Sociology: The Central Questions

Second Edition

Sociology: The Central Questions

Second Edition

William Kornblum
City University of New York, Graduate School and University Center

In collaboration with
Carolyn D. Smith

WADSWORTH
THOMSON LEARNING™

Australia • Canada • Mexico • Singapore • Spain •
United Kingdom • United States

WADSWORTH
THOMSON LEARNING™

Sociology Editor: Bryan Leake
Development Editor: Lisa Hensley
Editorial Assistant: Danette Cross
Technology Project Manager: Sarah Davis Packard
Marketing Manager: Laura Brennan
Marketing Assistant: Lisa Huebner
Project Editor: Claudia Gravier
Print/Media Buyer: Lisa Kelley
Permissions Editor: Caroline Robbins
Art Designer: David Beard
Photo Researcher: Lili Weiner
Copy Editor: Sandy Mann
Cover Printer: Lehigh Press, Inc.
Compositor: TSI Graphics
Printer: R. R. Donnelley, Willard

1 2 3 4 5 6 7 048 05 04 03 02 01

r For more information about our products, contact us at:
Thomson Learning Academic Resource Center
1-800-423-0563

For permission to use material from this text, contact us by:
Phone: 1-800-730-2214 **Fax:** 1-800-730-2215
Web: http://www.thomsonrights.com

Library of Congress Catalog Card Number: 00-111717
ISBN: 0-15-508562-x

Asia
Thomson Learning
60 Albert Street, #15-01
Albert Complex
Singapore 189969

Australia
Nelson Thomson Learning
102 Dodds Street
South Melbourne, Victoria 3205
Australia

Canada
Nelson Thomson Learning
1120 Birchmount Road
Toronto, Ontario M1K 5G4
Canada

Europe/Middle East/Africa
Thomson Learning
Berkshire House
168-173 High Holborn
London WC1 V7AA
United Kingdom

Latin America
Thomson Learning
Seneca, 53
Colonia Polanco
11560 Mexico D.F.
Mexico

Spain
Paraninfo Thomson Learning
Calle/Magallanes, 25
28015 Madrid, Spain

“Here’s a ‘central question.’ How can Kornblum’s *Sociology: The Central Questions* 2/e help me understand sociology and do well in my course?”

BEGINNING OF CHAPTER

Chapter Outlines let you see what you're going to read and what you'll be expected to understand when you're finished studying.

Central Questions help you organize your thoughts about a particular topic.

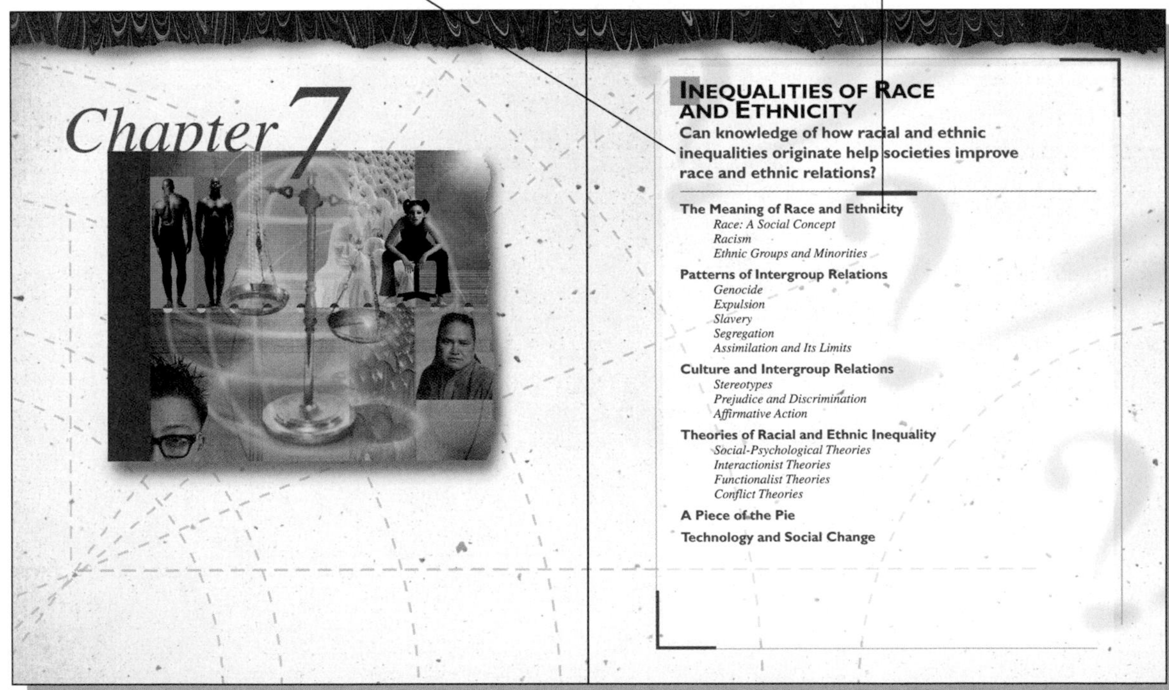

Chapter Opening Vignettes connect issues in sociology with events and people that affect all of us.

During the spring and summer of 2000, CBS Television made broadcasting history with its show *Survivor*. A so-called reality show centering on a group of Americans sent to live in a remote area such as a deserted island, the show is a combination soap opera, game show, and sports contest. The participants form a small society with its own subculture in the form of norms established by the show's producers. They must cooperate to perform the tasks required for survival, but each week they must also vote to eliminate one of their number. One contestant may win an endurance test (the sports element of the show) and thereby become immune from elimination during that week. The show was an almost instantaneous hit and was immediately copied by other networks. The final episode of the show's first season was one of the most heavily watched television events in history.

Sociologists and television critics were quick to point out that the show presented a scenario that bore no resemblance to reality. It was carefully produced, from the original casting of the contestants to the situations and interviews that were artfully edited, to heighten audience interest in the fate of each contestant. Sociologists also refuted the network's claim that the show was a unique social experiment. They noted that social scientists have conducted many similar but more carefully controlled social experiments, some of which will be discussed in this chapter.

Survivor does, however, demonstrate a well-known sociological phenomenon: When norms are structured to promote a certain outcome, people's behavior will conform to expectations and the outcome will be as predicted. The norms established by the show's participants promoted cooperation, but other norms made individual success, at the expense of others, more important. In the end, the person who won was the most skillful at creating coalitions favorable to himself while taking care to alienate as few other contestants as possible. The first season's winner, Richard Hatch, had well-honed skills in social interaction and professional experience in analyzing group behavior. He had a keen sociological imagination and the ability to apply it to achieve his own ends. We will see in this chapter how people develop such skills and under what conditions they use them for selfish purposes or for more altruistic ones. The process whereby a person masters the skills of interaction is part of the larger subject of socialization.

The "cast" of the first season of *Survivor*, shown here, became instant celebrities as their skills at interacting while promoting their own self-interests were broadcast throughout the world.

This chapter introduces some of the central issues in the study of how people become social beings. This may not seem to be a difficult process until one thinks about how many people have problems getting along well with others or how many conflicts we feel between the teachings of our parents and other social influences (from peers and others). Then there is the question of our biological endowments versus our social learning: How do these blend to create individuals who act in certain ways as social beings? If we look at the experiences of children growing up under conditions of severe social deprivation and lack of love, we find out more about the lasting influences of social learning and emotional support. The later parts of the chapter explore the ways in which people become social beings through their interactions in the family and in other types of groups. ■

WITHIN THE CHAPTER

Global Social Change boxes give you a worldwide view of important social issues and how they influence you.

GLOBAL SOCIAL CHANGE

Sociologists and Globalization

Globalization is widely discussed these days. It can mean different things to different writers, including "increased immigration, the feminization of the labor force, rising rents in global cities, the gentrification of urban neighborhoods, and the spread of informal work, from hot dog stand and gypsy cab to sweatshop seamstress and restaurant busboy" (Schaeffer, 1999, p. 1197). For sociologists, however, *globalization* is a term that guides a great deal of research and social action. Following are just two examples of how practicing sociologists are involved in understanding globalization.

Sociologist Saskia Sassen is one of the world's leading experts on the study of globalization. In her work, she argues that the expansion of foreign investment and the explosive growth in poor nations of industrial facilities owned by rich corporations are the main features of what we mean by globalization in its present form. In fact, Sassen and other sociologists who write about the subject know that the spread of markets for goods and services and the migration of people throughout the world have been going on since the age of exploration in the sixteenth century and before. Contemporary globalization, however, mostly has to do with international capitalism. Powerful corporations like IBM, Coca-Cola, General Electric, Sony, Nike, the major oil companies, and many of the world's largest banks and financial institutions increasingly have facilities located all over the world, not just in one nation.

These giant corporations encourage farmers and manufacturers in the regions where they operate to produce goods for export in the world markets. The export industries need workers to assemble garments and electronic components in factories and to pick flowers and fresh vegetables in the fields. In the cities where the management and financial centers of transnational firms are located, there is a parallel boom in demand for skilled workers who can operate computers and other sophisticated equipment; at the other end of the employment spectrum, there is increased demand for restaurant workers and less-skilled service employees to serve the needs of the more affluent. It is the expansion of demand for workers in skyscrapers, factories, and fields that drives migration, Sassen finds, not overpopulation, poverty, and economic stagnation, the traditional causes of emigration. Sassen's research on globalization shows the many ways in which international flows of people (migrants) and money also contribute to the growing gulf between the haves and the have-nots throughout the world.

On the other side of the world from the University of Chicago, where Sassen is based, Manjula Giri deals with different but related aspects of globalization. Giri, a sociologist and feminist activist in a small village in Nepal, received her Ph.D. in the United States but refused to be tempted by opportunities for an academic or research career. Instead, she returned to her native village, where she assists local women in their efforts to gain literacy, basic human rights, and greater equality with men. Together, these women have formed a village women's cooperative, often against the resistance of the village leaders. Giri has also been active in linking her local activities to national and international efforts to assert the rights and address the needs of poor women in the third world. She is active on all the women's committees of the United Nations, which are making important strides toward placing women's concerns on the agenda of global development.

Sassen and Giri are just two examples of how sociologists are involved in understanding and dealing with the impacts of globalization, but they represent the best the field has to offer. ■

Manjula Giri.

8

The Central Question: A Critical View boxes invite you to think critically about social issues and come to your own understanding of them.

THE CENTRAL QUESTION: A CRITICAL VIEW

Does culture account for the immense differences in the ways people throughout the world feel and behave?

Fauziya Kassindja, a young woman from the West African nation of Togo, escaped to the United States and was finally granted asylum after a long legal ordeal. She had created an international incident when she declared herself a political refugee on the ground that she was seeking to avoid genital mutilation (Dugger, 1996). In Togo and many other African societies, when a young woman reaches puberty she must undergo a ritual in which her clitoris and outer labia are cut away. This painful procedure, sometimes referred to as female circumcision, deprives her of much sexual pleasure and often leads to infection and reproductive problems later in life.

Kassindja appealed to the American authorities for legal asylum. The case created a number of precedents in immigration law and continues to generate much debate. Partly on the strength of the publicity surrounding this case, and partly owing to the efforts of other African women immigrants, in 1996 both the California state legislature and the U.S. Congress passed legislation banning clitorectomy in the United States. In international conventions on population, reproduction, and women's rights, however, women representing nations where clitorectomy is part of the culture often defend the practice. They may not personally support the norm of clitorectomy, but they resent the intolerance of Westerners, whose cultural norms (e.g., drinking alcoholic beverages or baring the female form) they find equally abhorrent.

Does the need to respect other cultures prevent one from taking a stand on practices like clitorectomy? Does avoidance of ethnocentrism mean that one must become a cultural relativist who "when in Rome, does as the Romans do"—even if it means violating one's own principles? Cultural relativism raises many controversial issues, especially in societies like the United States where people from different cultures are continually arriving, bringing with them different norms and values.

Many social scientists answer this question in both scientific and political terms. They argue that respect for other cultures does not prevent them from openly opposing cultural practices that perpetuate severe inequalities between women and men or that endanger women's health. Out of respect for colleagues from those cultures, they seek to persuade rather than pass laws that cannot be enforced. They seek to go beyond the issue of clitorectomy to the larger problems that prevent women in those cultures from speaking out against the practice themselves. So at world conferences the subject of clitorectomy is often subordinated to broader efforts to create policies that will improve women's status and power—for example, through greater access to education and modern forms of employment. At home, however, the same social scientists generally are highly supportive of legislation to protect women's rights to health and sexual freedom (Shweder, 2000).

Partly as a result of protests by feminists and human rights advocates, Fauziya Kassindja was freed from detention and granted asylum in the United States. She is shown here overcome by emotion upon her release from prison in York County, Pennsylvania, in 1996.

44

Mapping Social Change features give you an immediate visual representation of important social issues. Don't just read about it. See it!

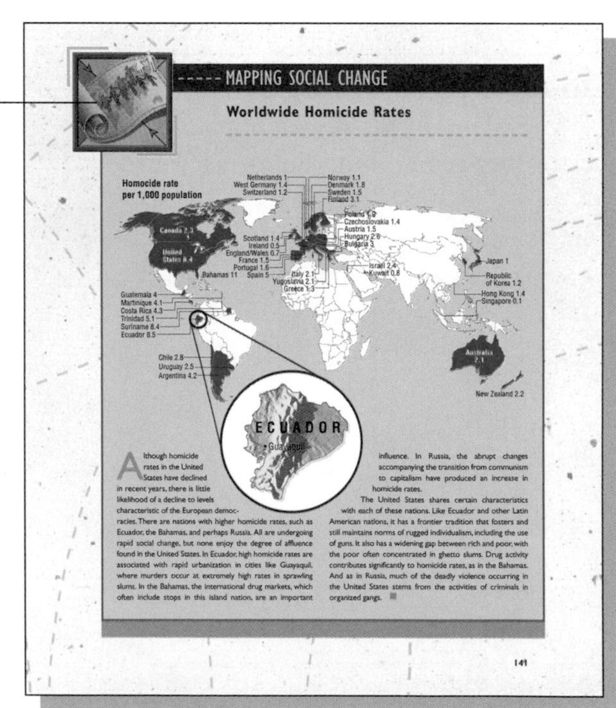

MAPPING SOCIAL CHANGE

Worldwide Homicide Rates

Although homicide rates in the United States have declined in recent years, there is little likelihood of a decline to levels characteristic of the European democracies. There are nations with higher homicide rates, such as Ecuador, the Bahamas, and perhaps Russia. All are undergoing rapid social change, but none enjoy the degree of affluence found in the United States. In Ecuador, high homicide rates are associated with rapid urbanization in cities like Guayaquil, where murders occur at extremely high rates in sprawling slums. In the Bahamas, the international drug markets, which often include stops in this island nation, are an important influence. In Russia, the abrupt changes accompanying the transition from communism to capitalism have produced an increase in homicide rates.

The United States shares certain characteristics with each of these nations. Like Ecuador and other Latin American nations, it has a frontier tradition that fosters and still maintains norms of rugged individualism, including the use of guns. It also has a widening gap between rich and poor, with the poor often concentrated in ghetto slums. Drug activity contributes significantly to homicide rates, as in the Bahamas. And as in Russia, much of the deadly violence occurring in the United States stems from the activities of criminals in organized gangs.

141

Sociology and Social Justice boxes refer back to the chapter opening vignette to give you a more complete picture now that you've finished reading the chapter.

SOCIOLOGY AND SOCIAL JUSTICE

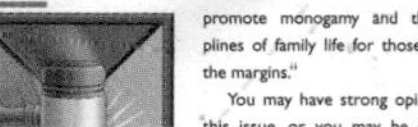

More than for blended families or even single-parent families, gay couples' desire to have families, adopt children, and live peacefully in communities of their choice is a fundamental issue of social justice. The status of gay and lesbian families is a "wedge" issue in American life. People with different opinions regarding the morality of homosexuality are driven apart by the political question of whether gay and lesbian marriages should be recognized as legal and thus qualify for full rights under marital and family laws. Ideological conservatives like William Bennett (1996), an outspoken critic of homosexual relationships, argue that broadening the definition of marriage to include same-sex unions would stretch it almost beyond recognition. Marriage, says Bennett, "is not an arbitrary construct which can be redefined simply by those who lay claim to it." On the other hand, defenders of homosexual marriage, such as Andrew Sullivan (1996), former editor of *The New Republic*, believe that permitting same-sex marriages would have a positive moral influence—"it would promote monogamy and the disciplines of family life for those cast on the margins."

You may have strong opinions on this issue, or you may be confused about what to think and how society should deal with the issue. How can a sociological perspective on the controversy help you make up your mind?

First we must look at the facts. There are many gay and lesbian people in this and other societies. It seems increasingly likely that homosexuality is not a mere lifestyle choice but a deeply felt aspect of a person's social being that cannot be changed easily, if at all. Moreover, it is clear that many gay and lesbian relationships are in fact marriages, judging from the couple's love and commitment to each other. How, then, can one limit membership in such fundamental institutions as marriage and the family only to heterosexuals?

But there are other sociological facts to consider. Clearly, there are parts of the nation where the vast majority of the population is bitterly opposed to homosexual marriage. Must norms that are morally offensive to antihomosexual majorities be forced upon them by the federal government? A sociologist would argue that such an action would only engender further hatred and perhaps violence. On the other hand, there are also parts of the nation where homosexual unions are more common, and they have gained support in the name of tolerance and fairness.

Many states and municipalities are likely to vote in favor of legal recognition of homosexual unions if given the opportunity. Most sociologists would argue, therefore, that as states and localities decide on the issue through democratic means, the norms will shift toward greater recognition of the rights of gay and lesbian marriages. ■

supported by the norms of a particular ethnic or religious community—neither partner is satisfied by the relationship. At the same time, neither feels that he or she can do anything to change the situation. Thus, the conflicts that might have produced change are reduced to indifference.

Today social scientists who study family interaction must deal with family structures that are more complex than the traditional nuclear family. Divorce and remarriage create many situations in which children have numerous sets of parental figures—parents and stepparents, grandparents and surrogate grandparents, and so on. These changes in family form result in new patterns of family interaction. For example, in a study of 2,000 children conducted over a five-year period, sociologist Frank Furstenberg found that 52% of children raised by their mothers do not see their fathers at all, not only because the fathers are absent by choice but because the opportunities for visits decrease as parents remarry or move away (cited in Aspaklaria, 1985). This produces situations in which parents strive to maintain long-distance relationships with their children and occasionally have brief, intense visits with them. Although this specific type of relationship may not be the most desirable, it appears that if parents and children can express love and affection even when the parents are divorced, the children's ability to feel good about themselves and to love others in their turn may not be impaired.

270

The Conflict Perspective

The conflict perspective on families and family interactions assumes that "social conflict is a basic element of human social life." Conflict exists "within all types of social interaction, and at all levels of social organization. This is as true of the family as it is of any other type of social entity" (Farrington & Chertok, 1993, p. 368). From this perspective, one can observe actual family interactions (the micro level) and ask why

271

Sociological Methods boxes allow you to see how sociologists conduct research.

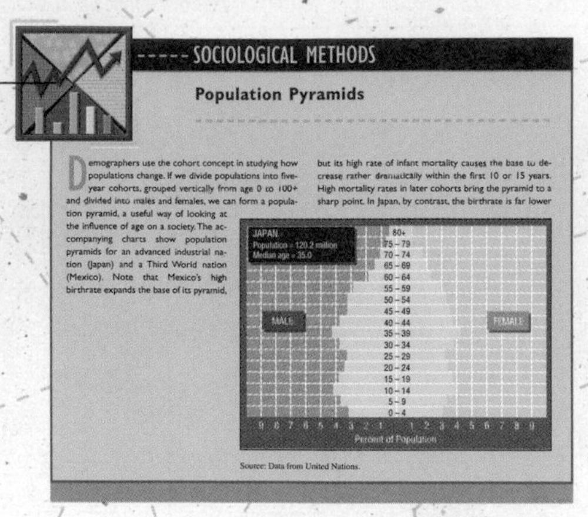

SOCIOLOGICAL METHODS

Population Pyramids

Demographers use the cohort concept in studying how populations change. If we divide populations into five-year cohorts, grouped vertically from age 0 to 100+ and divided into males and females, we can form a population pyramid, a useful way of looking at the influence of age on a society. The accompanying charts show population pyramids for an advanced industrial nation (Japan) and a Third World nation (Mexico). Note that Mexico's high birthrate expands the base of its pyramid, but its high rate of infant mortality causes the base to decrease rather dramatically within the first 10 or 15 years. High mortality rates in later cohorts bring the pyramid to a sharp point. In Japan, by contrast, the birthrate is far lower

Source: Data from United Nations.

colleges in coming decades. Between 1988 and 1993, for example, kindergarten enrollments rose by 7.2% and preschool enrollments by 14%, both trends that resulted in increasing enrollments in many school districts. This demographic effect, in turn, spurred increased debate over the investment of public funds in schools versus other uses, such as medical care for the poor and elderly (Gabriel, 1995a).

A Graying Population. As the parents of these children, the baby boom cohorts, move into middle age, the problems of their own parents, the aged, also take on increasing importance. Table 8.5 shows that the proportion of elderly people in the American population is increasing steadily. This increase—often referred to as the "graying" of America—will result in greater concern about the needs of the elderly and will augment the influence of the aged on American culture and social institutions.

Demographers and biologists disagree over estimates of how many very elderly people there will be in the U.S. population during the twenty-first century. Most Census Bureau demographers estimate that the population over age 85 will increase from about 3.3 million persons (mainly women) in 1990 to about 18.7 million in 2080. But demographers working with the U.S. Institute on Aging argue that there may be no biological limits on life expectancy. With advances in

238

Technology and Social Change sections show you how developments in technology shape our social world.

272 CHAPTER 9 The Family

conflict occurs, what the issues are, and how they are resolved. At a more macro level, one can ask how conditions of inequality and class conflict influence actual families or the laws and policies governing the family as a central institution in society.

Sociologists who study poverty argue that most families are poor for the obvious reason that they do not earn or receive enough income. But critics of liberal social-welfare policies argue that self-destructive behaviors of family members themselves, such as dropping out of school, drug use, and out-of-wedlock childbearing, are causes of poverty. Social-welfare policies that are too generous or take the form of entitlements that do not have to be earned are said to encourage behaviors that cause poverty.

Social scientists who study class conflict and poverty policies, notably Herbert Gans (1995), Christopher Jencks (1993), and Michael Katz (1983), point out that early in the twentieth century a large proportion of impoverished families were composed of elderly couples who did not have Social Security or Medicaid, two programs that have drastically reduced poverty among elderly Americans. Yet at the time that these policies were being debated, many members of the upper classes opposed them, viewing them as undemocratic transfers of wealth to less-deserving members of society. The situation today is similar, as members of the upper classes oppose adequate funding for day care and other policies that can help move adult family members from dependence on public support to employment at decent jobs.

The Functionalist Perspective

From a functionalist perspective, the family evolves in both form and function in response to changes in the larger social environment. As societies undergo such major changes as industrialization and urbanization, the family must adapt to the effects of those changes. Functionalist theorists have called attention to the loss of family functions that occurs as other social institutions like schools, corporations, and social-welfare agencies perform functions that were previously reserved for the family. We have discussed numerous examples of this trend in earlier chapters—the tendency of families to have fewer children as the demand for agricultural labor decreased, the changing composition of households as people were required to seek work away from their families of orientation, the increasing number of dual-income households, and so on. The functionalist explanation of these changes is that as the division of labor becomes more complex and as new, more specialized institutions arise, the family too must become a more specialized institution. Thus, modern families no longer perform certain functions that used to be within their domain, but they do play an increasingly vital part in early-childhood socialization, in the emotional lives of their members, and in preparing older children for adult roles in the economic institutions of industrial societies (Parsons & Bales, 1955).

TECHNOLOGY AND SOCIAL CHANGE

The family as a social institution is immensely influenced by technological change. Think of the impact of electricity and all the labor-saving devices made possible by electrification. Washing machines, sewing machines, vacuum cleaners, and refrigerators, to name only a few, revolutionized the way home labor was carried out. Much of the drudgery was taken out of housework. Women had more disposable time to seek education and enter the labor force. Men had fewer excuses for shirking housework (although most continued to do so). The family car made the family vacation a significant experience of childhood. The telephone made it far easier for families to stay in touch over great distances, just as today many families use the cellular telephone to coordinate pick-ups and gatherings even when they are out of the home (McGaw, 1987).

But technology has problematic aspects as well. Television, one of the most powerful innovations of the second half of the twentieth century, continues to have vast impacts on family life. The medium is often used as an "electronic baby-sitter," which means that children spend time immersed in television programs at the expense of family interactions that might better develop their emotional and verbal skills. The Internet can be a valuable learning tool, but it can also be a source of family conflict when children download information their parents disapprove of. Overall, technologies tend to increase the flexibility and adaptability of family life so that individual members can participate in many activities outside the family and still fulfill family expectations. But within individual families it is also true that particular technologies, like the family car or the Internet, may become contentious issues and areas of family conflict.

END OF CHAPTER

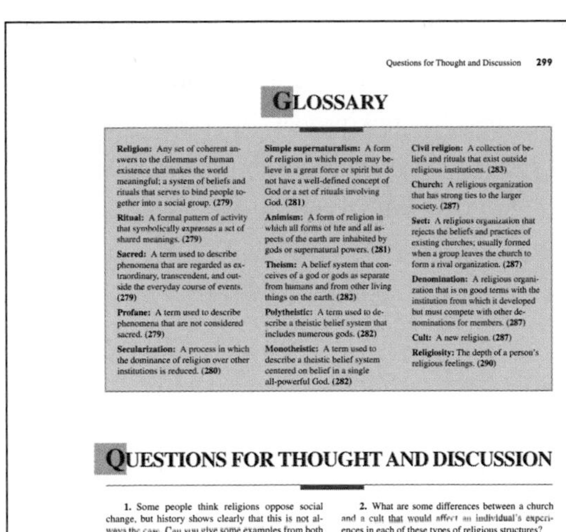

Questions for Thought and Discussion 299

GLOSSARY

Religion: Any set of coherent answers to the dilemmas of human existence that makes the world meaningful; a system of beliefs and rituals that serves to bind people together into a social group. **(279)**

Ritual: A formal pattern of activity that symbolically expresses a set of shared meanings. **(279)**

Sacred: A term used to describe phenomena that are regarded as extraordinary, transcendent, and outside the everyday course of events. **(279)**

Profane: A term used to describe phenomena that are not considered sacred. **(279)**

Secularization: A process in which the dominance of religion over other institutions is reduced. **(280)**

Simple supernaturalism: A form of religion in which people may believe in a great force or spirit but do not have a well-defined concept of God or a set of rituals involving God. **(281)**

Animism: A form of religion in which all forms of life and all aspects of the earth are inhabited by gods or supernatural powers. **(281)**

Theism: A belief system that conceives of a god or gods as separate from humans and from other living things on the earth. **(282)**

Polytheistic: A term used to describe a theistic belief system that includes numerous gods. **(282)**

Monotheistic: A term used to describe a theistic belief system centered on belief in a single all-powerful God. **(282)**

Civil religion: A collection of beliefs and rituals that exist outside religious institutions. **(283)**

Church: A religious organization that has strong ties to the larger society. **(287)**

Sect: A religious organization that rejects the beliefs and practices of existing churches; usually formed when a group leaves the church to form a rival organization. **(287)**

Denomination: A religious organization that is on good terms with the institution from which it developed but must compete with other denominations for members. **(287)**

Cult: A new religion. **(287)**

Religiosity: The depth of a person's religious feelings. **(290)**

QUESTIONS FOR THOUGHT AND DISCUSSION

1. Some people think religions oppose social change, but history shows clearly that this is not always the case. Can you give some examples from both the present and the past in support of this idea?

2. What are some differences between a church and a cult that would affect an individual's experiences in each of these types of religious structures?

Questions for Thought and Discussion give you a chance to explore issues in more depth.

Digging Deeper provides more references if you want to read in more detail about a particular topic or if you need to "kick-start" a paper or project.

Exploring Sociology on the Internet provides dozens of URLs for Web sites of interest.

300 CHAPTER 10 Religion

DIGGING DEEPER

Books

Religion, Deviance, and Social Control (Rodney Stark & William Sims Bainbridge, Routledge, 1997). An excellent source of material on religion, religious cults, and religious leadership.

One God: Peoples of the Book (Edith S. Engel & Henry W. Engel, eds.; Pilgrim Press, 1990). An introduction to the major monotheistic religions with a message of peace, commonality, and openness.

"The Sociology of Religion" (Robert Wuthnow; in Neil Smelser, ed., *Handbook of Sociology:* Sage Publications, 1988). An extremely useful review of current trends in the sociology of religion by one of the field's leading scholars. Includes an excellent bibliography of classic and recent sociological research on religion.

Base Communities and Social Change in Brazil (W. E. Hewitt; University of Nebraska Press, 1991). A fine case study of the influential Catholic ecclesiastical base communities movement in Latin America, based primarily on research in São Paulo, Brazil. A good example of empirical research on religion and social change at the community level.

Religion: The Social Context, 3rd ed. (Meredith B. McGuire; Wadsworth, 1992). A comprehensive text on the sociology of religion with excellent material on religion and social change.

Baptist Battles (Nancy Tatom Ammerman; Rutgers University Press, 1990). A seminal study of the conflicts that arise within a major Protestant denomination over issues of morality and religious practice.

Journals

Journal for the Scientific Study of Religion. Available in most college and university libraries; publishes recent research on religious practices, religiosity, and changes in religious institutions.

Other Sources

The Encyclopedia of American Religions, 4th ed. (Gale Research). Contains useful descriptions of religions in America; covers beliefs, organization, distribution in the population, and other aspects.

Exploring Sociology on the Internet 301

EXPLORING SOCIOLOGY ON THE INTERNET

Boston College Center for International Higher Education
www.bc.edu
Provides links to research centers and international news and library information dealing with the Catholic religion throughout the world.

The Anti-Defamation League
www.adl.org
Monitors hate activities in the United States and throughout the world; includes addresses of other useful research sites.

Ontario Consultants on Religious Tolerance
www.religioustolerance.org
Explores many religions and states; does not promote or denounce a specific belief. The site does list religious beliefs and news topics for over 20 denominations.

Andrew Greeley: Author, Priest, Sociologist
www.agreeley.com
Offers insights into Andrew Greeley's recent writings. The site includes previews of his recently published works and articles addressing recent issues in the Catholic religion and others.

PREFACE

Sociology, like many social sciences, addresses itself to a few central questions. This text explores those questions and examines the ways sociologists ask how human societies are changing. Understanding social change requires that we know about existing social conditions. To study how changes occur in societies and in individual lives, we must also know and appreciate what social forces act to prevent change and maintain social stability.

Unlike many other fields, however, sociology deals with subjects that seem entirely familiar to us. All of us have opinions about society, about the government, about our families, about our neighbors, about our town or society, our religion, our job, and about other aspects of our lives. Sociologists ask penetrating questions about these aspects of human social life, aspects that we often take for granted or argue about on the basis of our own experiences and prejudices. Instead of relying on their personal opinions, however, sociologists answer these questions by gathering facts according to the rules of scientific inquiry.

Knowledge of the methods of sociology, its theoretical perspectives, and the facts obtained through research gives sociologists a powerful advantage in debates about society and social change, especially when others are relying solely on opinion. That is why sociologists are sought out by businesses and political agencies to apply their methods and the facts that they discover to social issues of all kinds.

Sociology: The Central Questions, second edition, emphasizes how sociologists raise questions about social life and seek to answer them through their research. Each chapter opens with a central question; from that question arise the many more specific questions that researchers ask about particular areas of social life. For example, how do we become social beings? Is this process guided by our genes? Does it occur through learning in social situations? How are the rules of social life established and maintained in different societies with different cultures? Can we eliminate various kinds of social inequality? These are some of the central questions that sociologists ask and questions you will explore while using this textbook.

In order to discuss sociological questions or interpret sociological research, it is necessary to understand some of the basic concepts sociologists use in describing social life. This text introduces, defines, and applies many of the core concepts that sociologists use in their research and writing. When you use these sociological concepts, you will gain insight into your own social world and how it is changing. We trust that the many sociological insights in this book will serve you well in the years to come.

DISTINCTIVE ASPECTS OF THIS TEXT

Sociology: The Central Questions, second edition, is designed to serve as a tool for learning and appreciating sociology. Each chapter applies a specific set of instructional principles, each of which is intended to help you master sociological insights and stimulate your sociological imagination. These principles are the following:

- Exercising a global perspective: Because we live in an increasingly global society, a global perspective is incorporated throughout the textbook.
- Observing social change: A key to understanding

sociology is observing social change among individuals, groups, and societies around the world.

- Applying sociology: By applying sociology to everyday life, your comprehension is made more complete.
- Learning through visual presentation: Visual elements maximize learning by reflecting contemporary cultural assumptions and tastes.
- Incorporating diversity in theory and research: The research of women and minorities is contextually presented and balanced.
- Presenting accessible scholarship: The scholarship is presented in an accessible style, promoting comprehension of the material.

The teaching philosophy just described is incorporated in a variety of special features and pedagogical aids as well as in the body of the text. They include the following:

FEATURES

- *NEW TO THIS EDITION—*
  Each chapter begins with an ***introduction based on current events*** or situations that illustrate the questions to be addressed in the chapter; these are designed to capture students' attention and introduce them to the subject matter.
- *NEW TO THIS EDITION—*
  The Central Question: A Critical View boxes ask students to think critically about the central question addressed in the chapter—why it is important, and why it challenges our commonsense perceptions of social issues.
- *NEW TO THIS EDITION—*
  Sociology and Social Justice boxes build on the examples presented at the beginning of the chapter, relating them to problems of social justice and inequity; the use of photographs is intended to build students' ability to analyze visual cues for social meaning.
- *NEW TO THIS EDITION—*
  Technology and Social Change sections at the end of each chapter discuss the impact of new technologies on individuals, communities, and societies.
- ***Sociological Methods*** boxes apply contemporary research methods to topics discussed throughout the text. By integrating methodology, rather than confining it to a single chapter, the importance of research is reinforced.
- ***Global Social Change*** boxes present extensive examples of important worldwide changes, reinforcing the theme of social change and the text book's global perspective.
- ***Mapping Social Change*** uses maps and photos to depict the distribution of a condition or characteristic throughout the world or in a particular region.

PEDAGOGICAL AIDS

- ***Chapter summaries*** offer a concise rendering of the key concepts and relationships presented in the chapter.
- ***Digging Deeper*** presents sources that may be consulted for further information about topics covered in the chapter and includes listings of books, journals, and other sources.
- *NEW TO THIS EDITION—*
  Exploring Sociology on the Internet provides URLs of Web sites relevant to chapter topics.
- Other end-of-chapter elements include ***questions for thought and discussion*** and a ***glossary.***

ANCILLARIES FOR THE INSTRUCTOR

Instructor's Manual

Written by William Kornblum and Carolyn D. Smith, the Instructor's Manual includes discussions of the central questions; instructional goals; teaching suggestions that explain the distinctive features and central concepts of each chapter; topics for discussion; tips on using the tables, charts, and special features; pointers on teaching sociology across the disciplines; and lecture outlines.

Test Bank

Prepared by Ron Hammond of Utah Valley State College, the thoroughly revised test bank includes more than 2,000 multiple-choice, true/false, and short-answer questions.

Computerized Test Bank

Available in Macintosh and Windows formats, EXAMaster+ software allows you to create tests using fewer keystrokes. Easy to follow screens provide thorough step-by-step thorough test construction guidelines.

Web Site

This content-rich Web site contains Instructor Resources such as an instructor bulletin board, access to the Class Act course management system, a syllabus generator, downloadable Power Point slides for lecture preparation, and a downloadable Instructor's Manual.

ANCILLARIES FOR THE STUDENT

Study Guide

Written by Carolyn D. Smith and William Kornblum, the Study Guide allows for a self-paced review of the material. Each chapter begins with an outline and learning objectives, followed by a fill-in-the-blank review, a matching exercise in which key terms are matched with their definitions, a self-test consisting of multiple-choice and true/false questions, a section on analyzing tables and charts, and a reading comprehension exercise.

■ *NEW TO THIS EDITION—*

Integrated Activities Manual

Written by William Kornblum and Carolyn D. Smith, this new workbook contains a variety of exercises and activities designed to broaden and deepen the student's understanding of the material presented in the text.

Web Site

The book-specific Web site contains Student Resources such as an online glossary, links to related sites, Web activities, self-assessment quizzes, and a student bulletin board for discussion.

ACKNOWLEDGMENTS

I am especially grateful for the comments gleaned from reviewers of the manuscript, both in its first and second edition: Debbie Abowitz, Bucknell University; Robin Brown, Southern Union State Community College; Carole M. Carroll, Middle Tennessee State University; Ione DeOllos, Ball State University; E. Douglas Farley, Niagara County Community College; James E. Floyd, Macon College; Allen Furr, University of Louisville; David Greenwald, Bloomsburg University; Ron Hammond, Utah Valley State College; William J. Kinney, University of Saint Thomas; Phillip Kunz, Brigham Young University; Ronald R. Matson, Wichita State University; and Elizabeth Meyer, Pennsylvania College of Technology. I gratefully acknowledge the following instructors who reviewed content for this text: Glenn Currier, El Centro College; Michael J. Fraleigh, Bryant College; Ruby C. Lewis, Dekalb College—Central Campus; Anthony J. Mendonca, Community College of Allegheny County—Allegheny Campus; Hans Pieper, University of Evansville; Kanwal Prashar, Rock Valley Community College; George Primov, University of Miami; Ellen Rosengarten, Sinclair Community College; Matthew Smith-Lahrman, Dixie College; Ed Vaughn, University of Missouri; and Thomas J. Yacovone, Los Angeles Valley College. All of the reviewer comments were read carefully and thoughtfully considered.

My invaluable collaborator, Carolyn Smith, an author and professional textbook editor, made it possible for an overcommitted sociologist to stay on course and meet deadlines and complicated production schedules. At Harcourt, I have received excellent advice and support from Earl McPeek, publisher; Bryan Leake, acquistions editor; and Lisa Hensley, developmental editor. This text owes much to the creative design of David Beard and the skills of photo and rights editor Caroline Robbins. I appreciate the dedicated work of Claudia Gravier, project editor, and Holly Lewerenz, production manager, who guided this text through many hurdles.

William Kornblum

July 2001

ABOUT THE AUTHOR

William Kornblum is a professor of sociology at the Graduate School of the City University of New York, where he helps train future instructors and researchers in the social sciences. He also teaches undergraduates at various campuses of the City University, including Queens College, Hunter College, and City College.

A specialist in urban and community studies, Kornblum began his teaching career with the Peace Corps in the early 1960s, when he taught physics and chemistry in French-speaking West Africa. He received his doctorate in sociology from the University of Chicago in 1971. He has also taught at the University of Washington at Seattle and worked as a research sociologist for the U.S. Department of the Interior. At the CUNY Graduate School, he directs research on youth and employment and on urban policy. With his longtime research partner, Terry Williams, he co-authored *The Uptown Kids,* a sociological portrait of teenagers and young adults growing up in high-rise public housing projects. He was also the principal investigator of Project TELL, a longitudinal study of the ways in which home computers can improve the life chances of young people at risk of dropping out of school.

The author's other publications include *Blue Collar Community,* a study of the steel-making community of South Chicago; *Growing Up Poor* (with Terry Williams), a study of teenagers growing up in different low-income communities in the United States; and *Social Problems,* a comprehensive textbook about social problems and social policies in the United States.

CONTENTS IN BRIEF

CONTENTS

Chapter 4

Chapter 5

Chapter 6

Chapter 7

Chapter 8

Chapter 9

Chapter 10

Chapter 11

Chapter 12

Chapter 13

Chapter 14

Chapter 1

SOCIOLOGY, THE SCIENCE OF SOCIETY
Can sociology help us achieve better societies?

Sociology, the Human Science
The Sociological Imagination
Developing a Sociological Eye
The Social Environment

From Social Thought to Social Science
The Great European Sociologists
The Rise of Modern Sociology

Major Sociological Perspectives
Functionalism
Interactionism
Conflict Theory
The Multidimensional View of Society

Applying the Sociological Imagination
Formulating Research Questions
Reviewing the Literature

The Basic Methods
Observation
Experiments
Survey Research
Research Ethics and the Rights of Respondents

Analyzing the Data
Correlations
Mapping Social Data

Theories and Perspectives

Technology and Social Change

On June 30, 2000, with no prior warning, the prominent Egyptian sociologist Saad Eddin Ibrahim was arrested and imprisoned, charged with "defaming Egypt by his criticism of the Egyptian government." Ibrahim, a professor at Cairo's American University and founder of the Ibn Kaldoun Center for Development Studies, had recently written an article that was critical of the efforts many prominent Egyptian politicians were making to pass along their influence to their sons, a tendency hardly limited to Egypt or the Middle East. The authorities were also angered by a documentary film that the Center had begun distributing, which informed Egyptians of their voting rights and encouraged them to vote in the upcoming parliamentary elections. At this writing, the sociologist remains in prison. His situation carries some profound messages about what sociologists do and why sociology is sometimes seen as a subversive science.

Sociological research often shows how individual lives are shaped by larger social forces such as changes in political regimes, and changes in the ways people make their livings. Pointing out the effects of these changes and who is responsible for them may not be pleasing to those in power. As often occurs in nations throughout the world, "the government has convinced the people that Egypt and the government are one and the same thing," said the chief of Cairo's legal aid organization, which is defending Ibrahim. "If someone criticizes the government, then it is an insult against Egypt. If any crisis erupts, they do not solve it, but they punish the one who reported it" (quoted in Sachs, 2000).

According to sociologist Saad Ibrahim, voting in elections is necessary but not sufficient to maintain a democracy. Freedom of speech and protection of the rights of minorities are also fundamental to the rule of law in democracies.

In the United States, it may be less common for the sociological messenger to be punished for bringing bad news, but even here there are many examples of how sociological research can be controversial. This is especially true when the research reveals inequalities that favor some groups in society and deny social justice and equality of opportunity to others. Even the seemingly mundane issues sociology addresses, such as changes in the family, can lead to controversial viewpoints. Living a decent life and raising a family—seemingly modest goals—are fraught with difficulty in our increasingly complex and changing world. Relationships that seemed secure may shatter under the stress of unemployment, relocation, or differences in values among family members. Therefore, to study sociology is to try to understand how larger social forces are influencing our own lives. Sociology provides powerful intellectual tools that can help make sense of complicated issues: how populations grow, why poverty and crime are so difficult to eliminate, how nations can establish the rule of law rather than the rule of brute force.

This chapter begins with an overview of what sociology is and how it developed out of earlier forms of social thought. Of particular importance are the major sociological perspectives—powerful sets of concepts that sociologists use to analyze societies and social changes. Each sociological perspective has emerged from generations of research and social thought. None of these perspectives by itself is sufficient to explain complex social phenomena like wars or divorce, but when used creatively in combination, they provide enticing answers to many of the central questions of the field. These perspectives will reappear in later chapters.

The second half of the chapter introduces basic methods of sociological research, particularly the relationship of empirical research to sociological hunches and ideas. Examples of how research methods are put into action occur throughout the text. We return to issues of sociology and social justice at the end of this chapter. ■

SOCIOLOGY, THE HUMAN SCIENCE

Sociology is the scientific study of human societies and human behavior in the many groups that make up a society. Sociologists ask difficult, sometimes embarrassing, questions about human life in order to explore the consequences of cataclysmic events like those that shut down factories or enslave an entire people. To understand the possible futures of people who confront drastic changes, sociologists continually seek knowledge about what holds societies together and what makes them bend under the impact of major forces.

The Sociological Imagination

Whether you want to change the world, understand other cultures and societies, or merely find your own path to happiness, sociology will help you learn to ask questions that go beyond "common sense" and clichés. To do this, you must develop your sociological imagination—your ability to perceive how social forces influence your own life and those of the people around you.

Most people confuse social forces and personal troubles. If the economy falters and people become unemployed, they often blame themselves for failing; if money worries destroy a marriage, each spouse blames the other. People see crime and blame "human nature"; they see success and praise individual achievement. But this tendency to view life as a series of individual mistakes or successes obscures the fact that large-scale social forces also shape people's lives, often in ways that they cannot control. And the habit of seeing events mainly in terms of how they affect individuals blinds people to the possibility of improving the way their society is organized.

According to sociologist C. Wright Mills, who coined the term **sociological imagination,** people often believe that their lives can be explained mainly in terms of their personal successes or failures. They fail to see the links between their own individual biographies and the course of human history. "The facts of contemporary history," Mills cautions, "are also facts about the success and the failure of individual men and women" (1959, p. 3). He continues:

> When a society is industrialized, a peasant becomes a worker; a feudal lord is liquidated or becomes a businessman. When classes rise or fall, a man is employed or unemployed; when the rate of investment goes up or down, a man takes new heart or goes broke. When wars happen, an insurance salesman becomes a rocket launcher; a store clerk, a radar man; a wife lives alone; a child grows up without a father. (1959, p. 3)

According to Mills, neither a person's biography nor the history of a society can be understood unless one takes into account the influence of each on the other. The social forces of history—for example, war, depression or recession, increases in population, changes in production and consumption, and many other social conditions—become the forces that influence individuals to behave in new ways. But those new ways of behavior themselves become social forces and, in turn, shape history.

Developing a Sociological Eye

It takes more than just imagination to *see* the social forces around us. Often we cannot easily see the workings of social forces like population change, inequality, or racism. With some training in what to look for, however, we can develop an eye for social forces and phenomena and the social patterns that shape our lives.

Suppose, for example, that a female friend of yours has committed suicide. In trying to explain the horrible event, you and your friends all come up with reasons that deal with her internal state at the time of her suicide—depression, severe worry, great disappointment, and so on. Applying the sociological imagination to the event requires that you also ask questions about the person in a social context: How much contact did she have with friends and family, and did those contacts change in some way? Was she becoming more isolated? Was there a history of suicide in the family that would serve as a model for this tragic "solution" to the problems of her life? Were there any other social factors that might have contributed to her dire action, such as heavy debt or sudden loss of a job? The point is that we can never examine internal states without also wondering what larger social influences bear on the individual.

As sociologist Kai Erikson explains it, "There is a pattern in the way people grow up, become adults, choose occupations, form families, and raise children. There is a pattern in the way they become ill, commit crimes, compose music, or think thoughts." There are even patterns in the way people "make common cause with some of their fellow humans, and pattern in the way they exploit and abuse and even slaughter others of their fellow human beings as if they were not even of the same root species" (1997, p. 5). The study of these patterns and their consequences leads us to *see* the workings of social forces all around us. But if one major insight of sociology is that human life is patterned, another is that "those patterns are often imposed on the powerless by the powerful" (p. 5). Some of the major patterns of social life that we will explore in this book are the result of consensus and cooperation; others are the product of coercion and force. All of them will reveal the endlessly fascinating ways in which people in groups and as individuals attempt to cope with the personal challenges presented by these patterns of social life.

Through a variety of methods, including the analysis of visual material such as photographs, the analysis of statistical data, and comparisons among different cultures and societies, this book will help you develop a keener eye for important social patterns and their causes. Learning to keep an eye out for the sociological details that reveal social patterns and influences is part of how one develops a sociological imagination.

The Social Environment

The knowledge sociologists gather covers a vast range. Sociologists study religious behavior; conduct in the military; the behavior of workers and managers in industry; the activities of voluntary associations like parent-teacher groups and political parties; the changing relationships between men and women or between aging adults and their parents; the behavior of groups in cities and neighborhoods; the activities of gangs, criminals, and judges; differences in the behaviors of entire social classes; the way cities grow and change; the fate of entire societies during and after revolutions; and a host of other subjects. But ensuring that the information gathered is reliable and precise and using it to build theories of social cohesion and social change are among the most important challenges faced by the science of sociology.

As in any science, there are many debates in sociology about the appropriate ways to study social life and about which theories best explain social phenomena. Most sociologists, however, would agree with the following position:

> Human actions are limited or determined by "environment." Human beings become what they are at any given moment not by their own free decisions, taken rationally and in full knowledge of the conditions, but under the pressure of circumstances which delimit their range of choice and which also fix their objectives and the standards by which they make choices. (Shils, 1985, p. 805)

This statement expresses a core idea of sociology: Individual choice is never entirely free but is always determined to some extent by a person's environment. In sociology, *environment* refers to all the expectations and incentives established by other people in a person's social world. For the sociologist, therefore, the environment within which an individual's biography unfolds is a set of people, groups, and organizations, all with their own ways of thinking and acting. Certainly each individual has unique choices to make in life, but the social world that a person was born into—be it an urban ghetto, a comfortable suburb, or an immigrant enclave in a strange city—determines to varying degrees what those choices will be.

Levels of Social Reality. Sociologists look at behaviors in many environments, from the intimate glances of lovers to the complex coordination of a space shuttle launch. Thus, for purposes of analysis, we often speak of social behavior as occurring at three different levels of complexity: micro, macro, and middle. (See the study chart below.)

STUDY CHART

Levels of Sociological Analysis

Analytical Level		Social Behaviors Studied	Typical Questions
	Macro	Revolutions; intercontinental migrations; emergence of new institutions.	How are entire societies or institutions changing?
	Middle	Relations in bureaucracies; social movements; participation in communities, organizations, tribes.	How does bureaucracy affect personality? Do all social movements go through similar stages?
	Micro	Interaction in small groups; self-image; enactment of roles.	How do people create and take roles in groups? How are group structures created?

The **micro level** of sociological observation focuses on the patterns of interaction among a few people. One example is Erving Goffman's 1972 study, "Territories of the Self," which examined routine behaviors of everyday life. Goffman's research showed how seemingly insignificant ways of acting in public actually carry significant meanings. He categorized some of the ways in which we use objects as "markers" to claim a personal space:

> Markers are of various kinds. There are "central markers," being objects that announce a territorial claim, the territory radiating outward from it, as when sunglasses and lotion claim a beach chair, or a purse claims a seat on an airliner. . . . There are "boundary markers," objects that mark the line between two adjacent territories. The bar used in supermarket checkout counters to separate one customer's batch of articles from the next is an example. (pp. 41–42)

Some sociologists, however, deal almost exclusively with a larger scale or **macro level** of analysis This term refers to major changes in whole societies, such as revolutions or wars. (See the Global Social Change box on page 8.) One example of macrosociological analysis is the study of how the shift from heavy manufacturing to high-tech industries has affected the way workers earn their livings. Another example is the study of how the settlement of the American West gave rise to the beliefs and actions that drove Native Americans onto reservations.

Middle-level social phenomena are those that occur in communities or organizations such as businesses and voluntary associations. Middle-level social forms are smaller than entire societies but larger than micro-level social forms. When a mother living in poverty, for example, wonders how new welfare laws will affect her children and the other families in her community, she is wondering how changes at the macro level of society will affect life at the middle and micro levels—that is, in the community and in the homes of its residents.

These three levels of sociological analysis are helpful in understanding the experiences of immigrants who have recently arrived in your community. Macro-level social forces, like war or overpopulation, may account for the influx of immigrants. Middle-level social forces, such as available jobs or the presence of family members who immigrated earlier, may explain why certain ethnic groups are concentrated in particular communities within the United States. At the micro level of analysis there will be important differences in the way immigrant and native-born people interact on a daily basis, especially at first.

FROM SOCIAL THOUGHT TO SOCIAL SCIENCE

The roots of modern sociology can be found in the work of the philosophers and scientists of the Great Enlightenment, which had its origins in the scientific discoveries of the seventeenth century. That pivotal century began with Galileo's "heretical" proof that the earth was not the center of the universe; it ended with the publication of Isaac Newton's *Principia Mathematica.* Newton, often credited with the founding of modern science, not only discovered the laws of gravity and motion but, in developing the calculus, also provided later generations with the mathematical tools whereby further scientific discoveries could be made.

Hard on the heels of these discoveries came a theory of human progress that paved the way for a "science of humanity." Francis Bacon of England, René Descartes and Blaise Pascal of France, and Gottfried Wilhelm Leibniz of Germany were among the philosophers who recognized the social importance of scientific discoveries. Their writings emphasized the idea of progress guided by reason instead of by the notion that the human condition was ordained by God and could not be improved through human actions (Bury, 1932; Nisbet, 1969).

The rise of science began a transformation of the social order that continues today. But the vehicle of social change was not science itself. Rather, the modern era of rapid social change is a product of the many new social ideas that captured people's imagination during the eighteenth century. The revolutions that occurred in America and France, and much earlier in England, all resulted in part from social movements unleashed by the triumphs of science and reason. The ideas of human rights (all humans, not just the elite), of democracy versus rule by an absolute monarch, of self-government, and of applying reason and science to human affairs in general all arose during this period.

No longer could the Scriptures or the classics of ancient Greece and Rome be consulted for easy answers to age-old questions. Rather, it was becoming evident that new answers could be discovered through the use of the **scientific method**—repeated observation, careful description, formulation of theories based on possible explanations, and the gathering of additional data regarding the questions that followed from

GLOBAL SOCIAL CHANGE

Sociologists and Globalization

Globalization is widely discussed these days. It can mean different things to different writers, including "increased immigration, the feminization of the labor force, rising rents in global cities, the gentrification of urban neighborhoods, and the spread of informal work, from hot dog stand and gypsy cab to sweatshop seamstress and restaurant busboy" (Schaeffer, 1999, p. 1197). For sociologists, however, *globalization* is a term that guides a great deal of research and social action. Following are just two examples of how practicing sociologists are involved in understanding globalization.

Sociologist Saskia Sassen is one of the world's leading experts on the study of globalization. In her work, she argues that the expansion of foreign investment and the explosive growth in poor nations of industrial facilities owned by rich corporations are the main features of what we mean by globalization in its present form. In fact, Sassen and other sociologists who write about the subject know that the spread of markets for goods and services and the migration of people throughout the world have been going on since the age of exploration in the sixteenth century and before. Contemporary globalization, however, mostly has to do with international capitalism. Powerful corporations like IBM, Coca-Cola, General Electric, Sony, Nike, the major oil companies, and many of the world's largest banks and financial institutions increasingly have facilities located all over the world, not just in one nation.

These giant corporations encourage farmers and manufacturers in the regions where they operate to produce goods for export in the world markets. The export industries need workers to assemble garments and electronic components in factories and to pick flowers and fresh vegetables in the fields. In the cities where the management and financial centers of transnational firms are located, there is a parallel boom in demand for skilled workers who can operate computers and other sophisticated equipment; at the other end of the employment spectrum, there is increased demand for restaurant workers and less-skilled service employees to serve the needs of the more affluent. It is the expansion of demand for workers in skyscrapers, factories, and fields that drives migration, Sassen finds, not overpopulation, poverty, and economic stagnation, the traditional causes of emigration. Sassen's research on globalization shows the many ways in which international flows of people (migrants) and money also contribute to the growing gulf between the haves and the have-nots throughout the world.

On the other side of the world from the University of Chicago, where Sassen is based, Manjula Giri deals with different but related aspects of globalization. Giri, a sociologist and feminist activist in a small village in Nepal, received her Ph.D. in the United States but refused to be tempted by opportunities for an academic or research career. Instead, she returned to her native village, where she assists local women in their efforts to gain literacy, basic human rights, and greater equality with men. Together, these women have formed a village women's cooperative, often against the resistance of the village leaders. Giri has also been active in linking her local activities to national and international efforts to assert the rights and address the needs of poor women in the third world. She is active on all the women's committees of the United Nations, which are making important strides toward placing women's concerns on the agenda of global development.

Sassen and Giri are just two examples of how sociologists are involved in understanding and dealing with the impacts of globalization, but they represent the best the field has to offer. ■

Manjula Giri.

those theories. The idea that these same methods could be used to create a science of human society led to the birth of sociology. The French philosopher Auguste Comte coined the term *sociology,* confident that it would soon take its rightful place beside the reigning science of physics. Comte believed that the study of social stability and change was the most important subject for sociology to tackle. He made some of the earliest attempts to apply scientific methods to the study of social life. As Comte put it, "from the wretched inhabitants of Tierra del Fuego to the most advanced nations of Western Europe," there is such a great diversity of societies that comparisons among them will yield much insight into why they differ and how they change (1971/1854, p. 48).

The Great European Sociologists

In the nineteenth century, many philosophers and historians began to specialize in the study of social conditions and change. They saw that revolutions and civil wars had drastically disrupted more traditional ways of life. The scientific and technological innovations of the industrial revolution—railroads, the telegraph, steam power, and the like—were "shrinking" the earth by weakening the isolating effects of distance. Above all, the emergence of powerful new forms of social organization, especially capitalist enterprises—factories, banks, insurance companies, and the like—were creating new human needs and new ways of satisfying them. In the interests of expanding their markets and carrying out the "civilizing mission" of the European powers, entire regions of Africa, Asia, and Latin America were brought under the colonial dominion of the imperialist European nations, particularly England, France, and Germany. In response to these changes, the idea of creating a more systematic science of society, as Comte had urged, began taking hold of scholars' imaginations, and an increasing number of philosophers and historians began thinking of their research and writing as "sociology."

The most influential of the late nineteenth-century sociologists tended to think in macrosociological terms. Karl Marx, a German exile living in England, was both highly appreciative and extremely critical of the capitalist societies of his day. He predicted major upheavals arising from conflicts between the owners of wealth and the impoverished workers, and he believed that violent revolutions would transform society. The French sociologist Émile Durkheim disagreed, believing that social change would result from population growth and from changes in how work and community life were organized. Another pioneering European theorist, Max Weber, was the first to understand the overwhelming importance of bureaucratic forms of social organization in modern societies and to point out their increasing dominance in the lives of individuals.

Each of these early sociologists based his theories on detailed reviews of the history of entire societies. But to become a science rather than merely a branch of philosophy, sociology had to build on the research of its founders. The twentieth century brought changes of such magnitude at every level of society that sociologists were in increasing demand. Their mission was to gain new information about the scope and meaning of social change.

The Rise of Modern Sociology

We credit the European social thinkers and philosophers with creating sociology, but nowhere did the new science find more fertile ground for development than in North America. By the beginning of the twentieth century, sociology was rapidly acquiring new adherents in the United States and Canada, partly owing to the influence of European sociologists like Marx and Durkheim, but also because of the rapid social changes occurring at the time. Immigrants were arriving in waves, and population and industry were growing rapidly. There were race riots, strikes and labor strife, crusades against crime and vice and alcohol, and demands by women for the right to vote. These and many other changes caused American sociology to take a new turn. There was also a growing need for knowledge about exactly what changes were occurring and who was affected by them. Sociologists in North America therefore began to search for facts about changing social conditions—that is, the empirical investigation of social issues.

Empirical information refers to carefully gathered, unbiased data regarding social conditions and behavior. In general, modern sociology is distinguished by its relentless and systematic search for empirical data to answer questions about society. Journalists also seek the facts about social conditions, but their purpose is to present them as "stories" that will attract reader interest. Thus, because journalists cannot dwell on one subject very long, they are usually limited to citing examples and quoting experts whose opinions may or may not be based on empirical evidence.

In contrast, sociologists study a situation in depth, and when they do not have enough facts, they are likely to say something like: "That is an empirical question. Let's see what the research tells us, and if the answers are inconclusive, we will do more research." Evidence based on measurable effects and outcomes is required before one can make an informed decision about an issue.

In order to gain empirical information about social conditions, dedicated individuals undertook numerous "social surveys." Jacob Riis's account of life on New York City's Lower East Side (1890); W. E. B. DuBois's survey of African Americans in Philadelphia (1967/1899); Emily Balch's depiction of living conditions among Slavic miners and steelworkers in the Pittsburgh area (1910); Jane Addams's famous *Hull House Maps and Papers* (1895), which described the lives of her neighbors in Chicago's West Side slum area; and other carefully documented surveys of the living conditions of people experiencing the effects of rapid industrialization and urbanization left an enduring mark on American sociology.

By the late 1920s, the United States was the world leader in sociology, and the two great centers of American sociological research were the University of Chicago and Columbia University. At these universities and others influenced by them, two distinct approaches to the study of society evolved. The *Chicago School* sociologists emphasized the relationship between the individual and society. The major East Coast universities were more strongly influenced by European sociology and tended toward macro-level analyses of societies undergoing the change from feudalism to capitalism.

The sociology department at the University of Chicago (the oldest in the nation) was under the leadership of Robert Park and his younger colleague, Ernest Burgess. Park's main contribution to sociology was the development of an agenda for research that used the city as a "social laboratory." He favored an approach in which facts concerning what was occurring among people in their communities (the micro and middle levels) would be collected within a broader theoretical framework. That framework attempted to link macro-level changes in society, such as industrialization and the growth of urban populations, to patterns of settlement in cities and to how people actually lived in cities.

The Chicago School was distinctive for its emphasis on the relationships among social order, social disorganization, and the distribution of populations in space and time. Park and Burgess called this approach **human ecology.** Human ecologists seek to discover how populations organize to survive and prosper and how groups that are organized in different ways compete and cooperate. They also look for forms of social organization that may emerge as a group adjusts to life in new surroundings.

Modern ecological theories also consider the relationship between humans and their natural environment. We will see in later chapters that the way people earn their livings, the resources they use, the energy they consume, and their efforts to control pollution all have far-reaching consequences for individuals and for the society in which they live. These patterns of use and consumption also have a growing impact on the entire planet, and consequently ecological problems are becoming an increasingly important area of sociological research. (See the Mapping Social Change box on pages 12–13.)

MAJOR SOCIOLOGICAL PERSPECTIVES

Sociological *perspectives* are based on problems of human society, such as population size, conflict between populations, how people become part of a society, and other issues that will be addressed throughout this book. Although human ecology remains an important sociological perspective, it is by no means the only one employed by modern sociologists. Other perspectives, discussed in the following sections, guide empirical description and help explain social stability and social change. (See the study chart on page 11.)

Functionalism

When we speak of the family, the army, or the farm, we generally have in mind an entity marked by certain specific functions, tasks, and types of behavior. Family members nurture one another, armies fight, and farm workers plant and harvest. Individual interactions may determine how well a given person performs these various tasks, but it is the larger organization—the family, the army, the farm—that establishes the specific ways in which individuals do the work of that organization. In this sense, the organization, which outlives its members, has its own existence.

The **functionalist** perspective in sociology asks how society carries out the functions necessary to maintain social order, feed masses of people, defend

STUDY CHART

Major Sociological Perspectives

Perspective		Description	Generates Questions About . . .	Applications
	Interactionism	Studies how social structures are created in the course of human interaction.	How people behave in intimate groups; how symbols and communication shape perceptions; how social roles are learned and society is "constructed" through interaction.	Education practice, courtroom procedure, therapy.
	Functionalism	Asks how societies carry out the functions they must perform; views the structures of society as a system designed to carry out those functions.	How society is structured and how social structures work together as a system to perform the major functions of society.	Study of formal organizations, development of social policies, management science.
	Conflict Theory	Holds that power is just as important as shared values in holding society together; conflict is also responsible for social change.	How power affects the distribution of scarce resources and how conflict changes society.	Study of politics, social movements, corporate power structure.

against attackers, produce the next generation, and so on. From this perspective, the many groups and organizations that make up a society form the structure of human society. This social structure is a complex system designed to carry out the essential functions of human life. The function of the family, for example, is to raise and nurture a new generation to eventually replace the older generations; the function of the military is to defend the society; the function of schools is to teach the next generation the beliefs and skills they will need to maintain the society in the future; and a major function of religion is to develop shared ideas of morality.

When a society is functioning well, its major parts are said to be "well integrated" and in equilibrium. But periods of rapid social change can throw social structures out of equilibrium. When that happens, entire ways of life lose their purpose or function and the structures of society become poorly integrated. Formerly useful functions become "dysfunctional."

Consider the following example: In agrarian societies, in which most people work the land, families typically include three generations, with every member of the family living in close proximity to the others. High demand for labor and lack of machinery make early marriages and large families highly functional in such a society. But when agriculture becomes mechanized and the demand for farmhands has decreased, some families continue to produce large numbers of children. As adults, those children may migrate to cities, where they continue to value large families and to have many children. But if jobs are limited, they may be unemployed and their children may grow up in poverty. Such a family can be said to be poorly integrated with the needs of the society. The large family has become dysfunctional: It no longer contributes to the well-being of groups or individuals.

Interactionism

Interactionism is the sociological perspective that views social order and social change as resulting from the immense variety of repeated interactions among individuals and groups. Families, committees, corporations, armies, entire societies—indeed, all social forms—are a result of interpersonal behavior in which people communicate, give and take, share, compete, and so on. If there were no exchange of goods, information, love, and all the rest, obviously there could be no social life at all.

The interactionist perspective usually analyzes social life at the micro level of interpersonal relationships, but it does not limit itself to this level. It also looks at how middle- and macro-level phenomena result from micro-level behaviors or, conversely, how middle- and macro-level influences shape interactions among individuals. From the interactionist perspective, for example, a family is a product of interactions among a set of individuals who define themselves as family members. But each person's understanding of how a family ought to behave is a product of middle- and macro-level forces—religious teachings about

MAPPING SOCIAL CHANGE

Worldwide Calorie Consumption and Areas of Famine

From a biological perspective, the success of any species is measured by how well it meets the broad requirements of population growth and maintenance. Every day the world's 5 billion people seek and obtain enough food to convert into bodily energy, a minimum of perhaps 1,500 calories a day for survival at starvation levels (although this varies greatly with climate and other factors). More than 70% of the world's population is inadequately nourished (that is, obtains fewer than 2,500 calories a day), often while engaging in hard physical labor, whereas a smaller proportion, including most (but by no means all) of the North American population, lives comfortably well above the daily minimum.

Ecological theories help explain malnutrition and starvation in some areas of the world. In Somalia, for example, the combination of a desert environment and warring political factions produces social instability that leads to persistent famine. The problem is not so much that food is unavailable as that supplies cannot be delivered or distributed effectively. ■

These Iraqi children are searching in a dump for things to eat or sell. Scenes of children scavenging through garbage dumps are common throughout the world's poorest nations.

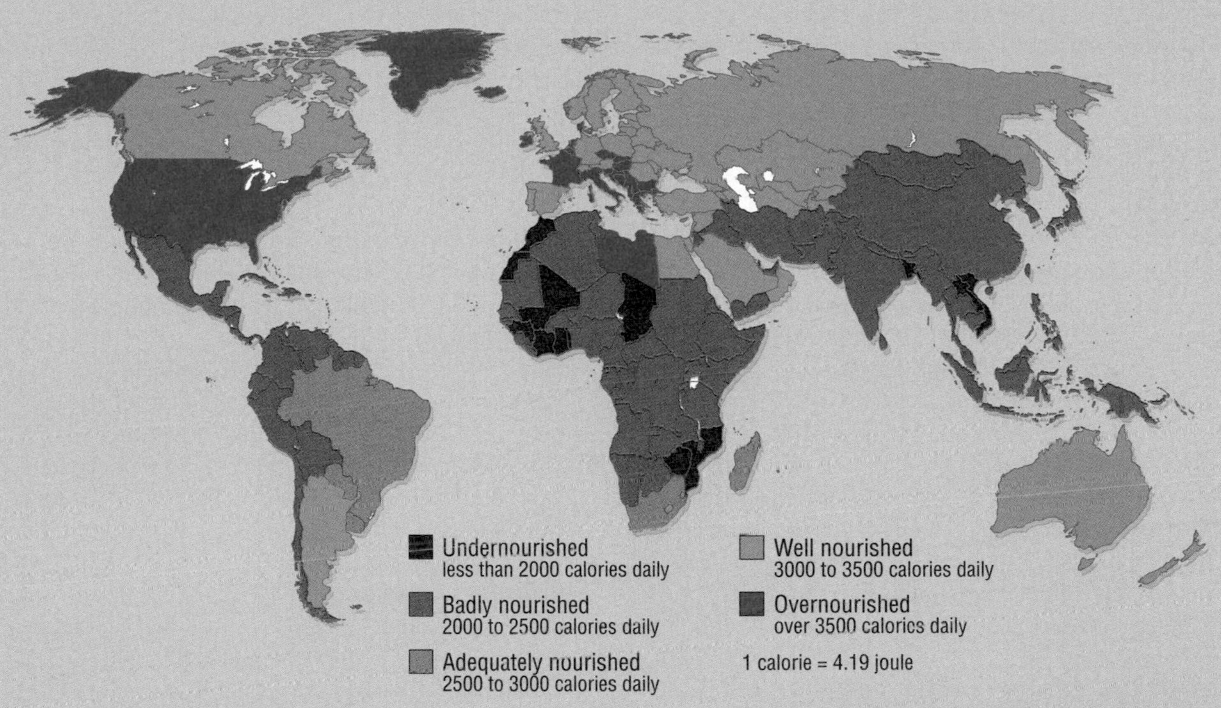

Hunger in the Developing World, by Region

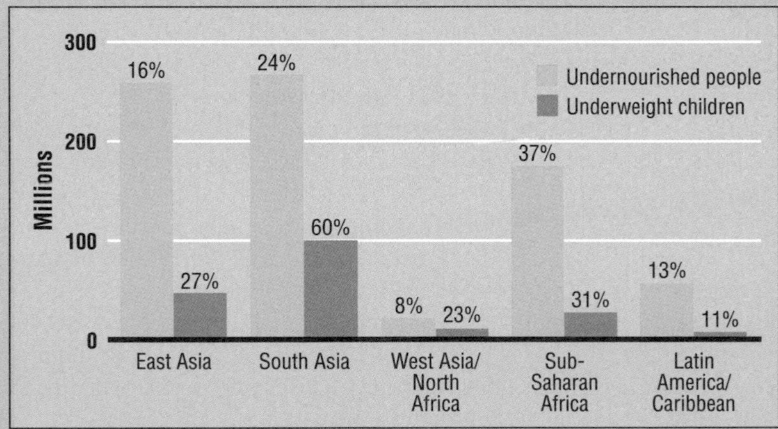

Conway, 1997

family life, laws dealing with education or child support, and so on. And these forces are always changing. You may, for example, have experienced the consequences of changing values that cause older and younger family members to feel differently about whether a couple should live together before getting married. In sum, the interactionist perspective insists that we look carefully at how individuals interact, how they interpret their own and other people's actions, and how the consequences of those actions affect the larger social group (Blumer, 1969b; Frank, 1988).

The general framework of interactionism contains at least two major and quite different sets of issues. One set has to do with the problems of exchange and choice: How can social order exist and groups or societies maintain stability when people have selfish motives for being in groups—that is, when they are seeking to gain as much personal advantage as they can? The second set of issues involves how people actually manage to communicate their values and how they arrive at mutual understandings. Research and explanations of the first problem fall under the heading of "rational choice" (or exchange theory), while the second issue is addressed by the study of "symbolic interaction." In recent years, these two areas of inquiry have emerged as quite different, yet increasingly related, aspects of the study of interaction.

Rational Choice—The Sociological View. Adam Smith, whose famous work *The Wealth of Nations* (1910/1776) became the basis for most subsequent economic thought, believed that individuals always seek to maximize their pleasure and minimize their pain. If over time they are allowed to make the best possible choices for themselves, they will also produce an affluent and just society. They will serve others, even when they are unaware that they are doing so, in order to increase their own benefit.

Although this theory, known as *utilitarianism,* is usually associated with economics, it can be applied to a variety of sociological issues. Often this rational-choice view of interaction is referred to as *exchange theory* because it focuses on what people get out of interactions and what they in turn contribute to a relationship or to a larger group. In every interaction something is exchanged. It may be time or attention, friendship, material values (e.g., wages or possessions), or less easily calculated values like esteem or allegiance. The larger the number of interacting members, the more complex the types of exchanges that occur among them. When people perceive an interaction as being one-way, they may begin to feel that they are being exploited or treated unfairly and may leave the relationship or quit the group (Homans, 1961). In industry, for example, underpaid workers may form a union, insist on collective bargaining, or even go on strike. But in doing so each worker will weigh the potential benefits against potential losses in pay, esteem, friendship, and so forth. The choices are not always easy, nor are the motivations always obvious. When many values are involved, the rational calculation of benefits and costs becomes even more difficult.

Rational-choice models of behavior prompt us to look at *patterns* of behavior to see how they conform to and depart from normal expectations of personal profit and loss. But those models do not always identify the underlying values. How we learn what to value in the first place, how we communicate our choices and intentions, and how we learn new values through interaction are all subjects that require other concepts besides those found in rational-choice theories of behavior. Such questions lead us toward research on how human interaction is actually carried out and understood by people in their daily lives.

Symbolic Interactionism. When we make choices about our interactions with other people, we may be said to be acting rationally. But other forces also shape our behavior. For example, you may select a particular college course because the instructor is rumored to be good. But what does "good" mean—an easy grader? friendly? humorous? The choice can be complicated, and you may not be aware of everything that goes into your decision. You might even choose a course as much to be with certain people as to be in that particular class.

Our choices tell other people about us: what we like, what we want to become, and so on. Indeed, how people choose to dress, how they carry themselves (body language), and how they speak convey a lot of information. Communication can occur without speaking (or when someone says one thing but actually means something else). But words are of great importance, too. Sociologists refer to all these aspects of behavior as *symbolic interaction.* From the symbolic-interactionist perspective, "society itself tends to be seen as a mosaic of little scenes and dramas in which people make indications to themselves and others, respond to those indications, align their actions, and so build identities and social structures" (Rock, 1985, p. 844; see also Goffman, 1959; Hughes, 1958).

Symbolic interactionists note how social life is "constructed" through mundane acts of social communication. For example, in all the choices students make—joining friendship groups, learning their school's informal rules, challenging those rules—the student social order, or "college culture," is actually "constructed." Erving Goffman, whose work was mentioned earlier, is known for his research on these processes. He applied the symbolic-interactionist perspective to the study of everyday interactions, to daily life in asylums and gambling houses, and to behavior in streets and public places. His work examines how people behave in social situations and how their "performances" are rated by others.

Note that the term *social construction* refers to the way people's interactions in their daily lives actually mark the nature of the groups and organizations they participate in. Every classroom, for example, is more or less similar in terms of the roles people perform—that is, the roles of teacher and student—but the way a teacher conceives of that role (for example, how informal he or she is when interacting with students) significantly affects the way actual classroom interactions take place. The symbolic-interactionist perspective helps us understand that the way we function in society's groups and organizations is not determined in advance but is subject to change as a result of the way we actually play out our roles. But this perspective often neglects the importance of power in shaping interactions and in determining the outcome of so many decisions, large and small.

Conflict Theory

The workings of power in human relations, especially in situations of conflict between groups and individuals, is the primary concern of conflict theory. Sociologists who take this perspective argue that conflict occurs as often as harmony. They believe that both the interactionist functionalist and perspectives fail to pay enough attention to the effects of power and conflict in society. In the twentieth century alone, two world wars and many civil wars disrupted the lives of millions of people. Almost as devastating was the Great Depression of the 1930s, the most severe economic slump in modern history. Worst of all were the nightmares of the Nazi Holocaust and the purges of Stalinist Russia, in which more than 20 million people were exterminated.

The world wars, the Depression, and the Holocaust shocked and demoralized the entire world. They also called into question the optimism of nineteenth- and early twentieth-century social philosophers, many of whom believed in the promise of progress through modern science and technology. Between 1914 and the end of World War II, modern ideas and technologies were used for horrible purposes. Bewildered intellectuals and political leaders turned to sociology to find some explanation for those horrors.

One explanation was provided by Marxian theory. Marx believed that the cause of conflict in modern times could be found in the rise of capitalism and the exploitation and domination that came with it. For example, in the early period of industrial capitalism, workers had to work twelve hours a day, six days a week; in less developed areas of the world, large populations were virtually enslaved by colonial powers.

For Marx, conflict lay at the core of capitalism, especially conflict between different economic classes. He argued that the division of people into different classes, defined by how they make a living and how much wealth they control, will always produce conflict. Marx believed that class conflict would eventually destroy or at least vastly modify capitalism. His perspective is at the heart of what has come to be known as **conflict theory.**

In the 1960s, when protests against racism and segregation, the Vietnam War, pollution of the environment, and discrimination against women each in turn became the focus of a major social movement, the conflict perspective gained prominence. It was not possible to explain so many major social movements with theories that emphasized how the social system would function if it were in a state of equilibrium. Even Marxian theory did a poor job of predicting the protest movements of the 1960s and their effects on American society. The environmental movement and the women's movement, for example, were not based on economic inequalities alone, nor were the people who joined them necessarily exploited workers. Sociologists studying the role of conflict in social change therefore had to go beyond the Marxian view. Many turned to the writings of the German sociologist Georg Simmel (1904), who argued that conflict is necessary as a basis for the formation of alliances. According to Simmel, conflict is one means whereby a "web of group affiliations" is constructed. The continual shifting of alliances within this web of social groups can help explain why certain individuals become involved in social movements and how much power those movements are able to acquire.

The concept of **power** holds a central place in conflict theory. From the functionalist perspective, society holds together because its members share the same basic rules of behavior. Conflict theorists point out that the role of power is just as important as the influence of shared beliefs in explaining why society does not disintegrate into chaos. Power is the ability of an individual or group to change the behavior of others. A nation's government usually controls the use of force (a form of power) to maintain social order. For sociologists who study conflict and power, the important questions are who benefits from the exercise of power and who loses.

The Multidimensional View of Society

Each of the sociological perspectives leads to different questions and different kinds of observations about social life. From the ecological perspective come questions about how populations can exist and flourish in various natural and social environments. From the interactionist perspective come questions about how people get along and behave in all kinds of groups and organizations. Functionalism asks questions about how society is structured and how it works as a social system. And the conflict perspective asks how power is used to maintain order and how conflict changes society. These different perspectives developed as sociologists asked different questions about society. In contemporary sociology, each perspective continues to stimulate relatively distinct research based on the types of questions being asked. Yet a great deal of research combines the insights of different perspectives in ways that vastly increase the power of the resulting analysis.

This brings us to the subject of how sociologists actually go about conducting their research. In the rest of this chapter, some of the basic sociological research methods will be introduced. An understanding of these methods will help you recognize the way different sociological writings address or utilize the major perspectives.

APPLYING THE SOCIOLOGICAL IMAGINATION

As citizens of a great democracy, we are encouraged to form opinions about the social issues of our time. We hear arguments about crime, poverty, education, and many other important questions. Political candidates are eager to tell us what to believe and what society should do about social issues. But often the claims and recommendations we hear are contradictory. Conflicting explanations for problems like poverty and welfare dependency may suggest very different solutions. It is difficult to decide what should be done. Sociology brings scientific methods to bear on these debates.

Many people associate social-scientific research with questionnaires and opinion surveys. Sociologists throughout the world do use these research techniques, but they also use many other methods to explore social conditions. In the following sections, we will review the most common research tools used by sociologists and how the findings lead to changes in theories and to new research.

Anyone conducting scientific research, regardless of the subject or method, will go through many intellectual and emotional ups and downs before the work is done. Research that originally appears easy to conduct soon becomes more complicated. Research questions appear less well-defined than at the outset. This can be just as true in biology or chemistry as it is in sociology or another social science. Nevertheless, in all research projects certain basic steps must be completed, although not always in the order listed here:

1. *Deciding on the problem.* At first, the problem to be studied may simply be a subject or topic of interest, but eventually it must be reduced to a specific research question or questions in order to provide a focus for the rest of the work.
2. *Reviewing the literature.* Usually others have conducted research on the same topic. Find their reports and use them to determine what you can accomplish through your own research.
3. *Formulating research questions.* The work of others often points to questions that have not been answered. Your own interest, time, resources, and available methods help determine what specific questions within your broad topic you can actually tackle.
4. *Selecting a method.* Different questions require different types of data—which in turn suggest different methods of data collection and analysis. Some of those methods are described in detail in this chapter.
5. *Analyzing the data.* Data are analyzed at each stage of the research, not just when preparing the final report.

THE CENTRAL QUESTION: A CRITICAL VIEW

Can sociology help us achieve better societies?

Sociologist Joyce Ladner is living proof that we can use sociology to achieve better societies. She is also a person who knows the value of a critical sociological view of the world she lives in. Her city, Washington, D.C., ranks among the lowest in the United States on all the critical indicators of social well-being. Rates of infant mortality, AIDS infection, homicide, and many other indicators show that the nation's capital is a sociological "basket case." Ladner points out that great poverty exists in the shadow of the White House and the Capitol, in part because Washington is a capital city and not a major city within a state. The local lawmakers depend on Congress for funds for health, education, and other public services, and often the legislators are more concerned about bringing benefits to their constituents at home than about addressing the social issues that plague the District of Columbia. Patterns of local political corruption have also contributed to the poor social conditions in the district.

Joyce Ladner

A sociologist at Howard University, Ladner was appointed to the DC Financial Control Board, which was established to deal with the financial and social crisis in the nation's capital. Ladner had her work cut out for her. The District's schools were dilapidated, funding was scarce, and she often had to confront residents who were angry about reductions in public services.

Ladner points out that during her years on the Board her sociological background was invaluable. "Sociology helps me understand the players," she observes. "It gives me the big picture, the framework" (quoted in ASA, 1997, p. 1).

Joyce Ladner grew up in Mississippi during the era of segregation. She joined the civil rights movement and developed a lifelong commitment to justice that she has applied to issues in child welfare and public education. Sociology allows her to see "the importance of structures and how they work." She knows that some structures must be changed and that change takes hard work. "We're making progress. But change is slow, and we're making important changes not yet seen" (quoted in ASA, 1997, p. 8).

As you study sociology, look for ways in which its concepts and research findings may help you sort out the confusions you see among people around you. There is a great deal of ignorance about how societies and organizations work and change. After this course, see if you have gained confidence that you and others with sociological insight can make a difference. ■

In this chapter we act as if this step-by-step procedure is always followed by researchers in the social sciences. But do not assume that this procedure can be applied to every research question. Research does not always progress easily from one step to another. Often earlier steps must be repeated. In addition, some research questions require that we devise new ways of conducting research or new combinations of existing research methods.

Formulating Research Questions

A good deal of information is required when researching a particular social issue. General questions about societies or social behavior have to be formulated so that responses produce measurable, meaningful data. Sociologists convert a broad question about social change into specific questions to be addressed in an empirical study—one that gathers evidence to describe behavior and to prove or disprove explanations of why that behavior occurs. These explanations are usually stated in the form of a **hypothesis,** a statement that expresses an informed (or "educated") guess regarding the possible relationship between two or more phenomena.

Consider the following: In a pioneering study, Émile Durkheim explored the possibility that suicide could be explained as much by social variables like rates of marriage—that is, social integration—as by individual psychological variables like depression. If this view is correct, he reasoned, the rates of suicide among various populations should vary along with measures of social integration. He therefore formulated these hypotheses:

- Suicide rates should be higher for unmarried people.
- Suicide rates should be higher for people without children.
- Suicide rates should be higher for people with higher levels of education. (Education emphasizes individual achievement, which weakens group ties.)
- Suicide rates should be higher in Protestant than in Catholic communities. (Protestantism places more stress on individual achievement than Catholicism does, and this in turn weakens group ties.)

Each of these hypotheses specifies a relationship between two variables that can be tested—that is, proved true or false—through empirical observation. In sociology, **variables** are characteristics of individuals, groups, or entire societies that can vary from one case to another.

In the hypotheses just presented, suicide rates, the proportion of people who are married or unmarried, the level of education, and the number of children in a household are all social variables. But what we are trying to explain is the rate of suicide for a society or nation. **Dependent variables** depend on the workings of other variables in the society (religion, marriage, education, etc.). They are the variables that a hypothesis seeks to explain. Other variables in hypotheses are called **independent variables.** They are the variables that the researcher believes cause a change in another variable (i.e., the dependent variable).

Reviewing the Literature

There is no need to conduct new research if the answers sought are already available in the "literature"—existing books or journal articles, published statistics, photos, and other materials. Most sociological research, therefore, begins in the library with a "review of the literature." However, it takes some imagination just to think of the kinds of studies to look for. The various sociological perspectives described earlier can help organize the search, especially when the perspectives are framed as questions.

Who? How Many? Where? In researching almost any subject involving human behavior, it is helpful to ask who is involved and in what numbers and where the behavior in question occurs. This is a way of phrasing the ecological perspective, which suggests why that perspective is helpful in beginning research on an issue. The ecological perspective gives rise to two types of studies: community studies and demographic studies. *Community studies* portray the typical day-to-day life of a particular population, while *demographic studies* provide counts of people in various relevant population categories.

If you were doing a research project on suicide, for example, you could read any number of community studies about why certain people are attracted to this fatal behavior. Donna Gaines's study of teenagers and suicide in a suburban New Jersey community is an excellent example of such a study. In *Teenage Wasteland: Suburbia's Dead End Kids* (1992), Gaines shows why some teenagers become so alienated and despondent that they try to take their own lives. For more ecological data, which would yield quantitative measures of how many suicides occur in different regions of the United States, you could begin by consulting the *Statistical Abstract of the United States,* published annually by the U.S. Bureau of the Census.

How Is the Situation Defined? Studies that take an interactionist approach tend to look at how people in a particular social category interact with one another and how they view their condition. Such studies allow one to see a social world, such as that of chronic welfare recipients, from the inside.

If, for example, your research dealt with assisted suicide, studies done from an interactionist perspective would help you see the dire choices facing people who contemplate taking their own lives with the aid of someone else. Lonny Shavelson's *A Chosen Death* (1998) offers moving and detailed accounts of how adults who are considering assisted suicide perceive their situation.

What Groups or Organizations Are Involved? This question stems from the functionalist perspective. It asks how society is organized to deal with a social issue or problem. Functionalist sociologists are concerned with how social policies actually function, as opposed to how they are supposed to function.

In searching the literature for research on groups or organizations, you would naturally look for titles that point to specific organizations. For example, Wanda Yvonne Johnson's *Youth Suicide: The School's Role in Prevention and Response* (1999) is an analysis of how schools need to function to identify kids at risk of suicidal behavior and, when prevention fails, how they can help children, parents, and teachers cope with the suicide of a peer.

Who Has the Power? This is a key question arising from the conflict perspective. Studies that take a conflict approach frequently criticize the operation of existing institutions. In searching for research on suicide from a conflict perspective, you would look for titles that somehow emphasize struggle or conflict. An example is Sue Woodman's *Last Rights: The Struggle Over the Right to Die* (1998), which shows that the controversy is dominated by powerful institutions like major churches and government agencies.

THE BASIC METHODS

Once the researcher has specified a question, developed hypotheses, and reviewed the literature, the next step is to decide on the method or methods to be used in conducting the actual research. Sociological research methods are the techniques an investigator uses to systematically gather information, or data, to help answer a question about some aspect of society. The variety of methods is vast, with the choice depending largely on the type of question being asked. The most frequently used methods are observation, experiments, and surveys.

Observation

Participant Observation. Much sociological research requires direct observation of the people being studied. A community study requires lengthy observation of a particular group and is rarely successful unless the sociologist also participates directly in the daily life of the people being observed. This method is known as **participant observation.** In this kind of observation, the sociologist attempts to be both an objective observer of events and an actual participant in the social milieu under study—not an easy task for even the most experienced researcher. The observer must record his or her observations and interactions in *field notes,* which supply the descriptive data that will be used in the analysis and writing phases of the study.

An excellent example of a study based on participant observation is one conducted by Douglas A. Harper, detailed in his book *Good Company* (1982). Harper spent months riding freight trains and living in hobo "jungles." His goal was to describe how hoboes actually live in a world considered deviant by most members of society. Following is Harper's description of the experience of getting into the world of the hobo:

> I'd been in the yards a couple of days, peeking around, asking questions and making plans. Every time I've gone back to the freights I've had to cross an emotional hurdle—they seem too big, too fast, too dangerous, and too illegal—and I get used to the idea by spending a few days in the yards, testing the waters. . . . I was shifting back into a tramp world for the fourth or fifth time. I'd made cross-country trips on freights and I'd spent some weeks the winter before living on Boston's skid row. These experiences were trips into a life I ordinarily did not lead. (1982, p. 3)

This kind of research describes the quality of life of the people involved. For that reason, it is often called *qualitative* research in order to distinguish it from the *quantitative* research methods that will be considered shortly. James Coleman observes that in qualitative research "we report a stream of action in which the interlinking of events suggests how the [social] system functions." (1964, p. 222)

Qualitative research carried out "in the field" is the best method for analyzing the processes of human interaction. A shortcoming of this approach, however, is that it is usually based on a single community or social system, which makes it difficult to generalize the findings to other social settings. Thus, community studies and other types of qualitative studies are most often used for exploratory research, and the findings serve as a basis for generating hypotheses for further research.

Unobtrusive Measures. Observation can employ numerous other techniques. One such technique is **unobtrusive measures,** observational techniques that measure the effects of behavior with minimal intrusion into actual social settings. Following are two examples:

- Chinese jade dealers have used the pupil dilation of their customers as a measure of the client's interest in particular stones. (Webb et al., 1966, p. 2)
- To study changes in the diet of Americans, investigators sifted through more than 2,000 household trash collections to measure the amount of fat being trimmed from meat packages. (Rathje, 1993)

Observations like these can be transformed into useful measures of the variables under study. Their nature is limited only by the researcher's creativity, and they intrude far less into people's lives than do interviews or participant observation.

Experiments

Although for both moral and practical reasons sociologists do not have many opportunities to perform experiments, there is much literature in the social sciences, especially social psychology (an interdisciplinary science that draws ideas and researchers from both sociology and psychology) that is based on experiments. The two experimental models used most frequently by social scientists are the controlled experiment, which is conducted in a laboratory, and the field experiment, which is often used to test public policies.

Controlled Experiments. The **controlled experiment** allows the researcher to manipulate an independent variable to observe and measure changes in a dependent variable. The experimenter forms two groups: an **experimental group** that will experience a change in the independent variable (the "treatment") and a **control group** that will not experience the treatment but whose behavior will be compared with that of the experimental group. (The control group is similar to the experimental group in every other way.) This type of experiment is especially characteristic of studies at the micro level of sociological research.

Consider the following example: Which line in Figure 1.1(b) appears to match the line in Figure 1.1(a) most closely? Could anything persuade you that a line other than the one you have selected is the correct choice? This simple diagram formed the basis of a famous series of experiments conducted by Solomon Asch in the early 1950s. These experiments showed that the opinion of the majority can have an extremely powerful influence on that of an individual.

Asch's control group consisted of subjects[1] who were seated together in a room but were allowed to make their judgments independently. When they looked at sets of lines like those in Figure 1.1, the

[1]The term *subject* refers to a person who participates in a controlled experiment.

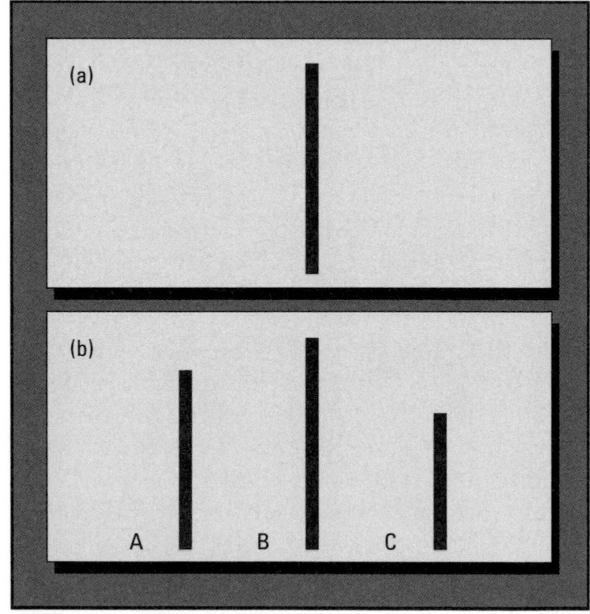

FIGURE 1.1
Lines Used in the Asch Experiment on Conformity
Cards like these were used in the Asch experiment. Subjects were asked to judge the lengths of various lines by comparing them with the three lines on the bottom card. The line on the top card quite obviously matches line B on the bottom card; all of the judgments were this simple.

subjects in this group invariably matched the correct lines, just as you no doubt have. But in the experimental group, a different result was produced by the introduction of an independent variable: group pressure.

Asch's experimental group consisted of subjects who were asked to announce their decisions aloud in the group setting. Each subject was brought into a room with eight other "subjects" who were actually confederates of the experimenter. When the lines were flashed on a screen, those subjects all chose a line that was not the matching one. When it was the real subject's turn to choose, he or she was faced with the unanimous opinion of a majority of "subjects" who had picked the wrong line. Thirty-two percent of the real subjects went along with the majority and chose the wrong line as well. And even among the subjects Asch called "independent"—the 68% of the real subjects who gave the correct response despite the pressure of the majority—there was a great deal of variation. Some gave the correct response at all times, whereas others gave it only part of the time (Asch, 1966).

Field Experiments. **Field experiments** are used extensively in evaluating public programs that address specific social problems. In these experiments, there is usually a "treatment group" of people who participate in the program and a control group of people who do not. In one example of this type of experiment, Angelo Atondo, Mauro Chavez, and Richard Regua of California's Evergreen Valley College attempted to test the theory that if students at risk of failure are linked up with adult mentors from their own communities their chances of success will improve. This popular theory stems in part from an influential article in the *Harvard Business Review* titled "Everyone Who Makes It Has a Mentor" (Collins & Scott, 1978).

Researchers have found little evidence to prove that mentors actually make a difference. Unless one can compare students who have mentors with similar students who do not, it is impossible to say with any scientific certainty—that is, beyond individual stories of success or failure—whether the mentors make a difference. The Evergreen Valley College researchers therefore assigned 115 entering Latino students with low English proficiency to Latino mentors with excellent English skills. Their control group was composed of a comparable number of Latino students with similar characteristics, but without mentors. Both groups attended the same classes and took the same exams. At the end of the semester, 89% of the students with mentors passed the freshman English course, compared to only 46% of the students in the control group. Later, more of the students with mentors went on to four-year colleges. While much more controlled research of this nature needs to be done before we can fully understand how mentors help students overcome educational problems, this example shows how a field experiment can advance our understanding of a social issue like success and failure in school (Bashi, 1991).

Survey Research

A sociological survey asks people to give precise information about their behavior, their attitudes, and at times the behavior and attitudes of others (e.g., family members). There is a world of difference between the modern sociological survey and the "social surveys" conducted around the turn of the twentieth century. Those surveys attempted to present an unbiased, factual account of the social conditions of a specific community; their findings could not be applied to other groups. In contrast, today's survey techniques make it possible to generalize from a small sample to an entire population. An example is election polls. A sample of 1,500 respondents can be used to predict how the entire electorate will vote in an election. Done properly, the modern survey is one of the most powerful tools available to social scientists.

The most ambitious and heavily used sociological surveys are national censuses. A *national census* is a full enumeration of all the members of the society, where they live and work, their family composition, age distribution, educational level, and other related data. Without these measures, a nation cannot plan intelligently for the needs of its people. The United States conducts a national census every 10 years.

Once a national census has gathered the basic demographic and ecological facts about a nation's people, it is possible to add more information through the use of smaller and far less costly **sample surveys.** In the United States, this is done regularly through the Census Bureau's Current Population Survey (CPS). From the CPS, we get monthly estimates of employment and unemployment, poverty, births, deaths, marriages, divorces, social insurance and welfare, and other indicators of the social well-being of the American people.

Opinion Polls. Another type of survey research, pioneered by sociologists early in the twentieth century, is the opinion poll. Today opinion polling is a

FIGURE 1.2
Public Support for Environmental Action, 1973–1996

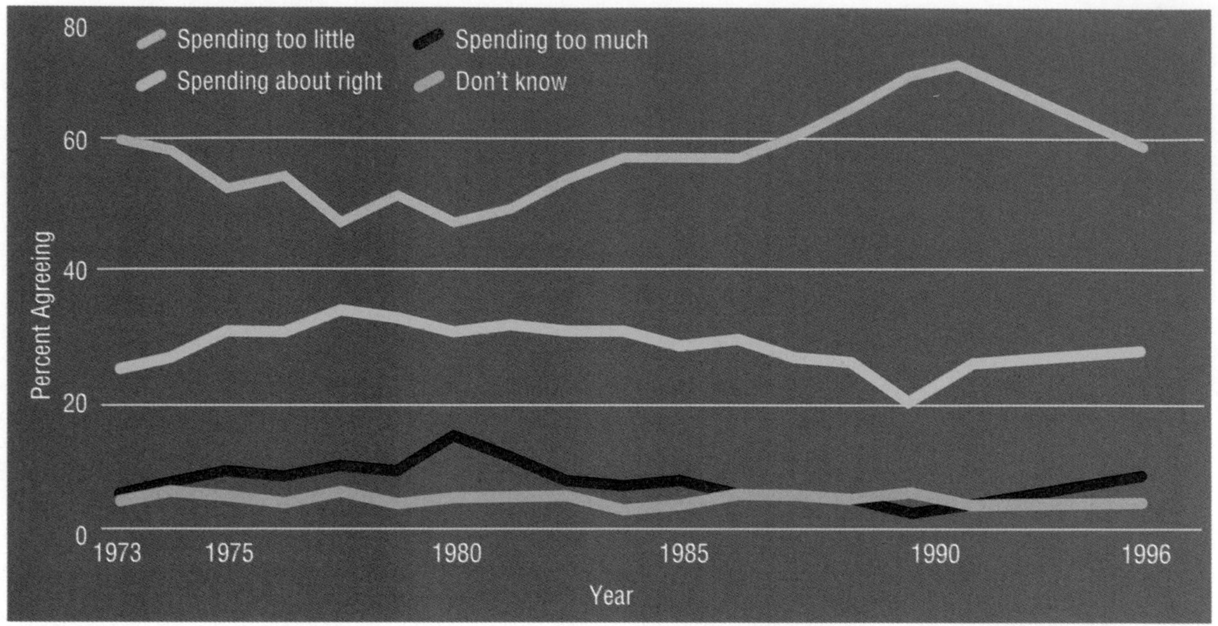

Source: NORC, 1996.

$2.5 billion industry (Rothenberg, 1990). Opinion polls help corporations make decisions about their products; political candidates use polls to measure voter support; and elected officials use polls to monitor public opinion on key political issues.

A good example of an opinion survey is the General Social Survey (GSS) conducted by the National Opinion Research Center (NORC). To get a better idea of how opinion surveys work, check the series of results from the GSS on environmental issues. Figure 1.2 shows how Americans responded to questions about levels of government spending for environmental preservation and regulations. Since those questions are asked almost every year (most recently in 1996), the responses give us a strong indication of trends in public opinion on this issue. Clearly, a majority of people in the United States favor environmental preservation (and spending for such programs).

The NORC General Social Survey is conducted each year, if possible, among a nationwide random sample of at least 1,500 adults. The term **sample** refers to a set of respondents selected from a specific population. When it is impossible or extremely costly to try to interview everyone in a population, social scientists often use samples to arrive at estimates of how certain phenomena are distributed in the population. In presidential elections, for example, it would be impossible to ask all potential voters what they think about the candidates, or at least it would be impossible to do so in a timely and cost-effective way. By asking a small, randomly chosen sample of people who represent the potential voters in terms of age, gender, region of the country, and so on, we can determine what proportion of that sample supports which candidate. The answers from that sample can then be generalized to the entire population. This can be done only if the respondents are chosen at random to ensure that they represent the larger population accurately and that no bias has entered into their selection. The statistics of sampling allows us to know within a specific margin of error how confident we can be that the findings from the sample would correspond to the opinions of the population as a whole. The larger the sample size, the smaller the margin of error will be.

The first step in selecting a sample is to define the population to be sampled. The next step is to establish rules for the *random* selection of respondents, with the goal of ensuring that within the specified population, everyone has an equal chance of being selected. A sample in which all potential respondents do not have the same probability of being included is called a *biased sample.* To avoid bias, respondents must be selected by some process of random sampling.[2] Only a random sample can be considered truly representative of the target population.

Before leaving the subject of opinion polling, one additional point should be made. Polls are subject to *sampling error.* A sampling error of plus or minus three percentage points, for example, would be normal in a national sample of 1,500 to 2,000 respondents. This is important information for anyone who reads and interprets poll data. It means that a difference of 3% or less between the percentages of American adults expressing certain opinions could be due to chance rather than to a real difference in the distribution of opinions. In other words, it is possible that just by chance a higher percentage of people with a certain opinion was included in the sample than is actually the case in the total population. The possibility of sampling error is more critical when there is a very small difference between two sets of responses—for example, when 51% of the sample favors one presidential candidate and 49% favors another. In that case, survey analysts would conclude that the election is "too close to call," as was the case in the 2000 presidential election.

Research Ethics and the Rights of Respondents

When asking questions in a survey, the researcher must be aware of the rights of respondents. Much sociological research deals with people's personal lives and inner thoughts. Although most research seems relatively innocent, there are times when the questions asked or the behaviors witnessed may be embarrassing, or worse. The federal government now requires that research involving human subjects be monitored by "human subjects review panels" at all research institutions that receive federal funding. The researcher must take precautions to protect the fundamental rights of human subjects, including privacy, informed consent, and confidentiality.

The right of **privacy** can be defined as "the right of the individual to define for himself, with only extraordinary exceptions in the interest of society, when and on what terms his acts should be revealed to the general public" (Westin, 1967, p. 373). **Confidentiality** is closely related to privacy. When a respondent is told that information will remain confidential, the researcher may not pass it on to anyone else in a form that can be traced to that respondent.

[2]Random sampling is accomplished by a variety of statistical techniques that are discussed in more advanced courses.

Informed consent refers to statements made to respondents (usually before any questions are asked) regarding what is being asked and how the information they supply will be used. It includes the assurance that their participation is voluntary—that is, that there is no compulsion to answer questions or give information.

The ethics of social research require careful attention. Before undertaking research in which you plan to ask people questions of any kind, be sure you have thought out the ethical issues and checked with your instructor about the ethics of your methods of collecting data and presenting the data to the public.

ANALYZING THE DATA

Survey research normally is designed to generate numerical data regarding how certain variables are distributed in the population under study. To understand how such data are presented and analyzed, refer to the Sociological Methods box on pages 24–25, which discusses some of the basic techniques of quantitative data presentation and analysis.

Correlations

The term **correlation** refers to a specific relationship between two variables: As one varies in some way, so does the other. Although the calculation of correlation is covered in statistics courses, for our purposes it will be helpful to know that the measure of correlation between two variables, termed the *correlation coefficient,* can vary between +1.0 and –1.0, with 0 representing no measured correlation at all. A correlation coefficient of +1 would indicate that the variables are positively and perfectly related—a change in one variable produces an equivalent change *in the same direction* (increase or decrease) in the other variable. A correlation coefficient of –1, on the other hand, would mean that the variables are perfectly inversely related—a change in one produces an equivalent change *in the opposite direction* in the other. In reality, most variables are not perfectly correlated, and correlation coefficients usually fall somewhere between these two extremes.

Although correlation can be very useful in the analysis of relationships among variables, it must not be confused with causation. Often a strong correlation that seems to indicate causality is in fact a spurious or misleading relationship. Take the example of storks and babies. The fable that storks bring babies seemed to be based on a statistical reality: In rural Holland

SOCIOLOGICAL METHODS

Understanding Tables

On approaching a statistical table, the first step is to read the title carefully. The title should state exactly what information is presented in the table, including the *units of analysis*—that is, the entity (e.g., individual, family, group) to which a measure applies. In Tables A and B, the units of analysis are households; in Table C, they are married-couple households.

Frequency Distributions

As you begin to study a table, make sure you understand what kind of information is being presented. The numbers in Table A are **frequency distributions.** For each year (1970 and 1998), they show how various types of households were distributed in the U.S. population. Frequency distributions indicate how many observations fall within each category of a variable. Thus, within the category "Married-Couple Family" of the variable "Type of Household," Table A indicates that there were 44,728 such households in the United States in 1970 and that the total increased to 54,317 by 1998.

Table A can tell us a great deal about social change in the United States. For example, social scientists use the term *household* to designate all the people residing at a given address, provided that the address is not a hospital, school, jail, army barracks, or other "residential institution." A household is not necessarily a married-couple family, as is evident from the categories listed in Table A. There are also many households in which a male, with or without his own children, or a female, with or without her own children, is the primary reporting adult. There are also millions of nonfamily households, in which the adults reporting to the census takers are not related (over 31 million in 1998). These households include roommates, cohabiting adults, nonfamily adults who do not consider themselves roommates, and many others.

In earlier times, it was considered unfortunate and a bit odd for a woman who was not a widow to head a household with children. It was assumed that a true family was composed of a male head, his wife, their children, and anyone related to them by blood who lived on their premises. The numbers in Table A show, however, that the number of female-headed households more than doubled between 1970 and 1998. The data also show that there was an increase in the number of family households (from 51 million in 1970 to almost 71 million in 1998) and almost as great a jump in the number of nonfamily households (from 12 million in 1970 to over 31 million in 1998).

Percent Analysis

Comparing the numbers for 1970 and 1998 in Table A can be misleading. Remember that these are absolute numbers. They show that the number of nonfamily households has increased and that the number of female-headed and male-headed households has also increased. However, the number of married-couple families has also increased, and so has the total number of households. How, then, can we evaluate the importance of these various changes?

To compare categories from one period to another, we need a way of taking into account changes in the overall size of the population. This can be done through **percent analysis,** as shown in the following example.

Table A shows that between 1970 and 1998 the number of male-headed households with one or more children rose from 341 to 1,798—about a fivefold increase. But how important is this increase in view of the increase in the total population between 1970 and 1998? By using the total number of households in each year as the base and calculating the percentage of each household type for that year, we can "hold constant" the effect of the increase in the total number of households. In the case of male-headed households with one or more children, we find that the increase as a percentage of all households (from 0.54% percent in 1970 to 1.75% in 1998) is not as important as the increase in absolute numbers would suggest. The calculation is as follows:

$$341{,}000 \;/\; 63{,}401{,}000 \times 100 = 0.54$$
$$1{,}798{,}000 \;/\; 102{,}528{,}000 \times 100 = 1.75$$

Table B presents each household type as a percentage of total households in that year. In this way, it eliminates the effect of the increase in overall population size between 1970 and 1998. The percentages in Table B reveal some significant changes that would not be evident from a comparison of absolute numbers. For example, although Table A seems to

TABLE A

Types of Households in the United States, 1970 and 1998 (in thousands)

Type of Household	1970	1998
All households	63,401	102,528
Family households	51,406	70,880
Without own children under 18	22,725	36,120
With own children under 18	28,732	34,760
Married-couple family	44,728	54,317
Without own children under 18	19,196	29,048
With own children under 18	25,532	25,269
Male householder	1,128	3,911
Without own children under 18	887	2,113
With own children under 18	341	1,798
Female householder	5,500	12,652
Without own children under 18	2,642	4,960
With own children under 18	2,858	7,693
Nonfamily households	11,945	31,648

Source: *Statistical Abstract,* 1999.

TABLE B

Types of Households in the United States, 1970 and 1998 (percentage of total households)

Type of Household	1970	1998
All households	100.0	100.0
Family households	81.2	69.1
Without own children under 18	35.8	35.2
With own children under 18	45.3	33.9
Married-couple family	70.5	53.0
Without own children under 18	30.3	28.3
With own children under 18	40.3	24.6
Male householder	1.9	3.8
Without own children under 18	1.4	2.1
With own children under 18	0.5	1.8
Female householder	8.7	12.3
Without own children under 18	4.2	4.8
With own children under 18	4.5	7.5
Nonfamily households	18.8	30.9

indicate that there has been an increase in married-couple families, Table B shows that when we take into consideration the overall increase in households, married couples actually declined as a proportion of the total, from 70.5% in 1970 to 53.0% in 1998. The proportion of female-headed families, especially those with children, increased during the period, as did the proportion of nonfamily households. But we cannot yet view these findings as definite. We must examine them carefully to be sure we are making the right kinds of comparisons. Look again at Table B. In the category of married-couple families, those with no children of their own decreased as a proportion of total households—from 30.3% in 1970 to 28.3% in 1998. But this apparent decrease is due to the overall increase in other types of households. It is still not clear whether, among married-couple families, those without children decreased in relation to those with children. Indeed, if we take only the married-couple households and compare categories within this category, as shown in Table C, we see that as a proportion of married-couple households, those without children actually increased from 42.9% to 53.5%.

TABLE C

Married-Couple Families, by Number of Own Children Under 18, 1970 and 1998 (percentage of total households)

	1970	1998
Married-couple families	100.0	100.0
Without own children under 18	42.9	53.5
With own children under 18	57.1	46.6

This example should convince you to pay close attention to the comparisons made in numerical tables. In this example, we see that married couples with children accounted for a smaller percentage of all married-couple households in 1998 than in 1970, but we would not have seen this without making the additional comparison presented in Table C.

SOCIOLOGY AND SOCIAL JUSTICE

Saad Ibrahim, the Egyptian sociologist who was jailed for supporting the democratic right to vote for the candidate of one's choice, is in good company among sociologists. In its first annual convention of the new millennium, members of the American Sociological Association honored a sociologist-social worker couple with the prestigious Distinguished Career Award for the Practice of Sociology. The recipients of the award, Frances Fox Piven and Richard Cloward, are famous for taking the lead in the fight for social justice. As fighters, they often make enemies, just as their Egyptian colleague did. So how does one explain the decision by the sociology profession's official organization to grant them this award? A definition of what is meant by social justice will help answer this question.

Social justice is a value that Americans hold dear even if they do not always understand its meaning. In the Pledge of Allegiance, for example, we pledge our fidelity to the flag and the nation and signal to ourselves and others that ours is a nation created under God, with liberty and justice for all. People who fight for social justice believe that we have not yet achieved justice for all. They believe that as long as some people are in prison largely because they are poor and could not afford adequate legal defense, we do not have social justice. As long as some people cannot achieve in school as well as others because their schools are deprived of adequate resources, we do not have social justice. As long as many people work for low wages in industries dominated by a small number of rich and powerful people, we do not have social justice. As long as some people are disadvantaged in society because of their skin color, gender, sexual preferences, religion, or immigrant status, we have yet to deliver on the promise of social justice.

By no means do all sociologists believe that members of their profession should be committed to fighting for social justice. Many believe that the sociologist's job is to present the facts about inequalities or injustices so that informed citizens and their leaders can act appropriately. Some social scientists, however, believe that it is not enough to present facts and interpretation. Action to correct injustices is also appropriate, they claim. Frances Fox Piven and Richard Cloward are two famous examples of social scientists who follow up their empirical work with direct action to combat social injustice. But the type of action they advocate is extremely important.

In their widely read books on the situation of poor people in the United States and elsewhere in the world, Piven and

until recent decades, there was a correlation between storks nesting in chimneys and the presence of babies in those households. The more storks, the more babies. In fact, however, the presence of babies in the home meant more fires in the fireplace and more heat going up the chimney to attract storks to nest there. Storks do not bring babies; babies, in effect, bring storks. But the real causal variable is heat, something that was not suggested in the original commonsense correlation.

Mapping Social Data

Data about individuals or households can be mapped in precise ways, providing a powerful tool for analyzing sociological data (Monnier, 1993). In a sense, the mapping of data is a form of correlation. The variable under study—poverty, crime, an illness, or any other variable—is mapped over a given territory, thereby correlating the variable with location.

Contemporary sociologists are making increasing use of sophisticated mapping techniques. For instance, see Figure 1.3, which maps the spatial distribution of white-collar workers in San Antonio, Texas. This example is from research by sociologist Michael White (1987), who studies the characteristics of neighborhoods in U.S. cities and towns. It shows that households with white-collar workers tend to be concentrated on the northern edges of the central city, extending outward into the northern suburbs. Of course, different cities will show different patterns of

Saad Ibrahim, the outspoken Egyptian sociologist and human rights activist, is shown here at a presentation he gave to students in Cairo after his release from jail for having criticized his government. The sociologist titled his lecture about his recent detention without trial "What I Did on My Summer Vacation."

Cloward have been in the forefront of the battle against poverty. They were highly critical of the old system of Aid to Dependent Children due to its stinginess and are even more vociferous critics of the current movement to end subsidies for poor families with children. While many political leaders and some social scientists applaud the policy of replacing cash payments to the poor with the demand that they work to receive supplemental benefits (workfare), Piven and Cloward (1993) point out that single mothers do not turn to welfare because they are pathologically dependent on handouts or unusually reluctant to work—they do so because they cannot get jobs that pay better than welfare. They assert that workfare programs seem better because they have been in effect only during a prolonged period of extremely high employment. Once the economy produces less jobs, the poor will be in worse shape than they were under the old system. This assertion can ultimately be proven only by empirical research. Unfortunately, if the facts show that Piven and Cloward are correct, it will be too late to help the millions of adults and children who have suffered unnecessarily.

Piven and Cloward (1996) believe in taking the lessons of their research to the streets, through protest organizations and social movements of poor people and those who sympathize with them. But such direct action, they argue, cannot be successful without positive change through electoral politics. This is why they were among the sponsors of the "motor voter" bill that Congress approved in 1995. The law, known officially as the National Voter Registration Act (NVRA), requires that states offer to register people to vote when they obtain or renew their driver's license or apply for benefits like Food Stamps, Medicaid, and disability services. These individuals may also register by mail. The law's supporters believe that the NVRA could do away with disparities in registration rates by age, income, and race. Approximately 1 million new voters have registered each month since the law went into effect, making it the most effective means of encouraging participation in the nation's political system since the nineteenth century. Thus, the fight for social justice, as Piven and Cloward conceive it, means that more people must get involved in democratic actions like political protests against injustice, but they need to back up those activities with their votes. ■

concentration. Such maps provide a useful way of showing how people's choices about where to live can shape an entire metropolitan region.

THEORIES AND PERSPECTIVES

The tables in the Sociological Methods box are valuable because they show how the nation's most important survey, the census, reveals fundamental changes in the structure of the population. But how do these facts about household composition relate to larger trends and issues? The census reveals trends, but it does not explain those trends unless an investigator armed with a set of hypotheses uses the facts to prove or disprove a theoretical point.

A **theory** is a set of interrelated concepts that seeks to explain the causes of an observable phenomenon. Some theories attempt to explain an extremely wide range of phenomena, while others limit their explanations to a narrower range. It is often said that sociology lacks highly predictive theories like those of the physical sciences, but this is debatable. If theories are judged by their ability to explain observable phenomena and predict future events, then sociology has some powerful theories.

At the beginning of the twentieth century, Émile Durkheim used his theory of social integration to

FIGURE 1.3
Percentage of White-Collar Workers in San Antonio, by Census Tract

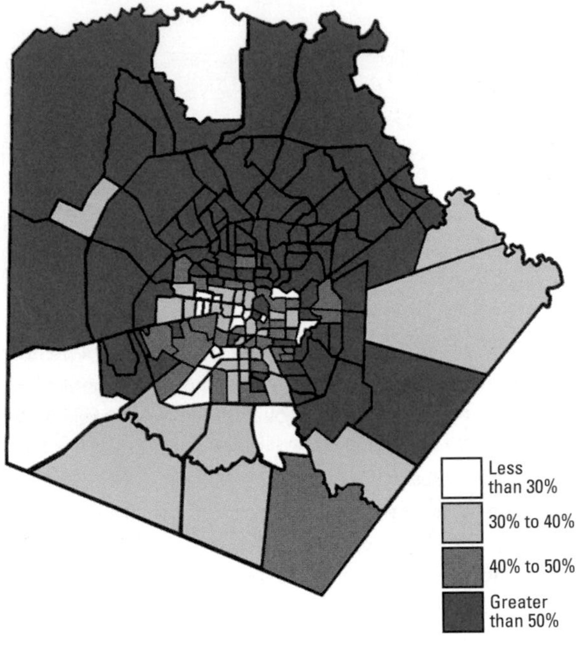

Source: White, 1987.

predict the social conditions that would increase rates of suicide. The same theory of social integration predicted the conditions that would lead to the rise of totalitarian regimes like that of the Nazis in Germany. Max Weber's theory of bureaucracy helps explain the experiences we encounter in all kinds of organizations. And Karl Marx's theory of class conflict is still the dominant explanation of the revolutions that occurred in feudal and early capitalist societies. No single sociological theory can explain all the complexities of human social life and social change, but in this respect, sociology is not very different from other social and physical sciences. The theory of relativity, for example, has not fully explained the physical forces that produced the universe, and different economic theories compete for the attention of policy makers.

To cope with the many levels of social explanation, sociologists employ **theoretical perspectives**—sets of interrelated theories that offer explanations for important aspects of social behavior. Like the methods of observation and analysis discussed in this chapter, theoretical perspectives are tools of sociological research. They provide a framework of ideas and explanations that helps make sense of data. The basic theoretical perspectives are the ones discussed earlier: interactionism, functionalism, and conflict theory. We also rely on the ecological perspective for descriptive data regarding communities and populations. At times, these perspectives offer competing explanations of social life; at other times, they seek to explain different aspects of society; and at still other times, they are combined in various ways.

TECHNOLOGY AND SOCIAL CHANGE

You do not have to become a professional sociologist for the study of sociology to make a difference in your life. As rapid social changes occur both in our immediate social environment and throughout the world, the need to anticipate and adjust to changes confronts us with many choices about how to live our lives. The person who has developed a sociological imagination and the ability to observe and study social relations has an advantage in adjusting to a changing world. This can readily be seen in the connections between sociology and technology.

Consider, for example, that the development of the personal computer was based to a large extent on sociological insights. It depended on insights into how computing technology could be moved from laboratories and corporations into the average home. Today the study of how home computers are actually used is a topic of great interest to sociologists, whose research is often consulted by the major technology corporations (Dunlop & Kling, 1991).

Throughout this book, you will see that sociological concepts and findings are powerful tools that can be applied in an almost infinite number of social situations. Sociological insights do not excuse individuals from responsibility for their actions, but they do make those actions far more comprehensible. In a world where some changes occur rapidly while other aspects of social life seem unchanging, it is important to gain some scientific knowledge about how cultures and societies shape our lives.

SUMMARY

Sociology is the scientific study of human societies and of human behavior in the groups that make up a society. The ability to see the world from this point of view has been described as the *sociological imagination.* Sociologists study social behavior at three levels of complexity: *micro, middle,* and *macro.*

Sociology developed in Europe in the nineteenth century. Among the outstanding early sociologists were Karl Marx, Émile Durkheim, and Max Weber. In the twentieth century, sociology developed most rapidly in North America. The approach known as *human ecology* developed under the leadership of Robert Park and Ernest Burgess; it emphasizes the relationships among social order, social disorganization, and the distribution of populations in space and time.

Functionalism is concerned primarily with the large-scale structures of society. *Interactionism* is a perspective that views social order and social change as resulting from all the repeated interactions among individuals and groups. *Conflict theory* emphasizes the role of conflict and power.

The basic steps of scientific research are deciding on the problem, reviewing the literature, formulating research questions, selecting a method, and analyzing the data. A *hypothesis* states a relationship between two *variables* that can be tested through empirical observation. Before collecting new data, a professional researcher reviews as much existing research as possible; this is known as a "review of the literature."

In *participant observation,* the researcher participates to some degree in the life of the people being observed. In a *controlled experiment,* the researcher establishes an *experimental group,* which will experience a "treatment," and a *control group,* which is similar to the experimental group in every way but will not experience the treatment. A sociological survey asks people to give precise information about their behavior and attitudes. Survey researchers must consider the rights of human subjects, especially *privacy, confidentiality,* and *informed consent.*

Sociological data are analyzed and presented in a variety of ways. These include statistical tables, *correlations,* and maps.

GLOSSARY

Sociology: The scientific study of human societies and human behavior in the groups that make up a society. **(5)**

Sociological imagination: According to C. Wright Mills, the ability to see how social conditions affect our lives. **(5)**

Micro-level sociology: An approach to the study of society that focuses on patterns of social interaction at the individual level. **(7)**

Macro-level sociology: An approach to the study of society that focuses on the major structures and institutions of society. **(7)**

Middle-level sociology: An approach to the study of society that focuses on relationships between social structures and the individual. **(7)**

Scientific method: The process by which theories and explanations are constructed through repeated observation and careful description. **(7)**

Human ecology: A sociological perspective that emphasizes the relationships among social order, social disorganization, and the distribution of populations in time and space. **(10)**

Functionalism: A sociological perspective that focuses on the ways in which a complex pattern of social structures and arrangements contributes to social order. **(10)**

Interactionism: A sociological perspective that views social order and social change as resulting from all the repeated interactions among individuals and groups. **(11)**

Conflict theory: A sociological perspective that emphasizes the role of conflict and power in society. **(15)**

Power: The ability of an individual or group to change the behavior of others. **(15)**

Hypothesis: A statement that specifies a relationship between two or more variables that can be tested through empirical observation. **(18)**

Variable: A characteristic of an individual, group, or society that can vary from one case to another. **(18)**

Dependent variable: The variable that a hypothesis seeks to explain. **(18)**

Independent variable: A variable that the researcher believes causes a change in another variable (i.e., the dependent variable). **(18)**

Participant observation: A form of observation in which the researcher participates to some degree in the lives of the people being observed. **(19)**

Unobtrusive measures: Observational techniques that measure behavior but intrude as little as possible into actual social settings. **(20)**

Controlled experiment: An experimental situation in which the researcher manipulates an independent variable in order to observe and measure changes in a dependent variable. **(20)**

Experimental group: In an experiment, the subjects who are exposed to a change in the independent variable. **(20)**

Control group: In an experiment, the subjects who do not experience a change in the independent variable. **(20)**

Field experiment: An experimental situation in which the researcher observes and studies subjects in their natural setting. **(21)**

Sample survey: A survey administered to a selection of respondents drawn from a specific population. **(21)**

Sample: A set of respondents selected from a specific population. **(22)**

Privacy: The right of a respondent to define when and on what terms his or her actions may be revealed to the general public. **(23)**

Confidentiality: The promise that the information provided to a researcher by a respondent will not appear in any way that can be traced to that respondent. **(23)**

Informed consent: The right of respondents to be informed of the purpose for which the information they supply will be used and to judge the degree of personal risk involved in answering questions, even when an assurance of confidentiality has been given. **(23)**

Correlation: A specific relationship between two variables. **(23)**

Frequency distribution: A classification of data that describes how many observations fall within each category of a variable. **(24)**

Percent analysis: A mathematical operation that transforms an absolute number into a proportion as a part of 100. **(24)**

Theory: A set of interrelated concepts that seek to explain the causes of an observable phenomenon. **(27)**

Theoretical perspective: A set of interrelated theories that offer explanations for important aspects of social behavior. **(28)**

QUESTIONS FOR THOUGHT AND DISCUSSION

1. Identify a social fact that pertains to your society (e.g., "The majority of Americans finish high school"). Why is this fact not explained by "human nature"? Is it likely to change in the near future? Has it changed in the past?

2. What are some of the methods used in the science of sociology? Why are there often ethical problems in applying the experimental method to the study of human groups and societies?

Digging Deeper

Books

The Sociological Imagination (C. Wright Mills; Oxford University Press, 1959). Still the best and most passionate statement of what the sociological imagination is, by the person who coined the phrase.

Classic Disputes in Sociology (J. A. Anderson et al., eds.; Allen & Unwin, 1987). A valuable introduction to the basic questions and debates in sociology.

The Handbook of Sociology (Neil J. Smelser, ed.; Sage Publications, 1988). A collection of reviews of the literature in major sociological fields.

The Basics of Social Research (Earl Babbie; Wadsworth, 1999). A popular primer on social research methods.

In the Field: Readings on the Field Research Experience, 2nd ed. (Carolyn D. Smith & William Kornblum, eds.; Praeger, 1996). A selection of personal accounts by a group of noted ethnographic researchers, designed to give students a sense of what it is actually like to conduct ethnographic research, especially participant observation.

The Practice of Social Research (Earl Babbie; Wadsworth, 1992). A practical text that introduces the ways in which sociologists actually formulate their research questions and conduct their research.

Journals

American Sociological Review. The journal of the American Sociological Association. Articles in this journal are often quite technical and may be somewhat advanced for the beginning student, but they offer a good perspective on current research.

American Journal of Sociology. The oldest journal in sociology; a treasure trove of articles going back to the early decades of the twentieth century. Consult the index for earlier papers, and recent issues for excellent new research.

Other Sources

Sociological Abstracts. A set of reviews of existing literature on a variety of social-scientific subjects. The abstracts are organized by topic and offer brief overviews of original research papers and other articles.

Census of the United States. An invaluable tool for all the social sciences, but one that requires a patient user. There are volumes for all states and cities in the United States. The tables show population totals and subtotals by sex, age, race, and occupation.

Exploring Sociology on the Internet

Sociology Department, Princeton University
www.princeton.edu/~sociolog/

A good example of how modern sociology departments are using the Internet. Offers links with other sociology departments on each continent, domestic and international research institutes, data archives, and Web pages for academic journals.

The Dead Sociologists' Society
http://diogenes.baylor.edu/www.providers/Larry_Ridener/dss/deadsoc.html

A page with a good sociological sense of humor. Provides links to pages devoted to some of the founders of sociology, such as Émile Durkheim, Karl Marx, Max Weber, and many others.

American Sociological Association
www.asanet.org

The official page of the American Sociological Association aims to serve sociologists in their work, advance sociology as a science and profession, and promote the use of sociology in society.

Society for Applied Sociology
www.indiana.edu/~appsoc/

The official site of the Society for Applied Sociology. Provides a forum for sociologists and others interested in applying sociological knowledge.

Chapter 2

CULTURE

Does culture account for the immense differences in the ways people throughout the world feel and behave?

A group of children in two schools in the Central American nation of Nicaragua have become the subject of one of the most important social-scientific discoveries of our time. The children, all of whom have been deaf since birth, were placed together in two schools in Managua, the nation's capital. Before the revolution of 1979, no formal education had been provided for deaf children. The revolutionary government created schools for hundreds of such children, who were then recruited from all parts of the small nation. And so began one of the most fascinating "natural" or unintended experiments in recent memory (Osborne, 1999).

Since their education had been entirely neglected, the children arrived in the city with only rudimentary ability to make pantomime gestures for common terms like "eat," "drink," and "ice cream." There are over 200 existing sign languages used by hearing-impaired people around the world, but the Nicaraguan children had not been exposed to any of them. Many of the pantomimes, or *mimicas* as they are called in Spanish, were highly individual. The children, isolated from teachers and other deaf children, had developed them alone or with their families. The rather inexperienced teachers in the new schools were frustrated because they had been taught old techniques of "finger spelling," which uses simple finger signs to make letters and spell out the words of spoken languages. But of course the children had no knowledge of Spanish or other spoken languages.

To their immense surprise, however, the instructors saw that the children, once placed together, began to build on each other's signs. "One child's gestures solidified into the community's word." The children also invented ways to link the words together in sentences. They were developing an entirely new sign language complete with grammatical structure and the ability to communicate the same complexities of thought and feeling that other humans do with verbal language. According to Steven Pinker, who studies how people acquire language, it is the first time "that we've actually seen a language being created out of thin air" (quoted in Osborne, 1999, p. 85; see also Pinker, 1994).

Anselmo Alemán, now 20, did not come to the school until he was 15, a relatively late start. He is now fluent in Nicaraguan Sign Language.

There are many important aspects of this story, but for sociologists, it is a striking example of the importance of language in human life. The children had language capabilities that could not be developed by their teachers. Instead, they used their own resources to develop their own language. With that language, they could then communicate with one another and with sign-language experts who came to study how this new language worked. Once the language rules the children had developed were known, teachers could use the new language to teach them how to read and understand Spanish. The children then had not only their own language but a way of entering the larger culture of their society. ■

THE MEANING OF CULTURE

In everyday speech, the word *culture* generally refers to pursuits like literature and music. But these forms of expression are only part of the definition of culture. To the social scientist, "a humble cooking pot is as much a cultural product as is a Beethoven sonata" (Kluckhohn, 1949, quoted in Ross, 1963, p. 96). Societies are populations that are organized to carry out the major functions of life. A society's culture consists of all the ways in which its members think about their society, understand its symbols and rules, and communicate among themselves.

Culture can be defined as all the modes of thought, behavior, and production that are handed down from one generation to the next by means of communicative interaction—speech, gestures, writing, building, and all other communication among humans—rather than by genetic transmission, or heredity. This definition encompasses a vast array of behaviors, technologies, religions, and so on—in other words, just about everything made or thought by humans. Notice how central language is in the definition of culture. Thinking, speaking, and writing are made possible by the ability to master language. Language is a cultural universal; that is, all people in all societies have language, although languages usually differ from one culture to another. When the deaf children of Nicaragua evolved their own language, they were also creating a culture. When that language and culture became intelligible to the children's teachers, the children could communicate with adults and thus begin to learn much more about their parents' culture than they were able to master before.

Among all the elements of culture that could be studied, sociologists are most interested in aspects that explain social organization and behavior. Thus, although they may analyze trends in movies and popular music (which significantly affect behavior in the modern world), they are even more likely to study aspects of culture that account for phenomena like the behavior of people in corporations or the conduct of scientific research.

Dimensions of Culture

The culture of any people, no matter how simple it may seem, is a complex set of behaviors and artifacts. People master their culture as children, and as adults they often do not notice how their culture shapes almost every aspect of their lives, from the way they express themselves to the food they eat to the way they think about what is right and wrong. A useful framework for thinking about culture was suggested by Robert Bierstedt (1963). He views culture as having three major dimensions: **ideas,** or ways of thinking that organize human consciousness; **norms,** or accepted ways of doing or carrying out ideas; and **material culture,** or patterns of possessing and using the products of culture. Figure 2.1 presents some examples of these three dimensions, along with two aspects of culture—*ideologies* and *technologies*—that combine elements from more than one dimension.

Ideas. As indicated in Figure 2.1, theories about how the physical world operates (scientific ideas), strongly held notions about what is right and wrong (values), and traditional beliefs, legends, and customs (folklore) are among the most important types of ideas found in any given culture. Of these, values are especially important because people feel so strongly about them and because they often undergo changes that result in social conflict (as we can see in contemporary debates about "a woman's right to control her reproductive destiny" versus "a fetus's right to life").

Values are socially shared ideas about what is right. Thus, for most people in North America, education is a value; that is, they conceive of it as a proper and good way to achieve social standing. Loyalty to friends and loved ones, patriotism, the importance of religion, the significance of material possessions—these and other values are commonly found in our culture and in many others, but of course there are wide differences in how people interpret these values and in the extent to which they adhere to them. In small tribal and peasant societies, there tends to be far greater consensus on values than in large, complex industrial societies. In a large and diverse society like ours, there is likely to be a good deal of conflict over values. Some people are satisfied with the way wealth and power are distributed, for example, while others are less satisfied with the status quo. Some people feel that it is desirable to attempt to improve one's own well-being and that society as a whole gains when everyone strives to be financially secure. Others assert that the value of individual gain conflicts with the values of community and social cohesion; too great a gap between the well-off and the less well-off creates suffering, envy, crime, and other problems (Etzioni, 1997).

Norms. From people's beliefs about what is right and good—that is, from their values—are derived the

FIGURE 2.1
Dimensions of Culture

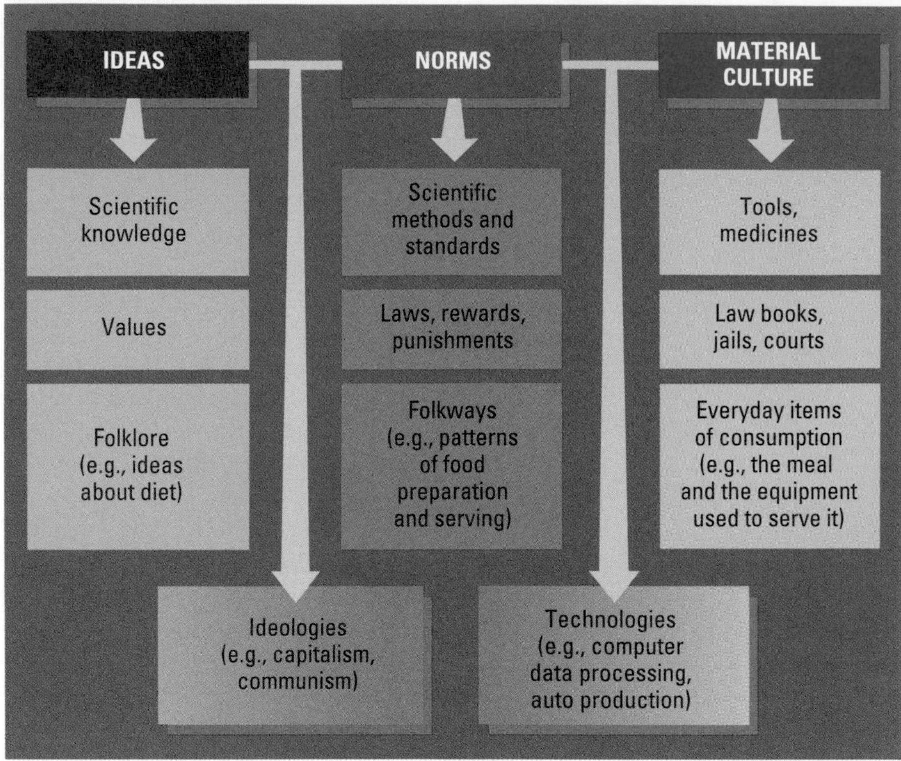

norms of a society. Values are more abstract than norms; they are the ideas that support or justify norms. Norms are specific rules of behavior or, as Robert Nisbet wrote, "the adjustments which human beings make to the surrounding environment. We may think of them as solutions to recurring problems or situations" (1970, p. 225). But norms involve more than behavior. Any given norm is supported by the idea that a particular behavior is correct and proper or incorrect and improper. "The moral order of society is a kind of tissue of 'oughts': negative ones which forbid certain actions and positive ones which [require certain] actions" (Nisbet, 1970, p. 226).

If we think about a complex aspect of everyday life like driving a car, it is evident that without norms life would be far more chaotic and dangerous. When we drive, we keep to the right, obey traffic lights and speed limits, and avoid reckless behavior that could cause accidents. These are among the many norms that allow the automobile to be such an essential article of North American culture.

Examples of norms are easy to find. Take the college or high school classroom. The classroom is organized according to norms of educational practice: There should be a textbook or books; there should be class discussion; there should be assignments, exams, and grades. Note that these norms correspond to actual behavior. Norms usually refer to behavior that we either approve or disapprove of. However, members of a culture often disagree about how a particular norm operates. Thus, in the case of classroom organization there is considerable debate about what educational practices are most effective. In Israel, for example, there is debate among orthodox and liberal Jews over whether boys and girls should be separated in classrooms, and similar debates divide fundamentalist and more secular Muslims in other Middle Eastern societies.

Laws are norms that are included in a society's official written codes of behavior. Laws are often developed by a specialized occupational group, such as priests in the ancient world and legislators, judges, and lawyers in the modern world. Some of the oldest

Creating a democratic society means more than instituting free elections and the norm of "majority rule." A young American lawyer working in Cambodia, where this photo was taken, explains: "When I came here in 1994 the courthouses were in ruins. They had to start a legal system literally from scratch. No more beating, no more cigarette burns, no more hitting with a rifle butt. You could clear out the prisons today if you reviewed the cases on the basis of procedural error" (quoted in Mydans, 1997, p. A1).

examples of laws include the Code of Hammurabi, the ancient Babylonian code that specified punishments equal to the gravity of the crime (e.g., an eye for an eye, a tooth for a tooth); the Ten Commandments, written by God, according to the Old Testament, on stone tablets at least 1,600 years before the birth of Christ; and the codes of Confucius, written and interpreted by the royal scribes of the ancient Chinese empire. In our time, laws are the special province of professional lawmakers, but all citizens participate in a network of laws and other written regulations that govern daily life. Basic behaviors like driving, going to school, getting married, and investing are greatly influenced by laws governing the conduct of these activities.

Ideologies. As indicated in Figure 2.1, **ideologies** combine two dimensions of culture; they are sets or systems of ideas and norms. Ideologies combine the values and norms that all the members of a society are expected to believe in and act upon without question. A classic study of the emergence of an ideology was Max Weber's (1974/1904) analysis of the link between Protestantism and capitalism *(The Protestant Ethic and the Spirit of Capitalism).* Weber noticed that the rise of Protestantism in Europe coincided with the rise of private enterprise, banking, and other aspects of capitalism. He also noticed that a majority of the most successful early capitalists were Protestants. Weber hypothesized that their religious values taught them that salvation depended not on good deeds or piety but on how they lived their entire lives and particularly on how well they adhered to the norms of their "callings" (occupations). As a result, the Protestants placed a high value on frugality and abstinence. To prove that they were worthy of salvation, they devoted themselves tirelessly to commerce and plowed their profits back into their firms. But Weber found that Catholics, who did not share these values, were less single-mindedly dedicated to their business ventures. They often spent their profits on good deeds rather than investing them in their businesses. Weber attempted to show how a set of religious values and norms combined with economic norms to create the ideology of capitalism.

A contemporary example of an ideology may be found among religious fundamentalists. Christian fundamentalists in the United States generally share a set of ideas and norms of behavior that include the value of prayer, the value of family and children, the negative value of abortion and secular humanism (an ethical system based on scientific knowledge rather than on religious teachings), and the belief in salvation and redemption for one's sins.

Material Culture. The third dimension of culture identified by Bierstedt is material culture. Material culture consists of all the things that a society produces. Mundane things like pots and pans or the wooden eating bowls of nonindustrial societies; immensely complex systems of things, such as the space shuttle; cherished items of religious worship like rosary beads or Indian fetish necklaces—all take their shape and purpose from the ideas of the culture that produces them. Members of societies that place a high value on science and efficiency are accustomed to seeing these values expressed in material objects. For example, we design our houses to conserve energy and create desirable combinations of view and privacy. We may take older forms like the ranch house or the Cape Cod bungalow and modify them to suit modern purposes; in this way, our houses combine tradition with usefulness. The particular form that appeals to us is usually a result of many different ideas, including what we know from our own upbringing, what we can afford, how much space we require, and the environmental conditions we anticipate.

Technologies. Technologies are another aspect of culture that spans two of Bierstedt's major dimensions. (See Figure 2.1.) **Technologies** are the things (material culture) and the norms for using them that are found in a given culture (Bierstedt, 1963; Ellul, 1964). Without the norms that govern their use, things are at best confusing and at worst useless or dangerous. In the United States, for example, new telecommunications technologies based on computers, modems, and networks like the Internet are a powerful means of communicating information over long distances. But norms for using these new technologies—such as norms about privacy, freedom of speech, and personal accountability—are only now developing, often out of serious conflicts about what content should and should not be tolerated on the networks.

Gun ownership is another case in which norms are very much in dispute. Most Americans agree that people should be free to own and use hunting weapons—but what about handguns, assault rifles, machine guns, and rocket launchers? What, if any, restrictions should there be on the distribution and use of such weapons? As we will see in later chapters, the debate about appropriate norms for controlling the distribution and possession of deadly firearms that are clearly not designed for hunting is an indication of a deeper cultural conflict over the right to bear arms versus the nation-state's monopoly over the use of force.

Norms and Social Control

Norms are what permit life in society to proceed in an orderly fashion without violence and chaos. Shared norms contribute enormously to a society's ability to regulate itself without resorting to the coercive force of armies and police. The term **social control** refers to the set of rules and understandings that control the behavior of individuals and groups in a culture. Park and Burgess (1921) noted that throughout the world there are certain basic norms that contribute to social control. "All groups have such 'commandments' as 'Honor thy father and mother,' 'Thou shalt not kill,' 'Thou shalt not steal'" (p. 787).

The wide array of norms that permit a society to achieve relatively peaceful social control is called its **normative order.** It is the normative order that creates what Morris Janowitz has termed "the capacity of a social group, including a whole society, to regulate itself" (1978, p. 3). This self-regulation requires a set of "higher moral principles" and the norms that express those principles. The norms define what makes a person a "good" member of the culture and society.

The most important norms in a culture are often taught as absolutes. The Ten Commandments, for example, are absolutes: "Thou shalt not kill," "Thou shalt not steal," and so on. Unfortunately, people do not always extend those norms to members of another culture. For example, the same explorers who swore to bring the values of Western civilization (including the Ten Commandments) to the "savage" Indians of the New World thought nothing of taking the Indians' land by force. Queen Elizabeth I of England could authorize agents like Sir Walter Raleigh to seize remote "heathen and barbarous" lands without viewing this act as a violation of the strongest norms of her own society (Jennings, 1975; Snipp, 1991). Protests by the Indians often resulted in violent death, but the murder of Indians and the theft of their land was rationalized by the notion that the Indians were an inferior people who would ultimately benefit from European influence. In the ideology of conquest and colonial rule, the Ten Commandments did not apply.

In the social sciences, punishments and rewards for adhering to or violating norms are known as **sanctions.** Cultural norms vary according to the degree of sanction associated with them. Rewards can range from a smile of gratitude to the Nobel Prize; punishments may vary in strength from a raised eyebrow to capital punishment. During the period of European colonial expansion, the murder of a native was far less strongly sanctioned than was the murder of another European.

The most strongly sanctioned norms are called **mores.**[1] They are norms that people consider vital to the continuation of human groups and societies, and therefore they figure prominently in a culture's sense of morality. **Folkways,** on the other hand, are far less strongly sanctioned. People often cannot explain why these norms exist, nor do they feel that they are essential to the continuation of the group or the society. Both terms—*mores* and *folkways*—were first used by William Graham Sumner in his study *Folkways* (1940/1907). Sumner also pointed out that laws are norms that have been enacted through the formal procedures of government. Laws often formalize the mores of a society by putting them into written form and interpreting them. But laws can also formalize folkways, as can be seen, for example, in laws governing the wearing of clothing in public places.

Sumner also pointed out that a person who violates mores is subject to severe moral indignation whereas one who violates folkways is not. People who violate folkways such as table manners or dress codes

[1]This term, pronounced "morays," is the plural of the Latin word *mos,* meaning "custom."

may be thought of as idiosyncratic or "flaky," but those who violate mores are branded as morally reprehensible. Thus, among prisoners there are norms for the treatment of other prisoners, based on the types of crimes they were convicted of. Rapists and child molesters, for example, are moral outcasts; their offenses are the most reprehensible. Such offenders are often beaten and tormented by other prisoners.

The Sociological Methods box on page 40 presents a typology of norms according to their degree of sanction and their mode of development—that is, whether they are formal or informal. Laws and other norms, such as company regulations and the rules of games and sports, are known as *formal norms.* They differ from *informal norms,* which grow out of everyday behavior and do not usually take the form of written rules, even though they too regulate our behavior. For example, when waiting to purchase a movie ticket, it is usually permissible to have one member of a small group save a place in line for the others, who may return later. And in a "pickup" basketball game, a player can call a foul and the opposing player usually cannot contest the call. Of course, there are times when such norms are disputed, depending on how the people involved define the situation. In the case of the basketball game, when the player on whom the foul is called disagrees with the call—and the score is extremely close—different definitions of the situation can lead to conflict.

CULTURE, EVOLUTION, AND HUMAN BEHAVIOR

Of all the species of living creatures on this planet, human beings are the most widely distributed. The early European explorers—Columbus, Magellan, Cook, da Gama, and many others—marveled at the discovery of human life thriving, more or less, in some of the earth's most inhospitable environments. Nevertheless, they were convinced that the "savages" they encountered were inferior to the more powerful Europeans; in fact, they had difficulty accepting the idea that the native peoples were fully human. In other words, they considered them biologically as well as culturally inferior.

Groups that are in conflict often accuse their enemies of being biologically inferior. The Nazis claimed that the Jews were a biologically inferior people. Serbian Orthodox Christians think of the Muslim Bosnians as innately inferior, just as some people in the United States believe that African Americans and other groups, like Puerto Ricans, are inferior. These and other examples of racial and ethnic prejudice will be discussed more fully in later chapters. Here we need to emphasize that differences in culture are frequently viewed (erroneously) as innate biological differences between distinct peoples.

Students of sociology must see beyond this tendency to confuse cultural and biological differences (Eisley, 1961). To do so, one needs to understand the difference between biological and cultural evolution.

Biological Versus Cultural Evolution

Charles Darwin's (1809–1882) revolutionary theory of **natural selection** is the central explanation of how living species evolve to adapt to their changing environments. It is based on the observation that unexpected physical changes, or *mutations,* in organisms occur more or less randomly from one generation to the next. When those mutations improve an individual organism's ability to survive in its environment, the new traits are "selected for"; that is, an individual that possesses those traits is more likely to survive, and hence more likely to reproduce and pass on its traits to the next generation, than individuals that lack the new traits. Over a few generations, such mutations can become so extensive that two species are created where formerly there was only one. This process of natural selection accounts for the great diversity of animal and plant life on the earth and for the ability of animals and plants to adapt to new environments.

Darwin's theory was based on his empirical observations of the natural world, especially those he had made as a scientist aboard the HMS Beagle during an extensive voyage of exploration in the early 1830s. On that voyage, Darwin had observed, in both fossils and living specimens, that some species of birds, turtles, and other animals had modified their physical form in ways that seemed to help them "fit" or "adapt" to their environment. It took him almost two decades of study and reflection to make sense of his observations. After all, much of what he had observed directly challenged the fundamental beliefs of most of the religious and scientific leaders of his day. For when Darwin considered all the information, and especially his observations of similar species on different islands, it became clear to him that God had not created all the living things on earth at once—instead, they had been evolving through natural selection over

SOCIOLOGICAL METHODS

Constructing a Typology of Norms

Typologies are ways of grouping observable phenomena into categories in order to identify regularities in what may appear to be a great variety of observations. For example, there are so many social norms that the average person has no hope of ever sorting them out without some kind of system for organizing them. In the table presented here, the subject is norms of various types. The sociologist constructs the types by comparing various dimensions along which norms may differ—such as whether they developed formally or informally or whether their degree of sanction (reward or punishment) is strong or weak.

		Relatively Weak	Relatively Strong
		Degree of Sanction	
Mode of Development	**Informal**	Folkways, fashions	Taboos, mores
	Formal	Misdemeanor laws, some rules, guidelines, civil rights laws	Capital-offense laws, felony laws

The norms listed in the Ten Commandments, for example, differ in the way they developed and are generally observed by society. The norm "Thou shalt not kill" not only is generally believed and passed along from one generation to the next but also is formally codified in laws. So is the commandment "Thou shalt not steal," which is an important part of the written legal code of our society. But the commandment "Remember the Sabbath day, to keep it holy" is not a written law in the United States, at least not in the federal statutes. In some states and communities, however, there are laws that specify that businesses must be closed on Sundays—even though not all religious groups recognize Sunday as the Sabbath. In addition, ideas about what behaviors are appropriate on the Sabbath are changing, and laws governing those behaviors are being challenged. These differences indicate that norms may differ according to whether they are informally taught to new generations or whether they are formal, written "laws of the land."

Another dimension along which norms may differ is the degree to which they are sanctioned—that is, the degree to which adherence is rewarded and violation is punished. The norm that men do not wear hats indoors is relatively weak. On the other hand, the norm that men and women do not casually display (or "flash") their genitals is strongly sanctioned.

Using these two comparative dimensions—mode of development (formal vs. informal) and degree of sanction (weak vs. strong)—we can create four categories: (1) norms that are informal and are weakly sanctioned (e.g., table manners, dress fashions); (2) norms that are informal but are strongly sanctioned (e.g., adultery); (3) norms that are part of the formal legal code but are weakly sanctioned (parking regulations, antismoking laws); and (4) norms that are formal laws and are strongly sanctioned (e.g., capital offenses like the murder of a police officer or treason in wartime).

By juxtaposing the two dimensions, each with its two categories, we create a fourfold classification of norms (see table). But as Max Weber observed, such a classification is an "ideal-typical" arrangement of observations in a form that accentuates some aspects and neglects others (Weber, 1949, in Gerth & Mills, 1958). These ideal types (folkways, mores and taboos, misdemeanor laws, and felonies or capital offenses) are useful because they establish standards against which to compare the norms of other cultures. For example, not all cultures treat the norm about the Sabbath the way Americans do. In parts of Israel or the Islamic lands, one could be arrested for violating the Sabbath laws. Typologies like this one also help identify areas of social change, as when one compares the way the Sabbath was treated in American laws early in the twentieth century, when stores were obliged to close on Sunday, with the way it is treated today, when Sunday is often viewed as another shopping day.

But the aspects of society described by ideal types are rarely so uncomplicated in real life. Even the norms that seem most formal and unambiguous, such as the prohibition against murder, become murky under some conditions, as in cases of self-defense or in war or in arguments about the death penalty and abortion. Studies of how actual behavior departs from the ideal-typical version invariably offer insights into how cultures and social structures are changing in the course of daily life. ■

many millions of years. And this theory could be applied to all other species, including humans.

The theory of natural selection had a dramatic impact on biological science but an even greater effect on the prevailing views of human society. In the late nineteenth and early twentieth centuries (and continuing to our own time), many social thinkers were profoundly influenced by what they thought were the lessons of biological evolution for human societies (Spencer, 1874; Sumner, 1963/1911). These **social Darwinists,** as they were called, concluded from Darwin's research on evolution that human inequality—the fact that some are rich and powerful while many others are poor and powerless—was itself the result of biological evolution. The first sociologist to become world famous, Herbert Spencer of England, coined the phrase "survival of the fittest," by which he meant that the people who are most successful at adapting to the environment in which they find themselves—that is, the better-educated, wealthier, more powerful people—are most likely to survive and to have children who will also be successful. It should be no surprise that the ideas of the social Darwinists were quite popular with powerful people in society: They seemed to imply that existing forms of inequality are justified by innate, genetically based differences in ability among humans.

Sociobiology. The tendency to explain social phenomena in terms of biological causes such as physiology or genes is known as *biological reductionism.* For example, theories proposing that there are genes that produce criminal behavior reduce the explanation of crime to biological causes. Some form of biological reductionism has emerged in every generation since Darwin's time.

The most recent version of biological reductionism is **sociobiology.** This term, coined by the Harvard biologist Edward O. Wilson (1975), refers to efforts to link genetic factors with the social behavior of animals. When applied to human societies, sociobiology has drawn severe criticism from both social scientists and biologists. Nevertheless, some sociologists support the sociobiological hypothesis that genes can explain certain aspects of human society and behavior (Caplan, 1978; Fox, 1989; Mascie-Taylor, 1990; Maxwell, 1991; Quadagno, 1979; Sahlins, 1976; van den Berghe, 1979). Because this hypothesis is so controversial, it deserves a closer look.

Consider, for example, the incest taboo, one of the strongest and most widespread norms in human life. The social scientist tends to explain the incest taboo as a cultural norm that is necessary for the existence of the family as a social institution. The family is an organized group with a need for well-defined statuses and roles. Should the different statuses within the family become confused, as would undoubtedly happen if sexual intimacy were permitted between children and their parents or between brothers and sisters, it would be difficult to maintain the family as a stable institution (Davis, 1939; Malinowski, 1927).

Sociobiology takes a different view. For the sociobiologist, the incest taboo develops from something other than cultural norms:

> [The incest taboo has] a deeper, more urgent cause, the heavy physiological penalty imposed by inbreeding. Several studies by human geneticists have demonstrated that even a moderate amount of inbreeding results in children who are diminished in overall body size, muscular coordination, and academic performance. More than 100 recessive genes have been discovered that cause hereditary disease . . . a condition vastly enhanced by inbreeding. (Wilson, 1979, p. 38)

Throughout most of the history of human evolution, sociobiologists point out, humans did not have any knowledge of genetics. Thus, "the 'gut feeling' that promotes . . . sanctions against incest is largely unconscious" (Wilson, 1979, p. 40). Individuals with a strong predisposition to avoid incest passed on more of their genes to the next generation because their children were less likely to suffer from the illnesses that result from inbreeding. And over many centuries of natural selection of individuals who did not inbreed, humans developed "an instinct [to avoid inbreeding] which is based on genes" (Wilson, 1979, p. 40).

This leap, from the observation of a strong and persistent norm like the incest taboo to the belief that certain human behaviors are genetically programmed, is an example of the sociobiological hypothesis regarding human nature. Sociobiologists have proposed a hypothesis in which not only the incest taboo but also aggression, homosexuality, and religious feelings are genetically programmed, and they believe that future discoveries by geneticists will prove their hypothesis correct.

Although it is true that genes set limits on human abilities and can be shown to influence many aspects of brain functioning, the hypothesis that genetic programming establishes complex forms of normative behavior is not supported by direct evidence: There is as yet no proof that such genes or sets of genes actually exist (Lewontin, 1992). Nevertheless, the rules of science require that we not reject the sociobiological hypothesis and that it remain an open area of investigation.

Cultural Evolution. The counterargument to sociobiology is that the human brain and other physical attributes are products of interaction between cultural and biological evolution and that in the past 100,000 years of human evolution there has been relatively little organic change in our species. Instead, the important developments in human life have occurred as a result of social and cultural changes. This widely accepted view, known as **cultural evolution,** denies that humans have innate instincts, such as an instinct to avoid incest. It argues instead that the great advantage of culture in human evolution is its creation of a basis for natural selection that would not be dependent on genes but would allow humans to adapt relatively quickly to any physical or social environment (Geertz, 1973).

Archaeological evidence shows that humans were using tools for primitive agriculture and making jewelry and personal adornments more than 30,000 years ago (Stevens, 1988). No doubt, they began using fire even earlier. Once humans could use fire and hand tools and simple weapons, the ability to make and use these items increased the survival chances of those who possessed these skills. This would create conditions for natural selection in which the traits being "selected for" were those that had to do with the manipulation of cultural and social aspects of life. These abilities (dexterity, linguistic ability, leadership, social skills, and so on) would in turn influence the further development of the human brain, again through the process of natural selection. Note that theories of cultural evolution do not deny the importance of natural selection. They emphasize, however, that the human traits that were selected for and became increasingly prevalent as innate, biological aspects of human individuals tended to be those that improved their ability to master and improve on their own cultures.

This view is supported by research on primate behavior showing that the higher primates use simple tools and that they teach this cultural technique to their young (Goodall, 1968; Schaller, 1964). For example, chimpanzees use sticks to probe for termites, and chew leaves to produce a pulp to be used as a sponge to draw water out of tree stumps (Peterson & Goodall, 1973). These and many other instances of rudimentary culture among animal species demonstrate that culture is not unique to humans; that is, we are not alone in using culture to aid in adaptation. What happened in the case of humans that did not happen (or at least has not happened yet) in any other species is that at a certain stage in human prehistory our ability to alter our cultures in response to changing conditions developed so quickly that the human species entered a new realm of social life. In other words, culture became self-generating, especially through the "breakthrough" developments of language, speech, and eventually writing. Once it had been freed from genetic constraints, culture had no limits.

LANGUAGE AND CULTURE

Perhaps the most significant of the inventions made possible by culture is language. In fact, the learning of culture takes place *through* language. From our enormous capacity to use language is derived our collective memory (myths, fables, sayings, ballads, and the like), as well as writing, art, and all the other media that shape human consciousness and store and transmit knowledge. Note that although the capacity to learn language appears to be innate (Chomsky, 1965, 1985), as we saw in the case of the Nicaraguan children, language does not occur outside a cultural setting. Indeed, it is the most universal dimension of human cultures.

What is unique about human language? Primatologists have shown that our closest evolutionary kin, the great apes (especially chimpanzees and pygmy chimpanzees), can learn language to some extent. Although their throats are not capable of producing the sounds that humans mold into language, apes do have the capacity to use language; that is, they can grasp the meanings of words as symbols for things and relationships. Through the use of sign language, or special languages using typewriters and other devices, apes can be taught a limited vocabulary (Zimmerman & Newman, 1995).

Fascinating as these experiments are, they only confirm the immense difference in communicative ability between humans and the apes. After months of training, an adult ape can use language with no more skill than an average 2½-year-old human infant (Zimmerman & Newman, 1995). On the basis of the research conducted so far, primate researchers conclude that an ape's innate ability to learn language is severely restricted. No amount of training can produce in apes the more advanced uses of language, including complex sentences containing abstract concepts, that are found in all normal humans regardless of their culture (Eibl-Eibesfeldt, 1989). For example, every human language allows its speakers to express an infinite number of thoughts and ideas that can persist even after their originators are gone. This property of human language, which is not shared by any other known species (Eisley, 1970; Rymer, 1992a, 1992b), allows human groups to transmit elements of their culture from one generation to the next.

So complete is the human reliance on language that it often seems as if language actually determines the possibilities for thought and action in any given culture. Perhaps we are unable to perceive phenomena for which we have no nouns or to engage in actions for which we have no verbs. This idea is expressed in the **linguistic-relativity hypothesis.** As developed by American linguists Edward Sapir and Benjamin Whorf in the 1930s, this hypothesis asserts that "a person's thoughts are controlled by inexorable laws or patterns of which he is unconscious. . . . His thinking itself is in a language—in English, in Sanskrit, in Chinese. And every language is a vast pattern-system, different from others" (Whorf, 1961, p. 135).

This observation was based on evidence from the social sciences, especially anthropology. For example, Margaret Mead's field research among the Arapesh of New Guinea had revealed that the Arapesh had no developed system of numbers. Theirs was a technologically simple society, and therefore complex numbering systems were not much use to them. They counted only "one," "two," "one and two," and "dog" (dog being the equivalent of four and probably based on the dog's four legs). To count seven objects, the Arapesh would say "dog and one and two"; 8 would be "two dog," and 24 would come out as "two dog, two dog, two dog." It is easy to see that in this small society one would quickly become tired of attempting to count much beyond 24 and would simply say "many" (Mead, 1971).

Other cultures have been found to have only a limited number of words for colors, and as a result they do not make some of the fine distinctions between colors that we do. And in a famous example, Whorf argued that many languages have ways of referring to time that are very different from those found in English and other Indo-European languages. In English we have verb tenses, which lead us to make sharp distinctions among past, present, and future time. In Vietnamese, in contrast, there are no separate forms of verbs to indicate different times; the phrase tôi đi về, for example, could mean "I'm going home" or "I went home" or "I will go home." The language of the Hopi Indians also lacks clear tenses, and it seemed to Whorf that this made it unlikely that the Hopi culture could develop the systems of timekeeping that are essential to modern science and technology.

Thus, in its most radical form, the linguistic-relativity hypothesis asserts that language determines the possibilities for a culture's norms, beliefs, and values. But there is little justification for this extreme version of the hypothesis. The Arapesh did not have a developed number system, but they could easily learn to count using the Western base-10 system. Once they were exposed to the money economy of the modern world, most isolated cultures formed words for the base-10 number system or else incorporated foreign words into their own vocabularies. They had no difficulty understanding the use of money, and this too became incorporated into so-called primitive cultures.

A more acceptable version of the linguistic-relativity hypothesis recognizes the mutual influences of culture and language. One does not determine the other. For example, someone living in Canada or the northern parts of the United States is likely to have a much larger vocabulary for talking about snow (*loose powder, packed powder, corn snow, slush,* and so on) than a person from an area where snow is rare. A person who loves to watch birds will have a much larger vocabulary about bird habitats and bird names than one who cares little about bird life. In order to share the world of bird-watchers or of winter sports fans, we will need to learn new ways of seeing and of talking about what we see. So, although the extreme version of the linguistic-relativity hypothesis is incorrect, it has been a valuable stimulus toward the development of a less-biased view of other cultures. We now understand that a culture's language expresses how the people of that culture perceive and understand the world and, at the same time, influences their perceptions and understandings.

CROSSING CULTURAL LINES

As our discussion of language has shown, until we study other cultures it is extremely difficult to view our own with detachment and objectivity. Unless we can see ourselves as others see us, we take for granted that our own cultural traits are natural and proper and that traits that differ from ours are unnatural and somehow wrong. Our ways of behaving in public, our food and dress and sports—all of our cultural traits—

THE CENTRAL QUESTION: A CRITICAL VIEW

Does culture account for the immense differences in the ways people throughout the world feel and behave?

Fauziya Kassindja, a young woman from the West African nation of Togo, escaped to the United States and was finally granted asylum after a long legal ordeal. She had created an international incident when she declared herself a political refugee on the ground that she was seeking to avoid genital mutilation (Dugger, 1996). In Togo and many other African societies, when a young woman reaches puberty she must undergo a ritual in which her clitoris and outer labia are cut away. This painful procedure, sometimes referred to as female circumcision, deprives her of much sexual pleasure and often leads to infection and reproductive problems later in life.

Kassindja appealed to the American authorities for legal asylum. The case created a number of precedents in immigration law and continues to generate much debate. Partly on the strength of the publicity surrounding this case, and partly owing to the efforts of other African women immigrants, in 1996 both the California state legislature and the U.S. Congress passed legislation banning clitorectomy in the United States. In international conventions on population, reproduction, and women's rights, however, women representing nations where clitorectomy is part of the culture often defend the practice. They may not personally support the norm of clitorectomy, but they resent the intolerance of Westerners, whose cultural norms (e.g., drinking alcoholic beverages or baring the female form) they find equally abhorrent.

Partly as a result of protests by feminists and human rights advocates, Fauziya Kassindja was freed from detention and granted asylum in the United States. She is shown here overcome by emotion upon her release from prison in York County, Pennsylvania, in 1996.

Does the need to respect other cultures prevent one from taking a stand on practices like clitorectomy? Does avoidance of ethnocentrism mean that one must become a cultural relativist who "when in Rome, does as the Romans do"—even if it means violating one's own principles? Cultural relativism raises many controversial issues, especially in societies like the United States where people from different cultures are continually arriving, bringing with them different norms and values.

Many social scientists answer this question in both scientific and political terms. They argue that respect for other cultures does not prevent them from openly opposing cultural practices that perpetuate severe inequalities between women and men or that endanger women's health. Out of respect for colleagues from those cultures, they seek to persuade rather than pass laws that cannot be enforced. They seek to go beyond the issue of clitorectomy to the larger problems that prevent women in those cultures from speaking out against the practice themselves. So at world conferences the subject of clitorectomy is often subordinated to broader efforts to create policies that will improve women's status and power—for example, through greater access to education and modern forms of employment. At home, however, the same social scientists generally are highly supportive of legislation to protect women's rights to health and sexual freedom (Shweder, 2000). ■

have become "internalized," so they seem almost instinctive. But once we understand another culture and how its members think and feel, we can look at our own traits from the perspective of that culture. This *cross-cultural perspective* has become an integral part of sociological analysis.

The ability to think in cross-cultural terms allows people to avoid the common tendency to disparage other cultures simply because they are different. However, most people live out their lives in a single culture. Indeed, they may go so far as to consider that culture superior to any other, an attitude that social scientists term **ethnocentrism.** This term refers to the tendency to judge other cultures as inferior in terms of one's own norms and values. The other culture is weighed against standards derived from the culture with which one is most familiar. The explorers' assumption that the Native Americans could benefit from the adoption of European cultural traits is an example of ethnocentric behavior. But such obvious ethnocentrism is not limited to historical examples. We still encounter it today—for example, in our use of the term *American* to refer to citizens of the United States and not to those of Canada and the South American nations. Another example is our tendency to judge other cultures by how well they supply their people with consumer goods rather than by how well they adhere to their own values.

To get along well in other parts of the world, a businessperson, politician, or scientist must be able to suspend judgment about other cultures, an approach that is termed **cultural relativity.** Cultural relativity entails the recognition that all cultures develop their own ways of dealing with the specific demands of their environments. This kind of understanding does not come automatically through the experience of living among members of other cultures. It is an acquired skill.

There are limits, however, to the value of cultural relativity. It is an essential attitude to adopt in understanding another culture, but it does not require that one avoid moral judgment entirely. We can, for example, attempt to understand the values and ideologies of the citizens who supported the Nazi regime, even though we abhor what that regime stood for. And we can suspend our outrage at racism in our own society long enough to understand the culture that produced racial hatred and fear. But as social scientists, it is also our task to evaluate the moral implications of a culture's norms and values and to condemn them when we see that they produce cruelty and suffering. To a large extent, worldwide outrage at South Africa's policies of apartheid (racial separation and enforced inferiority for black citizens) helped bring about the demise of those policies in the early 1990s.

Cultures whose norms negate the possibility that women can develop their full potential also invite critical evaluation. The Taliban regime in Afghanistan does not allow women to practice their professions or even appear outside the home unless their heads and bodies are entirely covered. In Afghanistan itself, educated women criticize these norms, which help maintain the power of male orthodox religious leaders. Elsewhere in the world, the issue of female circumcision is another example of cultural norms that are under fire for similar reasons.

In the world today there are an uncountable number of different cultures and variations within cultures. Every major nation hosts a variety of cultures, and there is invariably some conflict among them. We say, for example, that there is an "American culture," with its norms and values, its forms of entertainment, sport, justice, and so on. But just think of all the conflict about what those norms and values are. For example, is it acceptable to smoke? (See the Global Social Change box on pages 46–47.) Clearly, there are many cultural differences among us, just as there are among the people of most nations. For example, in France, which is often thought of as a Western nation with a rather homogeneous culture, there are many variations in forms of speech and food and even in ways of thinking from one region of the nation to another. In India, there are literally thousands of different cultures. Does this mean that it is impossible to establish shared meanings and rules among large numbers of people? Not at all. Indeed, in the modern world there are extremely strong cultures and sets of cultures known as civilizations that exert powerful forces for order even as they are subject to change.

CIVILIZATIONS AND CULTURAL CHANGE

Civilizations are advanced cultures. They usually have forms of expression in writing and the arts, powerful economic and political institutions, and innovative technologies, all of which strongly influence other cultures with which they come into contact. Some civilizations, like those of ancient Egypt and Rome, declined thousands of years ago

GLOBAL SOCIAL CHANGE

Changing Norms of Tobacco Use

What happens in a culture when a widespread norm actually encourages behavior that is destructive? What must be done to change such norms and behavior? What will occur in other societies to which the norm is spreading, and what are the moral responsibilities of the culture that is spreading the norm? Norms of tobacco use in the United States and throughout the world raise these questions with great urgency. The answers are swirling in the tobacco smoke that fewer Americans, but an increasing number of people in other nations, are consuming.

Tobacco was used by Native Americans for its pleasurable effects, as a medicinal substance, and for religious ceremonies (Ravenholt, 1990). Early European explorers and conquerors spread its use to Europe and other ocean trading societies (e.g., Turkey and Japan) during the sixteenth and seventeenth centuries. In England, pipe smoking was extremely popular in Queen Elizabeth's court in the late 1500s, but when James I assumed the throne after Elizabeth's death in 1603, the world witnessed its first great narcotics debate. The new king believed tobacco to be "a custome lothesome to the eye, hateful to the nose, harmfull to the braine, dangerous to the lungs" (quoted in Austin, 1978, p. 6). The king soon realized, however, that his efforts to ban tobacco use would not stem its spread among people who could afford it. He therefore decided to make tobacco a source of royal revenue by granting land in the Virginia colony to settlers who promised to grow tobacco as a cash crop whose importation he could control and tax.

Despite its high cost, the use of tobacco spread rapidly, especially in the form of snuff (powdered tobacco), which was popular among the European upper classes in the 1700s. Cigars and cigarettes were introduced to Europe and the United States in the early 1800s by soldiers returning from wars in the Middle East. During the American Civil War, smoking became popular among soldiers, who enjoyed a particularly American style of tobacco consumption, the chewing of a mixture of tobacco leaves and molasses. The invention of cigarette rolling machines in the late nineteenth century and their use in tobacco factories in Virginia and North Carolina made the supply of cigarettes almost unlimited and guaranteed that the economies of those states would be dominated by tobacco cultivation for many generations.

The Marlboro Man associates smoking with virility, freedom, and the open spaces of the American West, all of which appeal to a global audience of smokers.

During the twentieth century, spurred by modern advertising techniques and the development of mass media, the rate of cigarette consumption in the United States increased a hundredfold, from 2.5 billion cigarettes smoked in 1900 to a peak of 640 billion in 1988. The influence of American civilization throughout the world during this period also caused huge increases in worldwide cigarette consumption even as evidence of the health hazards of regular smoking mounted.

Although cigarette smoking is declining in the United States, smoking is on the rise elsewhere in the world. Cigarette smoking is now the most serious and widespread form of addiction in the world. It accounts for almost 5 million deaths annually in the United States, and the death toll from tobacco use is rising in other nations. In China, for example, cultivation, export, and consumption of tobacco are on the rise; given the size of the Chinese population, this represents an ominous trend for world health.

There is evidence in the declining curve of cigarette consumption in the United States that campaigns against smoking

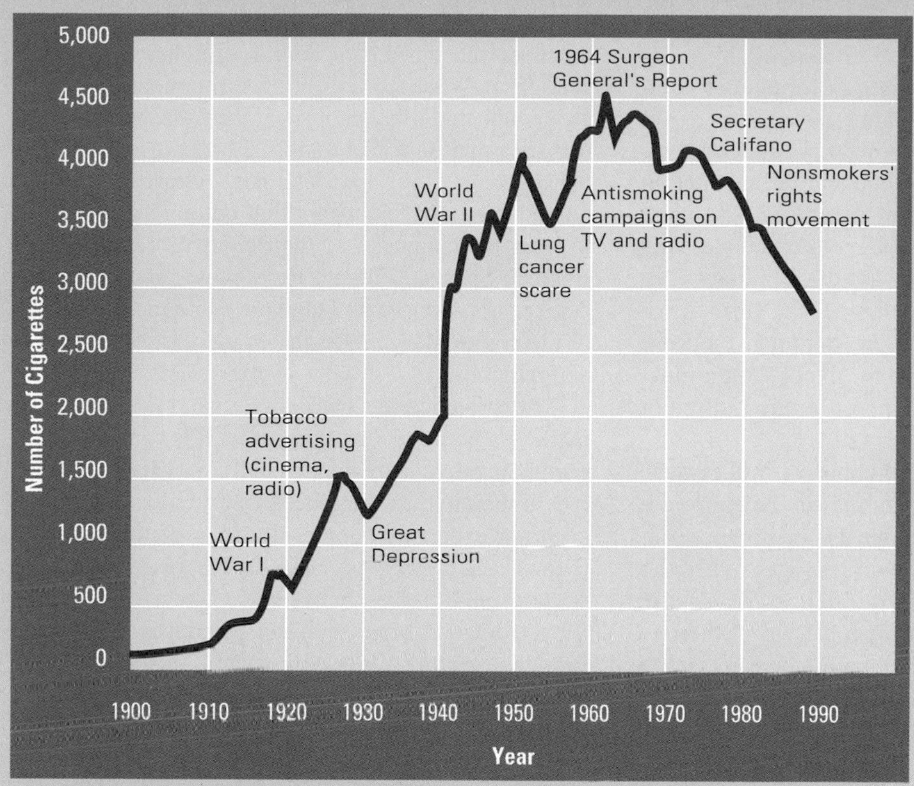

Number of Cigarettes Consumed Annually per Adult (age 18 and older), United States, 1900–1990

Source: Economic Research Service, U.S. Department of Agriculture.

Annual Adult per Capita Consumption of Manufactured Cigarettes, Selected Countries

Country	Consumption	Country	Consumption	Country	Consumption	Country	Consumption	Country	Consumption
Cuba	3,920	Australia	2,720	Portugal	1,730	Iraq	980	Nigeria	370
Greece	3,640	Korea, Republic of	2,660	Brazil	1,700	Thailand	900	Bolivia	330
Poland	3,300	Italy	2,460	Netherlands	1,690	Pakistan	660	Bangladesh	270
United States	3,270	France	2,400	China	1,590	Iran	620	India	160
Japan	3,270	Germany	2,360	Costa Rica	1,340	Senegal	610	Nepal	150
Canada	3,180	Turkey	1,970	Mexico	1,190	Kenya	550	Ethiopia	60
Switzerland	2,960	Colombia	1,920	Morocco	1,070	Laos	490		
Libya	2,850	Malaysia	1,840	Indonesia	1,050	Ghana	380		

Source: Ravenholt, 1990.

and concern about the hazards of tobacco use are rapidly changing the norms of smoking. Once considered chic and sexy, smoking is now increasingly seen as dangerous to the smoker and to others nearby. But levels of smoking in the United States remain extremely high (see table), suggesting that the norms established over a century of tobacco use are extremely difficult to reverse. And given the popularity of smoking elsewhere in the world, the prospects for turning smoking into a negatively rather than a positively sanctioned norm are dim.

and exist today mainly in museums and in the consciousness of scholars, artists, and scientists. Others are living civilizations with long histories, like those of China and India, that were conquered by other civilizations and are rising again in forms that combine the old with the new. Islamic civilization is yet another example of such a civilization; much of the unrest in the Islamic world is due to conflicts that pit orthodox leaders like the late Ayatollah Khomeini against Western-influenced leaders like the late King Hussein of Jordan. Then there are the dominant civilizations of North America, Europe, the former Soviet Union, and Japan. They are dominant because they compete on a world scale to "export" their ideas and their technology—in fact, their entire culture or "blueprint for living."

Like most of the principal concepts in the social sciences, the concept of civilization can be elusive. It is used in many different contexts, in popular language as well as in social-scientific usage. In popular speech, the word *civilization* is often used to make negative comparisons between people who adhere to the norms of polite conduct, and are therefore said to be "civilized," and those who are "uncouth" and act like "barbarians" or "savages." This is how the word was understood by the explorers of Columbus's time, and colonial conquest was justified in part as an effort to civilize the barbarians.

In the social sciences, the most common use of the term *civilization* stems from the study of changes in human society at the macro level, which often requires comparisons among major cultures. In this context, a **civilization** is "a cultural complex formed by the identical major cultural features of a number of particular societies. We might, for example, describe Western capitalism as a civilization, in which specific forms of science, technology, religion, art, and so on, are to be found in a number of distinct societies" (Bottomore, 1973, p. 130). Thus, Italy, France, Germany, the United States, Sweden, and many other nations that have made great contributions to Western civilization all have private corporations, and their normative orders, laws, and judicial systems are quite similar. Even though each nation may have a different language and differ in the way it organizes some aspects of social life (European and North American universities define academic degrees differently, for example), they all share similar norms and values and can all be said to represent Western civilization (Wallbank, 1996).

Through such processes as exploration and conquest, civilizations invariably spread beyond their original boundaries. Figure 2.2 shows how during much of the nineteenth century England spread its version of Western civilization throughout the world as it conquered tribal peoples and established colonies in Africa and Asia. Colonial rule brought *cultural imperialism,* the imposition of a new culture on the conquered peoples. This meant that colonial peoples had to learn the languages of their conquerors, especially English, Spanish, French, Portuguese, and Dutch (or Afrikaans). Along with language came the imposition of ideologies like Christianity in place of older beliefs and religions.

According to the historian Fernand Braudel, "The mark of a living civilization is that it is capable of exporting itself, of spreading its culture to distant places. It is impossible to imagine a true civilization which does not export its people, its ways of thinking and living" (1976/1949, p. 763). In his research on the contacts and clashes between the great civilizations surrounding the Mediterranean Sea during the 1500s, Braudel uses three important sociological concepts to explain the spread of civilizations around the world: acculturation, assimilation, and accommodation. (See the study chart on page 50.)

Acculturation

People from one civilization incorporate norms and values from other cultures into their own through a process called **acculturation.** Most acculturation occurs through intercultural contact and the borrowing or imitation of cultural norms. But there have been many instances of acculturation through cultural imperialism, in which one culture has been forced to adopt the language or other traits of a more dominant one. Thus, people in societies that were colonized in the nineteenth century were forced to learn the language of the conquering nation.

Aspects of our culture that we take for granted usually can be shown to have traveled a complicated route through other cultures to become part of our way of life. Braudel's study of the Mediterranean civilizations, for example, shows that many of the plants and foods that became part of life around the Mediterranean Sea, and later were imported to the New World, were themselves borrowed from other cultures and incorporated into those of the Mediterranean societies. The process of acculturation can be traced in

FIGURE 2.2

The British Empire in 1901

This map shows the extent of the British Empire in 1901. At the peak of its power, at the turn of the twentieth century, the empire had spread English civilization throughout the world. But its influence was strongest in peripheral, less-industrialized lands. In a mere half-century, most of the empire would crumble, but the effects of English civilization on language and social institutions throughout the world would endure.

Source: Eldridge, 1978.

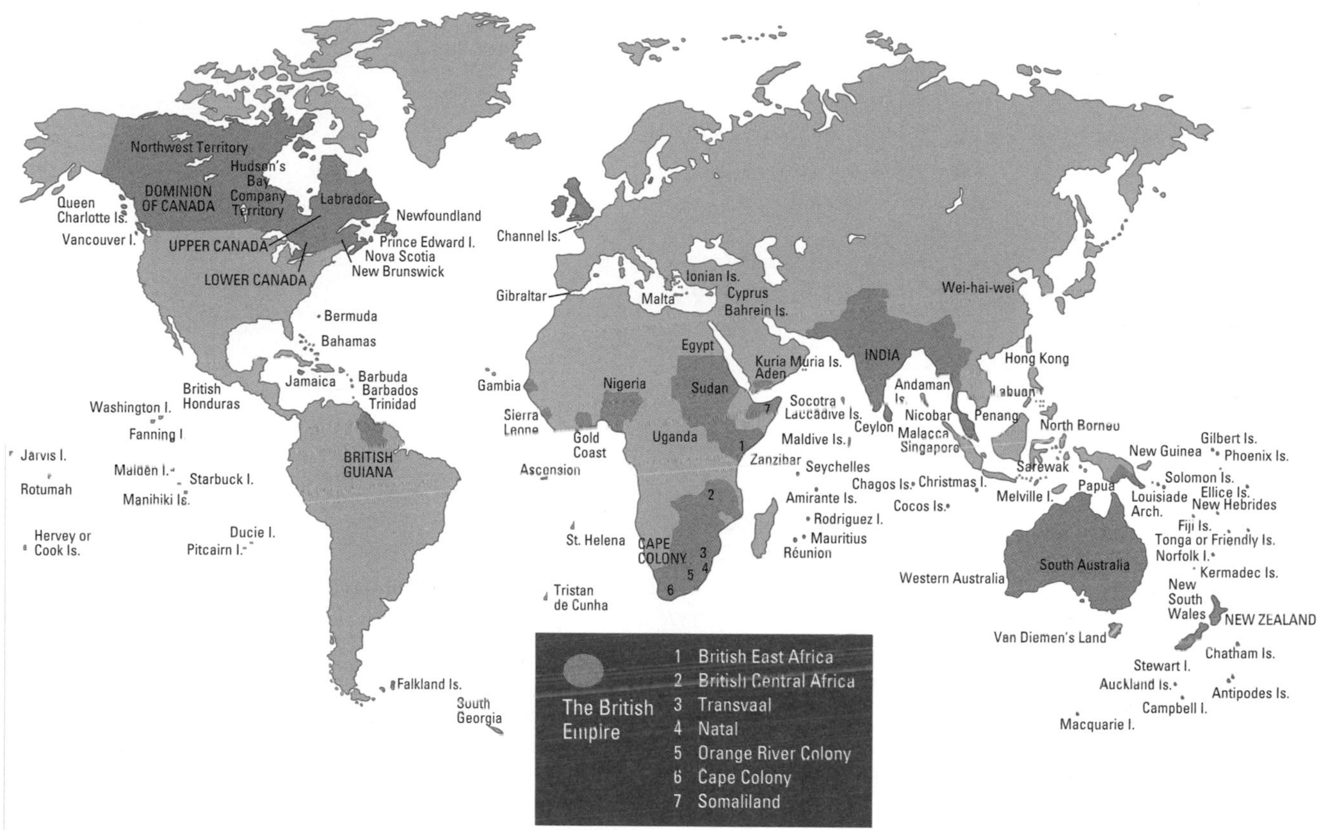

many other aspects of everyday life. As Ralph Linton has written:

> Our solid American citizen awakens in a bed built on a pattern which originated in the Near East. . . . He throws back covers made from cotton, domesticated in India. . . . He slips into his moccasins, invented by the Indians of the Eastern woodlands. . . . He takes off his pajamas, a garment invented in India, and washes with soap invented by the ancient Gauls. . . . He puts on garments whose form originally derived from the skin clothing of the nomads of the Asiatic steppes . . . and ties around his neck a strip of bright-colored cloth which is a vestigial survival of the shoulder shawls worn by the seventeenth-century Croatians. Before going out for breakfast he glances through the window, made of glass invented in Egypt, and if it is raining puts on overshoes made of rubber discovered by the Central American Indians and takes an umbrella, invented in southeastern Asia. (1936, p. 326)

The concept of acculturation can also be applied to how newcomers adopt the cultural ways of the host society. But acculturation is rarely a one-way process: At the same time that newcomers are becoming more like their hosts in values and behavior, they teach

STUDY CHART
Effects of Cultural Contact

Form of Contact	Description	Example
Acculturation	The process by which the members of a civilization incorporate norms and values from other cultures into their own.	Americans develop a taste for Italian, Chinese, and Mexican food.
Assimilation	The process by which culturally distinct groups within a civilization adopt the norms, values, and language of that civilization and gain equal statuses in its groups and institutions.	Spanish-speaking immigrants to the United States learn English and begin to move up the status hierarchy in education and jobs.
Accommodation	The process by which a smaller, less powerful society is able to preserve the major features of its culture even after prolonged contact with a larger, stronger culture.	Though conquered and forced onto reservations, Native Americans adapt by taking on many of the norms and values of the larger society while preserving aspects of their own culture.

members of the host society to use and appreciate aspects of their own culture. Because of the cultural diversity of its population, examples of acculturation are especially prevalent in the United States. For example, even small towns in the interior of the country have at least one Chinese restaurant, but often the food served in those restaurants is a highly acculturated form of the cuisine of China. Indeed, most of the things we think of as part of the American way of life, from hamburgers, pizza, and baseball to democracy and free enterprise, originally were aspects of other cultures. That they have become incorporated into American culture through acculturation, and in the process have become changed from their original forms, does not deny the fact of their "foreign" origin.

Assimilation and Subcultures

When culturally distinct groups within a larger civilization adopt the language, values, and norms of the host civilization and their acculturation enables them to assume equal statuses in the social groups and institutions of that civilization, we refer to that process as **assimilation.** When groups become assimilated into American society, for example, people often say that they have been "Americanized."

Assimilation has been a major issue for immigrant groups in North America. It is no surprise, then, that questions like the following are often asked: Will the various Hispanic peoples in America give up their language over time? Will American Jews marry members of other groups and lose their distinct identity? Will Italian Americans gradually forget their cultural heritage and come to think of themselves as "100 percent Americans" (Brimelow, 1995)? These are the kinds of questions that form the subject matter of racial and ethnic relations in pluralistic societies like those of the United States and the commonwealth of former Soviet republics. Both of these societies (which are also civilizations) are composed of a multitude of peoples, each of which once had its own culture but is under pressure to become assimilated into the dominant civilization.

When a culturally distinct people within a larger culture fails to assimilate fully or has not yet become fully assimilated, this group is known as a **subculture** within the larger culture. (The term is also applied to groups that have had significantly different experiences from those of most members of the society.) People who maintain their own subculture generally share many of the values and norms of the larger culture, but they retain certain rituals, values, traditions, and norms—and in some cases their own language—that set them apart. Thus, we speak of African American, Latino, Native American, and a host of other subcultures in the United States. As will be explained in Chapter 7, these subcultures are also known as *ethnic groups,* since their members have a sense of shared descent, a feeling of being "a people" with a history and a way of life that exists within a larger and more culturally diverse society.

Ethnic subcultures are created out of the experience of migration or invasion and subsequent adaptation to a host culture. But subcultures are also created out of the experience of people in complex societies who actively seek to create and maintain a way of life distinct from that of other members of their society (Fischer, 1987; Gans, 1976). For example, in large cities there are subcultures composed of artists and other people whose livelihood depends on the arts—theater people, rock musicians and record producers, visual artists, gallery owners, art critics, curators. The list could be expanded to include avant-garde artists who band together to find new forms of expression and perhaps to poke fun at conventional norms. When a subculture that challenges the accepted norms and values of the larger society establishes an alternative lifestyle, it is known as a **counterculture.** The hippies of the 1960s—along with members of New Left political groups, activists in the women's movement, and environmental activists—formed a counterculture that had a significant influence on American politics and foreign policy during the Vietnam War years (Roszak, 1969).

Accommodation and Resistance

Throughout history, many societies have withstood tremendous pressure to become assimilated into larger civilizations. But even greater numbers have been either wiped out or fully assimilated. Only a century ago, for example, one could still find many hundreds of hunter-gatherer societies throughout the world. Today, there are probably fewer than a hundred, and these live in the most isolated regions of the earth.

Larger and smaller societies do not usually develop ways of living together without the smaller one becoming extinct or totally assimilated into the larger one. But when the smaller, less powerful society is able to preserve the major features of its culture even after prolonged contact, **accommodation** is said to have occurred. For example, in the Islamic civilization of the Middle East before the creation of Israel in 1948, Jews and other non-Muslims usually found it rather easy to maintain their cultures within the larger Arab societies. Compare this pattern of accommodation with the experiences of the Jews in Spain, who were forced to leave in 1492 in one of the largest mass expulsions in history.

Accommodation requires that each side tolerate the existence of the other and even share territory and social institutions. The history of relations between Native Americans and European settlers in the Western Hemisphere is a complex story of resistance and accommodation. Throughout the period of conquest, expansion, and settlement by the Europeans there was continual resistance by the native peoples. This resistance took many forms, including refusal to adopt Christianity, to speak English or Spanish, to sell goods and services to the settlers, and to fight in the Europeans' wars. Resistance did not save Native Americans from death by disease, military conquest, or famine, but it did allow them to maintain their cultures and to borrow from the settlers the cultural customs that were most advantageous to them. For example, the Plains Indians adopted horses from the Spanish explorers, which completely changed their culture, and much later they borrowed trucks from American culture, which helped them adapt to modern ranching.

TECHNOLOGY AND SOCIAL CHANGE

A common feature of cultures in the contemporary world is that their norms are continually being challenged by new ideas. These new ideas often come through new channels of communication based on an ever-widening array of technologies. Early in the twentieth century, the telephone, radio, television, and movies

These French protestors against McDonald's carry signs saying, "Our stomach is not a garbage pail." Many Europeans are concerned about the growing influence of American-style fast food, considering it not only bad nutrition but also a form of cultural imperialism.

SOCIOLOGY AND SOCIAL JUSTICE

The story of the struggle by deaf children in Nicaragua to create a language, and of their teachers to master it in order to be able to teach them about their culture, is rich with sociological meaning. We should ask why these children did not have access to good teachers. Why was it necessary for them to evolve their own unique language? Part of the answer goes well beyond issues of language and culture to problems of social justice. The children were mainly from poor, rural backgrounds. The government was in the hands of rich and powerful landowners and their allies. The Nicaraguan revolution made it possible, finally, to create schools for these children. However inadequate those schools were, they allowed the children to live together with teachers who eventually discovered the rules of their sign language.

The story calls attention to how often a society's culture is not transmitted equitably to everyone. Not only in Central America but in the United States and elsewhere in the world, people find that access to their cultures is barred due to deficiencies in education or lack of access to communications media. A good example of how this occurs is found among deaf people in the United States.

The primary U.S. law that guarantees deaf people (along with others with major disabilities) equal rights in workplaces, schools, and public facilities was not passed until 1990. Passage of this law, the Americans With Disabilities Act, would not have occurred without the persistent demands and activities of disabled people and their families. Even now, with the law in effect, courts have been prone to narrow its

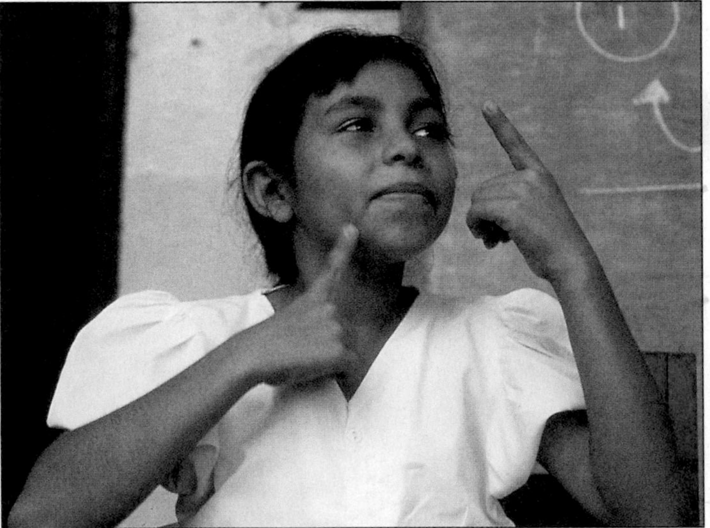

Nine-year-old Yuri Mejia signing the story of Babar.

scope and interpret it so as to restrict the situations to which it applies (Sheehan, 2000). Not having any language with which to learn one's culture, as was the case for Nicaragua's deaf children, is an extreme example of the cultural deprivation experienced by millions of people around the world. Globally, people's handicaps range from deafness and other physical conditions to the mere fact that they are women in societies where men control access to much of what is desired in the culture. Positive changes that will allow them greater access to all that their cultures have to offer is part of the global struggle for social justice (Krieger, 2000). ■

These demonstrators are protesting about the lack of handicapped access to bus facilities. Recently the courts have issued rulings narrowing and restricting the scope of laws governing handicapped access to public facilities, so that only through continual protest and mobilization of their allies can disabled people participate fully in their society and culture.

brought new values and new expectations about what human life could encompass to billions of formerly isolated villagers throughout the world. New forms of music like jazz and rock-and-roll became worldwide cultural forms through radio and then television.

These technologies helped spur what is known as "the revolution of rising expectations." People in many societies began questioning the status quo of inequality that was often supported by religious and political values, most often in the hands of men with wealth and power in their societies (da Sola Pool, 1983, 1990). A recent example is the adaptation of the popular television show *Who Wants to Be a Millionaire* for viewers in India. In poor areas such as the slums of Bombay, millions of people watch the show every night, dreaming of instant wealth.

Today the Internet and the World Wide Web, made possible by computers and telecommunications technologies, are bringing about new forms of cultural change. People all over the world can write to one another and exchange views and information almost instantaneously. They can share their opinions with people in countries where the ability to publicly express an opinion may be relatively new. They can call up Web sites that offer information about cultural norms or values they may never have imagined before. But sociologists often point to emerging problems with the way these new technologies are used. Is the Internet, which is dominated by English speakers, weakening the languages spoken by smaller segments of the world population? Is dramatically unequal access to technologies widening the already yawning gap between haves and have-nots throughout the world? Dominance of the Internet by American and some European cultures may indeed threaten the norms and values of other cultures, and these issues will continue to be on the frontiers of sociological research (Castells, 1999).

SUMMARY

Culture refers to all the modes of thought, behavior, and production that are handed down from one generation to the next by means of communicative interaction. It has three major dimensions: ideas, norms, and material culture. *Ideas* are the ways of thinking that organize human consciousness; they include *values,* or socially shared ideas about what is right. *Norms* are specific rules of behavior that are supported or justified by values; *laws* are norms that are included in a society's official written codes of behavior. A society's *material culture* consists of all the things it produces.

Ideologies combine ideas and norms; they are systems of values and norms that the members of a society are expected to believe in and act upon without question. *Technologies* combine norms and material culture; they are the things and the norms for using them that are found in a given culture.

A culture's array of norms, or *normative order,* constitutes a system of *social control. Sanctions* are rewards and punishments for adhering to or violating norms. Strongly sanctioned norms are *mores;* weakly sanctioned norms are *folkways.*

According to Darwin's theory of *natural selection,* mutations in organisms occur more or less randomly. Mutations that enable an individual organism to survive and reproduce are passed on to the next generation. Herbert Spencer and other *social Darwinists* attempted to apply Darwin's theory to humans' ability to adapt to social environments.

According to the theory of *sociobiology,* behaviors such as incest avoidance may be genetically programmed in human beings. A more widely accepted view is that cultural techniques allow humans to adapt to any physical or social environment.

The learning of culture is made possible by language. Human language allows its speakers to express thoughts and ideas that can persist even after their originators are gone. According to the *linguistic-relativity hypothesis,* language also determines the possibilities for a culture's norms, beliefs, and values.

Ethnocentrism is the notion that one's own culture is superior to any other. To understand other cultures, it is necessary to suspend judgment about those cultures, an approach termed *cultural relativity.*

A *civilization* is a cultural complex formed by the identical major cultural features of a number of particular societies. Civilizations invariably expand beyond their original boundaries. When people from one civilization incorporate norms and values from other cultures into their own, *acculturation* has occurred. The process by which culturally distinct groups within a larger civilization adopt the language, values, and norms of the host civilization and gain equal statuses in its institutions is termed *assimilation.* When a smaller, less powerful society is able to preserve its culture even after prolonged contact with a major civilization, *accommodation* has taken place.

GLOSSARY

Culture: All the modes of thought, behavior, and production that are handed down from one generation to the next by means of communicative interaction rather than by genetic transmission. **(35)**

Ideas: The ways of thinking that organize human consciousness. **(35)**

Norms: Specific rules of behavior. **(35)**

Material culture: Patterns of possessing and using the products of culture. **(35)**

Values: The ideas that support or justify norms. **(35)**

Laws: Norms that are written by specialists, collected in codes or manuals of behavior, and interpreted and applied by other specialists. **(36)**

Ideologies: Systems of values and norms that the members of a society are expected to believe in and act upon without question. **(37)**

Technologies: The products and the norms for using them that are found in a given culture. **(37)**

Social control: The set of rules and understandings that control the behavior of individuals and groups in a particular culture. **(38)**

Normative order: The array of norms that permit a society to achieve relatively peaceful social control. **(38)**

Sanctions: Rewards and punishments for abiding by or violating norms. **(38)**

Mores: Strongly sanctioned norms. **(38)**

Folkways: Weakly sanctioned norms. **(38)**

Natural selection: The relative success of organisms with specific genetic mutations in reproducing new generations with the new trait. **(39)**

Social Darwinism: The notion that people who are more successful at adapting to the environment in which they find themselves are more likely to survive and reproduce. **(41)**

Sociobiology: The hypothesis that all human behavior is determined by genetic factors. **(41)**

Cultural evolution: The process by which successful cultural adaptations are passed down from one generation to the next. **(42)**

Linguistic-relativity hypothesis: The belief that language determines the possibilities for thought and action in any given culture. **(43)**

Ethnocentrism: The tendency to judge other cultures as inferior to one's own. **(44)**

Cultural relativity: The recognition that all cultures develop their own ways of dealing with the specific demands of their environments. **(44)**

Civilization: A cultural complex formed by the identical major cultural features of a number of societies. **(48)**

Acculturation: The process by which the members of a civilization incorporate norms and values from other cultures into their own. **(48)**

Assimilation: The process by which culturally distinct groups in a larger civilization adopt the norms, values, and language of the host civilization and are able to gain equal statuses in its groups and institutions. **(50)**

Subculture: A group of people who hold many of the values and norms of the larger culture but also hold certain beliefs, values, or norms that set them apart from that culture. **(50)**

Counterculture: A subculture that challenges the accepted norms and values of the larger society and establishes an alternative lifestyle. **(51)**

Accommodation: The process by which a smaller, less-powerful society is able to preserve the major features of its culture even after prolonged contact with a larger, stronger culture. **(51)**

QUESTIONS FOR THOUGHT AND DISCUSSION

1. Name a norm of your society (e.g., smoking) and identify whether it is a formal or informal norm. How strongly is it sanctioned, and through what behaviors? Is the norm changing? What social forces are making it change?

2. Why is cultural relativity so important in studying cultures other than one's own? What are the problems that an uncritical use of cultural relativity may lead to?

DIGGING DEEPER

Books

The Interpretation of Cultures (Clifford Geertz; Basic Books, 1973). A set of essays that explore the many meanings of human culture, including a fine essay on the relationship between culture and human social and physical evolution.

Cultivating Differences: Symbolic Boundaries and the Making of Inequality (Michele Lamont & Marcel Fournier, eds.; University of Chicago Press, 1992). A collection of essays about how human cultures create and destroy boundaries among groups and societies.

Anthropology, 8th ed. (William A. Haviland; Harcourt Brace, 1997). An introductory textbook that covers in detail the cultures of nonindustrial peoples.

Journals

Daedalus. ***The journal of the American Academy of Arts and Sciences.*** It devotes entire issues to subjects of great importance in the world—such as AIDS, computers, violence, and nationalism—and contains some of the best writing on social and cultural change.

EXPLORING SOCIOLOGY ON THE INTERNET

The United Nations
www.un.org/
www.unsystem.org/

The UN's official home page and index page, presenting information about cultures and societies around the world.

International Red Cross
www.ifrc.org/

Includes links to many other groups concerned with humanitarian issues.

amazon.com
www.amazon.com/

The most successful online bookstore in the United States; a fine source to consult about changing aspects of culture anywhere in the world. Reviewers comment on recently published books, and readers add their comments via e-mail. You can find something here for almost any research need. If you are writing a paper about changing mores in the United States, for example, you could check out the latest book by Judith Martin (Miss Manners).

Amnesty International
www.amnesty.org/

A good place to start if you are looking for information about human rights and culture conflicts throughout the world. Information about the clashes between ethnic groups in Africa, the problems of leaders of opposition movements (e.g., Ken Saro-Wiwa), and the like can be found here or through links from this site.

The Southern Institute at Tulane University
www.tulane.edu/~so-inst/index.html

Provides access to discussion groups, databases, and research publications, especially on issues of culture and race in the United States and elsewhere in the world.

American Anthropology Association
www.ameranthassn.org/

The home page for professional anthropologists. An excellent place to look for information about the social-scientific study of cultures, especially those in other parts of the world.

Chapter 3

SOCIALIZATION

To what extent does society shape how individuals behave and what they can or will become?

During the spring and summer of 2000, CBS Television made broadcasting history with its show *Survivor.* A so-called reality show centering on a group of Americans sent to live in a remote area such as a deserted island, the show is a combination soap opera, game show, and sports contest. The participants form a small society with its own subculture in the form of norms established by the show's producers. They must cooperate to perform the tasks required for survival, but each week they must also vote to eliminate one of their number. One contestant may win an endurance test (the sports element of the show) and thereby become immune from elimination during that week. The show was an almost instantaneous hit and was immediately copied by other networks. The final episode of the show's first season was one of the most heavily watched television events in history.

Sociologists and television critics were quick to point out that the show presented a scenario that bore no resemblance to reality. It was carefully produced, from the original casting of the contestants to the situations and interviews that were artfully edited, to heighten audience interest in the fate of each contestant. Sociologists also refuted the network's claim that the show was a unique social experiment. They noted that social scientists have conducted many similar but more carefully controlled social experiments, some of which will be discussed in this chapter.

Survivor does, however, demonstrate a well-known sociological phenomenon: When norms are structured to promote a certain outcome, people's behavior will conform to expectations and the outcome will be as predicted. The norms established by the show's participants promoted cooperation, but other norms made individual success, at the expense of others, more important. In the end, the person who won was the most skillful at creating coalitions favorable to himself while taking care to alienate as few other contestants as possible. The first season's winner, Richard Hatch, had well-honed skills in social interaction and professional experience in analyzing group behavior. He had a keen sociological imagination and the ability to apply it to achieve his own ends. We will see in this chapter how people develop such skills and under what conditions they use them for selfish purposes or for more altruistic ones. The process whereby a person masters the skills of interaction is part of the larger subject of socialization.

The "cast" of the first season of *Survivor,* shown here, became instant celebrities as their skills at interacting while promoting their own self-interests were broadcast throughout the world.

This chapter introduces some of the central issues in the study of how people become social beings. This may not seem to be a difficult process until one thinks about how many people have problems getting along well with others or how many conflicts we feel between the teachings of our parents and other social influences (from peers and others). Then there is the question of our biological endowments versus our social learning: How do these blend to create individuals who act in certain ways as social beings? If we look at the experiences of children growing up under conditions of severe social deprivation and lack of love, we find out more about the lasting influences of social learning and emotional support. The later parts of the chapter explore the ways in which people become social beings through their interactions in the family and in other types of groups. ■

BECOMING A SOCIAL BEING

Socialization is the term sociologists use to describe the ways in which people learn to conform to their society's norms, values, and roles. People develop their own unique personalities as a result of the learning they gain from parents, siblings, relatives, peers, teachers, mentors, and all the other people who may influence them throughout their lives (Elkin & Handel, 1989). From the viewpoint of society as a whole, however, what is important about the process of socialization is that people learn to behave according to the norms of their culture. This process also makes possible the transmission of culture from one generation to the next. In this way, the culture is "reproduced" in the next generation (Danziger, 1971; Parsons & Bales, 1955).

Socialization occurs throughout life as an individual learns new norms in new groups and situations. However, for purposes of analysis, socialization can be divided into three major phases. The first phase is *primary socialization.* It refers to all the ways in which the newborn individual is molded into a social being—that is, into a growing person who can interact with others according to the expectations of society. Primary socialization occurs within the family and other intimate groups in the child's social environment. *Secondary socialization* occurs in later childhood and adolescence when the child enters school and comes under the influence of adults and peers outside the household and immediate family. During *adult socialization,* the third phase, the person learns the norms associated with new statuses such as spouse, journalist, programmer, grandparent, or nursing home patient (Danziger, 1971).

Nature and Nurture

Throughout recorded history, there have been intense debates over what aspects of behavior are "human nature" and what aspects can be intentionally shaped through nurture or socialization. During all the centuries of prescientific thought, the human body was thought to be influenced by the planets, the moon, and the sun, or by various forces originating within the body, especially in the brain, heart, and liver. In the ancient world and continuing into the Middle Ages, blood, bile, phlegm, and other bodily fluids were thought to control people's moods and affect their personalities. Human behavior and health could also be affected by evil spirits or witches. While these ideas are no longer dominant explanations of human behavior, none of them has ever entirely disappeared.

In the eighteenth century, a new and quite radical idea of the "natural man" emerged. The idea that humans inherently possess qualities such as wisdom and rationality, which are damaged in the process of socialization, took hold in the imaginations of many educated people. Social philosophers like Jean-Jacques Rousseau believed that if only human society could be improved, people raised in it would emerge with fewer emotional scars and limitations of spirit. This belief was based on the enthusiasm created by scientific discoveries, which would free humans from ignorance and superstition, and by new forms of social organization like democracy and capitalism, which would unleash new social forces that would produce wealth and destroy obsolete social forms like aristocracy.

In the United States, Thomas Jefferson applied these ideas in developing the educational and governmental institutions of the new nation. Much later, in the nineteenth century, Karl Marx and other social theorists applied the same basic idea of human perfectibility in their criticisms of capitalism. A revolutionary new society, they predicted, would overturn the worst effects of capitalism and finally realize Rousseau's promise that a superior society could produce superior people. As these examples indicate, many of the most renowned thinkers of the last two centuries have rejected the belief that nature places strict limitations on what humans can achieve.

The Freudian Revolution

Sigmund Freud was the first social scientist to develop a theory that addressed both the "nature" and "nurture" aspects of human existence (Nagel, 1994; Robinson, 1994). For Freud, the social self develops primarily in the family, wherein the infant is gradually forced to control its biological functions and needs: sucking, eating, defecation, genital stimulation, warmth, sleep, and so on. Freud shocked the strait-laced intellectuals of his day by arguing that infants have sexual urges and by showing that these aspects of the self are the primary targets of early socialization—the infant is taught in many ways to delay physical gratification and to channel its biological urges into socially accepted forms of behavior.

Freud's model of the personality is derived from his view of the socialization process. Freud divided the personality into three functional areas, or interrelated parts, that permit the self to function well in society.[1] The part from which the infant's unsocialized drives arise is termed the **id.** The moral codes of adults, especially parents, become incorporated into the part of the personality that Freud called the **superego.** Freud thought of this part of the personality as consisting of all the internalized norms, values, and feelings that are taught in the socialization process.

In addition to the id and the superego, the personality, as Freud described it, has a third vital element, the **ego.** The ego is a person's conception of him- or herself in relation to others, in contrast with the id, which represents self-centeredness in its purest form. To have a "strong ego" is to be self-confident in dealing with others and to be able to accept criticism. To have a "weak ego" is to need continual approval from others. The popular expression that someone "has a big ego" and demands constant attention actually signals a lack of ego strength in the Freudian sense.

In the growth of the personality, according to Freud, the formation of the ego or social self is critical, but it does not occur without a great deal of conflict. The conflict between the infant's basic biological urges and society's need for a socialized person becomes evident very early. Freud believed that the individual's major personality traits (level of security, fears and longings, ways of interacting with others) are formed in the conflict that occurs as the parents insist that the infant control its biological urges. This conflict, Freud believed, is most severe between the child and the same-sex parent. The infant wishes to receive pleasure, especially sexual stimulation, from the opposite-sex parent and, therefore, is competing with the same-sex parent. To become more attractive to the opposite-sex parent, the infant attempts to imitate the same-sex parent. Thus, for Freud, the same-sex parent is the most powerful socializing influence on the growing child, for reasons related to the biological differences and attractions between the child and the opposite-sex parent.

Contemporary sociologists who are influenced by Freud's biologically and socially based theories have used his concepts of same-sex attraction and modeling of the same-sex parent's behavior to explain differences between men and women. Alice Rossi (1977; Rossi & Rossi, 1990), for instance, argues that women's shared experience of menstruation and childbearing creates a strong bond between mothers and daughters. Nancy Chodorow (1978) claims that a woman's earliest experiences with her mother tend to convince her that she can be fulfilled by becoming a mother in her turn; thus, women are socialized from a very early age to "reproduce motherhood." Research on socialization has shown that men also are strongly influenced by the same-sex parent. Fathers often serve as models of behavior whom boys will emulate throughout their lives (Popenoe, 1996).

Freud's theory includes the idea that the conflicts of childhood reappear throughout life in ways that the individual cannot predict. The demands of the superego ("conscience") and the "childish" desires of the id are always threatening to disrupt the functioning of the ego, especially in families in which normal levels of conflict are either exaggerated or suppressed. Note, however, that Freud focused on the traditional family, consisting of mother, father, and children. The more families depart from this conventional form, the more the adequacy of Freudian socialization theory must be questioned.

Behaviorism

In the early decades of the twentieth century, Freud's theory was challenged by another branch of social-scientific thought, known as **behaviorism.** In contrast to Freud and others who saw many human qualities as innate or biologically determined (nature), behaviorists saw the individual as a *tabula rasa,* or blank slate, that could be written upon through socialization. In other words, individual behavior could be determined entirely through social processes (nurture). Behaviorism asserts that individual behavior is not determined by instincts or any other "hardware" in the individual's brain or glands. Rather, all behavior is learned.

Behaviorism traces its origins to the work of the Russian psychologist Ivan Pavlov (1927). Pavlov's experiments with dogs and humans revealed that behavior that had been thought to be entirely instinctual could in fact be shaped or **conditioned** by learning situations. Pavlov's dog, one of the most famous subjects in the history of psychology, was conditioned to salivate at the sound of a bell. The dog would normally salivate whenever food was presented to him. In his experiment, Pavlov rang a bell whenever the dog was fed. Soon the dog would salivate at the sound of the bell alone, thereby showing that salivation, which had always seemed to be a purely biological reflex, could be a conditioned, or learned, response as well.

The American psychologist John B. Watson carried on Pavlov's work with an equally famous series of

[1]Freud never expected that actual physical parts of the brain that correspond to the id, ego, and superego would be discovered. Instead, he was referring to aspects of the functioning personality that are observed in the individual.

Myths of children raised in the wild, like the story of Romulus and Remus, the legendary founders of Rome, have appeared throughout history. Social scientists believe that so-called "feral" children may have been abandoned by their parents because they suffered from mental impairments such as autism, a brain disorder that deprives children of certain emotional and linguistic capabilities. It is now understood that the symptoms of autism can be alleviated somewhat by extraordinary attention from parents and other adults.

experiments on "Little Albert," an 11-month-old boy. Watson conditioned Albert to fear baby toys that were thought to be inherently cute and cuddly, such as stuffed white rabbits. By presenting these objects to Albert and simultaneously frightening him with a loud noise (i.e., presenting a negative stimulus), Watson showed that the baby could be conditioned to fear any fuzzy white object, including Santa Claus's beard. He also showed that through the systematic presentation of white objects accompanied by positive stimuli he could extinguish Albert's fear and cause him to like white objects again.

On the basis of his findings, Watson wrote:

> Give me a dozen healthy infants, well-formed, and my own specified world to bring them up in and I'll guarantee to take any one at random and train him to become any type of specialist I might select—doctor, lawyer, artist, merchant-chief and, yes, even beggar-man and thief, regardless of his talents, penchants, tendencies, abilities, vocations, and race of his ancestors. (1930, p. 104)

For the behaviorist, in other words, nature is irrelevant and nurture all-important.

Behaviorists who followed Watson—the most famous being B. F. Skinner—developed even more effective ways of shaping individual behavior. Skinner and his followers reasoned that in order to avoid failures in socialization it is necessary to completely control all the learning that goes on in the child's social environment (Skinner, 1976). Sociologists are critical of the notion that it is possible to control the world of the developing person. They argue that while the behaviorists may show us how some types of social learning take place, psychological research often does not deal with real social environments. It has very little to say about what is actually learned in different social contexts, how it is learned (or not learned), and the influences of different social situations on the individual throughout life (Elkin & Handel, 1989). One type of situation that has been of interest to sociologists studying socialization processes is that of the child reared in extreme isolation.

Isolated Children

The idea that children might be raised apart from society, or reared by wolves or chimpanzees or some other social animal, has fascinated people since ancient times. Romulus and Remus, the legendary founders of Rome, were said to have been raised by a wolf. The story of Tarzan, a boy of noble birth who was abandoned in Africa and raised by apes, became a worldwide best-seller early in the twentieth century and has intrigued readers and movie audiences ever since. However, modern studies of children who have experienced extreme isolation cast doubt on the possibility that a truly unsocialized person can exist.

The discovery of a **feral** ("untamed") **child** always seems to promise new insights into the relationship between biological capabilities and socialization, or nature and nurture (Davis, 1947). Each case of a child raised in extreme isolation is looked upon as a natural experiment that might reveal the effects of lack of socialization on child development. Once the child has been brought under proper care, studies are carried out to determine how well he or she functions. Invariably those studies show that victims of severe isolation are able to learn, but that they do so far more slowly than children who have not been raised in isolation (Malson, 1972).

A famous case of social isolation in childhood is that of Genie, a girl born to a psychotic father and a blind and highly dependent mother. For the first 11 years of her life, Genie was strapped to a potty chair in an isolated room of the couple's suburban Los Angeles home. From birth, she had almost no contact with other people. She was not toilet-trained, and food was pushed toward her through a slit in the door of her room. When she was discovered by child welfare authorities after the mother told neighbors about the child's existence, the father committed suicide. Genie was placed in the custody of a team of medical personnel and child development researchers.

In the first few weeks after she was discovered, everyone who observed Genie was shocked. At first glance, she looked like a normal child, with dark hair and pink cheeks and a placid demeanor. Very quickly, however, it became clear that she was severely impaired. She walked awkwardly and was unable to dress herself. She had virtually no language ability—at most, she knew a few words, which she pronounced in an incomprehensible babble. She spat continuously and masturbated with no sense of social propriety. In short, she was a clear case of a child who had been deprived of social learning and in consequence was severely retarded in her individual and social development. She was alive, but she was not a social being in any real sense of the term. She had not developed a sense of self, nor did she have the basic ability to communicate that comes with language learning.

For years, researcher Susan Curtiss (1977) studied Genie's slow progress toward language learning. Curtiss showed that Genie could learn many more words than a retarded person would be expected to learn, but that she had great difficulty with the more complex rules of grammar that come naturally to a child who learns language in a social world. Genie's language remained in the shortened form that is characteristic of people who learn a language late in life. Most significant, Genie never mastered the language of social interaction. She had great difficulty with words such as *hello* and *thank you,* although she could make her wants and feelings known with nonverbal cues.

Extensive tests showed that in many ways Genie was highly intelligent. But her language abilities never advanced beyond those of a third grader. Genie gradually learned to adhere to social norms (e.g., she stopped spitting), but she never became a truly social being. Eventually, the scientists who worked with her came to the conclusion that the most severe deprivation, the one that was the primary cause of her inability to become fully social and to master language, was her lack of emotional learning and especially her feelings of loss and lack of love (Rymer, 1992a, 1992b).

The Need for Love

All studies of isolated children point to the undeniable need for nurturance in early childhood. They all show that extreme isolation is associated with profound retardation in the acquisition of language and social skills. However, they cannot establish causality, since it is always possible that the child may have been retarded at birth. Despite their lack of firm conclusions, studies of children reared in extreme isolation have pointed researchers in an important direction by suggesting that lack of parental attention can result in retardation and early death. This conclusion receives further support from studies of children reared in orphanages and other residential care facilities, which have shown that such children are more likely to develop emotional problems and to be retarded in their language development than comparable children reared by their parents (Goldfarb, 1945; Okun, 1996; Rutter, 1974).

In a series of studies that have become classics in the field of socialization and child development, the primate psychologist Harry Harlow showed that infant monkeys reared apart from other monkeys never learned how to interact with their mothers (Harlow & Harlow, 1962); that is, they could not refrain from aggressive behavior when they were brought into group situations. When females who had been reared apart from their mothers became mothers themselves, they tended to act in what Harlow could only describe as a "ghastly fashion" toward their young. In some cases, they even crushed their babies' heads with their teeth before handlers could intervene. Although it is risky to generalize from primate behavior to that of humans, these studies of the effects of lack of nurturance bear a striking resemblance to studies of child abuse in humans, which generally show that one of the best predictors of abuse is whether the parent was also abused as a child (Kempe & Helfer, 1980; Kempe & Kempe, 1978; Keniston, 1977; Polansky et al., 1981).

These findings confirm our intuitive knowledge that nurturance and parental love play a profound, though still incompletely understood, role in the development of the individual as a social being. These and related findings offer support for social policies that seek to enrich the socialization process with nurturance from other caring adults—for example, in early-education programs. Yet many researchers and policy makers remain convinced that biological traits place limits on what individuals can achieve, regardless of the kind of nurturance they receive. In recent years, proponents of this view have focused on how intelligence affects achievement. This is a particularly controversial question, especially when it is applied to supposed differences in intelligence test scores for different ethnic and racial groups.

A Sociological Summary

Neither extreme of the nature-nurture debate presents a complete picture of socialization. Nature may endow individuals with greater or lesser innate abilities, yet despite those differences, most people learn to function as social beings. And while humans have an infinite capacity to learn behaviors of all kinds, through socialization they learn the particular behaviors required to function within their own culture and society. Nevertheless, the nature-nurture debate will endure and will continue to stimulate new research on the biological and psychological bases of behavior. In the meantime, most sociological research will focus on the following hypotheses:

1. *The social environment can unleash or stifle human potential.* Genetic and other biological traits establish broad boundaries for individual achievement, but the environment in which a person is raised can cause his or her potential to be realized more or less fully within those boundaries.
2. *The social environment presents an ever-changing array of roles and expectations.* Through socialization, people learn to perform roles as members of a particular culture and society. As the society and its culture change, so do definitions of what makes a person well socialized. These changing definitions create endless possibilities for misunderstandings and conflict within the groups and institutions in which socialization takes place.

THE SOCIAL CONSTRUCTION OF THE SELF

The *self* is the outcome of socialization; it may be defined as "the capacity to represent to oneself what one wishes to communicate to others" (Elkin & Handel, 1989, p. 47). Genie, the isolated, unsocialized child discussed earlier, did not develop a self in early childhood. She did not learn to formulate in her own mind the words that would express her feelings to others. Through socialization, most children learn to convert cries of discomfort, hunger, or fear into socially understandable verbal symbols like "Want bottle" or "Go out now." These utterances show that the young child has learned to recognize his or her inner states and communicate them to others. "The child who can do this is on his or her way to becoming human, that is, to being simultaneously self-regulating and socially responsive" (Elkin & Handel, 1989, p. 47).

In sociology, the self is viewed as a social construct: It is produced or "constructed" through interaction with other people over a lifetime. Studies of how the self emerges therefore usually take an interactionist perspective.

Interactionist Models of the Self

The American sociologist Charles Horton Cooley was a major contributor to the interactionist view of the development of the self. In *Human Nature and the Social Order* (1956/1902), he introduced the concept of the "looking glass self." The looking glass self is the reflection of our self that we think we see in the behaviors of others toward us. We continually pay attention to the behavioral cues of others; we wonder whether they think we look good, are expressing ourselves well, are working hard enough, and so on. As we mature, the overall pattern of these reflections of other people's opinions becomes a dominant aspect of our own identities—that is, of how we conceive of ourselves. Cooley believed that through these processes we actually become the kind of person we believe others think we are.

Language, Culture, and the Self. Cooley's insight into the role of others in defining the self was the foundation for the view of the self proposed by George Herbert Mead. With Cooley, Mead believed strongly that the self is a social product. We are not

THE CENTRAL QUESTION: A CRITICAL VIEW

To what extent does society shape how individuals behave and what they can or will become?

Intelligence tests figure prominently in the work of social scientists who believe that genetic inheritance determines much if not most of what we achieve socially. But most research findings lead sociologists to question this belief. Still, the controversy continues, often with important implications for social policies that deal with education and other aspects of socialization.

A controversial study titled *The Bell Curve,* by Richard Herrnstein and Charles Murray (1994), presents the argument that intelligence is inherited and that genetic inheritance goes a long way toward explaining income inequality. Its authors claim that the growth in occupations requiring advanced education and technical skills is creating a demanding new economic environment in which many people are doomed to failure. They do not believe that efforts to address inequalities through preschool programs, improvements in public education, and the like will make a difference. The real obstacle, they argue, is intelligence—or rather the lack of it in vast numbers of people. Since by their reasoning intelligence is an innate trait that is distributed unequally among various subgroups of the population, social intervention cannot do much to equalize the economic effects of differences in intelligence.

Like many traits that vary from one person to another, such as height and weight, scores on intelligence tests are distributed in the shape of a "normal" or bell curve. Most people's IQ scores fall near the center of the distribution, while some are at either extreme. About these facts there is no argument. Murray and Herrnstein assert, however, that scores on the bell curve of intelligence are creating what they call "cognitive classes"—that is, categories of people whom they label "very bright," "bright," "normal," "dull," and "very dull" (see chart). They note that 5% of the U.S. population falls within the left and right extremes of the curve, and another 20% are in Class II and Class IV. By this reasoning, approximately 50 million residents of the United States are classified in the lower cognitive classes.

There has been much criticism of IQ as a single measure of intelligence. Many experts believe that intelligence is far too complex to be represented by a single measure like IQ (Gardner, 1983). Also, there is evidence of cultural and middle-class biases in the questions used to test IQ; examination of the test items reveals that they would be far more familiar to middle-class test takers than to those from a disadvantaged background. Moreover, just because IQ and poverty are correlated does not mean that IQ causes poverty. In Northern Ireland, for example, Catholic Irish individuals score lower on IQ tests than do Protestant Irish, but these differences in IQ are not found in the United States, where there are no differences in wealth or advantage between the two groups.

Do you think intelligence is all in the genes? Do you think it can be expressed as a single number, the IQ score? Or do you think, as many social scientists do, that intelligence in humans is multidimensional? Do you know people who seem to be able to perform brilliantly in social situations yet are not world beaters on tests, or whose bodies can perform astounding feats yet who are not necessarily great at math? Perhaps many of these differences are accounted for by a combination of biological endowments and opportunities to learn and practice. Before you accept the idea that IQ is destiny, please think long and carefully about these questions.

Defining the Cognitive Classes

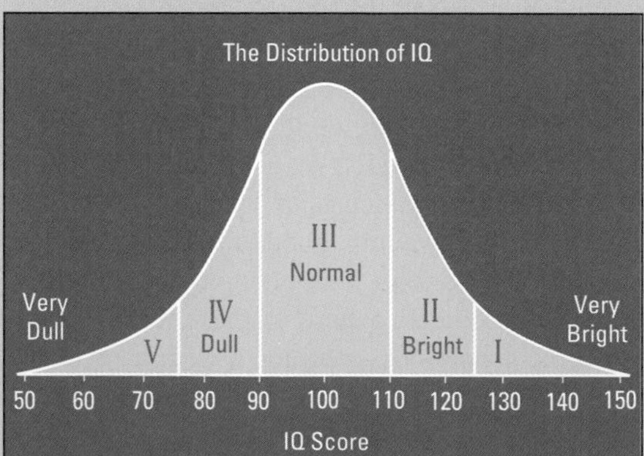

Caution: The labels imposed on the IQ curve and the scores used as boundaries between "cognitive classes" are those of Herrnstein and Murray and do not represent the thinking of many other social scientists.
Source: Herrnstein & Murray, 1994.

A century ago, long skirts and corsets were part of a larger set of cultural values and norms that required women to be subservient to men and to place men's needs above their own. Young women today are increasingly socialized by both parents to participate in society as independent individuals and not just as their father's obedient daughter or "the wife of so-and-so." This change has not occurred easily, and some people and groups still insist that women must conform to older values and norms.

born with selves that are "brought out" by socialization. Instead, we acquire a self by observing and assimilating the identities of others (Nisbet, 1970). The vehicle for this identification and assimilation is language. As Mead wrote, "There neither can be nor could have been any mind or thought without language; and the early stages of the development of language must have been prior to the development of mind or thought" (quoted in Truzzi, 1971, p. 272).

This view of the emergence of the self places culture at the center of the formation of the self. The kind of person we become is in large part a result of the cultural influences that surround us during socialization. Through interaction with people who are Catholic, for example, one takes on the language, the jokes, and the style of a person of that religion. If the father is a firefighter and the mother a nurse, certain attitudes about service to society and about illness and danger will carry over to the child. If the same child plays on sports teams with other children in the neighborhood, the norms and values of those children and their parents will become part of the child's experience and will be incorporated into his or her personality. Another child, growing up on a Sioux Indian reservation, would learn some of the same values—such as fair play, reward for achievement, and good citizenship—but would also learn the norms and values of the Sioux (e.g., reverence for one's ancestors and for the natural environment).

As each person learns the norms of his or her culture and its various ways of communicating—whether through language, dress, or gestures—and as each experiences the influences of a particular family and peer group, a unique self is formed. The self, thus, is a product of many influences and experiences; every person emerges with a personality of his or her own, and each has incorporated to varying degrees the values of the larger society and of a particular subculture.

Gender Socialization. Every culture has norms that specify behavior considered appropriate for men and women, boys and girls. From earliest infancy, we receive both explicit and implicit messages about behavior that is appropriate for boys or girls (Richmond-Abbott, 1992). **Gender socialization** refers to the specific aspects of socialization whereby people learn the behaviors and attitudes considered appropriate for males and females in their cultures. The Inuit boy learns how to hunt and fish, the girl how to skin animals and prepare furs, as gender-appropriate behaviors, but nowadays, each also learns to drive a snowmobile and take notes in class.

Throughout the world, there are major differences in what behaviors are considered appropriate for males and females. Indeed, one of the major areas of social change in the contemporary world is in the trend toward greater similarity in what is considered appropriate for males and females. Millions of women in traditional villages in Asia, Africa, and Latin America are experiencing dramatic changes in what is considered possible and appropriate for them as opposed to their brothers—although the latter are still far more likely to attend school and aspire to careers outside the family. In Europe and the United States as well, there have been revolutionary changes in the norms that are transmitted in gender socialization. We will see in subsequent chapters that in these societies there is still a great deal of inequality based on gender differences but that over the last 100 years or more there have been dramatic changes. At the turn of the twentieth century, for example, women were expected to live in a separate world of domestic chores and community activities. Today, women far more often assert their right to be active and competitive in the world outside the home and community.

Role Taking: The Significant Other and the Generalized Other. For Mead (1971/1934), two of the most important activities of childhood are play and games. In play, the child practices taking the roles of others. *Roles* are the behaviors associated with certain positions in society (e.g., first baseman, engineer, mother) and are discussed in more detail in Chapter 4. To "take" a role is to initiate and practice behavior that one has learned from watching others perform the role. If you watch preadolescent children play, you will see that they continually take or "try on" roles: "You be the mommy and I'll be the teacher, and you'll come to school to find out why . . ." or "Pretend I'm Michael Jordan and you're Reggie Miller and the score is tied in the fourth quarter . . ." They are reenacting the dramas of winning and losing or calling into question the behaviors of the schoolroom or trying to understand sickness and death.

This idea of **role taking** is central to the interactionist view of socialization. It refers to the way we look at social situations from the standpoint of another person from whom we seek a response. Mead believed that children develop this ability in three stages—the *preparatory, play,* and *game* stages—during which they gain their sense of self and learn to act as persons in society.

In the *preparatory stage,* the child attempts to mimic the behavior of people who are significant in his or her life. **Significant others** are people who loom large in our lives, people who appear to be directly involved in winning and losing, achieving and failing. They tend to be people after whom we model our behavior—or whose behavior we seek to avoid. In the preparatory stage, the child's significant others are those who respond to calls for help and shape social behavior like language.

In Mead's second stage, the *play stage,* children play at being others who are significant in their lives:

> They want to push the broom, carry the umbrella, put on the hat, and do all the other things they see their parents do, including saying what their parents say. The story is told of the 4-year-old playing "daddy" who put on his hat and his coat, said "good bye," and walked out the front door, only to return a few minutes later because he didn't know what to do next. He had taken as much of his father's work role as he could see and hear—the ritualized morning departure. (Elkin & Handel, 1989, p. 49)

As children grow in age and experience, they enter what Mead called the *game stage.* To take part in a game, a child must have already learned to become, in a symbolic sense, all the other participants in the game. Mead called this the ability to "take the role of the generalized other." Thus, in a baseball game,

> [The child] must know what everyone else is going to do in order to carry out his own play. He has to take all of these roles. They do not have to be present in consciousness at the same time, but at some moments he has to have three or four individuals present in his own attitude, such as the one who is going to throw the ball, the one who is going to catch it, and so on. (Mead, 1934, p. 151)

When we are able to take the role of the generalized other, we know that rules apply to us no matter who we are. We know, for example, that rules about not smoking in the school building apply equally to students, parents, and teachers and that those who violate the rules will not be excused because of their status (Mead, 1934).

The **generalized other** is a composite of all the roles of all the participants in the game. A person who participates in a game like baseball, for example, has developed the capacity for role taking and now (again using Mead's phrase) "takes the role of the generalized other." When little children play team games, they often have a hard time taking specific roles. As children mature, they become increasingly competent at games and team sports.

The generalized other represents the voice of society, which is internalized as "conscience." For some people, the generalized other demands perfection and strict adherence to every rule. For others, the generalized other may be extremely demanding where sports and other games are concerned but much more relaxed about achievement in school. For still others, the generalized other may insist on amassing large amounts of money or serving the community as the primary indicators of success. Within any given culture, such variations will be wide but will tend to follow certain easily recognized patterns.

Theories of Moral Development

Throughout life, people face a variety of moral dilemmas, and these have a significant effect on their personalities. Social scientists have devoted considerable study to the processes through which people develop concepts of morality. Among the best-known students of moral development are the Swiss child psychologist Jean Piaget and the American social psychologists Lawrence Kohlberg and Carol Gilligan.

Piaget stands with Freud as one of the most important and original researchers and writers on child development. In the 1920s, he became concerned with how children understand their environment, how they view their world, and how they develop their own personal philosophies. To discover the mental processes unique to children he used what was then an equally unique method: He spent long hours with a small number of children simply having conversations with them. These open-ended discussions were devoted to understanding how children actually think. In this way, Piaget discovered evidence for the existence of ideas that are quite foreign to the adult mind (Elkind, 1970). For example, the child gives inanimate objects human motives and tends to see everything as existing for human purposes. In this phase of his research, Piaget also described the egocentric aspect of the child's mental world, which is illustrated by the tendency to invent words and expect others to understand their meanings.

In the later phases of his research and writing, Piaget devoted his efforts to questions about children's moral reasoning—the way children interpret the rules of games and judge the consequences of their actions. He observed that children form absolute notions of right and wrong very early in life but that they often cannot understand the ambiguities of adult roles until they approach adolescence. This line of investigation was continued by the American social psychologist Lawrence Kohlberg, whose theory incorporates Piaget's views on the development of children's notions of morality.

Kohlberg's theory of moral development emphasizes the cognitive aspects of moral behavior. (By *cognitive,* we mean aspects of behavior that one thinks about and makes conscious choices about, rather than those that one engages in as a result of feelings or purely intuitive reactions.) In a study of 57 Chicago children that continued until the children were young adults, Kohlberg presented the children with moral dilemmas like the following:

> A husband is told that his wife needs a special kind of medicine if she is to survive a severe form of cancer. The medication is extremely expensive, and the husband can raise only half the needed funds. When he begs the inventor of the drug for a reduced price, he is rebuffed because the inventor wants to make a lot of money on his invention. The husband then considers stealing the medicine, and the child is asked whether the man should steal in order to save his wife. (Kohlberg & Gilligan, 1971)

On the basis of children's answers to such dilemmas at different ages, Kohlberg proposed a theory of moral development consisting of three stages: (1) *preconventional,* in which the child acts out of the desire for reward and the fear of punishment; (2) *conventional,* in which the child's decisions are based on an understanding of right and wrong as embodied in social rules or laws; and (3) *postconventional,* in which the individual develops a sense of relativity and can distinguish between social laws and moral principles. Very often subjects in the preconventional and conventional stages will immediately assume that stealing is wrong in the situation described, but postconventional thinking in older children will cause them to debate the fairness of rules against stealing, in view of the larger moral dilemma involved.

It should be noted that Kohlberg's studies involved children from secure American families. In those studies, the children were brought up in fairly similar circumstances, so the results could not reflect the impact of differences in such factors as poverty versus affluence or nurturance versus neglect. (See the Global Social Change box on pages 70–71.)

Gender and Moral Reasoning. Kohlberg's studies have been criticized for focusing too heavily on the behavior of boys and men and not exploring possible alternative lines of moral reasoning that may prevail among females (Gilligan et al., 1988; Wren, 1997). Pioneering work by social psychologist Carol

GLOBAL SOCIAL CHANGE

Growing Up on the Street: A Worldwide Trend

A boy playing on the railroad tracks lives in a shelter for homeless families in the Midwest. He is just one of the millions of homeless children in the world today. Throughout the world, there has been an alarming increase in the number of children living without adult supervision, usually in the slums and back alleys of major regional and national capitals. Social-scientific knowledge helps us understand the causes of this dangerous trend and documents the negative consequences of the socialization of homeless children on the streets. Although the chances of reversing this trend are not good, sociologists and other professionals with social-scientific training are addressing the problem by attempting to create alternative institutions that can fill the socialization needs of these children.

Homeless children often make play spaces out of industrial facilities. Chris, the boy who took this photo, noted that for him, "the best part of the shelter is the train track."

Children without supervision spend much of their time out of school and on the streets. Typically they are orphans, children of homeless families, runaways with parents or guardians elsewhere, or children who have been lost during natural or social catastrophes such as earthquakes and civil wars.

The causes of homelessness among children vary from one continent to another. In Africa and Indonesia, where rates of maternal death in childbirth may exceed 1,000 per 100,000 live births, large numbers of children are orphaned at birth. After several years in a public orphanage, these children often find their way to the streets, where they are socialized to street life by older, more experienced children. The AIDS epidemic in large parts of Africa plus famine and civil strife in the Sahel region (extending from Mauritania to Chad) also contribute to the growing number of wandering children finding their way into urban centers.

In Central and South America, the population of street children is growing as a consequence of poverty and abandonment as well as civil strife and political violence. In Guatemala, Nicaragua, Colombia, Brazil, and Peru—where civil strife in some rural areas has left thousands of children without parents—young, destitute migrants are flowing into regional urban centers. In Europe, similar problems are arising in Bosnia, Romania, and parts of the former Soviet Union, owing to civil wars and sudden shifts in family fortunes.

Gilligan, an early collaborator of Kohlberg, has produced an impressive body of evidence that demonstrates the propensity of females to make moral choices on the basis of a somewhat different line of reasoning from that generally followed by males. Gilligan's research, and that of others who have followed her lead, shows that females are more likely than males to base their moral judgments on considerations of caring as well as justice or law. More than males, females tend to look for solutions to moral dilemmas that also serve to maintain relationships. Caring solutions that consider the needs of both sides are therefore more often invoked by females.

A good example of this difference appears in the work of D. Kay Johnston (1988), in which adolescent boys and girls were presented with dilemmas taken from Aesop's fables. The young people were read a fable that presents a moral dilemma and then asked

In more economically developed areas of the world, orphaned infants tend to be placed in public orphanages. In affluent nations, however, homelessness among children is also on the upswing. In the United States, for example, high rates of crack abuse and incarceration of parents from high-poverty neighborhoods have caused a large share of the increase in homeless street children. As the chart shows, by 1990 almost 10% of U.S. children were living in households not headed by a parent. Of these, many thousands will leave orphanages and group homes to live on the streets.

Throughout the world, the socialization of street children is carried out by other children and adolescents. In order to survive, the homeless child needs protection from predatory adults. Most often the child finds the help of older children vital, but that help often comes at a high price in the form of sexual services, criminal activities, drug use, avoidance of school, and many other behaviors that will socialize the homeless young person for a life on the extreme margins of "respectable" society. To survive on the street, children must be members of an organized band or gang. Since legitimate employment is rarely available until they are older, they usually become adept at the hustles and scams of street life. They beg at busy intersections, scavenge garbage for food and fuel, find shelter where they can, and seek health care in the emergency rooms of hospitals. Their death rate, especially from violence and homicide, is extremely high.

In the United States, where there are more than 400,000 orphans and incarcerated young people living in institutions, the socialization that children experience in detention centers and juvenile correction centers often has the effect of preparing them to be more effective street hustlers and petty criminals when they are released. To date, the establishment of more successful socializing environments for homeless children has (with a few exceptions) been a matter of low priority. Left to survive by their wits, homeless street children scavenge, prostitute themselves, and beg—and as they grow older many turn to more ambitious and violent forms of crime.

Children Living in Households Not Headed by a Parent, United States

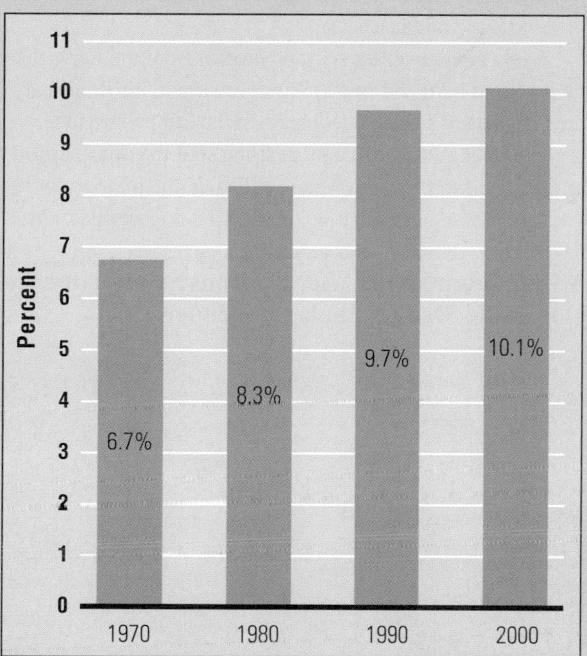

Source: Annie E. Casey Foundation, Center for the Study of Social Policy; *Statistical Abstract,* 2000.

what they understood the problem to be and how they would solve it. In the fable of the dog in the manger (see Table 3.1), the problem is clearly that the dog has taken sleeping space from the deserving ox. Some adolescents judged the situation purely in terms of which animal had the right to the space, and made statements like "It's [the ox's] ownership and nobody else had the right to it." Others sought a caring solution that would take into consideration both animals' needs, and made statements like "If there's enough hay, well, this is one way, split it. Like, if they could cooperate."

Table 3.1 shows that boys were more likely than girls to give solutions based on rights, while girls were more likely than boys to choose solutions that emphasized caring. Some subjects chose solutions that combined the two approaches. As Gilligan notes, "An innovative aspect of Johnston's design lay in the fact

TABLE 3.1

The Dog in the Manger

A dog, looking for a comfortable place to nap, came upon the empty stall of an ox. There it was quiet and cool and the hay was soft. The dog, who was very tired, curled up on the hay and was soon fast asleep.

A few hours later the ox lumbered in from the fields. He had worked hard and was looking forward to his dinner of hay. His heavy steps woke the dog, who jumped up in a great temper. As the ox came near the stall the dog snapped angrily, as if to bite him. Again and again the ox tried to reach his food, but each time he tried the dog stopped him.

Moral Orientation of Spontaneous Solution for the Dog in the Manger Fable, by Gender

Orientation	Female	Male
Rights (justice)	12	22
Response (caring)	15	5
Both	3	1

Source: Adapted from Johnston, 1988, p. 57.

that after the children had stated and solved the fable problems, she asked, 'Is there another way to think about this problem?' About half of the children . . . spontaneously switched orientation and solved the problem in the other mode" (Gilligan et al., 1988, p. xxi). On the basis of this and much subsequent research, Gilligan concludes that by age 11 most children can solve moral problems both in terms of rights (a justice approach) and in terms of response (a caring approach). The fact that a person adopts one approach in solving a problem does not mean that he or she does not know or appreciate others.

Gilligan and others who study moral development and gender point out that adolescence is a critical time in the development of morality and identity. However, in schools and elsewhere in society the message that comes across is that norms, values, and the most highly esteemed roles require that there be a "right way" to feel and think. Most often this right way is associated with the justice focus—and the caring focus is silenced, along with the voices of girls and others to whom it appears to be a valuable alternative mode of moral reasoning (Wren, 1997). In adolescent girls and many minority students, this form of silencing can be detrimental to the development of the self in social situations (Taylor, Gilligan, & Sullivan, 1995).

AGENCIES OF SOCIALIZATION

Agencies of socialization are the groups of people, along with the interactions that occur within those groups, that influence a person's social development throughout his or her lifetime (Elkin & Handel, 1989); **agents of socialization** are individuals, such as parents and teachers, who socialize others. The most familiar agencies of socialization are the family, schools, socializing agencies in the community, religion, the peer group, and the mass media. Later in life, adults may experience further socialization in the workplace, in universities, or in the military, to cite a few common examples. And people who wish to change self-destructive behaviors such as drug addiction or alcoholism may be resocialized in "12-step" programs like that of Alcoholics Anonymous. (Resocialization is discussed further in the next section.) Within all agencies of socialization, a number of processes are continually occurring that shape the individual's development. Among these are direct instruction, imitation or modeling of behavior, and reinforcement of particular behaviors (e.g., through rewards or punishments). Many of these processes appear in what sociologists call **anticipatory socialization.** Whenever an individual plays at a role he or she is likely to assume later in life, anticipatory socialization is taking place. By dressing up in adult clothing, for example, a child imitates the behavior of adults. The adolescent who attends a high school prom is being socialized in anticipation of a time when she or he will be expected to participate in formal social events. And a worker in a corporate setting may rehearse before coworkers a presentation he or she will later make to a potential client.

Another important aspect of agencies of socialization is that they must continually deal with social change. For example, as guns have become more readily available in large cities, peer groups have had to cope with the ever-present danger of violent death. As divorce and remarriage rates have risen, many families have had to deal with new structures that add much more complexity to family socialization patterns. As a result, the influence of social change on socialization patterns has become an important area of sociological research.

In the following pages, we briefly consider several agencies of socialization—the family, the schools, the community, religion, the peer group, and the mass media. Many of these agencies will be encountered in later chapters as well.

The Family

The family is the primary agency of socialization. It is the environment into which the child is born and in which his or her earliest experiences with other people occur—experiences that have a lasting influence on the personality. Family environments vary greatly, not only in terms of such key variables as parents' income and education but also in terms of living arrangements, urban versus rural residence, number of children, relations with kin, and so on. Much contemporary research centers on the effects of different family environments on the child's development. For example, Michael Lewis and Candice Feiring's intriguing study of 117 families, "Some American Families at Dinner" (1982), shows that a typical 3-year-old child interacts regularly with a network of kin, friends, and other adults who may play a significant role in the child's early socialization. (See the Sociological Methods box on page 74.)

One of the most significant changes occurring in American society today is the increase in the proportion of children growing up in poor families (almost 40% of all children under age 18) or in no family at all. In the early 1990s, nearly 10% of all children in the United States were growing up in households with no parent present. Most of these children (76%) were being socialized in the homes of grandparents and relatives; the remainder were living in group homes and with nonrelatives (Gross, 1992). There has also been a steady increase in the number of single-parent families, of which the vast majority are headed by women.

But changes in the way families are organized or how they cope with changing social conditions are not a new phenomenon. The values and ways of parents are never entirely valid for their children, although the degree to which this is true depends on how much social change is experienced from one generation to the next. Socialization creates the personalities and channels the behaviors of the members of a society, but that socialization is never entirely finished. Thus, in *Manchild in the Promised Land,* his masterpiece about growing up in Harlem during the period of rapid migration of blacks from the South, Claude Brown (1966) wrote that his rural-born parents did not "seem to be ready for urban life." Their values and norms of behavior made no sense to their son, who had to survive on Harlem's mean streets:

> When I was a little boy, Mama and Dad would beat me and tell me, "You better be good," but I didn't know what being good was. To me it meant that they just wanted me to sit down and fold my hands or something crazy like that. Stay in front of the house, don't go anyplace, don't get into trouble. I didn't know what it meant, and I don't think they knew what it meant, because they couldn't ever tell me what they really wanted. The way I saw it, everything I was doing was good. If I stole something and didn't get caught, I was good. If I got into a fight with somebody, I tried to be good enough to beat him. If I broke into a place, I tried to be quiet and steal as much as I could. I was always trying to be good. They kept on beating me and talking about being good. And I just kept on doing what I was doing and kept on trying to do it good. (p. 279)

Brown's story is the biography of a young man whose parents' experiences with social change created a severe disjunction in their lives. This left them ill equipped to socialize their son for the demands of a new environment. And so he learned to survive on the streets, by becoming a thief and a gang fighter. But by his own account, later in his life he was greatly influenced by people who had studied the social sciences and created well-functioning institutions, special schools in particular, that could bring out his talents and socialize him for a more satisfying and constructive life than he had led as a child.

The Schools

For most of us, regardless of what our home life is like, teachers are generally the first agents of socialization we encounter who are not kin. In some cases, children are also influenced by agents of socialization in the church (e.g., ministers or rabbis), but for most children the school is the most important agency of socialization after the family. Children experience many opportunities to perform new roles in school (e.g., as student, teammate, etc.). No wonder, then, that in these first "public" appearances they tend to be highly sensitive to taunts or teasing by other children. Even very young children can become distraught when they feel that they are not wearing the "right" clothes.

Schools are institutions where differences between the values of the family and those of the larger society come into sharp focus. In some communities, for example, there may be great concern about any form of sex education in school. Indeed, much of the conflict over educational norms—what ought to be taught, whether there should be prayer in school, and

SOCIOLOGICAL METHODS

Measures of Central Tendency

The following table, taken from Lewis and Feiring's study of American families, presents information about the social network of the typical 3-year-old:

The 3-Year-Old Child's Social Network (*N* = 117)*

	X	Range
Number of relatives other than parents seen at least once a week	3.20	0–15
Number of child's friends seen at least once a week	4.43	0–13
Number of adults seen at least once a week	4.38	0–24

*N refers to the number of subjects in a study.

Source: Lewis & Feiring, 1982, p. 116.

To understand this table, you must know something about the common measures of central tendency. The one used in the table is the *mean,* represented by *x*. To arrive at this number, each child's total number of friends, relatives, and other adults seen at least once a week is added to the totals for the other children in the sample. The grand total is then divided by the number of children in the sample. The resulting number can be used to represent the entire sample.

The other common measures of central tendency are the *median* and the *mode*. The median is the number that divides a sample into two equal halves when all the numbers in the sample are arranged from lowest to highest. The mode is simply the score that occurs most frequently in the sample. Sociologists make a point of specifying which of these measures they are using. They try to avoid the term *average* in statistical tables, as it can refer to any one of these measures.

To illustrate all three measures of central tendency, let us use the following data from 10 of the children in Lewis and Feiring's sample:

Child	Number of Friends Seen at Least Once a Week
1	2
2	3
3	3
4	7
5	12
6	2
7	3
8	1
9	0
10	5
	38

In this example, the mode would be 3 and the median would also be 3, but the mean would be 3.8.

the like—stems from differences between the values of some families and the values that many educators wish to teach. Such conflicts point to the exceedingly difficult situation of schools in American society. Research on school–family and school–community relations has shown that the schools are expected to conserve the society's values (by teaching ideals of citizenship, morality, family values, and the like) and, at the same time, play a major role in dealing with innovation and change (by expanding the curriculum to include new knowledge, coping with children's perceptions of current events, addressing past patterns of discrimination, and the like) (Goslin, 1965; Meier, 1995).

The Community

Schools may be the most important agency of socialization outside the family, but there are a number of other significant agencies of socialization in most communities—including day-care centers, scout troops, recreation centers, and sports leagues. These agencies engage in many forms of socialization. Parades on Memorial Day and the Fourth of July, for example, reinforce values of citizenship and patriotism. Participation in team sports instills values of fair play, teamwork, and competitive spirit. The uniforms and equipment required for these activities stimulate shopping trips to suburban malls and downtown department stores, and those trips prepare children and adolescents for the time when they will be consumers (McKendrick, Brewer, & Plump, 1982; Newman, 1988).

Of all these agencies of socialization, the day-care center is perhaps the most controversial. Polls often show that Americans have doubts about the effects of day care on very young children, and scandals involving charges of child abuse by day-care workers intensify those fears. But when both parents work outside the home, or when single parents must work to support their children, day-care centers may play a critical role in socialization. Many studies have shown that good quality centers are not harmful to children and in some cases may be beneficial, but the norm requiring that mothers stay home to care for young children remains strong in many communities (Kammerman, 1986).

Religion

Religion may be involved in socialization in different ways throughout an individual's lifetime. For children and adults who attend church services, the shared expression of spirituality and the teaching of the religious leader can be an extremely powerful source of moral beliefs. And these beliefs often shape one's behavior outside religious institutions. During World War II, for example, the majority of people who risked their lives to rescue Jews, gypsies, gays, and political refugees had deep religious and moral convictions (Winik, 1996). Recent sociological studies show, too, that among contemporary parents, those who attend religious services are half as likely to divorce as parents who never attend religious services. Again, the moral influence of religion is highly associated with moral beliefs that may help married couples endure hard emotional times. To the degree that religious behavior reduces the likelihood of divorce, it strengthens the family, the most important agency of socialization (Clydesdale, 1997). Finally, among elderly people religion often becomes a major source of solace as they face the continual death of loved ones and their own eventual demise.

Despite its contributions to lifelong socialization, religion can also be the source of innumerable conflicts. Disputes over child rearing, sexual conduct, the role of prayer in education, and many more rancorous conflicts are brought to public awareness by people with differing religious backgrounds. Should gay men and women serve as spiritual leaders? Should same-sex marriages be performed in churches and synagogues? Is prayer in the classroom a violation of other people's right to the constitutionally guaranteed separation of church and state? These questions hardly begin to exhaust the list of controversies based on differences in the moral teachings of various religious communities. In an election held in Kansas in 2000, for example, voters rejected school board candidates committed to making creationism, the religious interpretation of the earth's creation in seven days, stand on equal footing with the scientific theory of evolution over millions of years. We return to these issues in Chapter 10, but note here that in the Middle East, India, and the United States, sociologists are among those in the forefront of efforts to find ways in which people with strong convictions on either side of these disputes can come together peacefully to resolve their differences or to work out compromises that prevent moral conflicts from worsening (Vondra, 1996).

The Peer Group

In the United States, the **peer group** tends to be the dominant agency of socialization in middle and late childhood. Peer groups are interacting groups of people of about the same age. Among adolescents, peer groups exert a strong influence on their members' attitudes and values. Studies confirm the high degree of importance adolescents and adults alike attach to their friendship groups. Adolescents typically acquire much of their identity from their peers and consequently find it difficult to deviate from the norms of behavior that their peer group establishes (Gans, 1962, 1984; Homans, 1950; McAndrew, 1985; Whyte, 1943). In fact, the peer group may become even more important than the family in the development of the individual's identity.

There is often a rather high level of conflict within families over the extent to which the peer group influences the adolescent to behave in ways that are not approved of by the family. Even where conflict is limited, the peer group usually provides the child's first experience with close friendships outside the family. The peer group becomes the child's age-specific subculture—that is, a circle of close friends of roughly the same age, often with shifting loyalties.

The peer group usually engages in a set of activities that are not related to adult society. The peer subculture may, for example, include games that adults no longer play and may have forgotten. British researchers Iona and Peter Opie (1969) conducted a classic study of the games children play. They identified hundreds of games that were known by children between the ages of 6 and 10 all over England. They also identified the common elements in those games, such as chasing, pretending, and seeking, and found that in many of the games the central problem is to choose an "It," a seeker or chaser. The children's efforts to avoid being "It" seemed to express their desire to avoid roles that make them different from the others.

When children's peer groups are faced with conflict and social change—for example, in communities where there are high levels of poverty and demographic change, with groups of culturally distinct people moving in and out—the group often organizes itself in gangs for self-defense and aggression. There are many such gangs in large American cities characterized by rapid immigration, widespread drug use, and the ready availability of cheap handguns. Under these conditions, both male and female peer groups may become extremely dangerous environments for socialization, yet these groups exert a strong attraction on adolescents who seek protection and companionship in what is perceived as a hostile world (Sanchez-Jankowski, 1991; Williams, 1989).

The Mass Media

The most controversial agency of socialization in American society is the mass media. In debates about the effects of the media on socialization, television comes under the greatest scrutiny because of the number of hours children spend in front of the "electronic baby-sitter." Estimates of how much television children watch differ, depending on the methodology used to conduct the study (especially since simply having the set turned on does not mean that the children are actually watching). Many studies have shown that the amount of television viewed varies, depending on whether the child comes from a poor home with few alternatives or a more affluent one where other activities are available. Children from poor homes in urban communities often have the set on for seven or more hours a day, as opposed to children from more affluent homes, where the set is turned on for three or four hours daily (Elkin & Handel, 1989; Wellins, 1990).

The effects of all this television viewing on children and adolescents are a subject of intensive research. In particular, the effects of seeing violent acts on television or listening to violent music like gangsta rap are hotly debated. There is no doubt that the amount of violence shown on television is immense. George Gerbner, one of the nation's foremost researchers on television violence, has been monitoring the number of violent acts shown on television. His data indicate that the average child will have seen about 100,000 acts of violence on television before graduating from elementary school—and the number is far higher for children from poor neighborhoods (Gerbner, 1990).

SOCIALIZATION THROUGH THE LIFE COURSE

Childhood socialization is the primary influence on the individual. Regardless of what happens to a person later in life, the early childhood experiences that shaped the social self will continue to influence that person's attitudes and behavior. But people are also affected by their *life course:* the set of roles they play over a lifetime and the way those roles change as a consequence of social change. The life course of a person who came of age at the time of World War II, for example, would typically have involved service in the armed forces for men (perhaps including combat duty) and work in a factory (or noncombatant military service) for women. The life course of children born in the 1960s or 1970s is far less likely to include a long or intense experience of warfare. Natural disasters, major changes in the educational system or in the political stability of their society—these and other possible changes can influence the roles people play during their life course.

Changes in the culture of a society can also significantly alter the roles people play during their lives. In the first half of the twentieth century, for example, the role of wife was less likely to include periods of single parenthood than it is now. Fifty years ago, the role of an elderly person was far more likely to involve relative poverty and dependence on younger family members. The need to adjust to changes in society and culture and the fact that so many people are able to make such adjustments (though often with great difficulty) are evidence that socialization is not finished in childhood but continues throughout the life course (Riley, Foner, & Waring, 1988; Sorenson, Weinert, & Sherrod, 1986).

Adult Socialization

Two of the influences that produce socialization after childhood are significant others (especially new friends) and occupational mobility (especially new jobs). When one moves into a new neighborhood, for example, one often makes new friends, and this may lead to new activities. A person who has always been uninvolved in politics may, through the influence of new friends, become committed to a social movement such as the women's movement. New friends can also introduce a person to dangerous or unlawful activities such as drug use. One's *core identity*—the part of the self formed in early childhood that does not change easily—may prevent one from being overly influenced by a given peer group later in life, or it may actually cause that influence to be stronger (Elkin & Handel, 1989).

Such factors as stress and changing physical health may also have an influence. In a study of how people change their patterns of behavior later in life, Marjorie Fiske and David A. Chiriboga (1990) found that everyday hassles and the boredom of a predictable existence have a greater impact on the changes adults seek in their lives than was indicated by earlier research.

In any new activity, the newcomer or recruit must learn a new set of norms associated with the roles of the organization. Socialization associated with a new job, for example, often requires that a person learn new words and technical terms. The individual also needs to interact with a new peer group, usually composed of people whose work brings them together for long hours. And in most jobs the person will eventually be faced with choices between loyalty to peers and loyalty to those higher in the organizational hierarchy who can confer benefits like raises and promotions. In these and most other adult socialization experiences, the choices a person makes will be highly influenced by the individual's core identity, which continually shapes his or her responses to new situations and challenges (Rosow, 1965).

Resocialization

Many adults and even adolescents experience the need to correct certain patterns of prior social learning that they and others find detrimental. This need usually cannot be met through the normal processes of lifelong socialization, especially since the individual's negative behaviors may be leading toward a personal crisis and causing immense pain for family members and friends. Such a condition often responds to specific efforts at resocialization.

Resocialization is a process whereby individuals undergo intense, deliberate socialization designed to change major beliefs and behaviors. It is often aimed at changing behaviors like excessive drinking, drug use, and overeating—particularly common in affluent societies, where individuals are exposed to a great deal of choices and many pleasurable stimuli. Perhaps the most widespread and pernicious among the addictions is alcoholism, which affects millions of people in North America and is involved in much of the marital strife that leads to family breakup. Alcoholics Anonymous is a voluntary group program that uses techniques of resocialization and peer example to help alcoholics reject old behaviors and learn new, more positive ones that do not involve drinking.

Much successful resocialization takes place in what are often called **total institutions** (Goffman, 1961). These are settings where people undergoing resocialization are isolated from the larger society and are under the control of a specialized staff whose members themselves may have experienced the same process of resocialization. All aspects of the inmates' daily lives are controlled, and supervision is constant. Drug treatment centers, for example, are often set up as total institutions for the resocialization of addicts. Even the smallest rewards, like extra time alone or more freedom to walk around the grounds, are controlled by the staff and bestowed only on inmates who have made progress toward the resocialization goals.

In the resocialization process, the staff first attempts to tear down the individual's former sense of self. This stage may include various forms of degradation and abasement in which the individual is forced to reject the undesired thoughts and behaviors. In the next stage, the staff rewards the individual's attempts to build a new sense of self that conforms to the goals of the resocialization process (e.g., a sober person who accepts responsibility for previous wrongs and achieves new interests and a new awareness of personal strengths). However, while total institutions can be extremely powerful environments for resocialization, they sometimes run the risk of creating people who need to remain in that controlling environment—that is, people who have developed a need to be "institutionalized."

Erikson on Lifelong Socialization

The theories of social psychologist Erik Erikson are especially relevant to the study of adult socialization and resocialization. Erikson agreed with Freud that a person's sense of identity is shaped by early childhood experiences. But Erikson also focused on the many changes occurring throughout life that can shape a person's sense of self and ability to perform social roles successfully. He demonstrated, for example, that combat experiences can produce damaged identities because soldiers often feel guilty about not having done enough for their fallen comrades (this is sometimes called "survivor guilt"). For Freud, in contrast, war-produced mental illness was always related to problems that the soldier had experienced in early childhood.

In *Childhood and Society* (1963), Erikson's central work on the formation of the self, the concept of identification takes center stage. **Identification** is the social process whereby the individual chooses other people as models and attempts to imitate their behavior in particular roles. Erikson noted that identification with these role models occurs throughout the life course. He pointed out that even older people seek role models who can help them through difficult life transitions.

For Erikson, every phase of life requires additional socialization to resolve the new conflicts that inevitably present themselves. Table 3.2 presents the basic conflicts that each person must resolve throughout life and indicates how a positive role model can help in resolving those conflicts. (The major theories of socialization are summarized in the study chart on page 79.)

Resocialization at any age usually requires a strong relationship between the individual and the role model. In the military or in a law enforcement agency, for example, the new recruit is usually paired with a more experienced person who offers advice and can show the recruit the essential "tricks" of the new role—"how it's really done." Thus, in Alcoholics Anonymous (AA), once the self-confessed "drunk" faces up to that negative identity, he or she is taught new ways of thinking, feeling, and acting that do not involve drinking. An important step in this process is the development of a relationship with a *sponsor,* a person who has been through the same experiences and has made a successful recovery. In fact, one of AA's aims is to enable the resocialized alcoholic to eventually perform as a sponsor—that is, as a role model—for someone else who is going through the same process.

TECHNOLOGY AND SOCIAL CHANGE

Socialization practices are deeply affected by changing technologies. In North America and most of Europe, the past two generations have been raised with television in the home, often in the children's bedrooms. This is in marked contrast to people who were raised in the first half of the twentieth century, before the advent of television. Television exposes young people to far more adult imagery, including sexuality and violence, than was available to earlier generations. It also competes with reading, homework, hobbies, and physical activity to help produce a more sedentary and overweight population of consumer-citizens. Now the computer and the World Wide Web offer some competition to television in more affluent homes, but they also present many issues for parents to deal with (Greenman, 1999).

Like television, the networked computer brings the world into the home and potentially exposes children and adolescents to a great deal more cultural diversity than might otherwise be the case. But the new technology may also expose children to pornography and other temptations of the adult world. Ideally, parents want to guide their children's exposure to television and Internet programming in order to select educational and morally useful material. But with more and more parents in the labor force and children spending time alone or with peers by the television or computer, guiding children becomes more difficult in many households.

Technological "fixes" have been developed to help solve this problem. In particular, the television V-chip and filtering software for home computers offer parents a means to control children's exposure to undesirable material. Somewhat surprisingly, however, although

70% of parents say they would like to use the V-chip, less then 40% actually do (Greenman, 1999). As a result, political leaders are urging citizens to take a more active role in seeking out measures to filter programming and to become more attentive and active in monitoring their children's uses of the media (Clinton, 1999).

TABLE 3.2
Erikson's View of Lifelong Socialization

Stage of Life	Conflict	Successful Resolution in Old Age*
Infancy	Trust vs. mistrust	Appreciation of interdependence and relatedness.
Early Childhood	Autonomy vs. shame in the development of the will to be a social actor	Acceptance of the cycle of life, from integration to disintegration.
Play Age	Initiative vs. guilt in the development of a sense of purpose	Humor; empathy; resilience.
School Age	Industry vs. inferiority in the quest for competence	Humility; acceptance of the course of one's life and unfulfilled hopes.
Adolescence	Identity vs. confusion; struggles over fidelity to parents or friends	A sense of the complexity of life; merger of sensory, logical, and aesthetic perception.
Early Adulthood	Intimacy vs. isolation in the quest for love	A sense of the complexity of relationships; value of tenderness and loving freely.
Adulthood	Generativity vs. stagnation in interpersonal relationships	*Caritas* (caring for others) and *agape* (empathy and concern).
Old Age	Integrity vs. despair	Wisdom and a sense of integrity strong enough to withstand physical disintegration.

*Erikson did not believe that elderly people necessarily resolve these conflicts entirely, but when they do, the results are as shown here.

STUDY CHART
Theories of Socialization

Theorist	Description of Theory
Sigmund Freud	Through socialization, the infant is gradually forced to channel its biological urges into socially acceptable forms of behavior. Major personality traits are formed in the conflict between the child and its parents, especially the same-sex parent.
George Herbert Mead	Socialization is a process in which the self emerges out of interaction with others, not only in early infancy but throughout life. Role taking is central to socialization; role-taking ability develops through interaction during childhood in the preparatory, play, and game stages.
Jean Piaget	Children develop definite awareness of moral issues at an early age but cannot deal with moral ambiguities until they mature further. This insight is incorporated in Kohlberg's preconventional, conventional, and postconventional stages of moral development.
Erik Erikson	Throughout the life course, the individual must resolve a series of conflicts that shape the person's sense of self and ability to perform social roles successfully. Central to this process is identification, in which the individual chooses other people as role models.
Carol Gilligan	Children tend to develop different ways of resolving moral dilemmas. Some (more often male) tend to rely on strict rules of right and wrong, while others (most often but not always female) tend to make judgments based on notions of fairness and cooperation. In societies where male voices are dominant, issues of cooperation and fairness are passed over, to the detriment of female socialization for leadership and achievement.

SOCIOLOGY AND SOCIAL JUSTICE

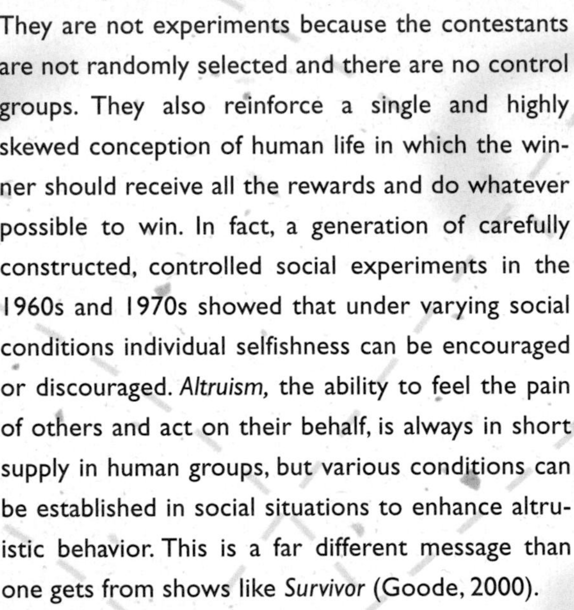

Survivor and similar "reality-based" television shows are heavily criticized by sociologists and other social scientists. Far from being "the ultimate social experiment," as they are often advertised as being, they are far removed from science and offer no new insights on human behavior. They are not experiments because the contestants are not randomly selected and there are no control groups. They also reinforce a single and highly skewed conception of human life in which the winner should receive all the rewards and do whatever possible to win. In fact, a generation of carefully constructed, controlled social experiments in the 1960s and 1970s showed that under varying social conditions individual selfishness can be encouraged or discouraged. *Altruism,* the ability to feel the pain of others and act on their behalf, is always in short supply in human groups, but various conditions can be established in social situations to enhance altruistic behavior. This is a far different message than one gets from shows like *Survivor* (Goode, 2000).

Shown here are photos from perhaps the best and most influential social-scientific experiment on how norms structure conflict or cooperation (Durlaf, 1999; Sherif, 1988). The research is known as the Robbers Cave experiment and was conducted by social psychologist Muzafer Sherif and associates in the mid-1950s. Sherif divided 30 randomly selected preadolescent boys, who had applied for a university-sponsored camp experience, into two groups (also by random assignment), which were named the Eagles and the Rattlers. At first, the groups were bunked on opposite sides of the camp and were unaware of each other's presence. The leaders who emerged in each group were boys with good communication skills and high levels of altruism. Then the groups were brought into contact. They played competitive games like softball and tug-of-war, always against the other group. The experimenters emphasized winning by keeping track of how well each group did. The competition caused new leaders to emerge, boys who were more athletic and more focused on winning, even if feelings got hurt in the process. Gradually the groups grew to hate each other and began raiding each other's bunks. Fights started.

Then the experimenters created a water emergency in the camp. The two groups had to cooperate for the good of the larger community. Animosities decreased, cooperation increased, and the most aggressive leaders were replaced with ones who were better able to lead cooperative efforts. Unlike the message conveyed by *Survivor,* which seems to indicate that life is a struggle to win out over others, the lesson of the Robbers Cave experiment is that the structure of the groups and the norms established for them influenced leadership and caused the behavior of the children to be either competitive and cutthroat or cooperative and altruistic. ■

Eagles' strategy in a second tug-of-war: sitting down to dig in while Rattlers (standing) exhaust themselves.

Successful effort to start the truck: Both groups have tug-of-war against the vehicle.

SUMMARY

Socialization refers to the ways in which people learn to conform to their society's norms, values, and roles. Primary socialization occurs in infancy, secondary socialization in childhood and adolescence; adult socialization refers to the learning of norms associated with new statuses.

Sigmund Freud believed that the personality develops out of socialization processes through which the infant is forced to control its bodily urges. In his model of the personality, unsocialized drives arise from the *id* and the moral codes of elders are incorporated into the *superego.* Conflicts between the id and the superego continually threaten to disrupt the functioning of the *ego* or social self.

Behaviorism asserts that all behavior is learned. It originated in the work of Ivan Pavlov, who showed that behavior could be shaped or *conditioned* by learning situations. This line of research was continued by John B. Watson and B. F. Skinner.

Feral children have experienced extreme isolation or have been reared outside human society. Studies of such children show that normal development requires the presence of other humans. Other studies have found that children who lack the attention and love of adults are more likely to develop emotional problems and to be retarded in their language development.

An early interactionist theory of personality development was Charles Horton Cooley's concept of the looking glass self. George Herbert Mead emphasized the idea of *role taking,* or looking at social situations from the standpoint of another person. Children develop this ability by mimicking the behavior of *significant others;* eventually they can take the role of the *generalized other*—that is, shape their participation according to the roles of other participants.

Lawrence Kohlberg proposed a sequence of stages of moral development in which the child's moral reasoning evolves from emphasis on reward and punishment to the ability to distinguish between social laws and moral principles.

Agencies of socialization are the groups of people, along with the interactions that occur within those groups, that influence a person's social development. The primary agencies of socialization are the family, schools, agencies in the community, religion, the *peer group,* and the media.

Socialization can occur throughout life. A person's core identity shapes his or her responses to new situations and challenges. Sometimes people undergo *resocialization* to correct patterns of social learning that they and others find detrimental.

Erik Erikson's theory of personality formation is based on the concept of *identification,* in which the individual chooses adults as models and attempts to imitate their behavior.

GLOSSARY

Socialization: The processes whereby we learn to behave according to the norms of our culture. **(61)**

Id: According to Freud, the part of the human personality from which all innate drives arise. **(62)**

Superego: According to Freud, the part of the human personality that internalizes the moral codes of adults. **(62)**

Ego: According to Freud, the part of the human personality that is the individual's conception of himself or herself in relation to others. **(62)**

Behaviorism: A theory that states that all behavior is learned and that this learning occurs through the process known as conditioning. **(62)**

Conditioning: The shaping of behavior through reward and punishment. **(62)**

Feral child: A child reared outside human society. **(63)**

Gender socialization: The specific aspects of socialization whereby people learn the behaviors and attitudes considered appropriate for males and females in their cultures. **(67)**

Role taking: Trying to look at social situations from the standpoint of another person from whom one seeks a response. **(68)**

Significant other: Any person who is important to an individual. **(68)**

Generalized other: A person's internalized conception of the expectations and attitudes held by society. **(68)**

Agencies of socialization: The groups of people, along with the interactions that occur within those groups, that influence a person's social development. **(72)**

Agents of socialization: Individuals who socialize others. **(72)**

Anticipatory socialization: Socialization that prepares an individual for a role he or she is likely to assume later in life. **(72)**

Peer group: An interacting group of people of about the same age that has a significant influence on the norms and values of its members. **(75)**

Resocialization: Socialization whose goal is to correct patterns of social learning that are considered detrimental. **(77)**

Total institution: A setting in which people undergoing resocialization are isolated from the larger society under the control of a specialized staff. **(77)**

Identification: The social process whereby an individual chooses role models and attempts to imitate their behavior. **(78)**

QUESTIONS FOR THOUGHT AND DISCUSSION

1. What is role taking, and why does it have to be learned? When sociologists talk about "taking the role of the other," what does that mean and how is it relevant to the role taking that people engage in throughout their lives? Can you think of an example in which people refuse to take the roles that society seems to be offering or imposing on them?

2. What objections can you make to the behaviorism of Watson or Skinner? Does behaviorism help explain what occurs when people are resocialized?

Digging Deeper

Books

The Child in the Family and the Community, 2nd ed. (Janet Gonzalez-Mena; Merrill, 1998). A valuable review of research and theory about child socialization in families, with a strong emphasis on cross-cultural differences and multicultural societies.

Child, Family, Community: Socialization and Support, 3rd ed. (Roberta Berns; Harcourt Brace, 1993). A basic text in the field of socialization, with a good review of relevant sociological research.

The Child and Society, 5th ed. (Frederick Elkin & Gerald Handel; Random House, 1989). A basic text that offers in-depth coverage of most aspects of childhood socialization.

Not in Our Genes (R. C. Lewontin, Steven Rose, & Leon Kamin; Pantheon, 1984). A strong but fair critique of biological thinking in the field of human development.

In a Different Voice (Carol Gilligan; Harvard University Press, 1982); ***Mapping the Moral Domain*** (Carol Gilligan et al.; Harvard University Press, 1988); and ***Between Voice and Silence: Women and Girls, Race and Relationship*** (Jill McLean Taylor, Carol Gilligan, and Amy M. Sullivan; Harvard University Press, 1995). In these important volumes of her research, both solo and collaborative, the pioneering social psychologist Carol Gilligan presents her argument that classic research and theory on moral development are biased toward male development and hence neglect important aspects of socialization and personality development.

Journals

Social Psychology Quarterly. A journal of the American Sociological Association that is devoted to recent research on "the processes and products of social interaction."

Sociological Studies of Child Development. An annual review, begun in 1986, of sociological research on socialization and child development. Represents many of the newest trends in this active area of sociological research.

Journal of Health and Social Behavior. A journal of the American Sociological Association that specializes in "sociological approaches to the definition and analysis of problems bearing on human health and illness." Often contains important articles on child welfare, socialization, and problems of health and development through the life cycle.

Other Sources

The Kids Count Data Book. (Annie E. Casey Foundation). An annual statistical review of the conditions under which children in the United States are being socialized; uses the best available data to measure the educational, social, economic, and physical well-being of children.

Basic Handbook of Child Psychiatry (Joseph D. Noshpitz, ed.; Basic Books, 1979). A two-volume work that contains many excellent articles about human development. The focus is interdisciplinary, although there is also an emphasis on abnormal socialization and treatment.

Statistical Yearbook. Published annually by the United Nations. This is a valuable source of comparative data on infant mortality and other indicators of the health of children, as well as statistical material on education, literacy, and other measures of socialization.

EXPLORING SOCIOLOGY ON THE INTERNET

Foundation for Children
www.unicef.org/

Offers excellent resources and links to other Web sites dealing with child development throughout the world, including problems of child labor, slavery, and abuse.

Kaiser Family Foundation
www.kff.org/

Features research on many aspects of socialization, sexuality, health, and agencies of socialization, with links to other important sites.

Foster Parent Home Page
http://fostercare.org/FPHP/

Includes debates on orphanages, surveys on child abuse and neglect, and analyses of media coverage of foster care.

The Annie E. Casey Foundation
www.aecf.org

Publishes *Kids Count,* an annual review of the situation of children in the United States and elsewhere in the world. A valuable starting point for research on the problems of socialization.

Trinity University, San Antonio, Texas
www.trinity.edu/~mkearl/

Examines issues in socialization patterns and processes; offers a number of links to related Web sites.

Gender and Society, Trinity College
www.Trinity.edu/~mkearl/gender.html

Features much current thinking on gender socialization and related topics, with links to fine sites with even more resources on gender socialization.

American Psychological Association
www.apa.org/pubinfo

Provides links to topics like work and family as they relate to recent study and research. Recent topics include a discussion of childhood socialization and the impact of television violence upon children's socialization habits.

Chapter 4

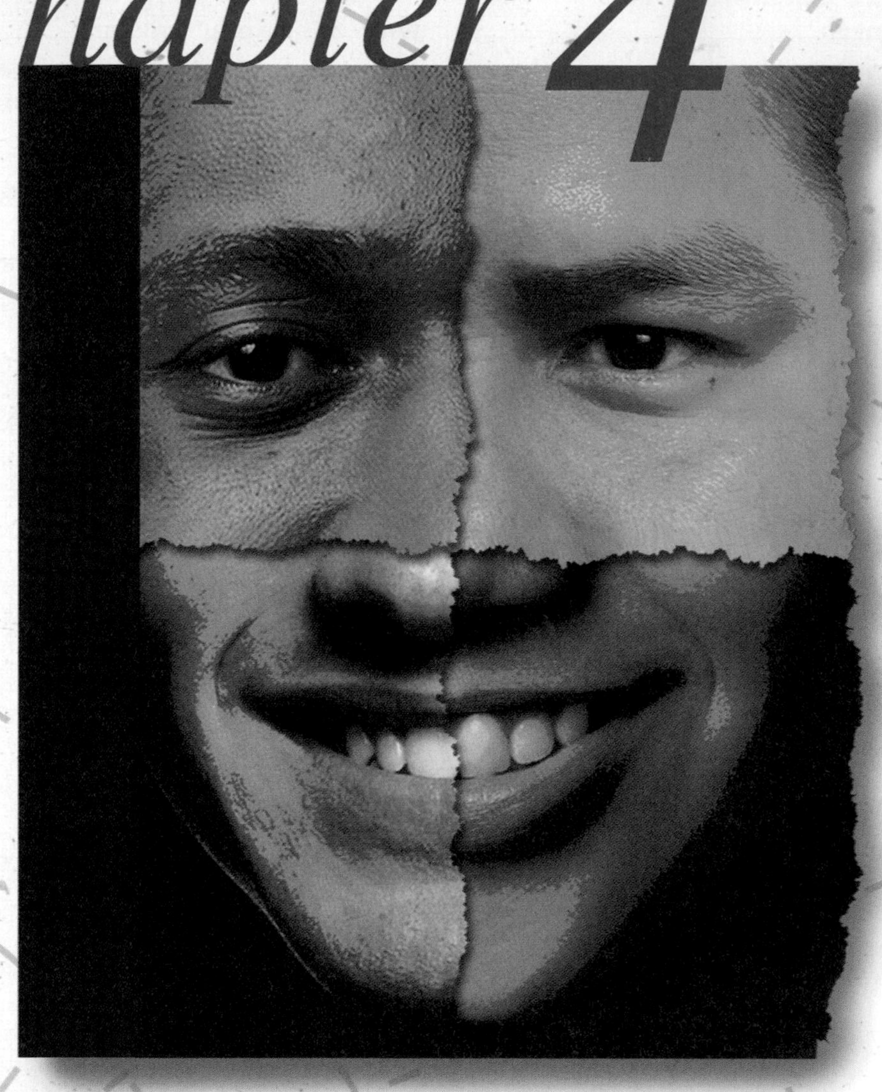

Social Structure and Group Interaction

Does social organization explain why human affairs do not dissolve into a "war of all against all"?

The Social Order

Society and Social Structure
Elements of Social Structure

The Evolution of Human Social Organization

The First Million Years: Hunting and Gathering
The Transition to Agriculture
The Industrial Revolution

Societies and Nations

The State
The Nation-State

Society and the Individual

Role and Status in Modern Societies

Social Groups

Characteristics of Social Groups
Types of Groups
Interaction in Groups
Interaction and Group Structure
Formal Organizations and Bureaucracy

Technology and Social Change

It's Saturday night in the emergency room where Lori Jones works as a triage nurse. The room is crowded with anxious parents, crying babies, and wheezing elderly patients. Behind Jones, inside the ER itself, paramedics have just brought in a man covered with blood. The patient is a big-bellied guy with a tattoo of an American eagle on his left arm and one of a tongue hanging out on his chest. He was rushed to the ER after crashing his car at 80 miles per hour.

After several years as a clinical specialist and triage nurse, Jones knows that to be happy in her roles and balance all the conflicting demands they make on her she must do some serious soul-searching. As she explains it, "Actually, I've never done anything really great here, never saved anyone's life. The paramedics have. Yet patients have been lucky to have me as a nurse. No, I've never done things that are dramatic. I mean, I was happy because patients took their medicine as directed for 10 days. That's a feat in itself."

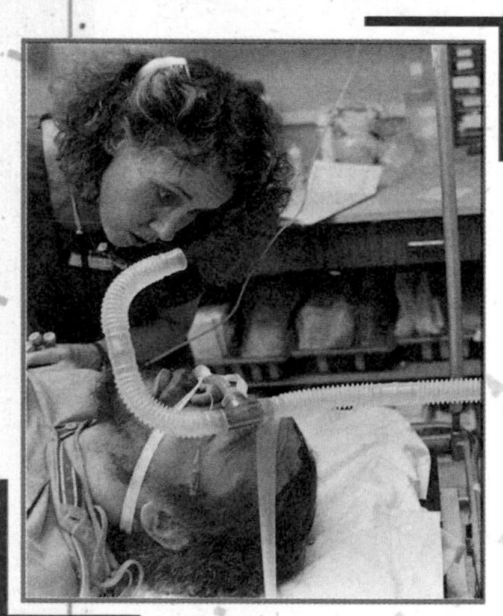

Jones also knows that the burdens of life in the ER are made even more difficult by the hospital's inner-city location, the need to provide services to so many poor people and minorities, and the fact that it is a public institution, dependent on a diminishing stream of tax revenues. "This is a city hospital on the edge," she observes, "losing money on indigent patients, and all around us is a massive system that has to raise more and more money, while politicians are threatening to close us down. So we are stretched incredibly thin, torn in so many different directions. It can be overwhelming."

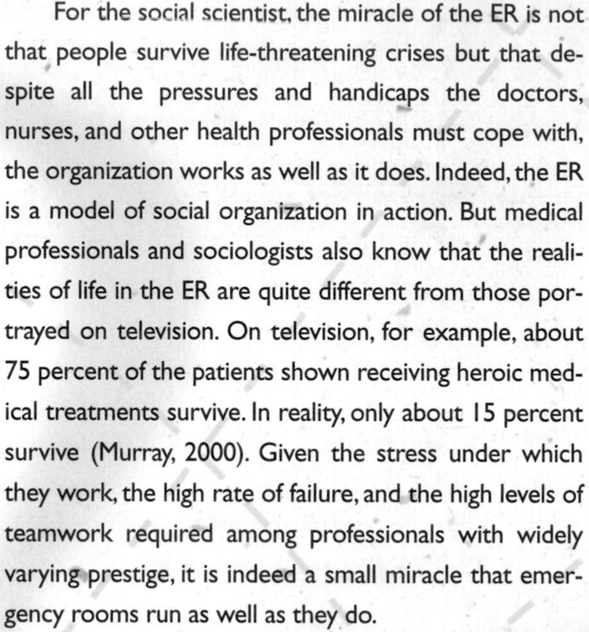

For the social scientist, the miracle of the ER is not that people survive life-threatening crises but that despite all the pressures and handicaps the doctors, nurses, and other health professionals must cope with, the organization works as well as it does. Indeed, the ER is a model of social organization in action. But medical professionals and sociologists also know that the realities of life in the ER are quite different from those portrayed on television. On television, for example, about 75 percent of the patients shown receiving heroic medical treatments survive. In reality, only about 15 percent survive (Murray, 2000). Given the stress under which they work, the high rate of failure, and the high levels of teamwork required among professionals with widely varying prestige, it is indeed a small miracle that emergency rooms run as well as they do.

This chapter is about the intricacies of social organization that we often take for granted but that make places like emergency rooms possible. It introduces many of the essential terms and concepts necessary for the study of how societies are organized and how they function. We define the basic concepts of social structure, including such key terms as *society, roles, statuses,* and *social institutions*. Then we examine some of the major turning points in the evolution of societies as human life became ever more complex and individuals became ever more interdependent. Two sections in the middle of the chapter deal with major concepts in the study of social structure. One focuses on the nation-state, which often contains numerous societies within its boundaries, and the other examines the way individuals "fit" into the modern social world through each society's system of roles and statuses.

The remaining sections deal with how social structures actually emerge through the social behavior of individuals in groups of all kinds. After analyzing variations in the form and composition of groups, we turn to some of the fascinating aspects of how people actually behave in different types of groups. ■

THE SOCIAL ORDER

In a hospital or a primary school or any place where people work together on shared tasks, smooth operations are often intermixed with periods of stress, pressure, or conflict, even among coworkers and friends. Under conditions of stress or conflict, we often blame ourselves or, even more frequently, some other person. We do not stop to look at the powerful influence of social organizations on our lives. One of the main goals of this chapter is to indicate the great diversity of human social structures and show how they influence individual behavior.

Of course, humans are not the only animals capable of social organization. We can learn a lot about life in human societies by comparing ourselves with other social animals. If you look closely at an ant colony or a beehive, for example, you can see a remarkable amount of organization. The nest or hive is a complete social world with workers, warriors, drones, a queen, and so on. Each individual ant or bee has something to do and seems to do it quite well. But the differences between human societies and those of social animals like bees are even more important than the similarities. Unlike a human, an individual social animal is able to perform only a certain number of innate (inborn) tasks. The human can learn an infinite number of tasks. The individual bee is born a worker or a queen. An individual human, in contrast, can be a nurse, a college student, a lawyer, a mother, a voter, a taxpayer, and on and on.

We often assume that we know how society works and how to steer our way through it. But there will be times, especially when we are learning to adapt to new social environments, when we will be unsure of what is expected of us and how we should perform. Worse still, there may be times when it seems that society itself is threatened, that its continued existence as we know it is endangered. We catch glimpses of the breakdown of society during riots or wars or severe economic recessions.

At earlier times in human history, plagues and famines were frequent reminders that people had little control over their own destinies. Today, plagues and famines still occur, but societies are more often faced with real or potential crises of their own making: war, genocide, environmental disasters, drug addiction, criminal violence. Thus, if we are to continue to exist and thrive as a species, it is vital that we study societies and social structures—how they hold together, how they change, and why they sometimes seem to fall apart.

Society and Social Structure

The term **society** refers to a population of people (or other social animals) that is organized in a cooperative manner to carry out the major functions of life, including reproduction, sustenance, shelter, and defense. This definition distinguishes between societies and populations. The notion of a population implies nothing about social organization, but the idea of a society stresses the *interrelationships* among the members of the population. In other words, a population can be any set of individuals that we decide to count or otherwise consider, such as the total number of people living between the Rio Grande and the Arctic Circle, whereas a society is a population that is organized in some way, such as the population of the United States or Canada or the Amish people of Pennsylvania. In the modern world, most societies are also (but not always) nation-states.

Social structure refers to the recurring patterns of behavior people create through their interactions and relationships. We say, for example, that the family has a structure in which there are parents and children and other relatives who interact in specific ways on a regular basis. The larger society usually requires that family members assume certain obligations toward one another. Parents are required to educate their children or send them to schools; children are required to obey their parents until they have reached an age at which they are no longer considered dependent. These requirements contribute to the structure of relationships that is characteristic of the family.

Throughout life, individuals maintain relationships in an enormous range of social structures, of which families are just one. There are many others. People may be members of relatively small groups like the friendship or peer group and the work group. They may also be members of larger structures like churches, business organizations, or public agencies. And they may participate in even more broadly based structures, such as political groups and party organizations, or interest groups like the National Rifle Association or Planned Parenthood. All of these social structures are composed of groups with different degrees of complexity and very different patterns of interaction. A military brigade, for example, is far more complicated than a barbershop quartet, and people behave quite differently in each. But both are social structures. In such structures, our time, our activities, and even our thoughts may be "structured" according to the needs and activities of the group.

Elements of Social Structure

Groups. The "building blocks" of societies are social groups. (See the study chart on page 91.) A **group** is any collection of people who interact on the basis of shared expectations regarding one another's behavior. One's immediate family is a group; so are a softball team, a seminar, a caucus, and so on. But a collection of people on a busy street—a crowd—is not a group unless for some reason its members begin to interact in a regular fashion. Usually a crowd is composed of many different kinds of groups—couples, families, groups of friends, and so on. They may be molded into a single group in response to an event that affects them all, such as a fire that creates the need for the orderly evacuation of a building.

Statuses. In every group there are socially defined positions known as **statuses.** Father, mother, son, daughter, teacher, student, and principal are all examples of familiar statuses in the family and the school. There are an infinite number of statuses in human societies. In a corporation, for example, the statuses range from president and chief executive officer to elevator operator and janitor. Between these two extremes there could be thousands of other statuses.[1] Moreover, the corporation can always create new statuses if the need arises. Thus, in the 1970s, when American society agreed to combat racism and sexism in business and government, many corporations invented the status of affirmative action director to provide equal opportunities for workers of both sexes and all racial and ethnic groups. In the 1990s, as the larger society became more deeply concerned about environmental pollution and drug abuse, corporations began to create new statuses like pollution control manager or drug counselor. Each of these statuses then became part of the "corporate structure."

Human societies rely heavily on the creation of statuses to adapt to new conditions like environmental pollution and drug abuse, and one can often observe this adaptation occurring in daily life. In the family, for example, it is increasingly common (though not universally condoned) for young adults to cohabit before marriage or after divorce. As a large-scale phenomenon, this is a relatively new trend in American society, so new that we have not defined very well the new status of the person who participates in such a relationship. Should such a person be called a *boyfriend* or *girlfriend, lover, mate,* or *significant other*? The awkwardness of these terms is due to the fact that this is an emerging status that our society has not yet fully accepted or defined.

This example highlights an essential point about human social structure: It is never fixed or perfectly formed but instead is always changing and adapting to new conditions. Often the process of change involves much conflict and uncertainty, and often there is little consensus about how one should perform in a given status. Should the president of a corporation be an aloof, aggressive leader who directs subordinates with little regard for their feelings? Or should the president show concern for employees' feelings and personal needs and perhaps thereby gain greater loyalty and motivation? This is just one of thousands of dilemmas arising from questions about how we should act in a given social status. To clarify our thinking about statuses in groups and the behaviors associated with those statuses, sociologists make a distinction between a *status* and a *role.*

Roles and Role Expectations. The way a society defines how an individual is to behave in a particular status is referred to as a **role.** If one thinks of any of the most common statuses—mother, father, student, teacher, and so on—a set of behaviors associated with that role will immediately come to mind ("takes care of children," "studies for exams," etc.). But think a bit more about that behavior and it will become obvious that different people perform the same role differently. That is why, for example, we may compare the behavior of one teacher with that of another. These different performances of particular roles are the product of **role expectations,** the society's expectations about how a role should be performed, together with the individual's perceptions of what is required in performing that role.

To appreciate the importance of role expectations you need only think of the mothers and fathers of your close friends: All hold the same statuses, but how different their behavior is! Part of that difference is due to personality—to psychological variables—but another part is due to how the individual mothers and fathers perceive what is expected of them in the statuses they hold. One mother and father may have been raised to believe that children should work to support the family. They will insist that their children get early job experience. Another couple may have been taught

[1] Like many other sociological terms, *status* has more than one meaning. It can refer to a person's rank in a social system and also to a person's prestige—that is, the esteem with which others regard him or her. The fundamental meaning is the one we use here.

STUDY CHART

Elements of Social Structure

Structural Element	Description	Example
Group	Any collection of people who interact on the basis of shared expectations regarding one another's behavior.	A discussion group; a Bible study class; a local union.
Status	A socially defined position in a group.	Orderly, practical nurse, registered nurse, resident, chief resident (all statuses in a hospital ward).
Role	The way a society defines how an individual is to behave in a particular status.	The doctor diagnoses and treats illnesses; the nurse provides care to patients under the doctor's supervision.
Role Expectations	A society's expectations about how a role should be performed, together with the individual's perceptions of what is required in performing that role.	A major league center fielder is expected to have a batting average over .300, drive in more than 75 runs, and cover the field with a minimum of errors.
Institution	A more or less stable structure of statuses and roles devoted to meeting the basic needs of people in a society.	The military is the primary institution devoted to providing national defense.

that childhood is too short and should be prolonged if possible. Other things being equal (e.g., adequate family income), that couple will not encourage their children to find jobs before they are more or less the same age the parents were when they went to work.

In sum, sociologists are interested in the way roles and statuses affect individual behavior. They also study how the lack of roles—for example, in the form of jobs—can affect people's lives. They do not deny the importance of personality, but they place greater emphasis on the influence of social structure as an explanation of individual behavior.

Organization in Groups. Groups vary considerably in the extent to which the statuses of their members are well or poorly defined. The family is an example of a group in which statuses are well defined. Although parents carry out their roles in different ways, the laws of their society define many of the obligations of parenthood, placing certain limits on what they can and cannot do. Other groups are much more informal. We may form groups for brief periods in buses, hallways, or doctors' offices. There are roles and statuses in these groups also, but they are variable and not well defined. All of the people riding on a bus are passengers, but they do not interact the way the members of a family do. At most, we expect civility and "small talk" from other passengers on a bus, but we demand affection and support from other members of our family.

Social change is constantly making the structure of groups more varied and complex. To continue with our earlier example, the family may seem to have well-defined statuses, but high rates of divorce and remarriage have increased the proportion of families in which one parent is a stepparent. Consider the difference in role expectations for "mother" or "father" in a situation in which each spouse must interact with a new spouse, a new set of children, and a new set of in-laws as well as an ex-spouse and children in the first family, plus the previous in-laws and his or her own parents. Balancing role expectations among the often competing demands of these groups can be a daunting task, one that was far less common when the norms of society made it difficult to divorce.

Groups are often connected with other groups to form a larger structure known as an *organization.* An army platoon, for example, includes the well-defined statuses of private, corporal, sergeant, and lieutenant, each with specific roles to play in training and combat.

Platoons are grouped together under the leadership of higher officers to form companies; this pattern is repeated at higher levels to create the battalion and the brigade.

Social Institutions. The structure of most of the important groups in society is determined by shared definitions of the statuses and roles of their members. When such statuses and roles are designed to perform major social functions, they are termed **institutions.** In popular usage, the word *institution* generally refers to a large bureaucratic organization like a university, hospital, or prison (usually with cafeterias that serve "institutional food"). But although the word has this meaning in everyday language, the sociological use of the term should not be confused with this meaning.

In sociology, an institution is a more or less stable structure of statuses and roles devoted to meeting the basic needs of people in a society. The family is an institution that controls reproduction and the training of new generations. The market is an institution that regulates the production and exchange of goods and services. The military is an institution that defends a society or expands its territory through conquest. Any particular family, corporation, or military unit is a group or organization within one of these institutions.

Within any given institution there are norms that specify how people in various statuses are to perform their roles. Thus, to be a general or a new recruit or a supplier of military hardware is to have a definite status in a military institution. But to carry out one's role in that status is to behave in accordance with a normative system—a set of mores and folkways that distinguish a particular institution from others. For example, in a well-known paper on becoming a military recruit, Sanford Dornbusch (1955) described how all the signs of a person's status in civilian life, from fashions in dress to the use of free time, are erased in boot camp. The new status of recruit must be earned by adhering to all the norms of military life. So it is with every social institution: Each has a specific set of norms to govern the behavior of people within that institution.

Institutions of religion, education, politics, and economics tend to become more complex over time as the society changes. It is helpful, therefore, to think of large-scale societies like the United States as having a number of institutional *sectors* (sets of closely related institutions). For example, the economy is an institutional sector that includes markets, corporations, and other economic institutions. Politics is an institutional sector containing legal, executive, legislative, and other political institutions. Each of these institutions, in turn, is composed of numerous groups and organizations. The Chicago Mercantile Exchange is a market organization; the House of Representatives is a legislative organization. In this book, as in much social-scientific writing, the term *institution* is used to refer to institutional sectors (e.g., religion) as well as to the institutions included within a sector (e.g., Buddhism or Christianity).

The history of human societies is marked by the emergence of new institutions like the university or the laboratory. In fact, a dominant feature of human societies is the continual creation of new social institutions, a feature that the social theorist Talcott Parsons (1951, 1966) labeled **differentiation.** By this term, Parsons meant the processes whereby sets of social activities performed by one social institution are divided among different institutions. In small-scale agrarian societies, for example, the family not only performs reproductive and training functions but is also the primary economic institution. As societies become larger and more complex, the processes of differentiation result in the emergence of new institutions designed to manage economic production (corporations), train new generations (schools), develop new technologies (science), or perform other important social activities.

THE EVOLUTION OF HUMAN SOCIAL ORGANIZATION

The ability of human populations to organize themselves to perform ever-greater tasks accounts in large part for the remarkable growth of our species over the last million years. But throughout most of human prehistory (that is, before the development of writing), human societies were relatively simple. They lacked the techniques of social organization that make societies so powerful today. At the end of the Neolithic period, about 8000 B.C., most people lived in small bands of related individuals. More complex social forms were developing in certain areas of the world, where concentrations of people were greatest, especially in the Middle East, East Africa, southern Europe, and a few fertile river basins in India, China, Latin America, and Central America. The momentous changes that occurred, and are still occurring, in our ability to use social organization to our own benefit are highlighted in the following brief account of the evolution of human societies.

The First Million Years: Hunting and Gathering

A human lifetime is no more than a twinkling in time. What are 70 or 80 years compared with the billions of years of the earth's existence? What is one generation compared with the millions of years of human social evolution? Yet many of us hope to leave some mark on society, perhaps to change it for the better, to ease some suffering, to increase productivity, or to fight racism and ignorance . . . what vaulting ambitions! And what a radical change from the worldview of our ancestors! The idea that people can shape their society—or even enjoy adequate shelter and ample meals—is widely accepted today. But for most of human history mere survival was the primary motivator of human action, and thus a fatalistic acceptance of human frailty in the face of overwhelming natural forces was the dominant worldview.

For most of the first million years of human evolution, human societies were developing from those of primates. Populations were small because humans, like other primates, lived on wild animals and plants. These sources of food are easily used up and their supply fluctuates greatly, and as a result, periods of starvation or gnawing hunger might alternate with bouts of gorging on sudden windfalls of game or berries. Thus, the hunting-and-gathering life that characterized the earliest human populations could support only extremely small societies; most human societies therefore had no more than about 60 members.

Recent archaeological evidence indicates that some hunting-and-gathering societies began to develop permanent settlements long before the advent of agriculture. The emergence of farming was one of the changes that accelerated human social evolution, but this new evidence shows that in parts of what is now Europe and the Middle East there were stable settlements of hunter-gatherers as early as 20,000 to 30,000 years ago. These rather large and complex societies were most firmly established at the end of the last Ice Age, some 12,000 to 13,000 years ago, in the area that is now Israel (Henry, 1989; Stevens, 1988).

Despite the slow pace of human evolution until about 35,000 years ago, some astonishing physical and social changes occurred during that long period, changes that enabled human life to take the forms it does today. Among these changes were the following:

- The development of an upright posture, freeing the hands for eventual use of tools, and an enlarged cerebral cortex, making possible vastly increased cognitive abilities and the development of language.
- Social control of sexuality through the development of the family and other kinship structures and the enforcement of the incest taboo.
- The establishment of the band of hunter-gatherers as the basic territorial unit of human society, coupled with the development of kinship structures that linked bands together into tribes. Within the band, the family became the primary economic unit, organizing the production and distribution of food and other necessities.

By the end of the last Ice Age, many aspects of this evolutionary process were more or less complete. Human societies had fully developed languages and a social structure based on the family and the band. To be cast out of the band for some wrongdoing—that is, to be considered a deviant person—usually meant total banishment from the society and eventual death, either by starvation or as a result of aggression by members of another society (Salisbury, 1962). But warfare and violence were not typical of early human societies. As anthropologist Marshall Sahlins has pointed out:

> Warfare is limited among hunters and gatherers. Indeed, many are reported to find the idea of war incomprehensible. A massive military effort would be difficult to sustain for technical and logistic reasons. But war is even further inhibited by the spread of a social relation—kinship—which in primitive society is often a synonym for "peace." (1960, p. 82)

On the other hand, one must not romanticize the hunter-gatherers. Their lives were far more subject to the pressures of adaptation to the natural environment than has been true in any subsequent form of society. Individual survival was usually subordinated to that of the group. If there were too many children to feed, some were killed or left to die; when the old became infirm or weak, they often chose death so as not to diminish the chances of the others. Thus, the frail Eskimo grandfather or grandmother wandered off into the snowy night to "meet the polar bear and the Great

Hunting-and-gathering societies still exist, but they tend to be found in remote and difficult environments where, as in the case of these Inuit Eskimos, it is still possible to pursue traditional occupations such as hunting and fishing. But like almost all the native peoples in the world today, the Inuit face pressures to become sedentary and adopt modern technologies and ways of life.

Spirit." Additional thousands of years of social evolution would pass before the idea that every person could and should survive, prosper, and die with dignity would even occur to our ancestors.

The Transition to Agriculture

For some time before the advent of plow-and-harvest agriculture in the Middle East and in the Far East, hunting-and-gathering societies were supplementing their diets with foods acquired through the domestication of plants and animals. In this way, they were able to avoid the alternation of periods of feast and famine caused by reliance on animal prey. Karl Marx was the first social theorist to observe that social revolutions like the shift to agriculture or industrial production are never merely the result of technological innovations such as the plow or the steam engine. The origins of new forms of society are to be found within the old ones. New social orders do not simply burst upon the scene but are created out of the problems faced by the old order. Thus, as some hunting-and-gathering societies experimented with domestication of animals and planting of crops, they were evolving into nomadic shepherding, or **pastoral,** societies in which bands followed flocks of animals. Others were developing into **horticultural** societies in which the women raised seed crops and the men combed the territory for game and fish.

As a result of these and other innovations, agriculture became the productive basis of human societies. Pastoral societies spread quickly throughout the uplands and grasslands of Africa, northern Asia, Europe, and the Western Hemisphere, and grain-producing societies arose in the fertile river valleys of Mesopotamia, India, China, and—somewhat later—Central and South America. Mixed societies of shepherds and marginal farmers wandered over the lands between the upland pastures and the lowland farms.

The First Large-Scale Societies. We have reached the beginning of recorded history (around 4000 B.C.), which was marked by the rise of the ancient civilizations of Sumer, Babylonia, China and Japan, Benin, and the Maya, Incas, and Aztecs. The detailed study of these societies is the province of archaeology, history, and classics. But sociologists need to know as much as possible about the earliest large-scale societies because many contemporary social institutions (e.g., government and religion) and most areas of severe social conflict (e.g., class and ethnic conflict) developed sometime in the agrarian epoch—between 3000 B.C. and A.D. 1600.

From the standpoint of social evolution, the following dimensions of agrarian societies are the most salient (Braidwood, 1967):

- Agriculture allows humans to escape from dependence on food sources over which they have no control. Agrarian societies produce surpluses that do not merely *permit* but *require* new classes of nonfood producers to exist. An example is the class of warriors, who defend the surplus or add to it through plunder.
- Agriculture requires an ever-larger supply of land, resulting in conflicts over territory and in wars with other agricultural or pastoral societies.
- The need to store and defend food surpluses and to house the nonagrarian classes results in new territorial units: villages and small cities.

New Social Structures. Freed from direct dependence on undomesticated species of plants and animals, agrarian societies developed far more complex social structures than were possible in simpler societies. Hunting-and-gathering societies divided labor primarily according to age and sex, but in agricultural societies labor was divided in different ways to perform more specialized tasks. Peoples who had been conquered in war might be enslaved and assigned the most difficult or least desirable work. Priests controlled the society's religious life, and from the priestly class there emerged a class of hereditary rulers who, as the society became larger and more complex, assumed the status of pharaoh or emperor. Artisans with special skills—in the making of armaments or buildings, for example—usually formed another class. And far more numerous than any of these classes were the tillers of the soil, the "common" agricultural workers and their dependents.

The process whereby the members of a society are sorted into different statuses and classes based on differences in wealth, power, and prestige is called **social stratification.** In general, societies may be **open,** so that a person who was not born into a particular status may gain entry to that status, or **closed,** with each status accessible only by birth. A key characteristic of agrarian societies is that their stratification systems became extremely rigid. Because the majority of the people were needed in the fields, there were few opportunities for people to advance from a lower level of society to a higher level. These therefore were closed societies.

The emergence of agrarian societies was based largely on the development of new, more efficient production technologies. Of these, the plow and irrigation were among the most important. The ancient agrarian empires of Egypt, Rome, and China are examples of societies in which irrigation made possible the production of large food surpluses, which in turn permitted the emergence of central governments led by pharaohs and emperors, priests and soldiers. Indeed, some sociologists have argued that because large-scale irrigation systems required a great deal of coordination, their development led to the evolution of imperial courts and such institutions as slavery, which coerced large numbers of agrarian workers into forced labor (Wittfogel, 1957).

By the close of the fifteenth century, before European colonial expansion changed the world map forever, the world's population was organized into a wide variety of societies with many different types of cultures and civilizations. The areas where the earliest civilizations emerged are to this day the most densely inhabited parts of the world. In 1500, these civilizations were marked by their ability to produce food surpluses through the use of the plow, domesticated animals, wheels, and carts. Above all, they were marked by the importance of towns and cities as centers of administration and religious practice. Although the Mayan and Incan civilizations did not have the wheel or the plow, their advanced systems of astronomy, art, and writing make a strong argument for their inclusion as world civilizations.

The explorations led by Portugal, Spain, and England in the sixteenth and seventeenth centuries resulted in conquest and European settlement in many parts of the Western Hemisphere and Africa. Conquest caused radical disruptions in the internal development of the world's cultures and civilizations. But even more drastic would be the changes caused by the industrial revolution of the eighteenth and nineteenth centuries.

These Minnesota Indians are harvesting wild rice in the traditional manner. In the contemporary world, agriculture remains as important as ever, but fewer and fewer people are needed to produce ever-greater crop yields.

The detail from a famous mural by Diego Rivera (left) shows what the Valley of Mexico looked like in 1519, before the conquistadors conquered the agrarian Aztec Empire. Today Mexico City, one of the world's largest metropolitan centers, occupies almost the entire valley. The social structure of the modern city is extremely complex because Mexico City is the nation's largest commercial center as well as its capital.

The Industrial Revolution

In 1650, when Holland, Spain, and England were the world's principal trading nations (Block, 1990), the population of England was approximately 10 million, of which about 90% earned a livelihood through farming of one kind or another. Just 200 years later, in 1850, the English population had soared to more than 30 million, with less than 20% at work in fields, barns, and granaries. England had become the world's first industrial society and the center of an empire that spanned the world.

A similar transformation occurred in the United States. In 1860, on the eve of the Civil War, there were about 30 million people in the United States. Ninety percent of that population, a people considered quite backward by the rapidly industrializing English, were farmers or people who worked in occupations directly related to farming. A mere 100 years later, in 1960, only about 8% of Americans were farmers or agricultural workers, yet they were able to produce enough to feed a population exceeding 200 million. Today, less than 3% of Americans work on farms and ranches. These dramatic changes in England and America—

Early in the industrial revolution, the dominant images of production were those of "satanic mills" belching smoke and dust. Today, the older urban industrial societies pride themselves on having made a transition to "postindustrial" production, which is dominated by information and highly technical forms of production.

and in other nations as well—occurred as a result of the industrial revolution.

The industrial revolution is often associated with innovations in energy production, especially the steam engine. But the shift from an agrarian to an industrial society did not happen simply as a result of technological advances. Rather, the industrial revolution was made possible by the rise of a new social order known as **capitalism.** This new way of organizing production originated in nation-states, which engaged in international trade, exploration, and warfare. Above all, the industrial revolution depended on the development of markets—social structures that would function to regulate the supply of and demand for goods and services throughout the world (Polanyi, 1944).

The transition from an agrarian to an industrial society affects every aspect of social life. It changes the structure of society in several ways, of which the following are among the most significant:

- The industrialization of agriculture allows many more people to be supported by each agrarian worker than ever before. Only a relatively small number of people live on the land; increasing numbers live in towns, cities, and suburbs.
- Industrial societies are generally far more receptive to social change than agrarian societies. One result is the emergence of new classes like industrial workers and scientific professionals (engineers, technicians) and new social movements like the women's movement and the movement for racial equality.
- Scientific, technical, and productive institutions produce both unparalleled wealth and unparalleled destructive capacities.
- The world "shrinks" as a result of innovations in transportation and communication. A "global society" develops, but at the same time the unevenness of industrialization leads to conflicts that threaten world peace.

We will return to the issues raised by industrialization throughout this book. For while many Western societies are fully industrialized, in many other parts of the world the industrial revolution is in full swing, and in still other, more remote areas older forms of agrarianism and tribalism continue to thrive. These basic differences in the social changes experienced by people in different regions of the world make it extremely difficult to sustain stable nations. This will become evident in the next section, where we explore the relationship between societies and nations.

SOCIETIES AND NATIONS

For most people in the world today, the social entity that represents society itself is the nation-state. The assumed correspondence between society and nation can be seen in the fact that expressions like "the United States" and "American society" are often used interchangeably. Moreover, most people think of their society in terms of national boundaries; thus, if you were asked to name the society of which you are a member, you would be more likely to say "the United States" than "California" or "Chicago" or "the University of Texas." Yet as we will see shortly, societies and nations are by no means the same thing.

The State

To understand the distinction between society and nation, we need to begin with the concept of the state. In a lecture at Munich University, Max Weber described a state as:

> a human community that (successfully) claims *the monopoly of the legitimate use of physical force* within a given territory. . . . The right to use physical force is ascribed to other institutions or to individuals only to the extent to which the state permits it. The state is considered the sole source of the "right" to use violence. (quoted in Gerth & Mills, 1958, p. 78; emphasis in original)

The **state** thus may be defined as a society's set of political institutions—that is, the groups and organizations that deal with questions of "who gets what, when, and how" (Lasswell, 1936). The **nation-state** is the largest territory within which those institutions can operate without having to face challenges to their sovereignty (their right to govern). Weber was careful to note that the state has a monopoly over the use of force, which under certain circumstances it grants to other agencies (e.g., state and municipal governments). But the state gains this right—the source of its power to influence the behavior of citizens—from the people themselves, from their belief that it is legitimate for the state to have this power. As we will see in later chapters, the concepts of power and legitimacy are essential to understanding the workings of the modern state; indeed, power is a basic concept at all levels of human social behavior. Here, however, we must examine the idea of a state as it operates in the concept of nation-state or, simply, nation.

The Nation-State

"One nation, under God, indivisible, with liberty and justice for all." We can say these words in our sleep, but we do not usually give much thought to the significance of what we are doing when we repeat the Pledge of Allegiance. Yet repeating the pledge is a highly significant action: It enhances the legitimacy of the state and thus helps create the nation—in this case, the nation known as the United States of America.

But do all the inhabitants of the United States of America think of themselves as members of one nation? To a large extent they do, and this is one of the greatest strengths of that nation. Yet at various times in American history, certain groups—African Americans, Native Americans, and the Amish, for example—have thought of themselves as separate peoples. And the idea of an "indivisible" nation was fought out in one of the bloodiest wars the world has ever known, the American Civil War of 1861–1865.

In the period since the Civil War, the United States has not experienced any real tests of its national solidarity. Countries like Canada, Lebanon, Zimbabwe, South Africa, Iran, Northern Ireland, and the nations that were formerly part of Yugoslavia—to name only a few—have been far less fortunate. For them, the issue of creating a national identity that can unite peoples who think of themselves as members of different societies remains a burning question.

Between 1945 and 1970, as peoples all over the world adjusted to the breakup of the European colonial empires, more than a hundred new states were created, many of them in the poorer regions of Africa and Asia. In these new nations, as in many of the nation-states that existed before the 1950s, the correspondence between national identity and society is often problematic, a situation that frequently produces social upheavals. In Nigeria, for example, the stability—indeed, the very existence—of the nation-state is endangered by animosities among the various societies included within its boundaries. Although the nation has rich oil resources and is the largest of the sub-Saharan nations, its per capita income fell to $250 in 1993 from $1,000 in 1980. This decline is due largely to conflicts among the major tribal groups and the failure of the nation-state's leaders to overcome those conflicts. Although there are scores of tribal and ethnic groups in the nation's population, the largest are the Hausa (21%), the Yoruba (20%), the Ibo (17%), and the Fulani (9%), each of which is afraid that the others might gain power at their expense (Olojede, 1995). (See the Mapping Social Change box on page 99.)

The lack of a clear match between society and nation can be seen in the case of entire groups who think of themselves as "a people" (e.g., the Ibo, the Jews, Native Americans, the Chechens in Russia) as well as in smaller groups. You and your friends are part of "American society," by which we mean the populations and social structures found within the territory claimed by the nation-state known as the United States of America. But you are also part of a local community with smaller structures and more "face-to-face" relationships. Indeed, this community level of society may have greater meaning for you in your daily life than society at the national level.

SOCIETY AND THE INDIVIDUAL

Imagine that you grew up in an agrarian village or a town in a slowly industrializing society and then came to live in the United States, as millions of people do each decade. A difficult part of that experience would be getting used to the impersonality of modern American society when compared with the close relationships you had with people you had known all your life in your native society. American society would seem to be composed of masses of strangers organized into highly impersonal categories. You would have to get used to being a shopper, an applicant, a depositor, a fan, a commuter, and so on, and you would have to shift from one to another of these roles several times in a day or even within a half hour.

Sociologists often describe this experience as a change from **gemeinschaft** (the close, personal relationships of small groups and communities) to **gesellschaft** (the well-organized but impersonal relationships of modern societies). These are German terms taken from the writings of the social theorist Ferdinand Tönnies (1957/1887). Complex industrial societies, Tönnies argued, have developed gesellschaft social structures like factories and office bureaucracies to such a degree that they tend to dominate day-to-day life in the modern world.

MAPPING SOCIAL CHANGE

Tribal Societies Within the Modern State of Nigeria

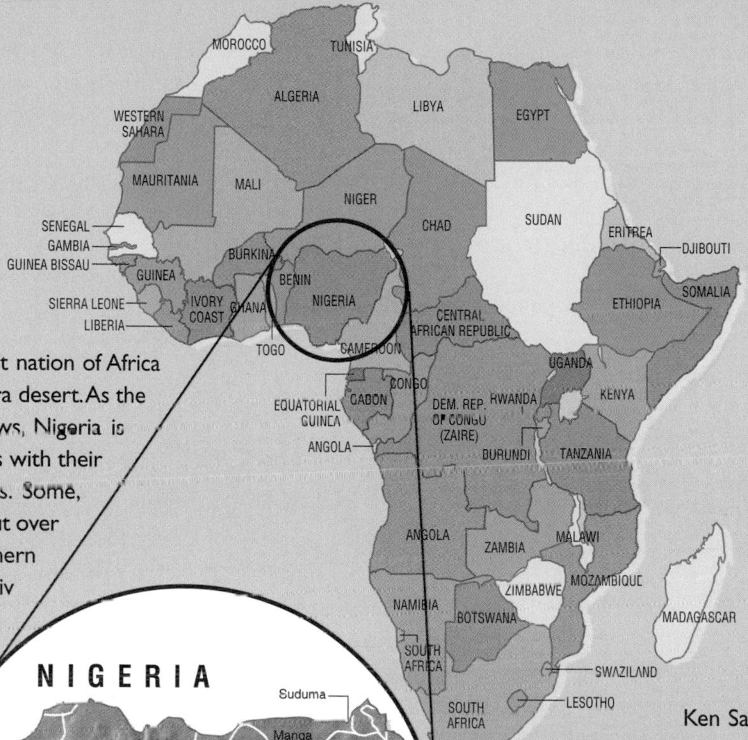

Nigeria is the largest nation of Africa south of the Sahara desert. As the detailed map shows, Nigeria is composed of many societies with their own cultures and languages. Some, like the Hausa, are spread out over vast territories in the northern savannas; others, like the Tiv and Ibo, are clustered in the more densely populated coastal plains and on the delta of the mighty Niger River. Nigeria was once considered a leader in the development of democratic political institutions. It had a lively cultural and intellectual life and produced some of Africa's best novelists and playwrights. Today, after many years of civil war and military coups, Nigeria is a military dictatorship. Conflict between its many societies, often over the disposal of oil revenues from the Niger delta region, and rampant political and economic corruption continually threaten the viability of this nation-state's democratic institutions.

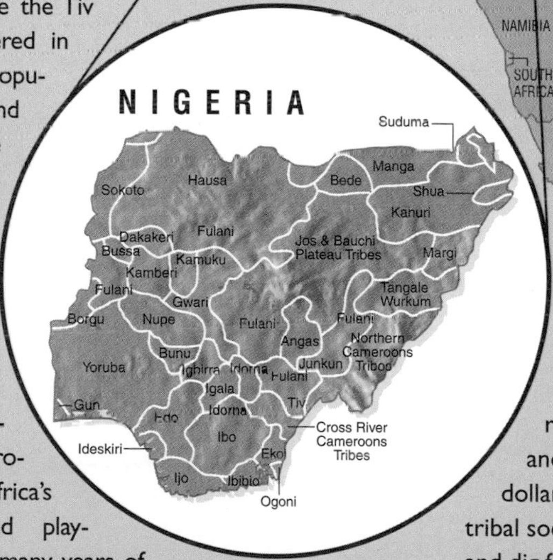

Ken Saro-Wiwa is considered to be a martyr of Nigeria's democratic hopes. Despite protests throughout the world, this celebrated writer, environmental activist, and political dissident was executed by the Nigerian dictatorship late in 1995. Saro-Wiwa was the founder and leader of the Movement for the Survival of the Ogoni People in Nigeria's oil-rich Niger delta region. Although the government and the Dutch Shell Corporation had taken billions of dollars' worth of oil from their lands, the Ogoni, a small tribal society in the Cross River region, still live in mud huts and dig for yams with bamboo sticks. Acid rain and oil pollution have destroyed many productive farms. Saro-Wiwa was attempting to organize a peaceful social movement to protest these conditions. His execution is an example of how environmental issues increasingly become confounded with intergroup politics in this land of tribal conflict. ■

THE CENTRAL QUESTION: A CRITICAL VIEW

Does social organization explain why human affairs do not dissolve into a "war of all against all"?

Role strain sometimes makes our social relations at work feel more like war than anything else. When everything seems to be going wrong, the normal operations of the workplace may fall apart in bickering and strife. Yet as the following case study illustrates, in such situations there is often a particular individual who is able to take the lead in finding solutions to the breakdown of social order.

At the beginning of Sherri's second month on the job, things were supposed to be getting easier, but they weren't. If anything, they were getting worse, especially the pressure during the lunch-hour madness. Lunch never seemed to go well. There were always more customers than she could handle. Too many tables to serve, busboys too slow, cooks yelling at her to pick up orders or write more clearly—things were so bad that she dreaded going to work. And now that chubby guy at table 14 was leering at her.

"Oh, Miss," he called out as Sherri shot by under a huge tray of burger plates. "Oh, Miss, why are you ignoring me? I haven't even gotten a menu." "Be right there, Sir," she muttered.

She dealt out the burger plates as fast as she could and dashed back into the kitchen to get the soups for another table. "Where've you been hiding?" the cook shouted at her from behind his stainless steel counter. "You better get that soup outta here before I have to heat it up again. Let's go. Move!"

Sherri flew through the swinging doors with the tepid soups. Out of the corner of her eye, she saw the chubby guy complaining to Karen, her supervisor, about the slow service. Karen shot her an angry glance. A party of six was waiting to be seated at one of her big tables, which needed to be cleaned off. It was too much. She felt the tears welling up in her eyes. Her makeup would be streaked. She quickly headed for the restroom.

On the way, Sherri felt a gentle touch on her shoulder. Betty, the veteran waitress who had broken her in when she started five weeks ago, smiled at her. "Take it easy, kid! It's not all that bad. My station is under control, so let me help you with the little fat guy and the others."

With her assured efficiency, Betty had Sherri's station under control in a matter of minutes. Later, after the rush was over, she took Sherri aside and showed her some tricks that would make the work easier, like warning the busboys when a table was almost ready to be cleaned up and working out a system for letting others know when she was having trouble. "You'll see," Betty assured her. "When you learn to assert yourself somewhat, and take care of emergencies first, and when you learn who you can ask for help, this job will be much easier."

In his extensive research on human relations in the restaurant industry, sociologist William F. Whyte (1984) found that high-pressure situations like the one Sherri faced often lead less-determined, more thin-skinned waitresses to quit. The ones who stay are emotionally tougher, better able to cope with the pressure. But he also showed that the intervention of tough yet sensitive pros like Betty often helps newcomers learn how to deal with the pressure, especially by learning the "tricks of the trade." However, not all newcomers are fortunate enough to have caring mentors to help them over the rough spots, and many quit before they can learn how to perform their roles well. ■

Role and Status in Modern Societies

Roles in Conflict. Gesellschaft forms of social organization are more complex than gemeinschaft forms in terms of the number of statuses people hold and, thus, the number of roles they must perform. One result of this greater complexity is that roles in secondary groups or associations often conflict with roles in primary groups like the family. Much of the stress of life in modern societies is due to the anxiety we experience as we attempt to balance the demands of our various roles. This anxiety is captured in the terms *role conflict* and *role strain.*

Role conflict occurs when, in order to perform one role well, a person must violate another important role. Parents who are also employees may experience this kind of conflict when their supervisors ask them to put in extra time, which cuts into the time they are able to spend with their children. An article in the *Harvard Business Review* (Schwartz, 1989) supporting the idea that corporations should develop a two-track system, one for executives who sacrifice family concerns for the company and another for those who take time off for family needs, stimulated intense controversy in both business and social-scientific circles. The so-called mommy track was attacked as discriminating against women who wish to succeed in business careers while raising their children.

Role strain occurs when people experience conflicting demands in an existing role or cannot meet the expectations of a new one. For example, law enforcement agents face role strain as they attempt to enforce laws favored by the majority while protecting the rights of minorities. Thus, federal rangers in the national forests and parks often experience role strain when they seek to balance the enforcement and community relations aspects of their role. Environmental laws may restrict the use of a wilderness area to hikers, but local residents may wish to drive their vehicles over old logging roads to reach isolated fishing or hunting grounds. When the rangers enforce the rules against local people (who are also their neighbors), they may experience severe role strain (Helvarg, 1994; Shanks, 1991).

Role strain in the form of anxiety over poor performance is at least as common as role strain caused by conflicting expectations. For example, the unemployed head of a family feels severe stress as a result of inability to provide for the family's needs (Bakke, 1933; Jahoda, 1982; Jahoda, Lazarsfeld, & Zeisel, 1971). The mother of a newborn baby often feels intense anxiety over how well she can care for a helpless infant, a feeling that is heightened if she herself is young and dependent on others (Mayfield, 1984).

Ascribed Versus Achieved Statuses. Role conflicts may occur in simpler societies, but they are far more common in societies undergoing rapid change. One reason for the relative lack of role conflict in simpler, more stable societies is that in such societies a person's statuses are likely to be determined by birth or tradition rather than by anything the person achieves through his or her own efforts. These **ascribed statuses** (peasant, aristocrat, slave, and so forth) usually cannot be changed and hence are less likely to be subject to different role expectations. Such statuses are found in industrial societies, too (statuses based on race or sex are examples), but they tend to become less important in modern institutions than **achieved statuses** like editor, professor, or Nobel Prize winner.

We expect to be able to achieve our occupational status, our marital and family statuses, and other statuses in the community and the larger society. Nevertheless, there is some tension in modern societies between the persistence of ascribed statuses and the ideal of achieved statuses. The empirical study of that tension and of efforts to replace ascribed with achieved statuses (e.g., through equality of educational opportunity) is another major area of sociological research. We will see in later chapters that many of the social movements studied by sociologists arose as different groups in a society—African Americans, wage workers, immigrants—organized to press for greater access to economic and social institutions like corporations and universities.

Master Statuses. One reason so many groups have had to organize to obtain social justice is related to the way statuses operate in many societies. Although any person may fill a variety of statuses, many people find that one of their statuses is more important than all the others. Such a status, which is termed a **master status,** can have very damaging effects (Hughes, 1945). An African American man, for example, may be a doctor, a father, and a leader in his church, but he may find that his status as a black American takes precedence over all of those other statuses. The same is often true for women. A woman may be a brilliant scientist and a leader in her community, and fill other statuses as well, but she may find that when she deals with men her status as a woman is more significant than any of the others.

The effects of a master status are also felt by people who have served jail terms, by members of various racial and ethnic groups, by Americans overseas (Smith, 1994), and by members of many other types of groups. For example, recent studies have shown that people who are noticeably overweight or obese often find that their status as a "fat person" is a master status that denies them opportunities to be appreciated for their performance in other statuses, such as student or worker. Too often, people with a stigmatizing master status find themselves avoiding social interaction, and they may feel excluded from many of the groups that make up the fabric of social life.

SOCIAL GROUPS

The social fabric of modern societies is composed of millions of groups. Some groups are as intimate as a pair of lovers. Others, like the modern corporation or university, are extremely large and are composed of many interrelated subgroups. People need to perform well in all of these groups, and to have this ability is to be considered successful by others. But the knowledge needed for success in a group with a formal structure of roles and statuses is quite different from the "smarts" needed for success in intimate groups.

In the rest of this chapter we analyze some important aspects of the way people behave in groups. We begin by describing the most common types of groups and showing how they can be linked together by overlapping memberships.

Characteristics of Social Groups

A **social group** is a set of two or more individuals who share a sense of common identity and belonging and who interact on a regular basis. Group members are recruited according to specific criteria of membership and are bound together by a set of membership rights and mutual obligations. Throughout this chapter, unless indicated otherwise, the terms *group* and *social group* are interchangeable.

Group identity, the sense of belonging to a group, means that members are aware of their participation in the group and know the identities of other members of the group. This also implies that group members have a sense of the boundaries of their group—that is, of who belongs and who does not belong. Groups themselves have a social structure that arises from repeated interaction among their members. In those interactions, the members form ideas about the status of each and the role each can play in the group (Hare et al., 1994; Holy, 1985; Homans, 1950). A skilled member of a work group, for example, may become the leader while an unskilled member may become recognized as the person others instruct and send on errands.

Through their interactions, group members develop feelings of attachment to one another. Groups whose members have strong positive attachments to one another are said to be highly cohesive, while those whose members are not very strongly attached to one another are said to lack cohesion. Groups also develop norms governing behavior in the group, and they generally have goals such as performing a task, playing a game, or making public policy. Finally, because groups are composed of interacting human beings, we must also recognize that all groups have the potential for conflict among their members; the resolution of such conflicts may be vital to continued group cohesion.

Types of Groups

Primary and Secondary Groups. If one walks along a beach on a nice summer day, the variety of groups one sees is representative of the range of social groups in the larger society. Many of the groups on the beach are composed of family members. Others are friendship groups, usually consisting of young men and women of about the same age. Some of these friendship groups are same-sex groups; others include both males and females. Then there are groups that form around activities like volleyball. Still other groups may be based on occupational ties; the lifeguards and their supervisors or the food service workers are examples of such groups.

Charles Horton Cooley defined **primary groups** as "those characterized by intimate face-to-face association and cooperation. They are primary in several senses, but chiefly in that they are fundamental in forming the social nature and ideals of the individual" (1909, p. 25). Out on the beach, the family groups are, of course, examples of primary groups. So are the groups of friends lying near one another on the sand. And among the food service workers or the lifeguards there are probably some smaller friendship groups or cliques within the larger occupational group.

The occupational groups serving the beachgoers are examples of **secondary groups.** The concept of a secondary group follows from Cooley's definition of primary groups, although he did not actually use the term (Dewey, 1948). Secondary groups are characterized by relationships that involve few aspects of their members' personalities. In such groups, the members' reasons for participation are usually limited to a small number of goals. The bonds of association in secondary groups are usually based on some form of contract, a written or unwritten agreement that specifies the scope of interaction within the group. All organizations and associations, including companies with employers and employees, are secondary groups.

Within most secondary groups, one can usually find a number of primary groups based on regular interaction and friendship. Scottish clans and Native American tribes, for example, are secondary associations that link individuals together through a network of kin relations. But within the clan or tribe there are many primary groups, often composed of men or women of roughly the same age.

As the number of people in a group increases, the number of possible relationships among the group's members increases at a greater rate. When there are two people in the group, there is just one relationship. But when a third person is added, there are three possible relationships (A–B, B–C, A–C). When a fourth person joins the group, there are six possible relationships; a group of only six people includes fourteen possible relationships. The Sociological Methods box on pages 104–105 describes this phenomenon in more detail.

Dyads and Triads. The basic principles of group structure and size were first described by the pioneering German sociologist Georg Simmel (1858–1918). Simmel perceived the need to study behavior in small and larger groups because groups are the basic units of life in all societies. Among his other contributions, he pointed out the significance of what he termed **dyads** (groups composed of only two people) and **triads** (groups composed of three people). He recognized that the strongest social bonds are formed between two people, be they best friends, lovers, or married couples. But he saw that dyads are also quite vulnerable to breakup; if the single relationship on which it is based ends, the dyad ends as well.

When the dyadic bond is strong (and the two who share it often jealously guard their intimacy), the introduction of a new person into the group frequently causes problems. Conventional wisdom has much to say about this matter ("Two's company, three's a crowd"), and if one observes children's play groups, it soon becomes clear that much of the conflict the children experience has to do with the desire to have an exclusive relationship with a best friend. In families, too, the shift from a dyad to a triad often creates problems. A couple experiencing a high level of conflict may believe that a baby will offer them a challenge they can meet together, thereby renewing their love for each other. At first, this may seem to be the case, but as the infant makes more demands on the couple's energy and time, the father may feel that the baby is depriving him of attention from his wife, and he may become jealous of the newcomer. For some couples, the addition of a child can actually increase the chance of divorce—resulting in a stable mother–child dyad and a single male.

Communities. At a level of social organization between the primary group and the larger institutions of the nation-state are communities of all descriptions. **Communities** are sets of primary and secondary groups in which the individual carries out important life functions such as raising a family, earning a living, finding shelter, and the like. Communities may be either **territorial** or **nonterritorial.** Both include primary and secondary groups. In general, territorial communities are contained within geographic boundaries, whereas nonterritorial communities are networks of associations that form around shared goals. When people speak of a "professional community," such as the medical or legal community, they are referring to a nonterritorial community.

Territorial communities are populations that function within a particular geographic area, and this is by far the more common meaning of the term as it is used both in everyday speech and in social-scientific writing (Loewy, 1993; Suttles, 1972). Territorial communities are usually composed of one or more *neighborhoods.* The neighborhood level of group contact includes primary groups (particularly families and peer groups) that form attachments on the basis of proximity—that is, as a result of living near one another.

In studies of how people form friendship groups in suburban neighborhoods, both Herbert Gans (1967, 1976) and Bennett Berger (1968) found that proximity tended to explain patterns of primary-group formation better than any other variable except social class. Thus, families that move into new suburban communities tend to find friends among others of the same social class who live near enough to allow ease of interaction and casual visiting; the same is true of students living in dorms on college campuses. On the international level, observers of the ethnic violence in

SOCIOLOGICAL METHODS

Diagramming Group Structure

A variety of techniques, collectively known as *sociometry,* are used to study the structure of groups. The three most frequently used sociometric methods are basic group diagrams, diagrams that indicate the valence of group bonds, and sociograms, which chart individuals' preferences in groups.

At their simplest, group diagrams show the number of members in a group. Each member is represented by the same symbol, and the presence of a relationship between two members is shown by a line. Diagram A shows how the number of possible relationships increases geometrically as the number of members in the group increases arithmetically.

In a slightly more sophisticated type of diagram, a bond between group members is shown only if it actually exists in the group's interactions. Where there is no relationship between two members, no line is drawn between them. In a peer group, it is usual for all the members to have a relationship, even if they do not all share the same strong feelings of friendship. But in work groups it is not uncommon for people to have relationships that are based on cooperation in carrying out specific tasks. If their work does not bring them together, they may not have any relationship at all.

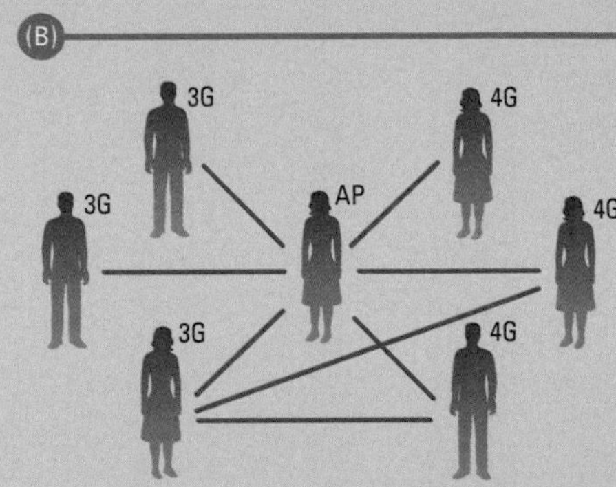

Diagram B represents a group in a school where the third grade (3G) and fourth grade (4G) teachers who teach

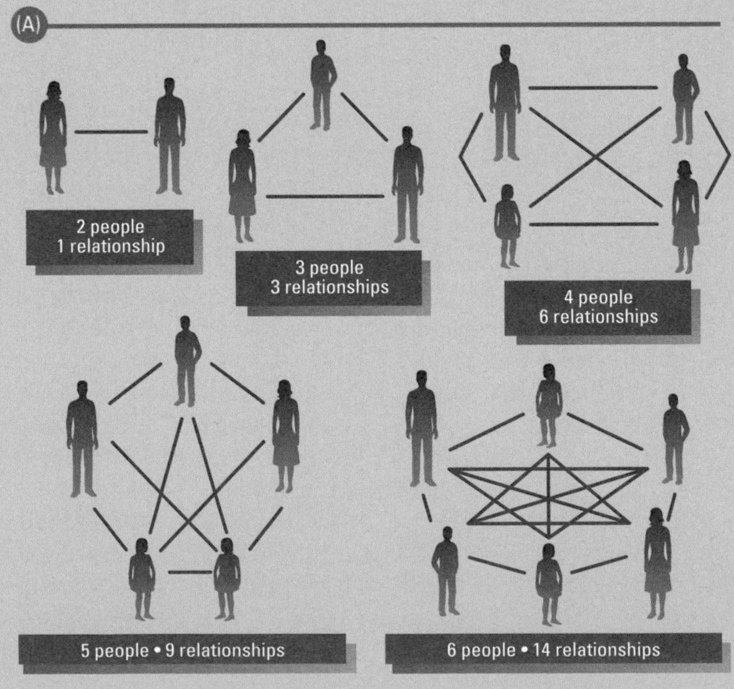

language arts are members of a committee convened by an assistant principal (AP). The diagram shows that three of the teachers share relationships (either through cooperation in projects at work or through friendship), while the other teachers are linked to one another only through their work with the assistant principal.

Valence is the feeling that exists between any two people in a group. It refers to the value they place on their relationship. The most basic valence in a relationship is positive or negative—that is, like or dislike. In most friendship groups, there is a balance of valences; otherwise the group will very likely break up. In diagram C, for example, Joe likes Dave and Marty and they reciprocate that liking, but Dave and Marty do not like each other. Joe must either convince the other two to like each other or choose between them.

(C)

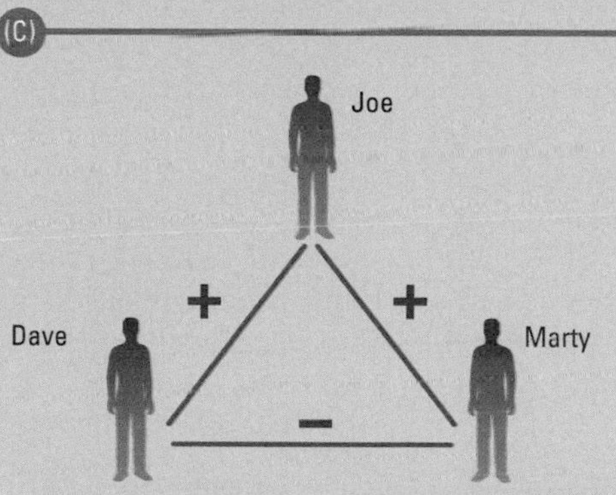

A difference in valence between two people can be indicated with a double line and two different signs. In diagram D, we see that Dave now likes Marty but Marty does not reciprocate. Perhaps Marty would like to have an exclusive relationship with Joe and views Dave as a threat. If Joe can show Marty that Dave likes him and if Dave continues to try to get along with him, the group is very likely to form a balanced triad, as shown in diagram E.

(D)

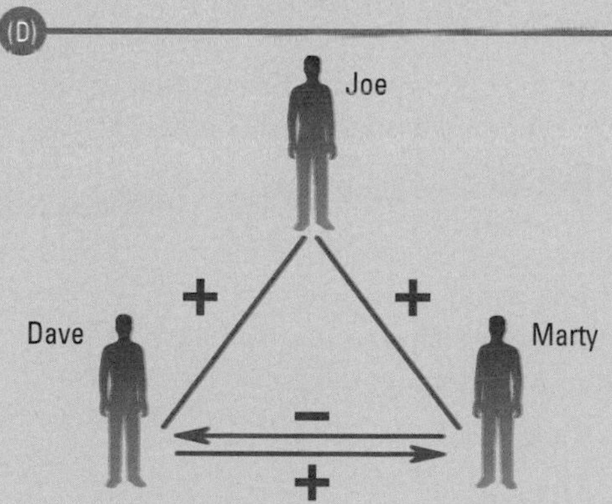

(E)

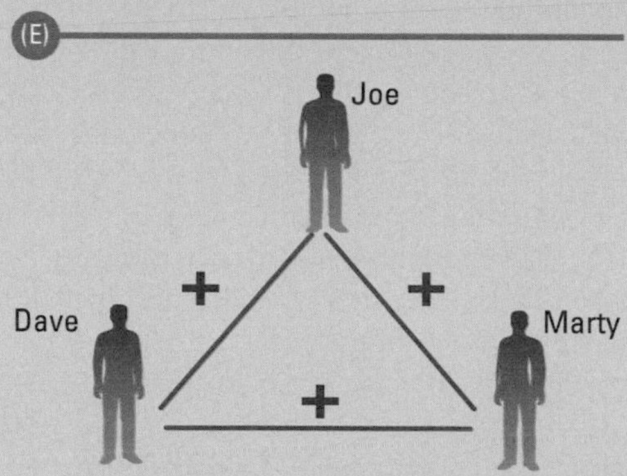

In a larger group, such as a school classroom or a club, the members express their preferences for each other in many ways. They call each other to talk about the events of the day or week; they gossip about each other; they give each other advice or criticism. These preferences result in a structure of cliques that exists within the larger group. When social scientists wish to chart such a structure, they simply ask all the members of the group to list the three or four individuals in the group with whom they would most like to spend time. By indicating each choice with a line, the researchers create a diagram known as a *sociogram*. Diagram F is a simple sociogram showing preferences among nine boys and girls. It reveals, among other things, that there is a clique centering on Dave and another centering on Latasha.

(F)

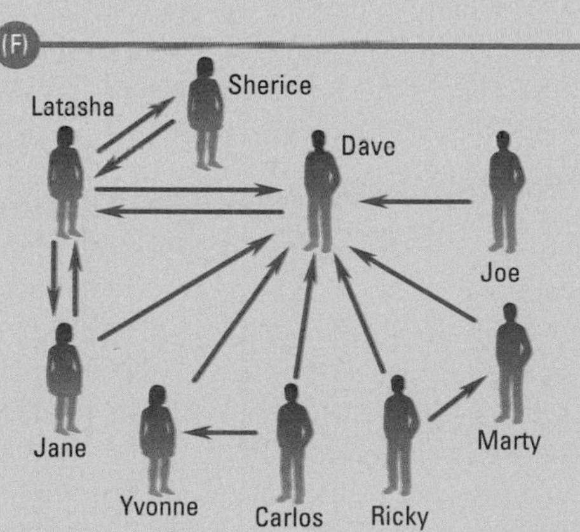

Rwanda report that where members of the Hutu and Tutsi tribes live close to each other and cooperate in bringing in the harvest, they are far less likely to bear grudges and resort to violence than members of the same tribes who do not interact in their daily lives (Bonner, 1994).

That physical propinquity and social-class similarities explain how friendships form may seem self-evident to us. Moving into new neighborhoods or into dorms is a constant feature of life in a large urban society. But the idea that mere propinquity, rather than kinship or tribal membership, can explain patterns of association would shock many people in simpler agrarian societies, where mobility is not a fact of everyday life. As Robert Park (1967/1926) noted, it is only under conditions of rapid and persistent social change that proximity helps explain why people form groups.

Networks. In modern societies, people form their deepest friendships in face-to-face groups, but these groups are, in turn, integrated into larger and more impersonal secondary-group structures that may extend well beyond the bounds of territorial communities. William F. Whyte opened *Street Corner Society* (1943), his study of street-corner peer groups, with this observation:

> The Nortons were Doc's gang. The group was brought together primarily by Doc, and it was built around Doc. When Doc was growing up, there was a kid's gang on Norton Street for every significant difference in age. There was a gang that averaged about three years older than Doc; there was Doc's gang, which included Nutsy, Danny, and a number of others; there was a group about three years younger, which included Joe Dodge and Frank Bonilli; and there was still a younger group, to which Carl and Tommy belonged. (p. 3)

The Nortons were active in a network of neighborhood-level peer groups. They would compete against other groups of boys of roughly the same age in a yearly round of baseball, bowling, and occasional interneighborhood fights (the latter occurring mainly in their younger teenage years). But the Nortons were also integrated into the community through some of its secondary associations. For example, at election time they were recruited by the local political party organization, and some were recruited by the racketeers who controlled their low-income neighborhood. Party organizations and organized-crime groups are examples of secondary associations that extend outside territorial communities.

In-Groups and Out-Groups. The "corner boys" studied by Whyte tended to be hostile toward similar groups from other street corners and even toward certain boys from their own territory. These distinctions between groups are common and are referred to as in-group–out-group distinctions. The **in-group** consists of one's own peers, whereas **out-groups** are those whom one considers to be outside the bounds of intimacy. Simmel observed that in-group–out-group distinctions can form around almost any quality, even one that many people would not consider meaningful at all. Thus, in a study of juvenile groups in a Chicago housing project, Gerald Suttles (1972) found that distinctions were made between boys who lived in lighter or darker brick buildings. Similarly, American children who have lived in foreign countries for several years may find themselves relegated to out-groups when they return to school in the United States (Smith, 1994).

In-group–out-group distinctions are usually based on such qualities as income, race, and religion. In "Cornerville," the community Whyte studied, group boundaries were based on educational background, particularly college versus noncollege education. One of Whyte's informants explained this distinction:

> In Cornerville the noncollege man has an inferiority complex. He hasn't had much education, and he has that feeling of inferiority. . . . Now the college man felt that way before he went to college, but when he is in college, he tries to throw it off. . . . Naturally, the noncollege man resents that. (Whyte, 1943, p. 79)

In-group–out-group distinctions often make it difficult for secondary associations to attract members from both groups. For example, when two ethnic or racial groups in a community make in-group–out-group distinctions, they often find themselves drifting toward different political parties or forming distinct factions within the same party rather than uniting to find solutions to common problems.

Interaction in Groups

"Human interaction," Herbert Blumer observed, "is mediated by the use of symbols, by interpretations, or by ascertaining the meaning of one another's actions"

(1969b, pp. 179–180). By "mediated" Blumer meant that we do not normally respond directly to the actions of another person. Instead, we react to our own interpretations of those actions. When we see other drivers moving into an intersection against the light, for example, we begin to feel that everyone is out for themselves and we would be fools to hold back. These interpretations are made in the interval between the stimulus (the other person's action) and the response (our own action).

What factors explain how people decide how to act in different kinds of groups? This question has been the subject of much research in the social sciences. Let us therefore take a brief look at some research on how people define their needs in social situations.

Definitions of the Situation. "Situations we define as real, are real in their consequences" stated W. I. Thomas (1971/1921), a pioneer in the study of social interaction. By this observation, which has come to be known as the "Thomas dictum," he meant that our understandings or definitions of what is occurring around us—whether they are correct or not—guide our subsequent actions. Thus, in a study of how dying patients are treated by medical groups in hospital emergency rooms, David Sudnow (1967) found that when a patient was brought in with no heartbeat, the person's age had a great deal to do with what happened next. The arrival of a younger patient who seemed to be dying would produce a frenzy of attempts to restart the heart. At times, the entire group of medical personnel would become involved. But an old person with no heartbeat was far more likely to be pronounced dead on arrival, with little or no mobilization of the medical group. The patient's aged appearance caused the members of the group to define the situation as one in which urgent and heroic efforts were not required.

Sudnow's study began by questioning how definitions of the situation account for the ways in which people interact. The emergency room teams that Sudnow observed were unofficially defining the situation in order to decide whether to apply heroic measures to heart attack victims. The subtle cues that define each situation are communicated through phrases, gestures, and other symbolic behavior, as well as through explicit evaluations of what was good and bad about a particular operation.

Ethnomethodology. Studies based on this interactionist perspective use a variety of techniques. One of these techniques is **ethnomethodology,** the study of the underlying rules of behavior that guide group interaction. This approach was used by Harold Garfinkel (1967) in studying how we use verbal formulas to create a flow of communication that we feel is normal and that we can understand. Garfinkel asked his students to engage in conversations with friends and family members in which they would violate some of the simple norms that most people follow in carrying on conversations. Following is an example:

> Victim [*waving cheerily*]: How are you?
>
> Student: How am I in regard to what? My health, my finances, my schoolwork, my peace of mind, my . . . ?
>
> Victim [*red in the face and suddenly out of control*]: Look! I was just trying to be polite. Frankly, I don't give a damn how you are. (Garfinkel, 1967, p. 44)

As this brief example shows, when the norms of spoken interaction that "entitle" us to continue or maintain a conversation are violated, the conversation often cannot continue because there is no longer a mutual area of interaction within which both parties tacitly agree to make sense.

The Dramaturgical Approach. Another technique used in research on interaction in groups is the **dramaturgical approach.** This approach is based on the recognition that much social interaction depends on how we wish to impress those who may be watching us. For example, Erving Goffman (1965) observed that people change their facial expressions just before entering a room in which they expect to find others who will greet them. Couples who are fighting when they are "backstage" often present themselves as models of friendship when they are "frontstage"—that is, when they are in the presence of other people. And many social environments, such as hotels, restaurants, and funeral parlors, are explicitly set up with a front and a back "stage" so that the public is spared the noisy and sometimes conflicted interaction occurring "behind the scenes." Using these and other strategies to "set a stage" for our own purposes is known as **impression management.**

In a well-known study that applied this dramaturgical view of group interaction, James Henslin and Mae Briggs (1971) drew upon Briggs's extensive experience as a gynecological nurse. Their research showed how doctors and nurses play roles that define the situation of a pelvic examination as unembarrassing and attempt to save each participant's "face" or self-image. According to Henslin and Briggs, the pelvic examination is carried out in a series of scenes. First, the doctor and the patient discuss the patient's condition. If the doctor decides that the patient needs a pelvic examination, the doctor will leave the room; this is the end of scene one. In scene two the nurse enters. Her role is to work with the patient to create a situation in which the majority of the patient is hidden behind sheets with only her depersonalized pelvic area exposed for clinical inspection. Through these preparations, the patient becomes symbolically distanced from the doctor, allowing the next scene, the pelvic examination itself, to be desexualized. The props, the stage setting (the examining room), and the language used help define the situation as a nonsexual encounter, thereby saving everyone involved from embarrassment.

Interaction and Group Structure

The examples just discussed involve decisions and strategies that individuals use in deciding how to behave in a given situation. But as members of groups we often explain our actions on the basis of our status in the group rather than on the basis of our individual expectations. "I must do this because I'm expected to lead," we might say, or "I can't decide that without talking to the others and especially to [the group's leader]." These explanations suggest that the structure of groups is a major factor in explaining behavior.

Six college students of the same sex are hired to participate in a small-group experiment. All are unknown to one another before the experiment. The specific task they are asked to perform is of little importance except that it must involve all the members of the group in a cooperative effort and must not be so difficult that they become frustrated or so easy that they become bored.

Initially, the group has no structure. All the subjects are peers, and all have come for the same reason: to earn some money for participating in the experiment. There are no predetermined statuses or roles. The students listen to a description of what they are to do and then are left alone to get the work under way. Actually they are being observed through a one-way mirror, and all their interactions are being recorded and counted.

Table 4.1 summarizes (or *aggregates*) the data obtained by counting all the interactions—including small utterances like "Oh" and "I see"—occurring in a group of six male subjects meeting for 18 one-hour sessions. This table shows who initiated each interaction and to whom it was directed, including interactions directed toward the group as a whole. The participants are ranked from high to low on this variable (i.e., initiating an interaction). Thus, subject 1 spoke 1,238 times to subject 2, spoke 961 times to subject 3, and so on (Homans, 1961). Subject 1 initiated 3,506 utterances to specific individuals (e.g., "Why don't you . . . ?") and 5,661 utterances to the group as a whole (e.g., "Why don't we . . . ?"), for a total of 9,167 utterances over all the sessions—more than any of the other participants. You can see that subject 6 spoke only about one ninth as often as

TABLE 4.1
Aggregate Matrix for 18 Sessions of Six-Man Groups

	Initiated to Individuals 1	2	3	4	5	6	Total to Individuals	To Group as Whole	Total Initiated
From Individuals 1		1,238	961	545	445	317	3,506	5,661	9,167
2	1,748		443	310	175	102	2,778	1,211	3,989
3	1,371	415		305	125	69	2,285	742	3,027
4	952	310	282		83	49	1,676	676	2,352
5	662	224	144	83		28	1,141	443	1,584
6	470	126	114	65	44		819	373	1,192
Total Received	5,203	2,313	1,944	1,306	872	565	12,205	9,106	21,311

Note: Differences between the total number of interactions and the number received are due to the fact that some utterances are answered indirectly by means of utterances directed at the entire group. The diagonal cells are empty because the experimenters did not count utterances by a subject to himself.

Source: *Social Behavior,* by George Caspar Homans. Copyright © 1961 by Harcourt Brace Jovanovich, Inc. Reprinted by permission of the publisher.

subject 1 and that most of what he said was directed to subject 1 (470 utterances). In fact, subject 1 received the most interaction from all the other participants, illustrating that people who initiate interactions are more likely to receive them. In later interviews, the experimenters also found that subject 1 was the most respected member of the group.

What makes the person who initiates and receives the most communications the most respected member of the group? The reasons vary with the specific tasks the group is performing, but if we assume competence at those tasks, usually the person at the center of the communication is spending a lot of time helping others as well as helping the group accomplish its goals. That person not only is concerned with his or her own performance but also takes pleasure in helping the others—giving approval, voicing criticism, making suggestions, and so on. In exchange for this help, the other members of the group give approval, respect, and allegiance to the central individual.

Is the person who initiates the most interactions respected simply because he or she talks a lot? Common sense tells us that this is far from true. People who talk a lot but have little to contribute soon find that no one is paying attention to them. Their rate of interaction then drops sharply. Table 4.1 cannot show such changes in interaction patterns, but they are frequently observed in small-group studies. Other patterns have also been identified. Robert F. Bales and Philip Slater (1955), for example, observed that the person who initiated the most interactions—both in getting tasks done and in supporting the suggestions of others—often came to be thought of as a leader. The group began to orient itself toward that person and to expect leadership from him or her. A second member of the group, usually the one who initiated the second-highest number of interactions, was often the best-liked person in the group.

Bales and his colleagues concluded that groups often develop both a "task leader" (or instrumental leader) and another leader, whom they called a "socioemotional leader" (or expressive leader). The former tends to adhere quite strictly to group norms and to take the lead in urging everyone to get the work done. Other members, however, are often left with ruffled feelings. The socioemotional leader typically is the person who eases the group through rough spots with jokes, encouragement, and attention to the group's emotional climate. Thus, most classes have a class clown whose antics help release the tension of test situations, and most teams have a respected member who can also joke around and get people to relax under pressure. At times these roles are performed by the same person, but generally as the group settles into a given task there is an informal division of labor between the two kinds of leaders.

As group size increases, or as the number of groups in an organization increases, the leaders of the larger organization must invest in ways of controlling the behavior of groups within it (Hechter, 1987). Frequently this investment takes the form of agents of control such as foremen, accountants, inspectors, and the like. These roles become part of the formal structure known as bureaucracy.

Formal Organizations and Bureaucracy

Informal organizations are groups whose norms and statuses are generally agreed upon but are not set down in writing. Usually such groups have leaders who help create and enforce the group's norms but have no formal leadership position in the group.

Formal organizations have explicit (often written) sets of norms, statuses, and roles that specify each member's relationships to the others and the conditions under which those relationships hold. Organization charts and job descriptions are typical of such organizations. Formal organizations take a wide variety of forms. For example, the New England town meeting is composed of residents of a town who gather to debate and discuss any issues the members wish to raise. The tenants' association of an apartment building is also composed of people who reside in a specific place, but its scope of action is usually limited to housing issues. Both of these are formal organizations, since there are rules defining who may participate and the scope and manner of that participation. As is true in many formal organizations, the members of town meetings and tenants' associations try to arrive at decisions through some form of democratic process—that is, by adhering to norms that allow the majority to run the organization but not to infringe on the rights of the minority.

A familiar type of formal organization is the **voluntary association.** People join groups like the Parent-Teacher Association (PTA) or the Rotary Club to pursue interests that they share with other members of the group. Voluntary associations are usually democratically run, at least in principle, and have rules and regulations as well as an administrative staff.

Churches, fraternal organizations, political clubs, and neighborhood improvement groups are examples of voluntary associations often found in American communities. Sociologists study these associations in order to understand how well or poorly people are integrated into their society.

Bureaucracies. Bureaucracies are another common type of formal organization. A **bureaucracy** is a specific structure of statuses and roles in which the power to influence the actions of others increases as one nears the top of the organization; this is in marked contrast to the democratic procedures used in other kinds of organizations. In popular speech, however, *bureaucracy* is often used as a negative term referring to a frustrating experience with rules and regulations and officials behind desks or counters. Terms like "red tape" or "pointy-headed bureaucrat" often reinforce the notion that bureaucracies are merely obstacles to individual freedom and creativity. Sociologists seek to understand the flaws and difficulties of running bureaucracies, but they also recognize that bureaucratic forms of organization are essential to contemporary life.

Bureaucracy is not limited to the public sector. Innovative "dot-com" businesses like Amazon.com are bureaucracies. So are older manufacturing corporations like Ford or General Electric. Voluntary associations usually have some of the elements of bureaucracies, but they are run as democratic structures in which power is based on majority rule rather than on executive orders, as is the case in pure bureaucracies. None of the goods and services we rely on could be obtained as easily as they are today without the activities of bureaucratic business administrations in which each person, performing his or her assigned role, carries out a set of designated tasks and reports to a superior with different tasks (Riggs, 1997).

We owe much of our understanding of bureaucracies to the work of Max Weber, who identified the following typical aspects of most bureaucratic organizations:

1. *Positions with clearly defined responsibilities:* "The regular activities required for the purposes of the organization are distributed in a fixed way as official duties."
2. *Positions ordered in a hierarchy:* The organization of offices "follows the principle of hierarchy; that is, each lower office is under the control and supervision of a higher one."
3. *Rules and precedents:* The functioning of the bureaucracy is governed "by a consistent system of abstract rules" and the "application of these rules to specific cases."
4. *Impersonality and impartiality:* "The ideal official conducts his office . . . in a spirit of formalistic impersonality . . . without hatred or passion, and hence without affection or enthusiasm."
5. *A career ladder:* Work in a bureaucracy "constitutes a career. There is a system of 'promotions' according to seniority, or to achievement, or both."
6. *The norm of efficiency:* "The purely bureaucratic type of administrative organization . . . is from a purely technical point of view, capable of attaining the highest degree of efficiency" (Weber, 1958/1922). (See the study chart on page 111.)

Weber observed that these characteristics of bureaucratic administrations were "ideal types." By this he meant that these features would be found in all true bureaucracies, but he also knew that in the real world there would always be problems or imperfections in achieving them. (See the Global Social Change box on page 112.) This is not just a minor sociological insight. If you understand the essential features of bureaucracies and why they are often difficult to achieve, you will have a definite advantage as you make your way through complex organizations like universities, businesses, military organizations, hospitals, and many other bureaucratic organizations. For example, when you apply for a job in an organization, there will generally be a "position description" that defines your function and responsibilities. But it will not always note the "gray areas" of your position that overlap with other positions and may cause confusion or conflict. Similarly, the hierarchical organization of offices, often referred to as the "chain of command," is associated with a cardinal rule of bureaucracies: The subordinate does not go over his or her supervisor's head and discuss work matters with a higher official, at least not without notifying the immediate supervisor. There must be major reasons for doing so, and one must understand that the consequences of "going outside the chain of command" can be serious—yet necessary in a real crisis. Every bureaucracy has its rules and precedents, but it also has ways in which people subvert those rules, sometimes in the interest of getting the work done. You need to understand this and

STUDY CHART

Characteristics of Bureaucracy

Characteristic	Example
Positions Clearly Defined	Job or position descriptions detail the responsibilities of each job, avoiding confusion about the duties of each jobholder.
Positions Ordered in a Hierarchy	Positions in the company or agency are ranked from top to bottom so that each position reports to another and supervisory responsibilities are clear.
Rules and Precedents	"The way we do things" in an organization is often written down, becoming "the book" of rules to be followed by its members.
Impersonality and Impartiality	At least in principle, actual performance on the job, rather than personal likes or favors, determines each individual's performance rating.
Career Ladder	New jobs are posted so that employees have a chance to move up in the organization on the basis of achievement, seniority, or both.
Norm of Efficiency	The company continually seeks to increase its efficiency by increasing productivity per employee; this can lead to layoffs or downsizing.

watch it in action so that you can work effectively in the organization.

Then there is the norm of impartiality and its violations. Success in moving up the career ladder is often a matter not only of ability to do the work but also of understanding the limitations of the norm of impersonality and impartiality. You will see people promoted not because of their ability or productivity but because they are personally connected with higher officials. If this form of "personal" rather than impersonal conduct pervades the organization, you will need to decide whether to play the game of cultivating personal relationships or move to another organization. In either case, you will find that to some degree friendships formed at work play a part in advancement, but the degree will vary greatly from one organization to another.

Do all these ways in which the "ideal typical" bureaucracy does not correspond to actual practice in organizations negate the importance of bureaucratic administration in the contemporary world? The answer is not a simple one. Weber believed that bureaucracy made human social life more "rational" than it had ever been in the past. Rules, impersonality, and the norm of efficiency are some of the ways in which bureaucracies "rationalize" human societies. By this Weber meant that society becomes dominated by groups organized so that the interactions of their members will maximize the group's efficiency. Once the group's goals have been set, the officials in a bureaucracy can seek the most efficient means of reaching those goals. All the less rational behaviors of human groups, such as magic and ritual, are avoided by groups organized as bureaucracies, which are based on science and rationality. Weber called this important change "the disenchantment of the world."

For this and other reasons, Weber had misgivings about the consequences of the increasing dominance of bureaucratic groups in modern societies. On one hand, his historical research on the evolution of bureaucratic forms of social organization convinced him that bureaucracies were essential to the increasing productivity of the modern world. Bureaucracies were also a requirement for the spread of the rule of law. Weber and subsequent researchers on bureaucracy inevitably found that without rule-abiding, professional administrators there is far less chance for powerless people to get a fair shake when pleading their cause. Bribery, nepotism (promotion of one's relatives instead of impartial promotion by merit), patronage (distribution of jobs in return for payments or favors rather than by merit), and many other forms of corruption would reign in government and business in the absence of strong bureaucratic administrations.

On the other hand, Weber saw that the increasing dominance of bureaucratic administrations in modern life created many obstacles to personal freedom. Inevitably, the growth of powerful bureaucratic organizations, whether private corporations or agencies of government, meant that more rules would be adopted and applied to control social behavior. In addition, people in a bureaucracy may increasingly look to their superiors for guidance and thus avoid taking full responsibility for their actions. An extreme example of this tendency is revealed in the history of Nazi officials

GLOBAL SOCIAL CHANGE

Kleptocracy and the Rule of Law

The ideal model of bureaucracy assumes leadership by high officials who set positive examples for all officials below them. Unfortunately, in the contemporary world there are innumerable examples of high officials, both appointed and elected, who use their authority to plunder the resources of their governments and societies. Notorious examples have been exposed in Haiti, Sierra Leone, Russia, Indonesia, Nigeria, and many other nations. The United States is hardly immune from corruption in high places, but it does not suffer from the pervasive systems of bribery and official corruption found in many other nations.

The term *kleptocracy* is often used to describe abuse of power for personal gain by high government and business officials in a nation (Lundahl & Silie, 1998). In the African nations of Liberia and Sierra Leone, for example, abuse of power by government officials became so widespread in the last three decades that citizens lost almost all faith in the ability of government to respond to any of their needs. Eventually, civil wars among ambitious army officers and government officials became the "normal" state of affairs. The rule of law broke down entirely, which means that citizens could no longer expect a minimum level of safety on the streets or that their property would be safe from marauding guerrilla troops. They also could not expect any grievance, civil or criminal, to be prosecuted fairly. Under these dire conditions, societies regress to a state of intense disorganization, distrust, violence, and chronic warfare.

While many Russians are experiencing downward mobility during the painful transition to a new social order, those close to the powerful or the criminal elites are doing quite well.

Elsewhere in the world, notably in Russia, the combination of kleptocrats in official bureaucracies and new "mafia" businesspeople also threatens peace and economic progress. Lack of trust in bank officials and in the ability of government officials to resist corruption pervades Russian society, so much so that Russia has the most U.S. currency in circulation of any nation outside the United States itself (Wickens, 1996). Russia's President Vladimir Putin was elected on the promise that as a former government police official he would take firm measures to combat kleptocracy and strengthen the rule of law. A major task for sociologists in Russia, therefore, is to monitor efforts to make good on this promise. ■

like Adolf Eichmann, who carried out Hitler's orders to exterminate Jews, Gypsies, the mentally ill, homosexuals, and many other groups during the Holocaust of World War II. These officials' pleas during later war crimes trials that they were merely carrying out orders from above illustrates the potential for excessive obedience to authority in bureaucratic organizations.

Bureaucracy and Obedience to Authority.

In Chapter 1, we discussed the experiments on conformity conducted by Solomon Asch. Asch's demonstration of the power of group pressure raised serious questions about most people's ability to resist such pressure. It also led to many other studies of conformity. None of those studies is more powerful in its implications and disturbing in its methods than the series of experiments on obedience to authority conducted by Stanley Milgram.

Milgram's study was designed to "take a close look at the act of obeying." As Milgram described it:

> Two people come to a psychology laboratory to take part in a study of memory and learning. One of them is designated as a "teacher" and the other as a "learner."
>
> The experimenter explains that the study is concerned with the effects of punishment on learning. The learner is conducted into a room, seated in a chair, his arms strapped to prevent excessive movement, and an electrode is attached to his wrist. He is told that he is to learn a list of word pairs; whenever he makes an error, he will receive electric shocks of increasing intensity.
>
> The real focus of the experiment is the teacher. After watching the learner being strapped into place, he is taken into the main experimental room and seated before an impressive shock generator. Its main feature is a horizontal line of 30 switches, ranging from 15 volts to 450 volts, in 15-volt increments. There are also verbal designations which range from SLIGHT SHOCK to DANGER—SEVERE SHOCK. The teacher is told that he is to administer the learning test to the man in the other room. When the learner responds correctly, the teacher moves on to the next item; when the other man gives an incorrect answer, the teacher is to give him an electric shock. He is to start at the lowest shock level (15 volts) and to increase the level each time the man makes an error, going through 30 volts, 45 volts, and so on. (1974, pp. 3–4)

The "learner" is an actor who pretends to suffer pain but receives no actual shock. The subject ("teacher") is a businessperson, an industrial worker, or a student, someone who has been recruited by a classified ad offering payment for spare-time work in a university laboratory.

Milgram was dismayed to discover that very large proportions of his subjects were willing to obey any order given by the experimenter. In the basic version of the experiment, in which the "learner" is in one room and the "teacher" in another from which the "learner" is visible but cannot be heard, 65% of the subjects administered the highest levels of shock, while the other 35% were obedient well into the "intense shock" levels.

Milgram used a functionalist argument to explain the high levels of obedience revealed in his experiment: In bureaucratic organizations, people seek approval by adhering to the rules, which often absolve them of moral responsibility for their actions. But he also explored the conditions under which conflict will take place—that is, the conditions under which the subject will rebel against the experimenter. As he placed the "teacher" in closer proximity to the "learner" in different variants of the experiment, he found that higher percentages of the "teachers" rebelled and refused to administer the maximum shock. Under all experimental conditions, however—even when the "teacher" was in direct contact with the "learner"—most of the "teachers" continued to administer high levels of shock. Their unquestioning obedience to authority is a disturbing verification of Weber's concern about the impact of bureaucracies on the individual's capacity to act according to moral principles.

When Milgram published the results of these experiments, he was criticized for deceiving his subjects and, in some cases, causing them undue stress. The controversy created by this study was one of the factors leading to the establishment of rules for the protection of human subjects in social-scientific research (see Chapter 1).

Commitment to Bureaucratic Groups.

Another question related to the impact of bureaucracy on individuals is how bureaucratic groups—which are based on unemotional and "rational" systems of recruitment, decision making, and reward—can sometimes attract strong commitment from their members. One possible answer is ideology: People believe in the goals and methods of the bureaucracy. Another explanation is that people within a bureaucracy form primary groups that function to maintain their commitment to the larger organization. There is considerable research evidence showing that both explanations are valid—both ideology and primary-group ties may operate to reinforce people's commitment to the organization.

An important study of this subject was Philip Selznick's (1952) analysis of what made the Bolsheviks so effective that they won out in the fierce competition for political dominance in post-czarist Russia. Selznick's data show that the ideologically based primary group was the key element in the Bolsheviks' organization. Small, secret "cells" of devoted communists were organized in neighborhoods, factories, army units, farms, and universities. Through years of tense and often dangerous political activity, the members became extremely devoted to one another and to their revolutionary cause. According to Selznick, this doubling of ideological and primary-group cohesion made the Bolsheviks themselves an "organizational weapon."

In the United States, there are radical militia groups that apparently use the same principle of creating secret "organizational weapons." But there are also corporations and charitable organizations that use small primary-group teams to implement innovations in computer technologies or send highly motivated groups of rescue workers to deal with refugee situations in Eastern Europe or Africa. Thus, as in any other field of knowledge, sociological insights into how social structures and groups operate under different circumstances can be used either for evil or for good.

There are parts of the world, especially in the most impoverished areas of Africa or in the politically embattled regions of the former Soviet empire, where the social order seems to have broken down. But when we look more closely at those regions, we see efforts being made to restore order and the rule of law. For example, one observer writes that Africa is on the brink either of escalating social disaster or renewed progress toward stability and modern development: "Forcing it toward the edge of collapse are the chaos makers, the warlords, the kleptocrats [politicians who pillage public funds]. Coaxing it toward long-term stability are the corruption fighters, the entrepreneurs, the democrats" (Goldberg, 1997, p. 36).

In addition to an understanding of social structures and group processes, those fighting for positive social change often require the help of other nations and major organizations, including international corporations and various branches of the United Nations. The struggles to restore peace in the former republics of Yugoslavia or in Haiti demonstrate the need for combinations of approaches at the macro and micro levels of social life. The nations of the world need to join together to act positively, but in the most troubled regions of the world, groups are seeking to develop social structures—courts, hospitals, refugee settlements, systems of delivering aid and fighting disease—that make orderly interaction possible.

People working in larger social structures—communities, states, nations—increasingly understand that to prevent the spread of war, disease, and intergroup hatreds, they must also build social structures to bridge the gaps between those who live in relative security and those who do not.

TECHNOLOGY AND SOCIAL CHANGE

Huge public bureaucracies are a feature of the contemporary social order in urban industrial societies. In the United States, about 3 million people work for the federal government in thousands of different jobs. In state, county, and municipal governments, there are more than 15 million employees. Critics of the sheer size of government employment continually point to government agencies with inflated payrolls as a reason why new policy initiatives should be carried out by private companies—even though they, too, are bureaucratic organizations (Weisberger, 1997). Sociologists and historians note that the rapid expansion of government employment has been associated with major wars, which, in turn, tended to expand the role of government. But peacetime programs have also contributed to the expansion of public and private bureaucracies. Policies that are extremely popular, such as social security and Medicare, require careful and efficient administration, all of which entail professional career bureaucrats. Regulations in the interest of transportation safety, environmental protection, law enforcement, and many other activities also entail bureaucratic administration.

But do these trends mean that expansion of bureaucratic employment is inevitable? And does it mean more "red tape" and needless delays? Today's administrators are turning to new strategies, often using information technologies, to increase administrative efficiency and slow the growth of bureaucratic employment. Each year, the U.S. Council for Excellence

SOCIOLOGY AND SOCIAL JUSTICE

The emergency room of a big-city hospital is not a place for the squeamish observer. The deadening calm of the night shift may suddenly be broken by the crackle of the paramedics' radio, announcing that they are bringing in a severely wounded gunshot victim or a baby with a deadly high fever or three bleeding gang fighters. Immediately the routine work of the ER is suspended in a rush of activity. Even the less seriously ill or injured patients in the waiting room intuitively realize that their needs must be set aside for the life-threatening emergency.

One reason hospital emergency rooms are such important places in the United States is that they are often the only source of medical care for new immigrants, poor families, and street populations (the homeless, drug addicts, prostitutes, and others). Our sense of social justice demands that people not be allowed to suffer without medical care or that people in extreme medical distress not be allowed to die on the streets. We also expect that in the case of emergencies of our own, there will be an emergency room nearby where we will receive immediate and highly professional care. All these expectations place difficult demands on the resources and staff of most emergency rooms, especially those in major urban centers. On many nights, the ability of the ER staff to function according to high professional standards is stretched to the limit.

Documentary photographer Eugene Richards (1989) spent many months conducting participant observation research in the emergency room of Denver General Hospital, known by local residents as the "knife and gun club." His photos capture the physically and emotionally draining work of the ER staff and the stark drama of the emergencies they deal with. They also show that the roles of the ER staff are well defined, so that even if the work stretches them to the limits of endurance they still know what they must do and how they must assist one another. Indeed, underneath its drama and emotion the ER is a marvel of human social organization, without which it could not be effective. ■

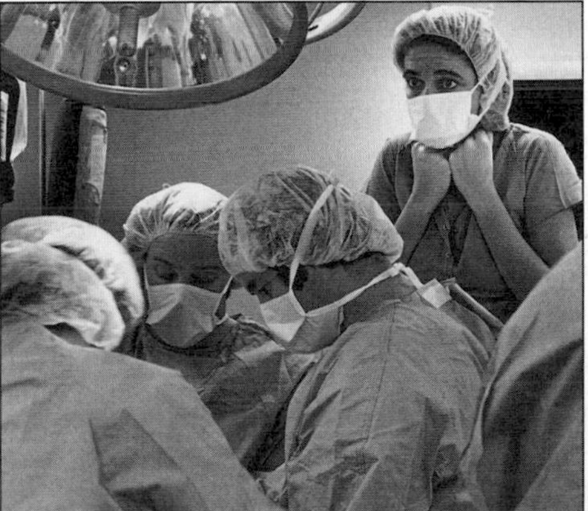

In a complex operation, the patient's life can depend on how well each participant's role is enacted in cooperation with all the others.

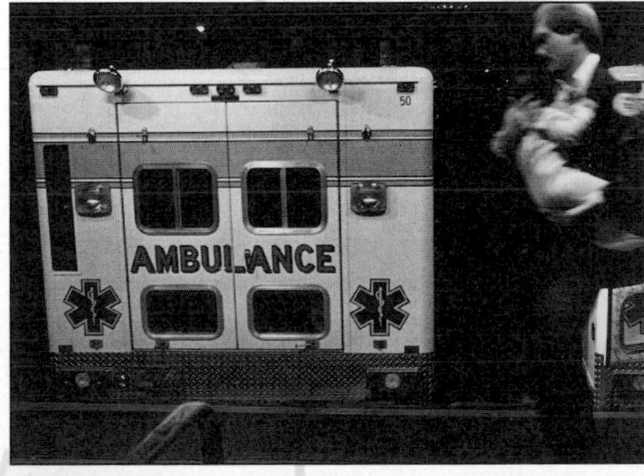

Paramedics are often involved in the most dramatic events that occur in the emergency room. While their status is lower than that of the doctors, the front-line aspect of their role gains them the respect of their colleagues.

in Government awards prizes to agencies that show the greatest innovation in speeding up their work and reducing the rate of growth in the numbers of their employees. The Internal Revenue Service, for example, is using computer and Internet technologies as well as older telephone technologies in highly innovative ways. All tax forms are now available on the Internet. People with modest earnings who may not have Internet access can file their tax returns by touch-tone phone. At the U.S. Patent Office, the Postal Service, and the Federal Aviation Administration, applications of computer technologies have dramatically improved these agencies' ability to perform their tasks with far lower personnel increases than would have been possible with older technologies. Of course, flight delays at airports are a continuing and highly frustrating situation, but much of that problem is due to unprecedented increases in air travel associated with the nation's relative affluence during the 1990s. Applications of technology to flight control and airline scheduling are no panacea, but without them the situation would be even worse.

Similarly, many political officials now realize that for the majority of citizens, encounters with government most often come in the form of frustrating experiences at the Department of Motor Vehicles, where long waits and stubborn officials often become synonymous with life in large bureaucratic administrations. Now many states are using computer technologies to streamline automobile registration and licensing procedures. A trip to the motor vehicle bureau is not something we relish, but it is becoming far less time-consuming and frustrating thanks to skillful applications of information technologies (Gergen, 2000).

SUMMARY

A *society* is a population that is organized in a cooperative manner to carry out the major functions of life. *Social structure* refers to the recurring patterns of behavior that create relationships among individuals and groups within a society.

Groups are collections of people who interact on the basis of shared expectations regarding one another's behavior. In every group, there are socially defined positions known as *statuses.* The way a society defines how an individual is to behave in a particular status is called a *role.* A social *institution* is a more or less stable structure of statuses and roles devoted to meeting the basic needs of people in a society.

The first million years of human social evolution were characterized by a hunting-and-gathering way of life. During that time, kinship structures developed and the band became the basic territorial unit of human society. The shift to agriculture occurred gradually as some human societies became *pastoral societies* based on the herding of animals while others evolved into *horticultural societies* based on the raising of seed crops. Large-scale agrarian societies evolved after the development of plow-and-harvest agriculture.

In agrarian societies, people produce food surpluses that can be used to feed new classes of non-food producers. These societies require increasing amounts of land, and this may lead to conflicts. The need to store and defend food supplies and to house non-food-producers results in the growth of villages and small cities.

The industrial revolution was spurred not only by technological advances but also by the rise of a new social order: *capitalism.* As a consequence of industrialization, fewer people live on the land and more live in cities and suburbs; greater openness to change results in the emergence of new classes and social movements, and scientific and technical advances produce great wealth and more rapid transportation and communication.

A *state* is a society's set of political structures; a *nation-state* is the territory within which those structures operate. The members of a society do not always think of themselves as members of a particular nation, and in extreme cases the lack of a clear match between society and nation can result in a civil war.

Adaptation to a more modern society involves a shift from *gemeinschaft* (close, personal relationships) to *gesellschaft* (well-organized but impersonal relationships). *Role conflict* refers to conflict between roles in secondary groups and roles in primary groups. *Role strain* occurs when a person experiences conflicting demands within a single role.

Ascribed statuses are determined by birth or tradition, whereas *achieved statuses* are determined by a person's own efforts. In more advanced societies, there is a tendency to replace ascribed statuses with achieved statuses.

A *primary group* is characterized by intimate, often face-to-face, association and cooperation. *Secondary groups* are characterized by relationships that involve few aspects of the personality; the members' reasons for participation are usually limited to a small number of goals. A group composed of only two people is a *dyad;* the addition of a third person forms a *triad* and reduces the stability of the group.

Territorial communities are contained within geographic boundaries and are usually composed of one or more neighborhoods. *Nonterritorial communities* are networks of associations formed around shared goals. A key factor in the formation of networks is in-group–out-group distinctions. Such distinctions are usually based on such qualities as income, race, or religion.

An important factor determining how people behave in a given instance is their definition of the situation. *Ethnomethodology* is the study of the underlying rules of behavior that guide group interaction. The *dramaturgical approach* views interaction as though it were taking place on a stage and unfolding in scenes.

Research on small groups has shown that they tend to develop two kinds of leaders: a "task leader" who keeps the group focused on its goals and a "socioemotional leader" who creates a positive emotional climate within the group.

Formal organizations have explicit sets of norms, statuses, and roles that specify each member's relationships to the others and the conditions under which those relationships hold. In a *voluntary association,* members pursue shared interests and arrive at decisions through some sort of democratic process. In a *bureaucracy,* members hold positions with clearly defined responsibilities, and those positions are ordered in a hierarchy.

One effect of the increasing dominance of bureaucracies in modern societies is the possibility that individuals will not take full responsibility for their actions. Commitment to bureaucratic organizations is greatest when it is supported by ideology or by strong primary-group attachments.

GLOSSARY

Society: A population that is organized in a cooperative manner to carry out the major functions of life. **(89)**

Social structure: The recurring patterns of behavior that create relationships among individuals and groups within a society. **(89)**

Group: A collection of people who interact with one another on the basis of shared expectations regarding the others' behavior. **(90)**

Status: A socially defined position in a group. **(90)**

Role: The way a society defines how an individual is to behave in a particular status. **(90)**

Role expectations: A society's expectations about how a role should be performed, together with the individual's perceptions of what is required in performing that role. **(90)**

Institution: A more or less stable structure of statuses and roles devoted to meeting the basic needs of people in a society. **(92)**

Differentiation: The processes whereby sets of social activities performed by one social institution are divided among different institutions. **(92)**

Pastoral society: A society whose primary means of subsistence is herding animals and moving them over a wide expanse of grazing land. **(94)**

Horticultural society: A society whose primary means of subsistence is raising crops, which it plants and cultivates, often developing an extensive system for watering the crops. **(94)**

Social stratification: The process whereby the members of a society are sorted into differen statuses. **(95)**

Open society: A society in which social mobility is possible for everyone. **(95)**

Closed society: A society in which social mobility does not exist. **(95)**

Capitalism: A system for organizing the production of goods and services that is based on markets, private property, and the business firm or company. **(97)**

State: A society's set of political structures. **(97)**

Nation-state: The largest territory within which a society's political structures can operate without having to face challenges to their sovereignty. **(97)**

Gemeinschaft: A term used to refer to the close, personal relationships of small groups and communities. **(98)**

Gesellschaft: A term used to refer to the well-organized but impersonal relationships among the members of modern societies. **(98)**

Role conflict: Conflict that occurs when in order to perform one role well a person must violate the expectations associated with another role. **(101)**

Role strain: Conflict that occurs when the expectations associated with a single role are contradictory. **(101)**

Ascribed status: A position or rank that is assigned to an individual at birth and cannot be changed. **(101)**

Achieved status: A position or rank that is earned through the efforts of the individual. **(101)**

Master status: A status that takes precedence over all of an individual's other statuses. **(101)**

Social group: A set of two or more individuals who share a sense of common identity and belonging and who interact on a regular basis. **(102)**

Primary group: A social group characterized by intimate, face-to-face associations. **(102)**

Secondary group: A social group whose members have a shared goal or purpose but are not bound together by strong emotional ties. **(102)**

Dyad: A group consisting of two people. **(103)**

Triad: a group consisting of three people. **(103)**

Community: A set of primary and secondary groups in which the individual carries out important life functions. **(103)**

Territorial community: A population that functions within a particular geographic area. **(103)**

Nonterritorial community: A network of relationships formed around shared goals. **(103)**

In-group: A social group to which an individual has a feeling of allegiance; usually, but not always, a primary group. **(106)**

Out-group: Any social group to which an individual does not have a feeling of allegiance; may be in competition or conflict with the in-group. **(106)**

Ethnomethodology: The study of the underlying rules of behavior that guide group interaction. **(107)**

Dramaturgical approach: An approach to research on interaction in groups that is based on the recognition that much social interaction depends on the desire to impress those who may be watching. **(107)**

Impression management: The strategies one uses to "set a stage" for one's own purposes. **(107)**

Informal organization: A group whose norms and statuses are generally agreed upon but are not set down in writing. **(109)**

Formal organization: A group that has an explicit, often written, set of norms, statuses, and roles that specify each member's relationships to the others and the conditions under which those relationships hold. **(109)**

Voluntary association: A formal organization whose members pursue shared interests and arrive at decisions through some sort of democratic process. **(109)**

Bureaucracy: A formal organization characterized by a clearly defined hierarchy with a commitment to rules, efficiency, and impersonality. **(110)**

QUESTIONS FOR THOUGHT AND DISCUSSION

1. Informal organizations almost always develop within the structure of statuses and obligations that define a formal organization. Can you give some examples of this? Can you explain why this is often a problem in organizations? How can knowledge of this tendency be useful to someone dealing with a large bureaucracy?

2. Explain what it means to say that democratic societies tend to place a higher value on achieved status than on ascribed status. Can you suggest some examples that support this point? Are there exceptions? (For example, does it help to be a member of the Kennedy family if one wishes to run for public office?)

DIGGING DEEPER

Books

The Division of Labor in Society (Émile Durkheim; Free Press, 1964). Durkheim's classic treatment of the impact of the rise of industrial urban societies. The famous preface to the second edition presents a succinct statement of how social change forever alters older forms of community.

Passions Within Reason: The Strategic Role of the Emotions (R. H. Frank; Norton, 1988). A highly readable treatment of a complex and often difficult subject: the relationships between emotion and social behavior. The emphasis is on rational behavior and its limits in the real world.

The Consequences of Modernity (Anthony Giddens; Stanford University Press, 1990). A thorough treatment of the elusive but centrally important concept of social structure.

The Presentation of Self in Everyday Life (Erving Goffman; Doubleday, 1959). One of the classics in the sociology of interaction at the micro level. A must-read for anyone wanting to gain greater insight into the dramaturgical approach to the analysis of social life.

Small Group Research: A Handbook (A. Paul Hare, Herbert H. Blumberg, Martin F. Davis, & Valerie Kent; Ablex, 1994). A comprehensive survey of classic and contemporary research on human behavior in small groups.

Big Structures, Large Processes, Huge Comparisons (Charles Tilly; Russell Sage, 1986). A highly readable account of the relationship among bureaucratic organizations, complex social structures, and social change.

The Rise of the West (William K. McNeill; University of Chicago Press, 1970). Not as slanted toward Western civilizations as its title may suggest, this masterful history of the world's major civilizations is a valuable source of material on the social structures and ideas that shaped the modern world.

The Great Transformation: The Political and Economic Origins of Our Time (Karl Polanyi; Beacon, 1944). Essential reading on the sociological implications of the shift to a market economy during the eighteenth and nineteenth centuries.

Journals

Administrative Science Quarterly. One of the principal journals in the field of organizations and behavior. A valuable source for theoretical and empirical studies of bureaucratic organizations.

Human Organization. Another good source of research on the problems and possibilities of efforts to improve human organizations. More diverse in its scope than ***Administrative Science Quarterly*** but also oriented toward economic organizations like firms and labor unions.

Other Sources

Historical Statistics of the U.S. from Colonial Times to 1970 (U.S. Bureau of the Census). An invaluable source of tables and charts showing changes in population, national origins, work, place of residence, and many other vital indicators of social change in the United States.

Statistical Yearbook (United Nations Educational, Scientific, and Cultural Organization [UNESCO]). Published annually, this is one of the best sources of comparative information about the world's nations and the trends occurring in them.

EXPLORING SOCIOLOGY ON THE INTERNET

Bureau of Labor Statistics
http://stats.bls.gov/blshome.html

Displays several pages and features on work and organizations in the U.S. economy, with links to international organizations like the International Labor Organization. Also provides links to many research reports and publications.

University of Michigan
www.umich.edu/asq

Offers many resources and links to other sites dealing with issues of social structure, bureaucracy, and administration.

U.S. Department of Labor
www.dol.gov

Offers a great deal of information and data about people in organizations, especially companies and corporations, as well as a link to the Economics and Statistics Administration, the source of much of the statistical, economic, and demographic information collected by the federal government.

The History Chart
www.hyperhistory.com/online n2/History n2/a.html

Offers a good schematic rendering of the history of the world from the perspective of human experience.

Sociological Timeline, University of Missouri
www.missouri.edu/~socbrent/timeline.htm

Provides a fascinating look at the development of sociology and its contributions to social change.

Society for the Study of Symbolic Interaction
www.sun.soci.niu.edu

A home page about symbolic interaction and the sociologists who use this method in their work, with links to related sites.

Ethnomethodology
www.bekkoame.or.ji

A highly creative Web site designed by an unofficial representative of the world of ethnomethodology, with many links to other fascinating sites.

Chapter 5

Deviance and Social Control

How do societies define what is deviant, determine who should be punished, and prevent deviance from threatening social stability?

Letizia Battaglia of Sicily felt that she had to do something about the outrageous behavior and criminal violence that surrounded her. Mafia violence threatened all the institutions of her society, from the family to the judicial system. Organized crime captains were in league with corrupt politicians. They were making real estate deals that were destroying many of the historic neighborhoods of her beloved city of Palermo. A skilled photographer and writer, Battaglia used these skills to document the murders. Her photos showed the devastating effects of organized crime violence on her city and on Sicilian society. Many of her photos were taken at the scenes of Mafia killings. Others, like that of the boy pretending to be a Mafia killer, revealed the widespread influence of crime and violence on the society's children and mothers. Always a social activist, Battaglia began appearing at public meetings to protest the destruction of historic parts of the city. Her work and her activism combined to form one of the most remarkable sociological explorations of crime and its effects available anywhere in the world.

This photo by Sicilian social activist Letizia Battaglia portrays a dramatic example of how some children are socialized into a deviant lifestyle.

Crime and corruption threaten the peace and the lives of masses of people in societies throughout the world. Sicily is not unique. In the nations of the former Soviet empire, crime is rampant. In many parts of Asia and Africa, collusion between politicians and criminals results in widespread theft, violence, and an almost complete breakdown of the rule of law. In the United States, crime is less of a danger to political and social stability than it is in nations like Sierra Leone or Russia, but the fact that we have over 1.8 million people in prison, among the highest incarceration rates in the industrialized world, shows that crime and punishment are important social issues here as well. And in the United States we cannot help but see that those who are poor and relatively powerless are far more likely to be judged and punished than people who can afford skilled legal counsel. This raises many questions about the meanings and definitions of deviance and crime in our own and other societies.

This chapter introduces the fascinating subjects of deviance, crime, and social control. After defining some of the key terms in the study of deviance, including criminal deviance, we turn to specific types of deviance and their impact on society. We then explore how different sociological perspectives explain deviance and crime. The final section deals with the social control of deviance and crime, which, as we will see, is not limited to control by the social institutions of police, courts, and prisons. ■

WHAT IS DEVIANCE?

Deviance, broadly defined, is behavior that violates the norms of a particular society. But because all of us violate norms to some degree at some time or other, we must distinguish between deviance and deviants. Deviance can be something as simple as dyeing one's hair purple or wearing outrageous clothing or becoming tipsy at an office party. Or it may be behavior over which the individual has little control, such as being homeless and living on the street, or it may consist of more strongly sanctioned departures from the society's norms—acts such as rape, mugging, and murder. Not all deviance is considered socially wrong, yet it can have negative effects for the individual. For example, "whistle blowers" who publicize illegal or harmful actions by their employers deviate from the norms of bureaucratic organizations and are often threatened with the loss of their jobs. Yet at the same time these employees benefit the public by calling attention to dangerous or illegal activities. As in the case of whistle blowers who are fired for reporting inconvenient truths, many deviant acts are defined as wrong simply because they offend people with enough power to define what is normative and what is deviant. In some Muslim countries, for example, a woman who has been raped is considered to have committed a deviant act and to have brought dishonor upon her family. She may be punished, even killed, with impunity by the men in her family, who are following the norms established by powerful Islamic religious leaders.

A deviant person is someone who violates or opposes a society's most valued norms, especially those that are valued by elite groups. Through such behavior, deviant individuals become disvalued people, and their disvalued behavior provokes hostile reactions (Davis, 1975; Goffman, 1963; Rubington & Weinberg, 1996; Sagarin, 1975). The word *deviant* therefore may be a label attached to a person or group. It may also refer to behavior that brings punishment to a person under certain conditions.

What are the conditions under which violations of norms are punished? This is an area of conduct in which there is often some uncertainty about what is legal and what is merely sleazy. Such questions reveal that deviance is not absolute. As sociologist Kai Erikson explains it, deviance "is not a property inherent in certain forms of behavior; it is a property conferred upon these forms by the audience which directly or indirectly witnesses them" (1962, p. 307). Some of us may believe that influence peddling is deviant, while others may believe that it is acceptable. Throughout this chapter and elsewhere in the text, we will see that power in social situations often explains more about who gets treated as deviant or criminal, and who gets punished, than the actual norms or formal laws of a society. Almost everywhere in the world there are people who have enough power to violate norms, even extremely strong ones against murder or major theft, without being labeled as deviant. There are also situations, such as the killing of people who appear to be threatening police officers or of gay people who are thought to be "coming on" to a hostile individual, in which murder can be officially defined as self-defense or merely a panic response. In understanding how such determinations are made, it is necessary to ask how social power has been used in defining the situation and outcomes of the controversy. Which of our views become the norm—and which are to be enforced through rewards and punishments is just as important as the behavior itself.

The ways in which a society prevents deviance and punishes deviants are known as **social control.** As we saw in Chapter 2, the norms of a culture, the means by which they are instilled in us through socialization, and the ways in which they are enforced in social institutions—the family, the schools, government agencies—establish a society's system of social control. In fact, social control can be thought of as all the ways in which a society establishes and enforces its cultural norms. It is "the capacity of a social group, including a whole society, to regulate itself" (Janowitz, 1978, p. 3).

The means used to prevent deviance and punish deviants are one dimension of social control. They include the police, prisons, mental hospitals, and other institutions responsible for applying social control, keeping order, and enforcing major norms. But if we had to rely entirely on official institutions to enforce norms, social order would probably be impossible to achieve (Bendle, 1999). In fact, the official institutions of social control deal mainly with the deviant individuals and groups that a society fears most. Less threatening forms of deviance are controlled through the everyday interactions of individuals, as when parents attempt to prevent their children from wearing their hair in the style of heavy metal or reggae musicians.

In offering graffiti artists a legitimate outlet for their work, authorities hope that they will voluntarily desist from defacing public property, making it less necessary to employ coercive forms of social control.

DIMENSIONS OF DEVIANCE

Deviance is an especially controversial topic: There is usually much disagreement not only about which behaviors are deviant but also about which behaviors should be strongly punished and which should be condoned or punished only mildly. The debate over whether abortion should be legal is a good example of such a disagreement. As Erikson noted, "Behavior which qualifies one man for prison may qualify another for sainthood since the quality of the act itself depends so much on the circumstances under which it was performed and the temper of the audience which witnessed it" (1966, pp. 5–6).

Consider, for example, former South African president Nelson Mandela. He was released from a maximum-security prison in 1990 after serving almost 30 years of a life sentence for his leadership of the movement to end apartheid, South Africa's racial caste system. The dominant white minority in South Africa regarded Mandela and other opponents of apartheid as criminals. But the black majority viewed him as a hero and a martyr; indeed, black South Africans revere Mandela the way people in the United States revere George Washington. Until the black majority began to mobilize world opinion and other nations began to enforce negative sanctions against South Africa, the white minority was able to define Mandela's opposition to apartheid as a criminal activity.

The power to define which acts are legal and which are illegal is an important dimension of deviance. In the case of black protest in South Africa, it is the single most important dimension. In the United States, the power of some groups to define certain acts as deviant helps explain why influence peddling is not a crime but cultivating marijuana is. The power to define what is deviant activity also plays a major role on the global stage. Despite years of extremely vigorous efforts to reduce the demand for illegal drugs in the United States, for example, Drug Enforcement Agency officials admit that both the demand and the supply remain high. Efforts to increase the cost of drugs through law enforcement seem to have little effect. Yet the United States continues to pressure Mexico, Colombia, and other Latin American nations to pursue antinarcotics programs even though these policies have contributed enormously to corruption and instability in those regions (Massing, 2000).

The fact that people in power or powerful nations can define deviant behavior and determine who is punished fails to explain differences in definitions of deviance in societies with different cultures.

Behavior that is regarded as normal in one society may be considered highly deviant in another. For example, in the United States the drinking of alcoholic beverages is considered normal, but in orthodox Islamic culture it is forbidden. Even within a society, members of certain social groups may behave in ways that are considered deviant by others. Thus, the official norms of the Catholic church do not permit the use of artificial methods of birth control, yet large numbers of Catholics use condoms, birth control pills, and other contraceptives. Similarly, the norms of Judaism require that the Sabbath be set aside for religious observance, yet many Jews work on the Sabbath and do not attend synagogue; in parts of Jerusalem they have occasionally been attacked by orthodox Jews for violating that norm. Clearly, differences in values are another important source of definitions of deviance and of disagreements about those definitions.

Another dimension of deviance has to do with attributes gained at birth (e.g., race or physical appearance) or other attributes that a person cannot control (e.g., having a convict as a parent), in contrast to actual behavior, which is usually voluntary. A criminal is deviant in ways that a mentally ill person is not, and it is criminal behavior that is most costly to society. Yet in many situations a mentally ill person is labeled as deviant, and this label may actually drive him or her toward criminality. A related issue concerns how people who deviate from generally accepted norms manage to survive in societies where they are considered outsiders. In fact, deviant people often form their own communities with their own norms and values, and these deviant subcultures sustain them in their conflicted relations with "normal" members of society.

An important point is that deviant subcultures, which engage in prostitution, gambling, drug use, and other "deviant" behaviors, could not exist were they not performing services and supplying products that people in the larger society secretly demand. It would be wise, therefore, not to draw the distinction between deviant and normal people too sharply. Many people deviate from one norm or another, and their deviations often create opportunities for others whose identities and occupations are already deviant.

In sum, three dimensions of social life—power, culture, and voluntary versus involuntary behavior—give rise to the major forces operating in any society to produce the forms of deviance that are typical of that society.

Deviance and Stigma

To narrow the range of phenomena we must deal with in discussing deviance, let us keep in mind Erving Goffman's (1963) distinction between stigma and deviance. "The term **stigma,**" Goffman stated, "refers to an attribute that is deeply discrediting" and that reduces the person "from a whole and usual person to a tainted and discounted one" (p. 3). People may be stigmatized because of mental illness, eccentricity, membership in a disvalued racial or nationality group, and the like. In some instances, their stigma is visible, as in the case of a disfigured person like the Elephant Man. Suffering from a disease that grossly distorted his face, the Elephant Man was rejected by society even though he was a highly intelligent person, and was treated badly because of the groundless fear that he aroused in others. In other cases, stigma is revealed only with growing acquaintance, as in the stigma attached to the children of convicts.

Stigma and Master Status. For the stigmatized person, the disqualifying trait defines the person's master status (Becker, 1963; Scull, 1988). (See Chapter 3.) A blind person, for example, may be an excellent musician and a caring parent, but the fact that he or she is blind will often outweigh these achieved statuses except in unusual cases like that of Stevie Wonder.

Stigmatized people deviate from some norm of "respectable" society, but they are not necessarily social deviants. The term deviant, Goffman argued, should be reserved for people "who are seen as declining voluntarily and openly to accept the social place accorded them, and who act irregularly and somewhat rebelliously in connection with our basic institutions" (1963, p. 143). The small group of students at Columbine High School in Littleton, Colorado, who were known as the "Trench Coat Mafia" were deviants in that they rejected the dominant norms of the school, which emphasized conventional success in academics and sports. They felt especially hostile toward the "jocks," who often teased them. Cliques are common in high schools, and when clique rivalries become heightened, group members may try to stigmatize members of other cliques by labeling them as "retards" or "faggots," for instance. The Trench Coat Mafia surely brought such stigma upon itself through its rejection of others. Members of the small clique were often picked on and harassed. They always felt as if they were on the outside looking in at the popular and successful students. "Most of the time, the members appeared to like it that way. As many cliques of young people do, the members played up their differentness. They wore army gear, black trench coats, and Nazi symbols. They spoke German to each other and were quite vocal about their fascination with Hitler and World War II" (Dority, 1999). But could anyone have imagined that two members of the disaffected clique, Dylan Klebold and Eric Harris, harbored such virulent hatred of other students that they would commit the ultimate of deviant acts, mass murder and suicide? This is a question that Littleton educators and parents will never be able to stop asking themselves. For professional social scientists, the tragedy emphasizes the importance of greater understanding of the processes whereby stigma and extreme deviance destroy lives and threaten communities.

According to the definition of stigma, the population of social deviants is smaller than that of stigmatized individuals; only some stigmatized behaviors are socially deviant. Deviant behaviors are characterized by denial of the social order through violation of the norms of permissible conduct. This point should be kept in mind as we continue our discussion of criminals and other people who are considered social deviants.

Deviance and Crime

Most of us, at some time in our lives, deviate from one or more norms. We double park or jaywalk or experiment with an illicit pleasure. We skip church services or disobey our parents or take far greater liberties with one or more of the Ten Commandments than we would ever admit publicly. The act of deviating does not make a person a deviant. Deviants are people who are socially recognized as continual violators of specific norms that people or groups with power define as important. Such norms are often the focus of political struggles. For example, homosexuals were considered criminal deviants in England and the United States until gay men and lesbian women began forcefully confronting the norms against homosexual behavior by insisting that their love is not deviant and that their relationships should be legitimated through legal sanctions (marriage) and economic rights (e.g., health insurance coverage for same-sex partners). The fact that some groups, such as fundamentalist Christians and orthodox Jews, continue to define homosexuality as deviant is further evidence that most norms are open to interpretation and change.

Much of the study of social deviance focuses on crime. **Crime** is usually defined as an act, or the omission of an act, for which the state can apply sanctions. Those sanctions are part of the criminal law—a set of written rules that prohibit certain acts and prescribe punishments to be meted out to violators (Kornblum & Julian, 2001). But the questions regarding which specific behaviors constitute crime and how the state should deal with these behaviors are often controversial.

In every society, there are some behaviors that almost everyone will agree are criminal and should be punished, and other behaviors that only some consider criminal. All societies punish murder and theft, for example, but there is far more variation in the treatment of adultery, prostitution, and pornography. Indeed, the largest number of "crimes" committed in the United States each year are so-called public order crimes such as public drunkenness, vagrancy, disorderly conduct, prostitution, gambling, drug addiction, and certain homosexual interactions. Many sociologists claim that these are victimless crimes because they generally cause no physical harm to anyone but the offenders themselves (Schur, 1973; Silberman, 1980). Not all social scientists agree with this view, however. Some point out that crimes like prostitution actually inflict damage on society because they are usually linked with an underworld that engages in far more serious and costly criminal activities (Wilson, 1977).

Measuring Crime. In the United States, crime is considered one of the nation's most serious social problems. The most serious, most frequently occurring, and most likely to be reported crimes are called index crimes by the Federal Bureau of Investigation because they are included in its crime index, a commonly used measure of crime rates. The FBI tabulates all the crimes reported to the thousands of municipal police jurisdictions throughout the nation. Its listing of trends in index crimes includes crimes against people—murder, forcible rape, aggravated assault, and robbery (in which the victim is confronted by the perpetrator)—and major crimes against property, which include burglary (mainly breaking and entering), larceny-theft (which includes many forms of embezzlement, check forging, etc.), auto theft, and arson (which is often committed to obtain insurance funds). While there are many other types of crime, these index crimes are those that the American public and law enforcement professionals feel are most threatening to public safety. Note, however, that the sale of illegal drugs, which accounts for almost half of the U.S. prison population, is not an FBI index crime.

It is important to understand that the FBI crime reports are highly problematic as estimates of actual crime rates. First, there may be differences in the quality of reporting by police administrations from one jurisdiction to another. Second, and most significant, the crimes enumerated are only those actually reported to the police. The numbers do not reflect crimes or crime attempts that go unreported.

To address the problem of unreported crime, the U.S. Department of Justice sponsors an annual victim survey in which large random samples of household members are asked about their experiences with crime. When the official crime reports are compared to the National Crime Victimization Survey, the crime rates in the latter are generally higher. This means that a great deal of crime (amounts vary for different types of crime) are unreported. Figure 5.1 shows that rates of violent crime have declined since 1973 and especially during the good economic times of the 1990s. But the figure also shows that the numbers of violent crimes reported to the police are lower than the numbers reported in the victim survey. Perhaps most significant, the number of actual arrests for violent crime is, on average, about half of the number of crimes recorded by the police.

Deviance and Changing Values

Definitions of crime and deviance are constantly changing as the society's values change. Almost every week, for example, there are reports in the media of men who have battered and even killed their wives, often after serving brief prison sentences for previous attacks on their spouses. Why, many ask, are law enforcement agencies unable to prevent and punish spouse abuse? Advocates of new laws and greater investment in prevention point out that until fairly recently a married woman was viewed as the husband's household possession. Although both husband and wife had sworn to "love, honor, and cherish" their partner, these norms were applied far more forcefully to women than to men. This double standard is changing as men and women become more equal before the law. In the minds of many judges and juries, however, this change is not yet complete, and men who are found guilty of abusing their spouses often receive relatively light sentences.

In any society, agreement on particular aspects of crime and deviance can range from weak (in cases in which there is much controversy) to strong (in cases in which there is little disagreement). Negative sanctions, or punishments, can also range from very weak to very strong. Capital punishment is the strongest sanction in the United States, with life imprisonment following close behind. Minor fines or the suggestion that a person undergo treatment for a behavior that is viewed as deviant are relatively weak sanctions. Nor are all sanctions are formal punishments meted out according to law. Some deviance can be punished by means of shunning or the "silent treatment," and milder infractions can be controlled by simply poking fun at the person in an attempt to change his or her behavior.

Table 5.1 uses these distinctions to construct a typology of deviance as it is generally viewed in the United States today. Table 5.2 suggests what such a typology would have looked like before the Civil War. Together, they highlight some of the continuities and changes in patterns of deviance in the United States over the past century. You will probably find points in them to argue with, which is further evidence of the difficulty of classifying deviant behaviors in a rapidly changing society.

FIGURE 5.1
Four Measures of Serious Violent Crime

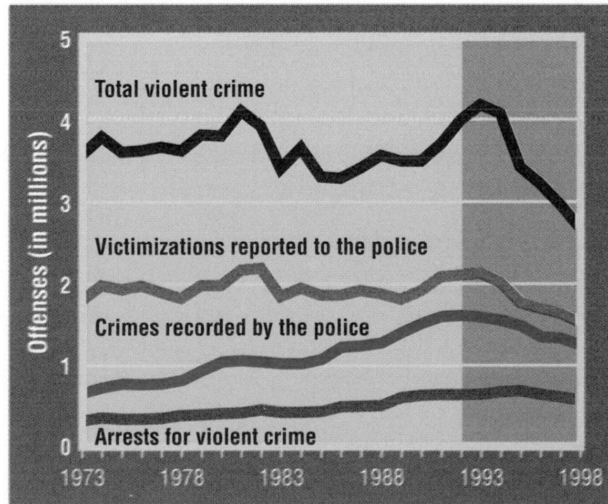

Source: Bureau of Justice Statistics, U.S. Department of Justice.

Deviant Subcultures and Markets

Even when the majority of the people in a society can be said to support a particular set of values, the society will contain many subcultures whose lifestyles are labeled deviant by the larger population. A deviant subculture includes a system of values, attitudes, behaviors, and lifestyles that are opposed to the dominant culture of the society in which it is found (Brake, 1980). The members of the subculture are also members of the larger society; they have families and friends outside the subculture with whom they share

TABLE 5.1
A Typology of Deviance in the United States

DEGREE OF CONSENSUS	Strength of Sanction: Weak	Strength of Sanction: Strong
Weak	• Recreational drugs • Homosexuality • Abortion	• Sale of whiskey during Prohibition • Prostitution • Abortion before 1973 Supreme Court ruling
Strong	• Schizophrenia • Driving while intoxicated • Public drunkenness • Corporate crime • Wife or child beating	• Major crimes (felonies) • Treason

TABLE 5.2
Deviance in the United States Before the Civil War

DEGREE OF CONSENSUS	Strength of Sanction: Weak	Strength of Sanction: Strong
Weak	• Killing Native Americans • Lynching African Americans • Prostitution • Wife or child beating	• Abortion • Indebtedness • Illegitimacy • Divorce • Adultery
Strong	• Political corruption • Corporate crime	• Major felonies against whites • Treason • Homosexuality

many values and norms. But within the subculture they may pursue values that are opposed to those of the larger or "mainstream" culture. Subcultures evolve their own rather insulated social worlds or communities, with local myths ("X is a slick dealer who once beat the tables in Vegas and was asked to leave the casino"), ways of measuring people's reputations ("he's a punk," "he's cool"), rituals and social routines, particular language or slang, formal and informal uniforms (e.g., the attire worn by bikers), and symbols of belonging such as tattoos (Simmons, 1985).

Subcultures of drug users and dealers provide many illustrations of the ways in which people who engage in deviant activities often develop their own vocabulary and norms. In his fascinating accounts of the subculture of crack dealers and crack-cocaine addicts, urban sociologist Terry Williams (1989, 1992) describes how everyone in the crack subculture—addicts, dealers, and hangers-on—used the language of the famous television series *Star Trek* in referring to the drug and its use. The drug itself was often called "Scotty," and smoking the drug was referred to as "beaming up."

Professional criminals who think of their illegal activities as occupations—including major drug dealers, bookmakers and gamblers, hired killers, and loan sharks—often become part of secret crime organizations. The best known of these (but hardly the only one) is the Mafia. Originally a Sicilian crime organization, the Mafia spread to the United States via Sicilian immigration in the late nineteenth century. Today, families associated with the Mafia are active in most large cities in North America and elsewhere and often extract millions of dollars from legitimate businesses in exchange for "protection." In recent years, there have been some successful crusades against the Mafia, but competing crime organizations are always ready to engage in lucrative criminal activities if the Mafia is weakened.

The subculture of organized crime is not unique to North America. In Japan, rates of homicide and mugging are small fractions of those in U.S. cities (Buruma, 1984), yet organized crime thrives there. The Yakuza, as the crime gangs are called, control many illegal businesses, just as they do in other countries, and they often resort to violence to discipline their members. As in the United States, organized crime in Japan is most directly associated with behaviors that are considered deviant, such as prostitution, gambling, and drug use, and a distinct subculture has emerged among those who engage in such behaviors.

Underground Markets. Many deviant subcultures flourish because they are closely related to enormous markets for illegal goods and services. Whenever a good or service that is considered valuable by some people is banned, there is the probability that an illegal "underground" market for that good or service, and the deviant behaviors associated with it, will flourish. Prohibition of whisky, wine, and beer in the United States early in the twentieth century produced huge markets for these goods that were largely controlled by criminal subcultures and greatly stimulated the rise of organized crime syndicates (Gusfield, 1987). When state governments raise taxes on tobacco to discourage smoking and generate revenues, opportunities for tobacco smuggling are created. Current prohibitions on sales of marijuana, opiates, and many other drugs do not necessarily decrease the demand for these substances, but they create the conditions for underground markets and the deviant subcultures (like prostitutes and pimps, or drug dealers and junkies) that are associated with them.

In sum, many deviant subcultures flourish because they provide opportunities for engaging in behavior that is pleasurable to some people but is considered deviant in "respectable" (conventional) society. Clearly, therefore, the line between what is normal and what is deviant is not nearly as distinct as one might believe from official descriptions of the norms of good conduct. On the other hand, one must recognize that membership in a deviant subculture, especially for those without money and power, often leads to exploitation and early death. In view of these problematic aspects of deviant subcultures, sociological and other explanations of deviant behavior take on special importance.

PERSPECTIVES ON DEVIANCE

Biological Explanations of Crime

Early in the twentieth century, the Italian criminologist Cesare Lombroso (1911) claimed to have proved that criminals were throwbacks to primitive, aggressive human types who could be recognized by physical features like a prominent forehead, shifty eyes, and red hair. Although Lombroso's theory has been thoroughly refuted by modern researchers, efforts to link body type with crime seem destined to reappear from time to time. In the 1940s, for example, the psychologist and physician William Sheldon announced that body type was correlated with crime. He believed that human beings could be classified into three types: ectomorphs, or thin people; endomorphs, or soft, fat people; and mesomorphs, or people with firm, well-defined muscles. The latter were most prone to crime, according to Sheldon, but he neglected to account for the possibility that mesomorphs simply might have led harder lives than less muscular individuals and hence were better equipped to commit crimes that require strength (Glueck & Glueck, 1950).

The more modern biological view of crime is represented in the influential work of social scientists

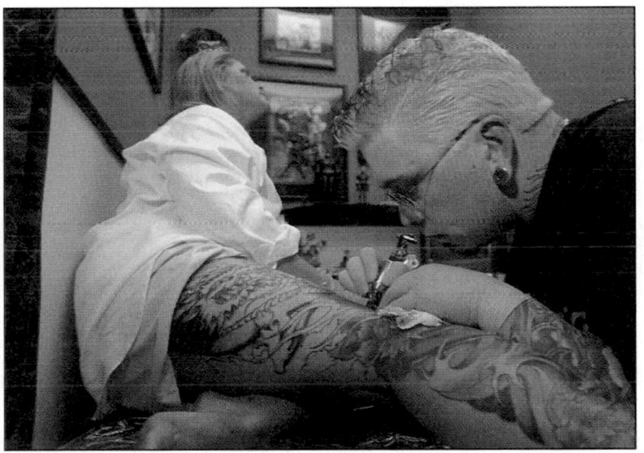

Tattoos are an example of how norms and sanctions shift with changes in culture. Early in the twentieth century, tattoos were considered somewhat exotic and deviant, a sign that one was a member of an outsider group such as bikers or prison inmates. Tattoos were long fashionable among sailors and other world travelers, who originally copied them from the peoples of the Pacific Islands. But a "proper" gentleman or lady never displayed a tattoo. Today, that norm is changing in Europe and North America as tattoos become part of what is considered fashionable adornment for young men and women.

THE CENTRAL QUESTION: A CRITICAL VIEW

How do societies define what is deviant, determine who should be punished, and prevent deviance from threatening social stability?

This central question about deviance helps organize our thinking about the subject as an academic discipline, but it can seem far removed from our own lives. Yet issues of deviance are never far from our own life experiences. Think about your own experiences in school and the way fellow students talked about each other. Undoubtedly there was gossip about "sluts" or "fatties" or "faggots," for example. Peers who acted in ways that violated norms of sexual conduct or desired body image were stigmatized as deviants. Anyone who has experienced such labeling and rejection will understand how powerful and lasting the experience of dealing with deviant labels can be.

Sociological approaches to deviance can help us deal with these issues. Take the problem of being overweight as an example. Being overweight is often treated as a form of deviance, especially when it invokes stigma and leads to sanctions such as discrimination against overweight people in hiring decisions. Obesity (as opposed to being somewhat overweight) is medically defined as a condition of body weight over 20% above normal, or as a body fat proportion above 30% for women and 25% for men (Bjorntorp, 1997). Obesity is a real danger to many people, for it puts them at risk for diabetes and other serious illnesses. Almost 30% of Americans are obese by this definition, the highest percentage in the urban industrial world. But over half of Americans, and an even greater proportion of women, believe that they are fat and that they deviate from a cultural ideal.

Few people can look like Brad Pitt or Julia Roberts. Today the average female model weighs 20% less than the average woman, and 85% of all American women weigh more than that model. It is likely that some of the very thin people, especially young women, you know are starving themselves to conform to an almost impossible cultural ideal of beauty, and yet it is the overweight person who is often seen as deviant and is blamed for his or her overindulgence. So how does a sociological perspective help us understand these issues?

First, sociologists point out that our culture bombards us with conflicting messages in the form of advertising, television shows, and other communications that exhort us to be thin and yet enjoy the bounty of the rich American diet, including double-scoop ice-cream cones, milk shakes, cakes and cookies, burgers and fries, thick steaks, beer, and potato chips. Few marketing messages in our consumer culture invite us to live moderate lives. Second, these cultural conflicts produce endless opportunities for making profits from dieting schemes, diet books, spas, and "fat farms." Extremely powerful corporations like Coca-Cola and McDonald's can

Richard J. Herrnstein and James Q. Wilson. Their research has led them to conclude that both biology and social environment play important roles in producing criminals. In commenting on a recent study by researchers at the U.S. Department of Justice reporting that more than one third of adult criminals in U.S. jails have at least one parent who is or was a criminal, Herrnstein argues that whatever determines criminality "is transmitted both genetically and environmentally. So kids brought up in criminal families get a double exposure" (quoted in Butterfield, 1992b, p. A16).

A study by Deborah W. Denno offers support for this view. In her research on 1,000 African American boys from poor neighborhoods of Philadelphia, Denno (1990) found that a disproportionate number of individuals who had become criminals had histories of childhood hyperactivity. This genetically transmitted condition makes it difficult for children to concentrate and succeed in school. It also can make it hard for them to hold jobs. Thus, Denno sees the causal chain as one of biological hyperactivity leading to social failure and then to crime.

But to many social scientists such studies are not convincing. Marvin E. Wolfgang, one of the nation's leading experts on criminality, points out that all the studies deal with people from poor neighborhoods,

spend almost unlimited funds on advertising that encourages heavy use of sugars and carbohydrates as "comfort foods" that Americans indulge in between meals.

A third sociological observation on the problems of deviant fatness points to a paradox. On a global scale, to be fatter than the norm is considered desirable and is almost always correlated with greater wealth. In China, for example, less than 1% of the population is overweight, and these people are almost always comparatively wealthy. In the United States, however, the reverse is true. Poorer, less-educated people are more likely to be overweight than wealthier and better-educated people. This is because poorer people tend to eat more fast foods, more starches (carbohydrates), more snacks between meals, and more comfort foods such as donuts and French fries, while wealthier people have more knowledge about nutrition and can afford to spend more on fish and salads and sorbets (Grunwald, 1995).

Does all this mean that to be overweight is not deviant? Does it mean that to be overweight is the fault of society and not an individual problem? In our society, we treat fatness, especially obesity, as a form of deviance, and this is yet another example of how society and culture create norms and, thus, deviance. The sociological point here is that it helps people as individuals to understand the larger social forces that together encourage the problem of fatness. The overweight person can often take steps to eat a more healthful diet and gradually lose weight. Our culture tends to favor extremes of eating and dieting, which are not especially helpful in addressing the overall problems that sociologists and other professionals describe. ■

where a high proportion of people will be sent to jail whether they are related or not and whether they were hyperactive as children or not. Hyperactivity is also found in children from wealthier families but is not so often associated with criminal careers because the families can afford special schools and programs to deal with the condition (Angier, 1995; Butterfield, 1992b).

As these criticisms by a leading criminologist suggest, there are many other, more sociological perspectives on why crime occurs, why some people commit crimes of various kinds, and why others are law-abiding citizens. It is to these major sociological perspectives that we now turn.

Social-Scientific Explanations of Deviance and Crime

The Functionalist Perspective. The influential American sociologist Robert K. Merton (1938) developed a typology of deviance based on how people adapt to the demands of their society. Merton's aim was to discover "how some social structures exert a definite pressure upon certain persons in society to engage in nonconformist rather than conformist conduct" (p. 672). "Among the elements of social and cultural structure," Merton continued, "two are important for our purposes. . . . The first consists of culturally defined goals, purposes, and interests. . . . The second . . . defines, regulates, and controls the acceptable modes of achieving these goals" (pp. 672–673).

Merton's explanation of deviance is based on Émile Durkheim's concept of **anomie,** or normlessness. In Merton's view, anomie results from the frustration and confusion people feel when what they have been taught to desire cannot be achieved by the legitimate means available to them. Merton believes that people in North America and other modern societies exhibit high levels of anomie because they are socialized to desire success in the form of material well-being and social prestige. For many, however, the means of attaining these culturally defined ends, such as hard work and saving, seem to be out of reach. This is especially true for people who experience rapid social changes, such as the closing of factories, and find themselves deprived of opportunities to attain what they have come to expect from life.

Merton's theory of deviance is a functionalist theory because it argues that for some members of society the conventional matching of socially approved goals and means of achieving these goals does not function satisfactorily. Their ability to achieve their goals is blocked. As a result, they feel various forms of pressure to either modify their goals or resort to deviant means of achieving conventional goals.

Consider some examples. Possession of money is a culturally defined goal. Work is a socially acceptable mode of achieving that goal; theft is not. Mating is also a culturally defined goal. Courtship and seduction are acceptable means of achieving it; kidnap and rape are not. But if theft and rape are unacceptable means of achieving culturally defined goals, why do they exist? According to Merton, the gap between culturally defined goals and acceptable means of achieving them causes feelings of anomie, which in turn make some people more likely to choose deviant strategies of various kinds.

Through socialization, we learn the goals and acceptable means of our society. Most of us would love to be rich or powerful or famous. We accept these goals of our culture. We also accept the legitimate means of achieving them: education, work, the electoral process, plastic surgery, acting school, and so on. We are conformists. But not everyone accepts either the cultural goals or the accepted means of achieving them. Some become "innovators" in that they explore (and often step over) the frontiers of acceptable goal-seeking behavior; others (e.g., hoboes) retreat into a life that rejects both the goals and the accepted means; some rebel and seek to change the goals and the institutions that support them; and still others reject the quest for these precious values while carrying out the rituals of social institutions. (In many bureaucracies, for example, one can find ritualists who have given up the quest for promotion yet insist on receiving deference from people below them in the hierarchy.) Merton's typology thus is based on whether people accept either the cultural goals of their society, the acceptable means of achieving them, or neither. Table 5.3 presents this typology, and Table 5.4 gives examples of all the deviant types in Merton's framework.

You may be wondering what there is in Merton's theory that explains why some poor people resort to crime while so many others do not. The answer lies in the way people who experience anomie gravitate toward criminal subcultures. For example, adolescents who choose to steal to obtain things that their parents cannot afford must learn new norms—they must learn how to steal successfully and must receive some approval from peers for their conduct (Cloward & Ohlin, 1960). As a result, they drift toward deviant peer groups. But this explanation presents a problem for the functionalist perspective: If there are deviant subcultures, the idea that there is a single culture whose goals and means are shared by all members of society is called into question. The presence of different subcultures within a society suggests another possible cause of deviant behaviors: conflict between groups.

Conflict Perspectives. Social-scientific explanations of crime and deviance that take a conflict perspective usually reflect one of two viewpoints. They look at deviance in terms of cultural conflict or as an extension of Marxist conflict theory.

In an influential essay, "Crime as an American Way of Life," Daniel Bell (1962) observed that at the turn of the twentieth century cultural conflict developed between the "Big City and the small-town conscience. Crime as a growing business was fed by the revenues from prostitution, liquor, and gambling that a wide-open urban society encouraged and that a middle-class Protestant ethos tried to suppress with a ferocity unmatched in any other civilized society" (p. 128). This example of conflict between the official morality of the dominant American culture and the norms of subcultures that do not condemn gambling, drinking, prostitution, and the like, encouraged the growth of criminal organizations. In the United States as in many other societies, such organizations thrive by supplying the needs of millions of people in cities,

TABLE 5.3
The Merton Typology—Modes of Adaptation by Individuals Within the Society or Group

Merton explained that (+) signifies "acceptance," (–) signifies "elimination," and (+/–) signifies "rejection and substitution of new goals and standards." The line separating rebellion from the other roles signifies that the individual no longer accepts the society's culture and structure; other individuals, although they may deviate, continue to accept the society's culture and structure.

Mode of Adaptation	Cultural Goals	Institutional Means
Conformity	+	+
Innovation	+	–
Ritualism	–	+
Retreatism	–	–
Rebellion	+/–	+/–

towns, and rural areas who appear to be law-abiding citizens yet engage in certain deviant behaviors.

The prohibition of alcoholic beverages in the United States from 1919 to 1933 is an example of how cultural conflict can lead to situations that encourage criminal activity. Prohibition has been interpreted as an effort by the nation's largely Protestant lawmakers to impose their version of morality on immigrant groups for whom consumption of alcohol was an important part of social life. Once they were passed, however, laws against the production and sale of alcoholic beverages created opportunities for illegal production, bootlegging (smuggling), and illegal drinking establishments (speakeasies). These in turn supported the rise of organized crime syndicates. The current laws against drugs like marijuana and cocaine have similar effects in that they lead to the clandestine production and supply of these illegal but widely used substances. Similar opportunities to make illegal profits arise when states pass laws raising the drinking age (Gusfield, 1966; Kornblum & Julian, 1998).

For Marxian students of social deviance, however, the cultural-conflict explanation is inadequate because it does not take into account the effects of power and class conflict. Marxian sociologists believe that situations like Prohibition do not occur just because of cultural conflict. They happen because the powerful classes in society (i.e., those who own and control the means of production) wish to control the working class and the poor so that they will produce more. From the Marxian perspective, as criminologist Richard Quinney (1980) points out, "crime is to be understood in terms of the development of capitalism" (p. 41), as it was in Marx's original analysis. From this perspective, most crime is essentially a form of class conflict—either the have-nots taking what they can from the ruling class, or the rich and their agents somehow taking what they can from the poor.

The economic "robber barons"—John Jacob Astor, John D. Rockefeller, J. P. Morgan, Leland Stanford, Andrew Carnegie, and many others—amassed huge fortunes in the period of booming industrial growth following the Civil War. But they often resorted to illegal means in pursuing the culturally approved goal of great wealth. Among other tactics, they used violence to drive settlers off land they had purchased or to break

TABLE 5.4
Examples of Social Roles Based on the Merton Typology

These examples are meant to illustrate possible outcomes, not stereotypes. Perceptions of approved goals and means may vary. In some cases, for example, the very rich, who feel that they could be even richer were it not for legal obstacles in their path, may choose to bend the rules. Also, a number of cultural values besides wealth shape the likelihood that a member of a given class will behave as he or she does.

Mode of Adaptation	The Poor	The Middle Class	The Rich
Conformity	The working poor	The suburban family	The wealthy civic leader
Innovation	The mugger	The embezzler	The stock manipulator
Ritualism	The chronic welfare recipient	The resigned bureaucrat	The hedonist
Retreatism	The wino or junkie	The skidding alcoholic	The bohemian
Rebellion	The bandit	The anarchist	The racist

strikes by workers, and they were not above manipulating prices in order to drive out competitors and monopolize the markets for steel, oil, coal, precious metals, food products, and numerous other goods. In Merton's typology of deviance, their actions would classify them as "innovators" (see Table 5.4), but from a Marxian viewpoint, they were merely carrying out "the logic of capitalism," which was based on the exploitation of the poor by the rich and powerful.

Marxian students of deviance point out that legal definitions of deviant behavior usually depend on the ability of the more powerful members of society to impose their will on the government and to protect their actions from legal sanctions. Thus, the crimes of the robber barons almost always went unpunished. Definitions of what is criminal and who should be punished are generally applied more forcefully to the poor and the working class (Bendle, 1999; Quinney, 1978).

Marx and his collaborator Friedrich Engels recognized that the working class (or *proletariat,* as they called it) would resort to individual crimes like robbery when driven to do so by unemployment and poverty, but they believed that the workers would be more likely to form associations aimed at destroying capitalism. The chronic poor, on the other hand, would form a class that Marx called the *lumpenproletariat,* people who were unable to get jobs in the capitalist system or were cast off for not working hard enough or for being injured or sick. Marx did not believe that members of this class would join forces with the proletariat. Instead, they would act as spies, informers, and thugs whose services could be purchased by the rich to be used against the workers. Marx agreed with other thinkers of his time who called the lumpenproletariat the "dangerous class" created by capitalism; from its ranks came thieves, prostitutes, gamblers, pickpockets, con artists, and contract murderers.

It can be helpful to think of crime and deviance as symptoms of the class struggles that occur in any society and to show that laws that define and punish criminal behaviors are often imposed by the powerful on those with less power. But to attribute crime as we know it to the workings of capitalism is to suggest that if capitalism were abolished, crime would vanish. This clearly is not the case. In China, for example, there are thousands of prisoners in a vast chain of prison camps. Many of those prisoners were convicted of the deviant act of opposing the dominant regime (Ning, 1995; Wu & Wakeman, 1995). And in Cuba homosexuality is considered a serious crime and is severely punished because the society's leaders think of it as an offense against masculinity and a symbol of a "decadent" subculture. As these examples indicate, societies with Marxian ideologies are hardly free from their own forms of deviance and crime.

The Interactionist Perspective. Functionalist theories explain deviance as a reaction to social dysfunction; conflict theories explain it as a product of deviant subcultures or of the type of class struggle that occurs in a society in a particular historical period. Neither of these approaches accounts very well for the issues of recruitment and production. *Recruitment,* in this context, refers to the reasons why some people become deviants while others in the same social situation do not. *Production* refers to the creation of new categories of deviance in a society. These processes are reflected in the interactionist perspective.

One aspect of the recruitment process—that of **differential association**—was addressed by the sociologist and criminologist Edwin H. Sutherland. In 1940, Sutherland published a paper, "White Collar Criminality," in which he argued that official crime statistics do not measure the many forms of crime that are not correlated with poverty. Outstanding among these are *white-collar crimes*—that is, the criminal behavior of people in business and professional positions. In Sutherland's words:

> White-collar criminality in business is expressed most frequently in the form of misrepresentation in financial statements of corporations, manipulation in stock exchange, commercial bribery, bribery of public officials directly or indirectly in order to secure favorable contracts and legislation, misrepresentation in advertising and salesmanship, embezzlement and misapplication of funds, short weights and measures and misgrading of commodities, tax frauds, misapplication of funds in receiverships and bankruptcies. These are what Al Capone called "the legitimate rackets." These and many others are found in abundance in the business world. (1940, pp. 2–3)

Sutherland was pointing out that an accurate statistical comparison of the crimes committed by the rich and the poor was not available. But his paper on white-collar crime also set forth an interactionist theory of crime and deviance:

> White-collar criminality, just as other systematic criminality, . . . is learned in direct or indirect association with those who already practice the behavior; and . . . those who learn this criminal

> behavior are segregated from frequent and intimate contacts with law-abiding behavior. Whether a person becomes a criminal or not is determined largely by the comparative frequency and intimacy of his contacts with the two types of behavior. This may be called the process of differential association. (pp. 10–11)

The concept of differential association offers an answer to some of the weaknesses of functionalist and conflict theories. Not only does it account for the prevalence of deviance in all social classes, but it also provides clues to how crime is learned in groups that are culturally distinct from the dominant society. For example, in the 1920s sociologist Clifford Shaw (1929) observed that some Chicago neighborhoods had consistently higher rates of juvenile delinquency than other neighborhoods. These were immigrant neighborhoods, but their high rates of delinquency persisted regardless of which immigrant groups lived there at any given time. Sutherland's theory explained this pattern by calling attention to the culture of deviance that had become part of the way of life of teenagers in those neighborhoods. According to Sutherland, the teenagers became delinquent because they interacted in groups whose culture legitimated crime. It was not a matter of teenage delinquents deviating from conventional norms because the approved means of achieving approved goals were closed to them. Rather, they acted as they did because the culture of their peer group made crime an acceptable means of achieving desired goals.

In an empirical study that tested Sutherland's theory, Walter Miller (1958) found that delinquency in areas with high rates of juvenile crime was in fact supported by the norms of lower-class teenage peer groups. In three years of careful observation, Miller found that delinquent groups had a set of well-defined values: trouble, toughness, smartness, excitement, fate, and autonomy. Whereas other groups felt that it was important to stay out of trouble, the delinquent groups viewed trouble—meaning fighting, drinking, and sexual adventures—as something to brag about, as long as they didn't get caught. Toughness, as shown by physical prowess or fearlessness; smartness, as evidenced by the ability to con or outsmart gullible "marks"; the excitement to be found in risking danger successfully; one's fate as demonstrated by luck or good fortune in avoiding capture; and the autonomy that crime seemed to provide in the form of independence from authorities—all were values of delinquent groups that distinguished them from nondelinquent groups in the same neighborhoods.

Remember, however, that not all deviants are people whose means of achieving success have been blocked or who are acting out some form of class struggle or have associated with a deviant group. For example, many alcoholics and drug users are not thought of as deviant, either because their behavior is not considered serious or because it is not witnessed by other people. From an interactionist perspective, the key question about such people is how their behavior is understood by others. The central concepts that attempt to answer this question are *labeling* and the idea of the *deviant career.*

According to symbolic-interactionist theory, deviance is produced by a process known as **labeling,** meaning a societal reaction to certain behaviors that labels the offender as a deviant. Most often, labeling is done by official agents of social control like the police, the courts, mental institutions, and schools, which distribute labels that stick, often for life, such as "troublemaker," "hustler," "kook," or "blockhead" (Becker, 1963; Gusfield, 1981; Kitsuse, 1962; Rubington & Weinberg, 1996; Schur, 1984). Following is Howard Becker's explanation of what occurs in the labeling process:

> Social groups create deviance by making the rules whose infraction constitutes deviance and by applying those rules to particular people and labeling them as outsiders. From this point of view, deviance is not a quality of the act the person commits, but rather a consequence of the application by others of rules and sanctions to an "offender." The deviant is one to whom that label has been successfully applied; deviant behavior is behavior that people so label. (1963, p. 9)

In a famous experiment that tested the effects of labeling, David Rosenhan (1973) and eight other researchers were admitted to a mental hospital after pretending to "hear voices." Each of these pseudopatients was diagnosed as schizophrenic. Before long, many of the real patients with whom the pseudopatients associated recognized them as normal, but the doctors continued to think of them as schizophrenic. As the pseudopatients waited in the lunch line, for example, they were said to be exhibiting "oral-acquisitive behavior." Gradually the pseudopatients were released with the diagnosis of schizophrenia in remission, but none was ever thought to be cured.

Rosenhan and his researchers also observed that not only did the diagnosis of schizophrenia label the patient for life, but the label itself became a justification for other forms of mistreatment. The doctors and hospital staff disregarded the patients' opinions, treated them as incompetent, and often punished them for infractions of minor rules. The hospital's social atmosphere was based on the powerlessness of the people who were labeled mentally ill.

Rosenhan's study accelerated the movement to reform mental institutions and to deinstitutionalize as many mental patients as possible. But studies of deinstitutionalized mental patients (many of whom are homeless), together with research on the problems of released convicts, have shown that the labels attached to people who have deviated become incorporated into their definitions of themselves as deviant. In this way, labeling at some stage of a person's development tends to steer that person into a community of other deviants, where he or she may become trapped in a deviant career (Bassuk, 1984).

In a classic empirical study of youth gangs, William J. Chambliss (1973) applied both the conflict and interactionist perspectives. For two years, Chambliss observed the Saints, a gang of boys from rather wealthy families, and the Roughnecks, a gang of poor boys from the same community. Both gangs engaged in car theft and joyriding, vandalism, dangerous practical jokes, and fighting; in fact, the Saints were involved in a larger overall number of incidents. But the Roughnecks were more frequently caught by authorities, described as "tough young criminals headed for trouble," and sent to reform school. The wealthier youths were rarely caught and were never labeled delinquent.

Chambliss observed that the Saints had access to cars and could commit their misdeeds in other communities, where they were not known. The Roughnecks hung out in their own community and performed many of their antisocial acts there. A far more important explanation, according to Chambliss, pertained to the relative influence of the boys' parents. The parents of the Saints argued that the boys' activities were normal youthful behavior; the boys were just "sowing their wild oats." The parents' social position enabled them to influence the way their children's behavior was perceived, an influence the parents of the Roughnecks did not have. Thus, the Roughnecks were caught and labeled and became increasingly committed to deviant careers, but the Saints escaped without serious sanction.

Reasonable as the labeling perspective appears, it has not always been borne out by empirical research. Some studies have found that people who have been labeled delinquent after being caught and convicted of serious offenses go on to commit other deviant acts, but other studies have found that labeling can lead to decreased deviance and a lower probability of further offenses (Rubington & Weinberg, 1996). (The various theoretical approaches to crime that we have discussed are summarized in the study chart on page 139.)

CRIME AND SOCIAL CONTROL

Sociologists often distinguish between the voluntary and coercive aspects of social control. Voluntary social control results from individual self-control and the ability of groups to convince their members of the goodness of their values and norms. Coercive social control is based on the use or threat of force (Janowitz, 1991). Every society depends on some mix of voluntary and coercive social control, but the more agreement there is among its members about values and norms, the more social control of behavior tends to be voluntary. Indeed, the vast majority of our actions are guided by the desire to behave in socially appropriate—that is, normative—ways.

At the beginning of the chapter we defined social control in broad terms as all the ways in which a society establishes and enforces its cultural norms. Certainly it is true that without socialization and the controlling actions of social groups like the family, schools, the military, and corporations there would be much more anomie, crime, and violence. Thus, social control can be voluntary as well as coercive. When neighborhoods organize to prevent crime, this is an example of voluntary social control of crime. When local, state, or national governments invoke laws and use their police powers to combat crime, they are resorting to more coercive forms of social control. In considering a society's means of controlling crime, sociologists also study the interactions between voluntary and coercive social control. Critical sociologists point out that the tendency to invoke police power against juvenile crimes, such as underage drinking or graffiti, may actually usurp less coercive forms of social control, so that, for example, neighbors become dependent on police patrols and fail to develop voluntary measures as well (O'Malley, 1996).

When the United States consisted mainly of small agrarian communities, social control of deviance and crime was carried out by the local institutions of the family or the church. Parents, for example, were expected to control their children; if they did not, they

STUDY CHART

Theoretical Perspectives on Deviance

Perspective	Description	Critique
Biological Theories	Crime and other forms of deviance are genetically determined.	There is evidence of effects of socialization.
Sociological Theories		
Functionalism	Deviance and crime result from the failure of social structures to function properly. Every society produces its own forms of deviance.	Does not explain why some people drift into deviant subcultures.
Conflict Theories		
Cultural conflict	Cultural conflict creates opportunities for deviance and criminal gain in deviant subcultures (e.g., prohibitions create opportunities for organized crime).	Explains a narrow range of phenomena.
Marxian theory	Capitalism produces poor and powerless masses who may resort to crime to survive. The rich employ their own agents to break laws and enhance their power and wealth.	Crime exists in societies that have sought to eliminate capitalism.
Interactionist Theories		
Differential association	Criminal careers result from recruitment into crime groups based on association and interaction with criminals.	Excellent explanation of recruitment but not as effective in explaining deviant careers.
Labeling	Deviance is created by groups that have the power to attach labels to others, marking particular people as outsiders. It is extremely difficult to shed a label once it has been acquired, and the labeled person tends to behave in the expected manner.	Not always supported by empirical evidence. People can also use labels (e.g., drunk) to change their behavior.

would lose the respect of other members of the community. But as the size, complexity, and diversity of societies increase, the ability of local institutions to control all of the society's members is diminished (Wirth, 1968/1938). Societies therefore develop specialized, more or less coercive institutions to deal with deviants. Courts, prisons, police forces, and social-welfare agencies grow as the influence of the community on the behavior of its members declines.

Capital Punishment—The Global Debate

Capital punishment provides an illustration of the issues raised by the use of coercive forms of social control. Only recently (during the twentieth century in some nations) has the death sentence come to be thought of as barbaric. In simpler societies and earlier civilizations, execution was more than a penalty carried out on an individual; it was also an occasion for a public ceremony. People attended beheadings and hangings, and hawkers sold them food and favors as they waited for the bloody pageant. The villain's death reaffirmed their common values, their solidarity as a people who could purge evil elements from their midst.

Today, by contrast, the value of capital punishment is a matter of heated debate throughout the world. In 1998, 1,625 prisoners were executed in 37 countries, four of which—China, the Democratic Republic of the Congo, the United States, and Iran—accounted for 80% of the total. All of the European nations have abolished capital punishment, and Russia has placed a moratorium on executions. Many countries decided to suspend or abolish capital punishment as a way of turning over a new leaf after long periods of political repression. This is true of South Africa, the former communist countries of eastern

Europe, and several Latin American nations (Argentina, Brazil, Peru, Nicaragua, and El Salvador; Jacot, 1999).

The United States differs from other advanced, urban industrial nations in that it has an extremely high rate of executions. But this is a relatively recent change. While most states in the United States were conducting executions during most of the twentieth century, there was mounting evidence from social-scientific research that challenged the fairness of the executions. Incidents of wrongful execution, along with clear evidence that poor people and minorities were far more likely to be executed than whites from more advantaged backgrounds, convinced the U.S. Supreme Court that capital punishment could be considered "cruel and unusual punishment" according to the Constitution (Radelet & Bedau, 1992; Wolfgang & Reidel, 1973). In 1972, in a five-to-four decision, the Court ruled that in the absence of clear specifications for when it might be used, the death sentence violated the Eighth Amendment. But Congress has since passed new legislation that legalized the death sentence, and the Court has not overruled it. As a result, capital punishment has been reinstated in many states; there are now between 15 and 30 executions a year in the United States, with Texas, Florida, Mississippi, and Arizona leading in numbers of executions carried out. Indeed, in many states the debate about capital punishment has shifted away from its morality to questions about technology, such as the use of lethal injections versus the electric chair.

But does capital punishment have the effect of deterring people from committing murder, as its advocates claim? Much of the evidence on this subject is negative. Little encouragement is found in comparisons among states that have the death penalty, states that have had it with interruptions, and states that have never had it. For example, despite its highly visible and frequent use of capital punishment, the murder rate in Texas is almost 7 per 100,000 people, while the rate in Massachusetts, which has not had capital punishment, is 2 per 100,000 people (*Statistical Abstract,* 2000). Yet in interviews with criminals charged with robbery, James Q. Wilson (1977) found evidence that fear of the death penalty discouraged them from carrying guns. There is still considerable opposition to the death penalty among people who feel that it represents cruel and unusual punishment and that innocent people are sometimes executed, but surveys indicate that the tide of public opinion has turned toward support for it. These surveys also show, for better or worse, that the public's increased concern about crime has led to greater emphasis on use of the death sentence as retribution for murders that have already been committed and less emphasis on the possible deterrent effect of capital punishment. (The Mapping Social Change box on page 141 compares homicide rates in the United States and other nations.)

The Boom in Prison Populations

Prisons are the social institutions with the primary responsibility for dealing with criminals. In recent years, the population of American men and women serving prison terms, on probation or parole, or awaiting sentencing has exploded to an extraordinary record high of over 5 million. Research on the causes of this prison population boom shows that drugs or alcohol led to the incarceration of four out of five inmates (Wren, 1998). The "war on drugs" campaign has resulted in an unprecedented use of coercive social control. (See the Global Social Change box on page 142.) The conservative criminologist John DiIulio has conducted research on how many prisoners are convicted for "drug only" offenses and has found that these prisoners account for at least 25% of the total. These offenders have never been charged with any other type of crime but are often serving long prison sentences and doing "hard time" (Butterfield, 1997). If current trends persist, sociologists estimate that within a decade more than 7.5 million people will be under some form of law enforcement surveillance in the United States (Kornblum & Julian, 2001).

The arguments used to support reliance on prison as the central institution of punishment and reform are mainly functionalist in nature. The functions of prisons are said to be deterrence, rehabilitation, and retribution (i.e., punishment; Goode, 1994; Hawkins, 1976). As Bruce Jackson, a well-known student of prison life, explains it, a prison is supposed to deter criminals from committing crimes ("its presence is supposed to keep those among us of weak moral strength from actions we might otherwise commit"); it is supposed to rehabilitate those who do commit crimes ("within its walls those who have, for whatever reason, transgressed society's norms are presumably shown the error of their ways and retooled so they can live outside in a more acceptable and satisfactory fashion"); and it is supposed to punish criminals—the only function that it clearly fulfills (1972, p. 248).

MAPPING SOCIAL CHANGE

Worldwide Homicide Rates

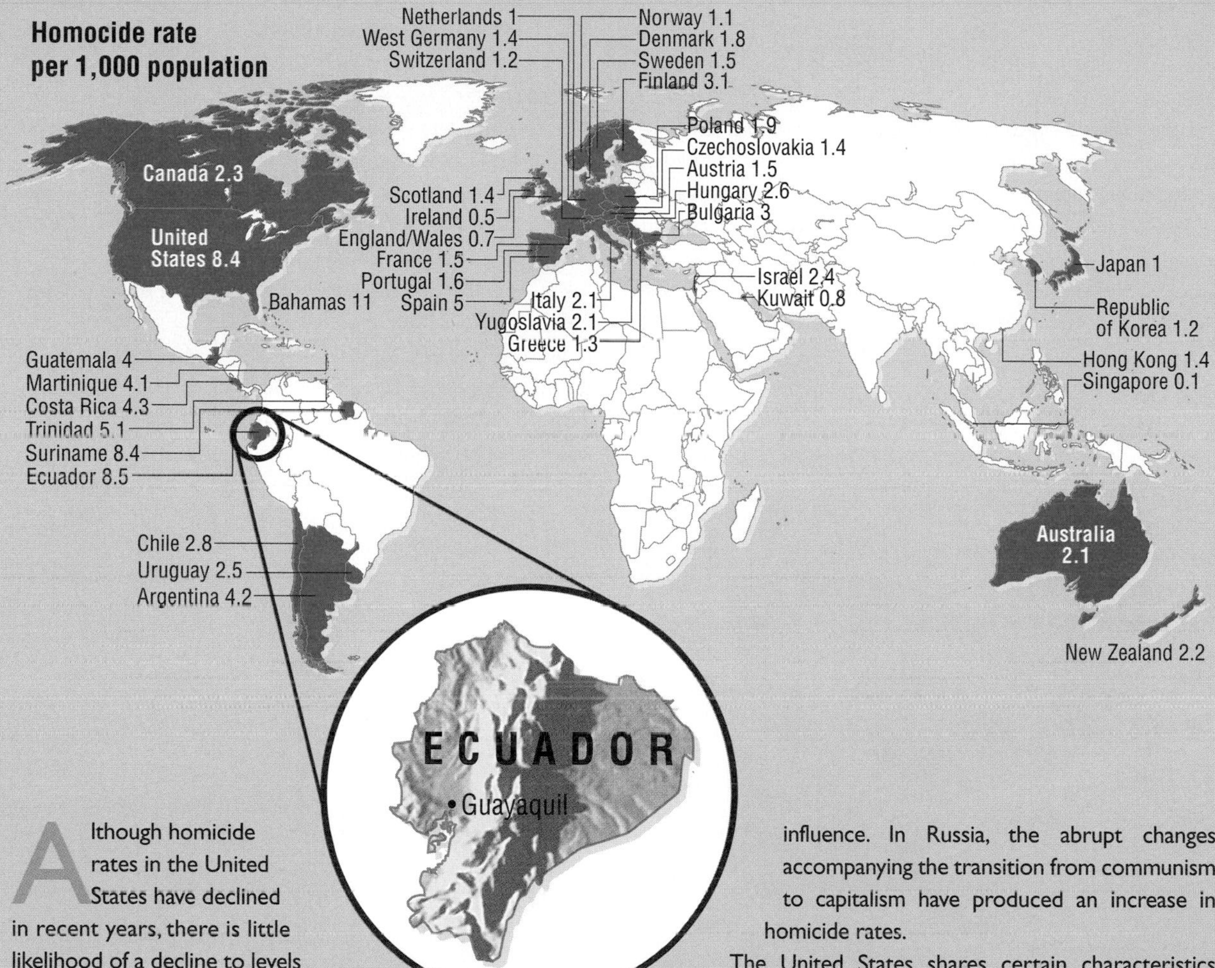

Although homicide rates in the United States have declined in recent years, there is little likelihood of a decline to levels characteristic of the European democracies. There are nations with higher homicide rates, such as Ecuador, the Bahamas, and perhaps Russia. All are undergoing rapid social change, but none enjoy the degree of affluence found in the United States. In Ecuador, high homicide rates are associated with rapid urbanization in cities like Guayaquil, where murders occur at extremely high rates in sprawling slums. In the Bahamas, the international drug markets, which often include stops in this island nation, are an important influence. In Russia, the abrupt changes accompanying the transition from communism to capitalism have produced an increase in homicide rates.

The United States shares certain characteristics with each of these nations. Like Ecuador and other Latin American nations, it has a frontier tradition that fosters and still maintains norms of rugged individualism, including the use of guns. It also has a widening gap between rich and poor, with the poor often concentrated in ghetto slums. Drug activity contributes significantly to homicide rates, as in the Bahamas. And as in Russia, much of the deadly violence occurring in the United States stems from the activities of criminals in organized gangs.

GLOBAL SOCIAL CHANGE

Impacts of the U.S. War on Drugs

During the summer of 2000, the United States and Colombia agreed on a $1.3 billion cooperative program to address the problem of Colombian narcotics cultivation and the massive traffic of cocaine and heroin into the United States. Faced with civil war and the threat of growing anarchy within its borders, the Colombian government reluctantly agreed that in return for large increases in U.S. aid and military advice it would crack down more strenuously on drug cultivators and on the guerrillas who depend on help from the drug lords to continue their decades-long struggle against the government (Golden, 2000).

The Colombian government also agreed to take the first steps toward testing the use of a powerful fungus that kills the coca and poppy plants that produce the substances used in making the drugs. The fungus was developed in the former Soviet Union when that nation was at war with Afghanistan and wished to destroy the opium poppy economy that helped support their opponents' war efforts. But the fungus may have unknown side effects on animals and other crops. Environmentalists in both the United States and Colombia strenuously oppose the policy.

One consequence of the American war on drugs is increased surveillance and searches at border crossings, in airports, and elsewhere.

U.S. intervention in Colombian affairs is only one of many examples of the global impact of the war on drugs. Mexico, Peru, Thailand, Cambodia, Turkey, and other nations that are substantial narcotics suppliers have all experienced the force of the war on drugs. Quite frequently, efforts to combat the cultivation and sale of drugs in poorer nations also have implicated some national leaders, who cannot resist the temptation to dip their own hands into the huge flow of money associated with the traffic in drugs.

The larger sociological point here is that drug prohibitions in rich nations create major and almost irresistible opportunities for profit in poorer nations. Campaigns against cultivators and profiteers can destabilize these nations, but they can also, as in the case of Mexico, contribute to greater resolve to combat corruption and crime. In 2000, Mexicans elected a president from an opposition party after decades of one-party rule, largely in the hope that the new president would address the serious problem of government corruption. The latter is often stimulated by the involvement of government officials in the Mexican–U.S. drug trade (Dillon, 2000). ■

FIGURE 5.2
Prisoners in Need of Treatment

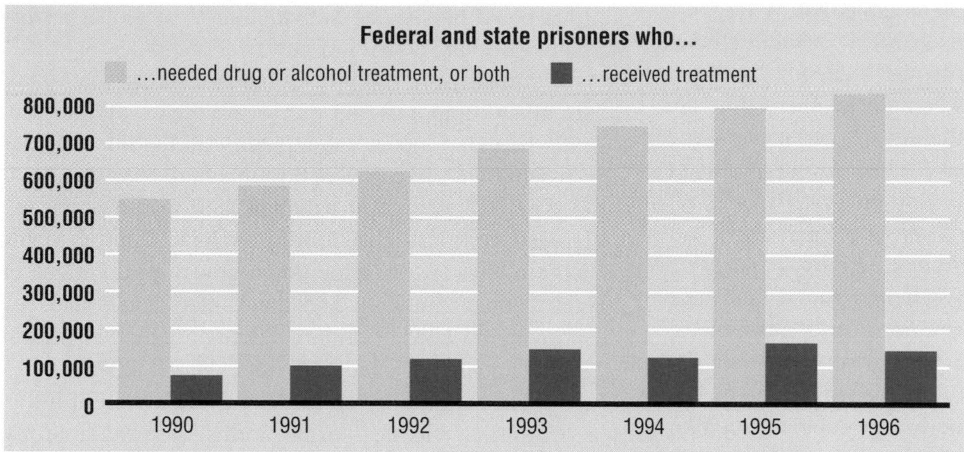

Source: The National Center on Addiction and Substance Abuse at Columbia University.

Schools for Crime? Sociologists who study prisons and other kinds of total institutions have found that new members of such organizations usually undergo a period of resocialization during which they are deprived of their former statuses through haircuts, uniforms, and the like (Goffman, 1961). (See Chapter 3.) The goal of this process is to socialize the inmates to behave in ways that suit the organization's needs. But in most such organizations there is also a strong inmate subculture consisting of norms that specify ways of resisting the officials in favor of such values as mutual aid and loyalty among the inmates. Thus, in Jackson's (1972) collection of case histories of criminals and prison inmates, a seasoned convict summed up the world of the total institution as follows: "A penitentiary is like a prisoner of war camp. The officials are the enemy and the inmates are captured. They're on one side and we're on the other" (p. 253).

Only relatively recently, mainly in the second half of the twentieth century, has the goal of rehabilitation—the effort to return the criminal to society as a law-abiding citizen—been taken seriously as a function of prisons. Critics of the prison system, many of whom argue from a conflict perspective, often claim that far from rehabilitating their inmates, prisons in fact function as "schools for crime." Failure to offer drug and alcohol treatment programs to more than one-third of the inmates in need of treatment (see Figure 5.2) is a glaring example of the failure of prisons to address the rehabilitation needs of their populations. As early as 1789, the English philosopher and economist Jeremy Bentham wrote that "prisons, with the exception of a small number, include every imaginable means of infecting both body and mind. . . . An ordinary prison is a school in which wickedness is taught by surer means than can ever be employed for the inculcation of virtue" (pp. 351–352). In the intervening years, many studies have attempted to show under what conditions this is true or not and what can be done to increase the deterrent effects of prison and prevent it from becoming a community that socializes criminals.

Yet there is much sociological research that defends prisons as a means of deterrence and a necessary form of retribution. James Q. Wilson (1977) holds to the functionalist position: Societies need the firm moral authority they gain from stigmatizing and punishing crime. He believes that prisoners should receive better forms of rehabilitation in prison and be guaranteed their rights as citizens once they are outside again. But, he states, "to destigmatize crime would be to lift from it the weight of moral judgment and to make crime simply a particular occupation or avocation which society has chosen to reward less (or perhaps more) than other pursuits" (p. 230). Some states, notably Louisiana, Alabama, and Florida, have taken the desire to get tough on crime to the extreme of reinstating chain gangs.

Recidivism

Whatever their sociological perspective, all students of the American prison system agree that by far the least successful aspect of prison life is rehabilitation. The term **recidivism** refers to the tendency for a person convicted of a crime to commit another crime after a term of imprisonment. Recidivists, or repeat offenders, are often people who have failed during the time they were incarcerated to develop alternative ways of life that would permit them to avoid the associates and situations that lead to criminal activity. Research conducted in the 1970s consistently found that the only effective rehabilitation programs in prisons, measured by lower rates of recidivism, involve job training and education (Martinson, 1972). Indeed, during the 1980s, as prison populations more than doubled, sociologists and criminologists found that men from very poor and minority backgrounds were beginning to regard prison as the only way they could obtain job training, health care, and other social services. Prisons, after all, provide more adequate housing than the large urban shelters for the homeless, and in prison one gets three meals a day and, in some cases, a chance to improve one's skills and self-confidence.

The conservative mood of the United States in the 1990s made a profound impact on rehabilitation programs in federal and state prisons. At this writing, there are bills in Congress that would deny prisoners any opportunity to take courses for high school or college credit. Some legislators, showing far more interest in punishment than in rehabilitation, support a proposal that would void all court-ordered efforts to correct inhumane conditions in prisons (Kunen, 1995). As one of the nation's leading researchers on crime and prisons has observed, in our society "we have an exaggerated belief in the efficacy of imprisonment. The real problem is that we make life really terrible for some people and then blame them when they become dangerous" (Norval Morris, quoted in Butterfield, 1992a).

TECHNOLOGY AND SOCIAL CHANGE

New technologies always seem to spawn new opportunities for both deviance and social control. The jet transport draws its hijackers; the telephone, its heavy breathers; the computer, its hackers. The computer, associated as it is with a burgeoning technological revolution in human communications, is especially vulnerable to deviant activities, from child pornography to the malicious development of computer viruses that could be seen as individual acts of global terrorism. Recently, a series of powerful computer "worms," intentionally created viruses that are spread via e-mail to other computers and that worm their way into certain files of a victim's computer with the intent of destroying files, have swept throughout the world. The "I Love You" worm, for example, began in Manila and spread from Asia to Europe and the United States within one day, shutting down government offices in the United Kingdom and the United States and destroying millions of dollars' worth of computer files in private companies and universities (Gold, 2000).

The availability on the World Wide Web of information that violates sexual norms—for example, by giving children easy access to explicit sexual materials—creates situations that demand greater social control of the technology. This demand, in turn, has led to the development of a wide variety of filtering software that permits parents or teachers to bar children's access to specific Web sites. Tracking software, on the other hand, allows people in control of computer technologies, at work or in school, to know at every moment what users under their authority are doing with the technologies. The widespread use of tracking software raises many questions about whether computers and the Internet are technologies of freedom or technologies of greater social control and perhaps authoritarianism (Greenberg, 1999).

Institutions that specialize in social control, specifically the police and penal institutions, are adapting computer technologies for a wide variety of applications. Electronic anklets that allow authorities to ensure that someone under house arrest remains within a designated area, mapping software that allows authorities to pinpoint areas where certain types of crime are occurring so they can intervene more effectively, computerized video surveillance and alarm systems that "patrol" wealthy neighborhoods—all are examples of how new technologies meet the demand for increased social control in societies where fear of crime is high. Other examples include drug-testing technologies that allow authorities to screen employees and students for drug use. On one hand, these applications of technology to social control provide protection against deviance and crime, but on the other hand they clearly increase the surveillance capability of those in power and thereby put the freedom of those under surveillance at risk.

SOCIOLOGY AND SOCIAL JUSTICE

Letizia Battaglia, shown here dancing with gypsies in the streets of her hometown of Palermo, is an award-winning photographer, municipal councilwoman, ecological activist, defender of women's and of human rights, and, most of all, fierce opponent of the Sicilian Mafia. As noted at the beginning of the chapter, Battaglia has used her photography to call attention to the ravages caused by the Mafia and especially to the suffering of Sicilian women, whose human potential is so often subordinated and stifled in a world of poverty and violence. But Battaglia would be the first to acknowledge that the successes she and other Sicilian activists have had in combating the Mafia depend on the activities of an aroused and courageous public. Following is her description of what happened in Palermo after the brutal bombing assassination of Judge Giovanni Falcone, his wife, and three bodyguards by the Mafia in 1992:

It began subtly in response to the Mafia's murder of Judge Giovanni Falcone. But word spread quickly: Whoever was against the Mafia should hang out a sheet. White. It was a gesture of great courage, because you saw one building where at first there wasn't anything, and then suddenly there was a sheet, and the person who lived there was saying, "I am against the Mafia. I live here. You can come and cut my throat—I will always be against you." On the next anniversary, we went into a risky neighborhood and we all walked together, laughing, saying beautiful things against the Mafia, because of the sweetness of life. And every now and then you saw a window open up a crack, and a woman would pretend to shake out the crumbs from a white tablecloth, and you would see the woman. We called out to her in force, and then she would withdraw inside, but she had done it! (Battaglia, 1999, p. 10) ■

Letizia Battaglia in a lighter moment.

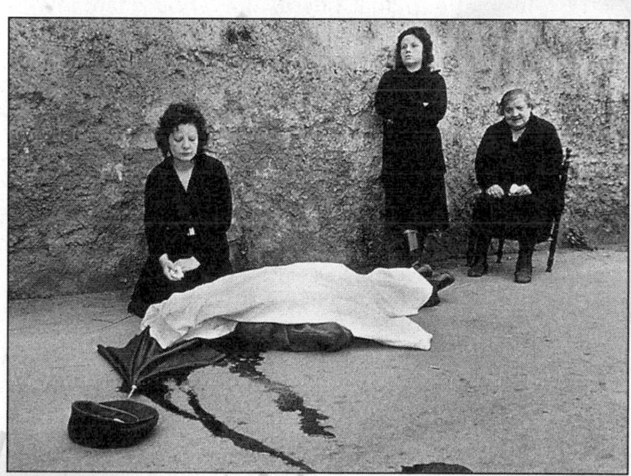

Letizia Battaglia's photos taken directly after Mafia killings have been extremely effective in stimulating local and international protests against organized crime.

SUMMARY

Deviance is behavior that violates the norms of a particular society. A person who violates or opposes a society's most valued norms is labeled as deviant. The ways in which a society prevents deviance and punishes deviants are known as *social control.* There is usually much disagreement not only about which behaviors are deviant but also about which ones should be condoned or punished only mildly. An important dimension of deviance is the power of some groups in society to define which acts are legal and which are illegal.

Deviance should be distinguished from *stigma.* A stigmatized person has some attribute that is deeply discrediting, but is not a social deviant—that is, someone who voluntarily violates social norms.

Crime is usually defined as an act, or the omission of an act, for which the state can apply sanctions. Those sanctions are part of the criminal law—a set of written rules that prohibit certain acts and prescribe punishments to be meted out to violators. As a culture's values and norms change, so do its notions of what kinds of behavior are deviant and how they should be sanctioned.

A deviant subculture includes a system of values, attitudes, behaviors, and lifestyles that are opposed to the dominant culture of the society in which it is found. Many deviant subcultures are harmful to society because they sustain criminal occupations.

Biological explanations of deviance relate criminality to physical features or body type. According to Robert Merton's theory (a functionalist theory of deviance), through socialization people learn what goals are approved of in their society and the approved means of achieving them. Those who do not accept the approved goals and legitimate means of achieving them are likely to engage in deviant behaviors.

Cultural conflict theories concentrate on the ways in which conflicting sets of norms result in situations that encourage criminal activity. Marxian theories place more emphasis on class conflict, explaining various types of crime in terms of the social-class position of those who commit them.

Edwin Sutherland's theory of *differential association* (an interactionist theory of deviance) holds that whether or not a person becomes deviant is determined by the extent of his or her association with criminal subcultures. Interactionists believe that deviance is produced by the process of *labeling,* in which the society's reaction to certain behaviors is to brand or label the offender as deviant.

The methods used to control crime change as societies become more complex. Larger, more diverse societies tend to develop standardized, more or less coercive institutions to deal with deviants. The primary functions of prisons are said to be deterrence, rehabilitation, and punishment. However, numerous studies have found that prisons are not successful in rehabilitating their inmates and in fact often serve as "schools for crime."

GLOSSARY

Deviance: Behavior that violates the norms of a particular society. **(125)**

Social control: The ways in which a society encourages conformity to its norms and prevents deviance. **(125)**

Stigma: An attribute or quality of an individual that is deeply discrediting. **(127)**

Crime: An act or omission of an act that is prohibited by law. **(128)**

Anomie: A state of normlessness. **(134)**

Differential association: A theory that explains deviance as a learned behavior that is determined by the extent of a person's association with individuals who engage in such behavior. **(136)**

Labeling: A theory that explains deviance as a societal reaction that brands or labels as deviant people who engage in certain behaviors. **(137)**

Recidivism: The tendency for a person convicted of a crime to commit another crime after a term of imprisonment. **(144)**

QUESTIONS FOR THOUGHT AND DISCUSSION

1. The argument that some types of crime actually increase the cohesion of a society by making people understand the threats to their social order is an example of functionalist sociological thought. Explain why this is so. What would a conflict theorist add to the discussion?

2. Sociologists often note that formal institutions of social control like the police and the courts are part of a culture's larger system of social control. Explain this statement. (Hint: Think about how the activities of agents of formal social control may be frustrated or supported by changes in a society's culture.)

DIGGING DEEPER

Books

Going Up the River: Travels in a Prison Nation (Joseph T. Hallerian; Random House, 2001). A bleak assessment of the rapid rise in the U.S. prison population, by a sociologically informed reporter for the *Wall Street Journal.*

Street Women (Eleanor Miller; Temple, 1986). An empirical study of how women become involved in crime and organize their lives as street hustlers.

The Outsiders (Howard Becker; Free Press, 1963). An original application of the interactionist and labeling perspectives to the study of deviance and crime, using empirical examples.

Stigma: Notes on the Management of Spoiled Identity (Erving Goffman; Prentice Hall, 1982). A classic study of how people who differ from the norm are often labeled as deviant; explores the consequences of such labeling.

Constructions of Deviance: Social Power, Context, and Interaction (Patricia A. Adler & Peter Adler, eds.; Wadsworth, 1994). A collection of contemporary articles on theories and research about many aspects of deviant behavior.

Other Sources

Crime in the United States (Uniform Crime Reports). An annual report on criminal offenses, arrests, and law enforcement employment. Data are reported by local police departments to the Federal Bureau of Investigation and published by the U.S. Department of Justice.

National Crime Survey. A continuing series of reports on the incidence of crime throughout the United States; published by the U.S. Department of Justice.

Understanding and Preventing Violence (Albert J. Reiss & Jerry A. Roth, eds.; National Academy Press, 1993). A summary volume of a six-volume report on violence in the United States. Contains excellent reviews of the literature on gun control, interpersonal violence, and the biological and social antecedents of violent behavior.

Journals

Social Justice. The journal of critical criminology, highly useful for commentary on social policy.

EXPLORING SOCIOLOGY ON THE INTERNET

U.S. Department of Justice
www.usdoj.gov/

A regularly updated home page with links to various agencies and projects.

Violence Against Women Office
www.usdoj.gov/vawo/

Offers information on the National Domestic Violence Hotline, copies of federal legislation and regulations, ongoing research reports and studies, and a Domestic Violence Awareness Manual targeted to federal employees.

Federal Bureau of Prisons
www.bop.gov/

Provides statistics on the federal prison population, both inmates and staff.

The Journal of Prisoners on Prison
www.synapse.net/~arrakis/jpp/jpp.html

A lively and innovative journal written for, by, and about inmates.

The National Gay and Lesbian Task Force
www.ngltf.org/gi.html

A "progressive civil rights organization" with a home page that tracks political issues of concern to lesbians and gays and suggests activities for those who want to advance their civil rights.

United States Federal Bureau of Investigation
www.fbi.gov

Provides links to 20 information sources, uniform crime report information, and publications and announcements; lists the FBI's most wanted fugitives.

Chapter One
www1.psi.net/chapterone/

Offers an online opportunity to purchase books from leading public policy organizations. Searches can be conducted using subjects, titles, or publishers.

Chapter 6

STRATIFICATION AND SOCIAL CLASS

Can the social-class inequalities that affect our lives so dramatically ever be reduced?

Ann Berrios is a secretary in a large and profitable U.S. corporation. She is also a keen observer of the way inequality touches her life and that of people like her, as we see from this editorial she wrote for an influential metropolitan newspaper:

> Politicians have been talking a lot about the prosperity that has transformed the country, and though I see it all around me, I have yet to feel it. My husband, a science professor, and I couldn't afford to have children for years. We waited until we could wait no longer. We lived 12 years in a neighborhood we were not wild about before we moved into our present house in a picturesque neighborhood in an excellent school district. It takes our two incomes to live where we do.
>
> The young mothers I see dress casually and expensively. They wave goodbye to their husbands as they go off to work, stroll with their baby carriages about the neighborhood, and later meet their older children at the bus stop. My children do not know what it is like to come home immediately after school. They stay for hours in after-school day care.

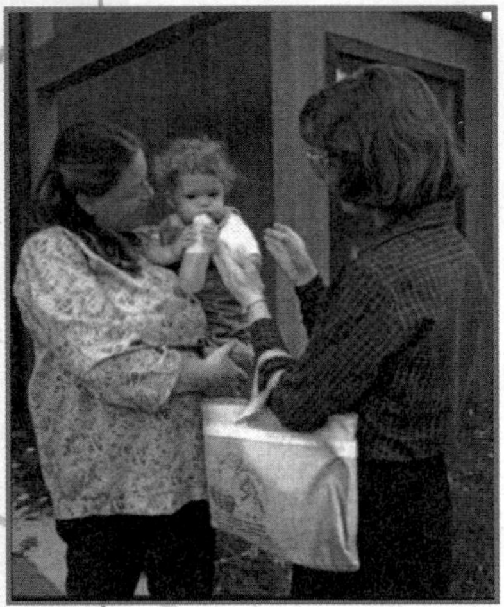

Ann has a bachelor's degree and a master's degree that she earned by taking one course at a time at night. Much as she would like to find a higher-paying job, she has had little success. "The market for people with résumés like mine," she observes, "does not seem to allow for upward mobility." Unemployment has reached record lows, but this secretary wonders, "If jobs were so abundant, shouldn't I be able to find one that's better than the one I have now?"

"I suspect that the real job opportunities are in places like Wal-Mart and Sears," she comments. "Near my house, all stores, great and small, post signs inviting applicants. Long lines at the check-out counters and the absence of personnel in the aisles attest to understaffing." At home, Ann and her husband pinch their pennies to pay the mortgage and save for their children's education. When she was a child, Ann remembers, her parents "could afford vacations to interesting places, private lessons and private schools for me and my sister. They even had money left over for contributions to their colleges and their church."

> My two children get only the lessons our school district offers. They have never been to Disney World, and my son is getting too old for it. Though, of course, it is possible to survive childhood without a trip to Disney World, it would be nice not to have to keep putting it off.
>
> We are not hungry. Our jobs provide us with excellent health and dental insurance. We tighten our belts and pay our bills. Like the coffee and pastry I buy each morning, the extras are not necessary, but they make life more palatable. It bugs me that at the height of my earning potential, I cannot afford them (Berrios, 2000, p. A27).

This chapter explores a variety of concepts, theories, and factual situations that help explain why economic and social inequalities exist in societies throughout the world. The first sections explain what sociologists mean by social stratification and how inequalities emerge from the way people earn their living in various types of societies. We then consider how social-class inequality is experienced in the United States. We continue with a discussion of the major sociological theories of stratification and inequality. The final section explores some of the reasons for the persistence of poverty and why it is so difficult to reduce inequality even in an affluent society like the United States. ■

CASTE, CLASS, AND SOCIAL MOBILITY

In countries like China, India, and the nations of Africa, continents away from the headquarters of the U.S. corporation where Ann Berrios works, the vast majority of people would consider the life of an American secretary quite luxurious. In these countries poverty and disease are constant reminders of the narrow gap between survival and death. But this observation does not diminish Ann's feelings of deprivation. Faced with inequalities of wealth and income, what matters most is with whom one compares one's situation. Everywhere in the world people divide themselves into those with more of what their culture deems important and those with less. In every society, people generally group themselves into different categories according to how they earn their living. This produces a system of social stratification that is unique to each society yet is similar in many ways to those of other societies.

Social strata are invisible strata, or layers, in a society that may be more or less closed to entry by people from outside any given layer. A society's system for dividing people into strata according to such attributes as wealth, power, race or ethnicity, gender, and prestige is its system of **social stratification.** Societies that maintain rigid boundaries between social strata are said to have **closed** stratification systems; societies in which the boundaries are easily crossed are said to be **open** societies.

In open societies, it is possible for some individuals and their families, and even entire communities, to move from one stratum to another; such movement is termed **social mobility.** A couple whose parents were unskilled workers may become educated, learn advanced job skills, and be able to afford a private house instead of renting a modest apartment as their parents did. Such a couple is said to experience **upward mobility.** If they have enough wealth to help their parents live comfortably and help other family members as well, the entire family may enjoy upward social mobility. If everyone with the same education and skills and the same occupation experiences greater prosperity and prestige, the entire occupational community is said to be upwardly mobile. But in an open society fortunes can also decline. People with advanced skills—in engineering or higher education, for example—may find that there are too many of them. They may not be able to afford the kind of housing, medical care, education for their children, and other benefits that they have come to expect. When this occurs, they are said to experience **downward mobility.**

The best examples of closed societies are found in **caste** societies. **Castes** are social strata into which people are born and in which they remain for life. As discussed in Chapter 4, membership in a caste is an **ascribed status**—a status acquired at birth—rather than an **achieved status**—one based on the efforts of the individual. Members of a particular caste have no hope of leaving that caste. Slaves and plantation owners, for example, formed a caste society in the United States before the Civil War. The slaves were captives; runaway slaves were pursued and returned to their masters. Their children were born into slavery. Plantation owners, on the other hand, had amassed great wealth, especially in the form of land, and this wealth was passed on to their children. On occasion, a plantation family might lose its wealth, or another family might acquire a plantation and the accompanying wealth and prestige, but this form of social mobility was rather infrequent and did not alter the caste nature of the plantation system.

Today, although the caste system is officially illegal, much of India remains influenced by caste-based inequalities. South Africa, on the other hand, is an example of a society that is moving away from a rigid caste system. Under apartheid, blacks in South Africa were a racially defined caste that was kept at the lowest rungs of the stratification system by violent repression and laws mandating racial segregation. Although this situation has been changing since the transition to majority African rule, the extreme poverty of the black population indicates that most aspects of the caste system are far from ended.

Classes, like castes, are social strata, but they are based primarily on economic criteria such as occupation, income, and wealth. England is famous for its social classes and for the extent to which social class defines how people are viewed and how they view themselves. Classes are generally open to entry by newcomers, at least to some extent, and in modern societies there tends to be a good deal of mobility between classes. Moreover, the classes of modern societies are not homogeneous—their members do not all share the same social rank. There are variations in people's material well-being and in how much prestige they are accorded by others. Within any given class, these variations produce groups, known as **status groups,** that are defined by how much honor or prestige they receive from the society in general.

The concept of status groups is illustrated by "high society." The nobility of England is one of the world's most prestigious status groups, despite the recent well-publicized marital troubles of the royal family. In the United States, people with names like Rockefeller, DuPont, Lowell, Roosevelt, and Harriman—of Western European Protestant descent—often have more prestige than people with just as much wealth who are of Italian, Jewish, or African American descent. In cities throughout the United States, these very rich, prestigious families, who form a status group as defined by wealth and reputation, also tend to form groups (e.g., "high society") that interact among themselves and play significant roles in their communities. The society pages of metropolitan newspapers devote most of their gossip to the philanthropic activities and private affairs of these families.

DIMENSIONS OF SOCIAL STRATIFICATION

Rankings from high to low are only one aspect of social stratification. The way people live (often referred to as their *lifestyle),* the work they do, the quality of their food and housing, the education they can provide for their children, and the way they use their leisure time are all shaped by their place in the stratification system.

The way people are grouped with respect to access to scarce resources determines their **life chances**—that is, the opportunities they will have or be denied throughout life: the kind of education and health care they will receive, the occupations that will be open to them, how they will spend their retirement years, and even where they will be buried. The place in a society's stratification system into which a person is born (be it a comfortable home with access to good schools, doctors, and places to relax, or a home that suffers from the grinding stress of poverty) has an enormous impact on what that person does and becomes throughout life. A poor child may overcome poverty and succeed, but the experience of struggling out of poverty will leave a permanent mark on his or her personality. And most people who are born poor will not attain affluence and leisure, even in the most open society.

Global Stratification and Life Chances

Out of almost 6 billion people living today, about four-fifths live in countries where the majority of the population is extremely poor. When we look at patterns of stratification and inequality around the world, therefore, the most basic questions concern whether people in different regions or nations are safe from the major threats to human existence (UNDP, 1997, 1999). These threats to survival are often analyzed in terms of seven categories:

Economic insecurity. The probability of losing access to the means of existence.

Food insecurity. The probability of losing access to food supplies adequate to sustain life.

Health insecurity. The probability of loss of life due to illnesses that are routinely corrected by contemporary health care (e.g., infant mortality, dysentery).

Personal insecurity. The probability of death or serious injury due to violence, workplace accidents, or other avoidable conditions.

Environmental insecurity. The probability of loss of livelihood or life due to ecological disasters, overuse of resources, or settlement in unstable environments such as flood plains, mountainsides, and the like.

Community and cultural insecurity. The probability that entire communities or cultures will be displaced, forced to migrate, or otherwise eliminated due to conflict, rapid social change, disaster, and so forth.

Political insecurity. The probability of civil war, genocide, or other forms of warfare and politically motivated violence.

These forms of insecurity are closely interrelated. Each society's structure of stratification and economic inequality creates conditions and inequalities that heighten some groups' insecurities while reducing those of more privileged groups. In general, the poorer an individual, family, or village community is, the more likely it is to experience one or more forms of insecurity. In his Nobel prize-winning research on famines, for example, the Indian social scientist Amartya Sen showed that the likelihood of death in famines throughout the world normally is not directly due to lack of food supplies in the famine-struck nation. Instead, for the most part, it is due to the victims' lack of entitlement to food, either because of extreme lack of money, lack of food or other goods to exchange, or lack of access to government emergency supplies. Sen shows that people suffering from extreme economic insecurity, such as beggars, landless farm laborers, the rural unemployed, and their families, have the least claim to whatever food supplies are available in cases of extreme shortage. In

FIGURE 6.1
Disparity of Income Between Rich and Poor, Selected Countries, 1820–1992
This graph traces the widening income gap between rich and poor countries over the last two centuries. These changes in the income of nations are measured in gross domestic product per capita (in 1990 U.S. dollars). Changes in particular nations, such as Argentina and Egypt, are shown by individual lines. The shading indicates the income range among the richest five countries (represented by dots), which in 1992 were the United States, Switzerland, Japan, Germany, and Denmark.

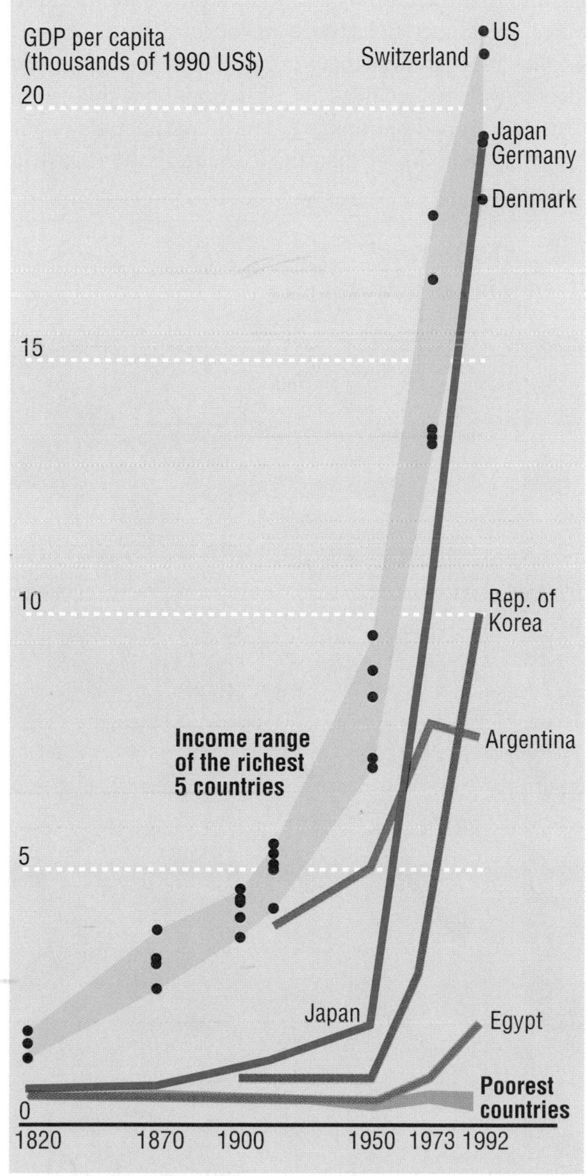

Source: UNDP, 1999.

consequence, mortality rates due to starvation and associated illnesses (dehydration, dysentery, infectious diseases, etc.) are highest among these least secure groups. Even when famines occur in the poorest nations, there is usually enough food to go around. However, because of the workings of the stratification system, food never gets to the people and places where it is most needed (Sen, 1981, 2000).

On a global scale, the gap in income between the poorest and the richest nations has continued to widen. In 1960, the richest 20% of the people in the richest countries had 30 times the income of the poorest 20%. By the late 1990s, that disparity had increased to 74 times as much. These changes represent a recent acceleration of a trend in world inequality that has been continuing for the past two centuries, as we see vividly in Figure 6.1. Figure 6.2 shows how dramatically the wealth (net worth) of the world's 200 richest people has increased in the past few years. In fact, the assets of the three richest people on earth are more than the combined gross domestic product (the value of all goods and services produced in a nation) of all the world's least-developed nations. The assets of the 200

FIGURE 6.2
Net Worth of the 200 Richest People

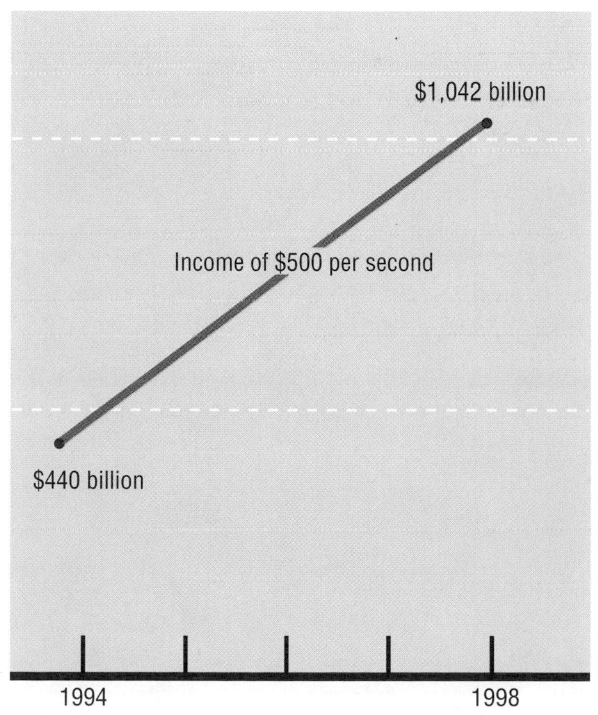

Source: UNDP, 1999.

richest people amount to more than the combined assets of 41% of the earth's population. A yearly contribution of 1% of the wealth of these individuals (about $7 billion) could provide universal access to primary education for everyone in the world (UNDP, 1999).

Global Inequality and Health. Of the various categories of insecurity just listed, health insecurity may have the greatest impact, especially among women and children. Health insecurity has many different meanings, especially if one lives in an affluent nation or a nation with high proportions of poor individuals and families. An estimated 20% of the people living in less-developed nations have no access at all to health services. And in the poorest, least-developed nations about 51% are estimated to have no access to health care at all (UNDP, 1997). In the United States, the quality of health care one receives is directly related to one's place in the stratification system, with poor families likely to experience the greatest health insecurity. But even those without any health insurance can find emergency health services. This is typically not true in the poorest nations, where health problems are most urgent.

Two indicators—infant mortality and life expectancy—are used as primary measures of efforts to reduce glaring inequalities between rich and poor. Nations that are investing their growing wealth in their populations should have decreasing rates of child mortality and increasing life expectancy. Figure 6.3 shows that despite recent declines in infant mortality throughout the world, high infant mortality rates persist in the least-developed nations. Because far more people in these nations are poor, their infant mortality rates are higher than those in more affluent nations. Figure 6.4 shows that improvements in life expectancy have occurred in all nations, but that in the least-developed nations average life expectancy is almost 20 years lower than the average in the industrial countries.

FIGURE 6.3
Trends in Infant Mortality (per 1,000 Live Births)

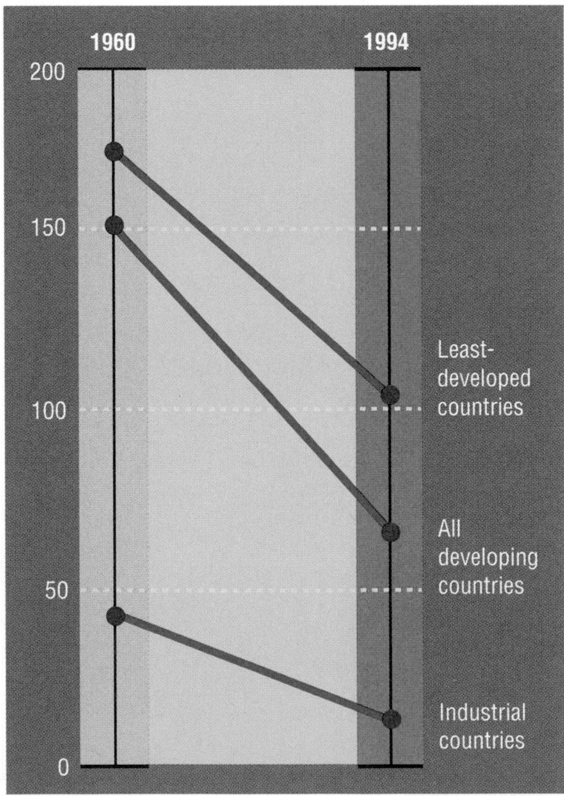

Source: UNDP, 1997.

FIGURE 6.4
Trends in Life Expectancy

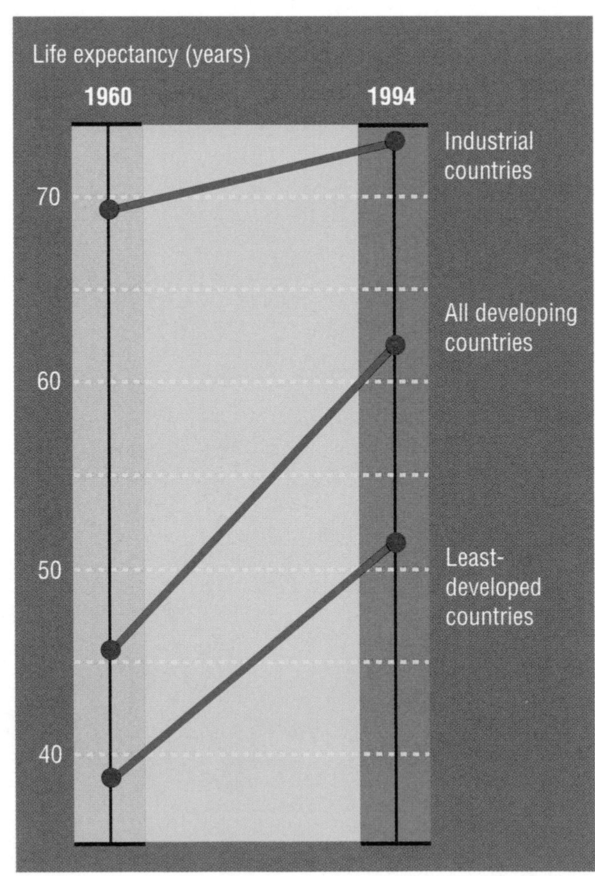

Source: UNDP, 1997.

Throughout the world, income is the single most important factor in determining survival. At the extreme of poverty, in Africa, with an annual per capita income of US $800, life expectancy is 52 years. At the other end of the income spectrum, in the United States, with an annual per capita income of US $20,000, average life expectancy is 75 years. There are, however, low-income regions that have invested in the health and safety of their populations (e.g., some states of India), and there are affluent regions that still rank relatively low on measures of health and survival (Mosley & Cowley, 1991; Sen, 1993).

In impoverished nations like Ethiopia and Bangladesh, the primary causes of death before old age are diseases such as amoebic dysentery (severe diarrheal diseases), which result from contaminated water supplies and poor sewage systems. In more affluent nations, the rate of death before age 60 increases for adolescents and young adults, especially males, owing to the greater risks they take in their leisure activities (dangerous sports, drug use, etc.) and the higher rates of interpersonal violence in some of these nations (especially the United States).

Stratification systems in many regions of the world tend to exclude women from educational opportunities. Female illiteracy and poverty rates are closely associated with high rates of infant mortality. Figure 6.5 dramatically illustrates this fact for selected nations. Note, for example, that in Bolivia women with no education are four times more likely to experience the death of children under five years of age than women with high school or higher levels of education.

Stratification and the Means of Existence

The principal forces that produce stratification are related to the ways in which people earn their living. The economic insecurity due, for example, to loss of work, loss of crops, or the fear that one's company may go out of business is less pervasive as one goes up through the levels of the stratification system. The rich can experience reversals of their economic fortunes, but these reversals are far less likely to produce the health crises, food scarcities, and other dire consequences of economic insecurity that people at lower levels of the stratification system routinely face.

In the nonindustrialized world, the majority of the people are small farmers or peasants. When they look up from their toil in the fields, they see members of higher social strata—the landlords, the moneylenders,

FIGURE 6.5
Child Mortality by Mother's Education, 1997–1998

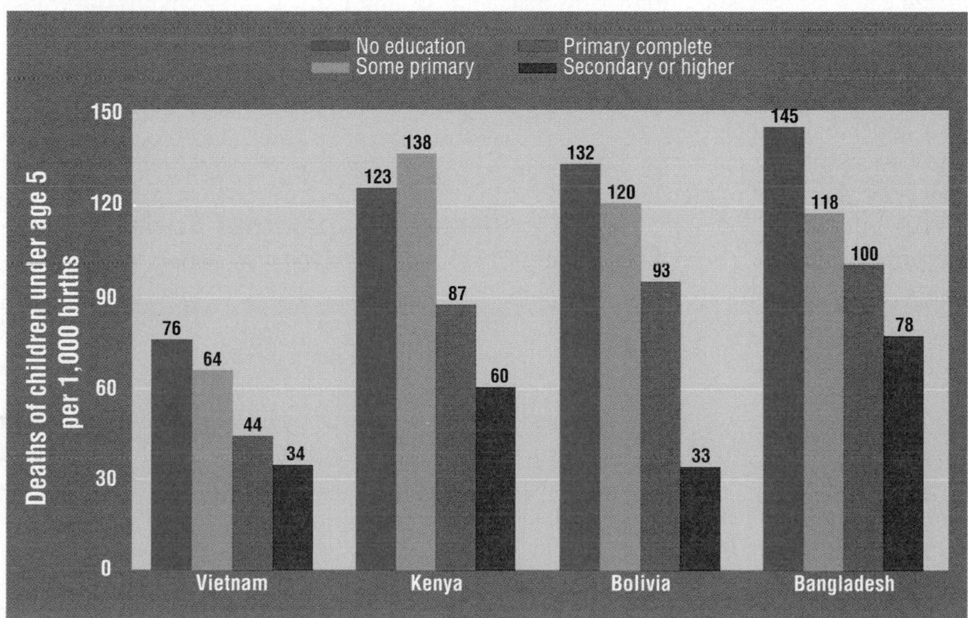

Source: Ratzen, Filerman, & LeSar, 2000.

the military chiefs, the religious leaders. These groups control the peasants' means of existence: the land and the resources needed to make it produce. Even when farmers or peasants are citizens of a modern nation-state with the right to vote and access to education, health care, and other benefits, when harvests are poor the landowners still take their full share, leaving less for the peasants. The system of stratification that determines the peasants' fate depends on how much the land can yield. In contrast, in the stratification systems of industrial societies, whether capitalist or socialist, most people are urban wage workers whose fates are determined by the managers of public or private firms. If the firms are no longer productive or consumers no longer desire their products, urban workers may lose their jobs and suffer economic hardship. (See the Global Social Change box on page 159.)

The stratification systems of the United States and Canada, where less than one-twentieth of the population works the land, are most relevant to understanding people's life chances in modern industrial societies. But to understand the conditions of life for most of the world's population we must study inequalities in rural villages, where well over two-thirds of the earth's people till the soil and fish the rivers and oceans.

Stratification in Rural Villages. In India and China, more than 1.2 billion people spend their lives coaxing an existence from the soil. Millions of other rural villagers squeeze a modest livelihood from the land in Central and South America, Africa, and Southeast Asia. Social divisions in these villages are based largely on land ownership and agrarian labor. The poorest farmers and agricultural laborers, who typically do not own land themselves, are most prone to extreme "income poverty," or lack of enough money to feed themselves and their families. A crude but useful definition of income poverty is the number of adults in a population who do not have incomes worth more than one U.S. dollar a day. At present, that number for the world as a whole is over 1 billion people, of whom the vast majority are landless peasants and farm laborers (UNDP, 2000).

In peasant societies, the farm family, which typically works a small plot of land, is the basic and most common productive group. Such families can be found in the villages of modern India. The Indian village reveals some important dimensions of stratification in Third World societies. For example, women are assigned to hard work in the fields and at the same time are expected to perform almost all of the household duties. This is true even in families that are well off. Men from the higher castes may be innovators, but women of all castes and male members of the "backward" castes do most of the productive work (Bhalla & Lapeyre, 1999).

In rural villages, the facts of daily life are determined largely by one's place in the local system of agricultural production. The poor peasant family, with little or no land, hovers on the edge of economic disaster. Work is endless for adults and children alike. Meals are meager; shelter is skimpy; there is not much time for play. Among the gentry, who have large land holdings and hired help to ease the burdens of work, there are "the finer things of life"—education, ample food and shelter, music and games to pass the time. Of course, wealth brings additional responsibilities. Participation in village or regional politics or in charity work takes time away from pleasure, at least for some. But the power that comes with such activities provides opportunities to amass still more wealth. Thus it is that, with few exceptions, the poor remain poor while the rich and powerful usually become richer and more powerful. And between the rich and the poor there are other strata—the middle peasants and the middle castes with skills to sell—whose members look longingly at the pleasures of the rich and console themselves with the fact that at least they are not as unfortunate as the humble poor in the strata below.

How does this scheme of social stratification compare with that found in industrial societies? We will see that there is more mobility in industrial societies but that, just as in rural societies, to be born into the lower strata is to be disadvantaged compared with people who are born into higher strata.

Stratification in Industrial Societies. The industrial revolution profoundly altered the stratification systems of rural societies. The mechanization of agriculture greatly decreased the number of people needed to work on the land, thereby largely eliminating the classes of peasants and farm laborers in some societies. This dimension of social change is often called **structural mobility:** An entire class is eliminated as a result of changes in the means of existence. As noted in Chapter 4, the industrial revolution transformed the United States from one in which almost 90% of the people worked in farming and related occupations into a nation in which less than 10% did so. Similar

GLOBAL SOCIAL CHANGE

Remittances and Migration

Throughout the world, able-bodied men and women are attracted to the opportunities for economic and social mobility offered in the industrialized nations. Poor Koreans seek to migrate to Japan; Chinese migrate throughout the cities of Southeast Asia and, if possible, to North America; Latin American agricultural and industrial workers may dream of finding their way into the United States and Canada; in Africa, workers dream of finding their way to the metropolitan centers of Europe. This hardly begins to describe the flow of "economic" immigrants in the contemporary global economy.

Everywhere, immigration by the poor is accompanied by hardships and sacrifice. Look at any kitchen in an American restaurant, for example, and you are likely to find recent immigrants, typically Mexicans or Caribbeans, washing dishes and doing the most menial tasks for minimum wages, and sometimes for less. Almost miraculously, however, these same immigrants are likely to be sending a portion of their meager wages to loved ones in the villages and towns they left behind. Indeed, the global flow of such funds, known as remittances (*remesas* in Spanish) has become a major source of income for many villages that "send" their men and women abroad. These funds hardly reverse the growing income gap between rich and poor nations, which has doubled in size in the past thirty years, but they often make an enormous difference in the home villages and regions (Jones, 1998).

Between North America and Latin America, the annual flow of remittances, often called migradollars, exceeds $9 billion annually. The accompanying table offers examples of the total volume of remittances sent from the United States by migrants who came from Mexico and selected Central American nations. The middle column represents the number of households that sent migrants and the average amount of migradollars these households received in a year from the hard work of their migrant family members. The table shows, for example, that over $3.4 billion was sent back to Mexico to more than 1.4 million households, which received an average of $2,427. These funds are keeping dreams alive and changing the politics and lifestyles of villages throughout Latin America and the world, but they are earned at an enormous cost to personal relationships (Perez-Lopez & Diaz, 1998).

Remittances by Immigrants to Home Households

	Remittances (millions of U.S. dollars)	Number (in thousands)	Remittances/ Household (in U.S. dollars)
Costa Rica	55	14.9	3,691
Guatemala	205	63.5	3,228
Honduras	49	30.8	1,591
El Salvador	324	133.5	2,427
Dominican Republic	315	137.7	2,288
Mexico	3,465	1,427.6	2,427
Total	4,413	1,808.0	2,441

Source: Perez-Lopez & Diaz, 1998.

changes took place in England and most of the European nations and are now taking place in many other parts of the world.

But structural mobility did not end with the industrial revolution. Today automation, foreign competition, and technological advances are creating new patterns of structural mobility. Older smokestack industries like steel and rubber have been steadily losing factories and jobs; newer industries based on information and communication technologies have been creating plants and jobs. However, the workers who have become "superfluous" as a result of the closing of their plants are not always willing to move to the new jobs, nor are they often trained to meet the demands of such jobs. Thus, structural mobility often leads to demands for social policies like job-training programs and unemployment insurance.

A second major change brought on by the industrial revolution was a tremendous increase in **spatial mobility** (or geographic mobility). This term refers to the movement of individuals, families, and larger groups from one location or community to another. The increase in spatial mobility resulted from the declining importance of the rural village and the increasing importance of city-centered institutions like markets, corporations, and governments. Increasingly, one's place of work became separate from one's place of residence. People's allegiance to local communities was weakened by their need to move, both within the city and to other parts of the nation. As a result, social strata began to span entire nations. Working-class people created similar communities everywhere, as did the middle classes and the rich (Dahrendorf, 1959; Janowitz, 1978).

Despite these immense changes, our relationship to the means of existence is still the main factor determining our position in our society's stratification system. We continue to define ourselves to one another first and foremost in terms of how we earn a living: "I am a professor; she is a doctor; he is a steelworker." Once we have dealt with the essentials of our existence—essentials that say a great deal about the nature and the quality of our daily lives—we talk about the things we like to do with our lives after work or after educating ourselves for future work.

Stratification and Culture

Why do people accept their "place" in a stratification system, especially when they are at or near the bottom? One answer is that they have no choice; they lack not only wealth and opportunities but also the power to change their situation. But lack of power does not prevent people from rebelling against inequality. Many also believe that their inferior place in the system is justified by their own failures or by the accident of their birth. If people who have good cause to rebel do not do so and instead support the existing stratification system, those who do wish to rebel may feel that their efforts will be fruitless.

Another reason people accept their place in a stratification system is that the system itself is part of their culture. Through socialization, we learn the cultural norms that justify our society's system of stratification. The rich learn how to act like rich people; the poor learn how to survive, and in so doing they tacitly accept being poor. Women and men, as well as the young and old, learn to accept their assigned places. Yet despite the powerful influence of socialization, at times large numbers of people rebel against their cultural conditioning. To understand their why this occurs, we need to examine the cultural foundations of stratification systems.

The Role of Ideology. You will recall from Chapter 2 that an ideology is a system of ideas and norms that all the members of a society are expected to believe in and act on without question. American workers, for example, have been taught that employment in private firms and the pursuit of individual gain are commendable. The ideology of American capitalism teaches that workers are free to sell their labor as individuals to the highest bidder. If they are dissatisfied with the pay or conditions of their work, they are free to find a better situation.

Every society appears to have ideologies that justify stratification and are used to socialize new generations to believe that existing patterns of inequality are legitimate. In the United States, for example, people love to hear versions of the "rags-to-riches" theme, as embodied in the stories of Horatio Alger. After the Civil War, Alger wrote numerous books based on this theme. In his first novel, *Ragged Dick,* the central character is a poor but honest boy who comes to the city looking for work. As he walks the streets, he sees a runaway carriage he leaps onto the horses and stops them. In the carriage, is a beautiful young woman who turns out to have a rich father. The father takes Dick into his business, where he proves his great motivation and becomes highly successful.

The castes of Hindu India are supported by religious ideology. The Rig-Veda taught that Hindu society was, by divine will, divided into four castes, of which the Brahmins were the highest because they were responsible for religious ceremonies and sacrifices (Majundar, 1951; McNeill, 1963). Over time, other castes with different tasks were added to the system as the division of labor progressed and new occupations developed. Still another powerful ideology had its origins in Europe before the spread of Christianity. Tribal peoples in what is now France and Germany associated their kings with gods, and that association became stronger in the feudal era (Dodgson, 1987).

In our own era, the civil rights movement in the United States, the movement to end apartheid in South Africa, the struggle of the Northern Irish Catholics for independence from Britain, and other social movements often invoke the ideology of radical Christianity. "We Shall Overcome," the theme song of the civil rights movement, was borrowed from the African American Baptist spiritual, "I Shall Overcome," and transformed into a moving song of hope and protest with religious overtones.

Stratification at the Micro Level. These relationships between religious ideologies and the stratification systems of civilizations are macro-level examples of how culture maintains stratification systems from one generation to the next. But we can also see the connection between culture and stratification in the micro-level interactions of daily life. The way we dress—whether we wear expensive designer clothes, off-the-rack apparel, or secondhand clothing from a thrift shop—says a great deal about our place in the stratification system. So does the way we speak, as anyone knows who has ever been told to get rid of a southern or Brooklyn accent in order to "get ahead." Our efforts to possess and display **status symbols**—material objects or behaviors that convey prestige—are encouraged by the billion-dollar advertising industry. Many other examples could be given, but here we will concentrate on two sets of norms that reinforce stratification at the micro level: the norms of deference and demeanor (Goldhamer & Shils, 1939).

By **deference** we mean the "appreciation an individual shows of another to that other" (Goffman, 1958, pp. 488–489). In popular speech, the word *deference* is often used to indicate how one person should behave in the presence of another who is of higher status. Formulas for showing deference differ greatly within different subcultures. Among teenage gang members, for example, there are ways of showing deference through greeting rituals. Not performing those rituals properly may cause a gang member to feel "dissed" (disrespected), which can lead to violence. These formulas for showing deference illustrate how our society's stratification system is experienced in everyday life. In the United States, for example, we learn to address judges as "Your Honor" rather than "Judge." In most European countries, with their histories of more rigid stratification, people who want to show deference go further and address the judge as "Your Excellence."

Our **demeanor** is the way we present ourselves—our body language, dress, speech, and manners. It conveys to others how much deference or respect we believe is due us. Here again, the interaction can be symmetrical. A college professor must make the first move toward informality in relations with students, for instance, but among professors of equal rank there is far more symmetry. The move toward informal demeanor, such as the use of first names, can be initiated by whoever feels most comfortable in his or her status. On the other hand, asymmetry in the use of names can be used to reinforce stratification, as in an office where the secretaries are addressed by their first names but address their supervisor by his or her last name plus a title such as Mrs., Dr., or Professor.

Deference, demeanor, and other ways in which we behave according to the micro-level norms of stratification further reinforce our sense of the correctness of those norms. Thus, in times of rebellion against a society's stratification system, those norms are explicitly rejected. During the 1960s, for example, students from wealthy families often wore jeans and tie-dyed shirts in social situations that would normally call for suits or dresses. Long hair and beards, refusal to wear a bra, and other symbolic acts in violation of generally accepted norms of demeanor were intended to communicate rejection of the society's stratification system.

Power, Authority, and Stratification

When the macro dimensions of social stratification change, these changes may be reflected in the behavior of people at the micro level, and, in turn, accelerate changes throughout the entire society. Thus, for example, the civil rights and women's movements of recent decades have altered the norms of demeanor for blacks and women. More accepted terms for African Americans have replaced the demeaning terms used in the past. Women prefer not to be called "girls" by men and especially by their supervisors, although some women continue to use the term among themselves. These may seem to be trivial matters, but when people demand respect in everyday interactions, they are also demonstrating their determination to create social change at a more macro level—in other words, to bring about a realignment of social power.

Max Weber defined **power** as "the probability that one actor within a social relationship will be in a position to carry out his own will despite resistance" (1947, p. 152). This definition is very general, however; it applies equally to a mugger with a gun and to an executive vice president ordering a secretary to get coffee for a visitor. But there is a big difference between the types of power used in these examples. In the first, illegitimate power is asserted through physical coercion. In the second, the secretary may not want to obey the vice president's orders yet recognizes that they are legitimate; that is, such orders are understood by everyone in the company to be within the vice president's power. This kind of power is called **authority.**

Even when power has been translated into authority, there remains the question of how authority originates and is maintained. This is a basic question in the study of stratification. As we saw earlier, the fact that people in lower strata accept their place in society requires that we examine not only the processes of socialization but also how power and authority are used to maintain existing relations among castes or classes. We turn, therefore, to a consideration of how stratification and inequality operate in the United States.

SOCIAL-CLASS INEQUALITY IN THE UNITED STATES

Americans experience far lower levels of economic, health, and personal insecurity than people in most other nations. But inequality and social stratification operate in powerful ways in American life as well. Income, wealth, education, and good jobs are desired by Americans just as they are elsewhere in the world. But access to these desired values is far from equal and tends to divide households and individuals by social class strata.

The basic and most readily available measures of inequality in any society are income, wealth, and educational attainment. Of these measures, income is the one that is most often used to give an initial view of social-class inequality. A common method of comparing incomes in a large population like that of the United States is to divide the population into fifths, or *quintiles,* on the basis of personal or household income. Table 6.1 divides the U.S. population into quintiles, going from the 20% with the least income to the 20% with the most income. The income share of all households in each fifth of the population is given as an average. Note that in the mid-1970s the top fifth of U.S. households earned 43% of all income while the bottom fifth earned only 4.4% of the total. By 1996, the gap had widened, and the richest fifth of households were taking home an average of $115,000 while the poorest earned an average of less than $9,000, or 3.7% of total income earned by Americans that year. Despite a recent booming

TABLE 6.1
How Households Divided the Nation's Income: 1976 and 1996 (in 1996 dollars)

1976			1996	
Share of All Income	Segment Average	Household Segments	Segment Average	Share of All Income
43.3%	$85,335	Top 20%	$115,514	49.0%
24.8	48,876	Second 20%	54,922	23.2
17.1	33,701	Middle 20%	35,486	15.1
10.4	20,496	Fourth 20%	21,097	9.0
4.4	8,672	Bottom 20%	8,596	3.7
100.0%	**$39,416**	**All households**	**$47,123**	**100.0%**
16.0%	**$126,131**	**Richest 5%**	**$201,684**	**21.4%**

Source: Hacker, 1997.

economy, the middle quintile of households saw their incomes decline from a 17.1% share to a 15.1% share of national income. By 1999 and 2000, as the economy continued to grow, the middle quintiles gained a bit more, but the master trend is that shown in the table. The table also indicates that the richest 5% of the U.S. population were the major winners during this period: By the mid-1990s, they were taking home over one-fifth of the nation's income.

Critics point out that these figures do not include medical and other noncash benefits received by many poorer Americans. They also note that these are pretax figures; taxes have the effect of reducing income disparities. There is some merit to these criticisms. Nevertheless, when public benefits and the effects of taxes are accounted for, the proportional results shown in Table 6.1 still hold (Gramlich, Kasten, & Sammartino, 1993; Hacker, 1997). Moreover, if one studies the distribution of wealth rather than of income alone, the picture is even more skewed in favor of the rich. If we measure wealth in terms of the net financial worth of households—that is, their total assets–only about 6% of American households have a net worth of $250,000 or more, whereas over 25% have a net worth of less than $5,000.

Recent studies show that the United States has replaced Great Britain and Ireland as the Western industrial nation with the largest gap between rich and poor (Hacker, 1997; Wolff, 1995). The percentage of total private wealth owned by the richest 1% of the U.S. population is rising quickly, while the opposite is true in Great Britain, which, unlike the United States, has never been thought of as an egalitarian society.

Figure 6.6 shows how net worth is distributed among some of the most significant aspects of household wealth. The richest 10% of families own homes valued at about 35% of the total worth of American homes. The same 10%, however, own almost 80% of other real estate, over 90% of securities (stocks and bonds), and about 60% of the money in bank accounts (Hacker, 1997). Remember, too, that inequality expressed in dollars of income or possession of property and stocks translates into large differences in what people can spend on health care, shelter, clothing, education, books, movies, trips to parks and museums, vacations, and other necessities and comforts.

The distribution of educational attainment is more nearly equal than the distribution of wealth and income. **Educational attainment,** or number of years of school completed, has become more equal among major population groups in the United States, but important inequalities remain. The educational attainment of the nation's black population, for example, has risen dramatically in the past generation. In 1960, only 20% of African Americans (who were age 25 or older at the time) completed four years of high school or more, while the comparable figure for whites was 43%. Today, about 77% of African Americans have completed four years of high school or more, compared to about 84% of whites. However, only about

FIGURE 6.6
Proportion of American Families Owning Various Forms of Wealth

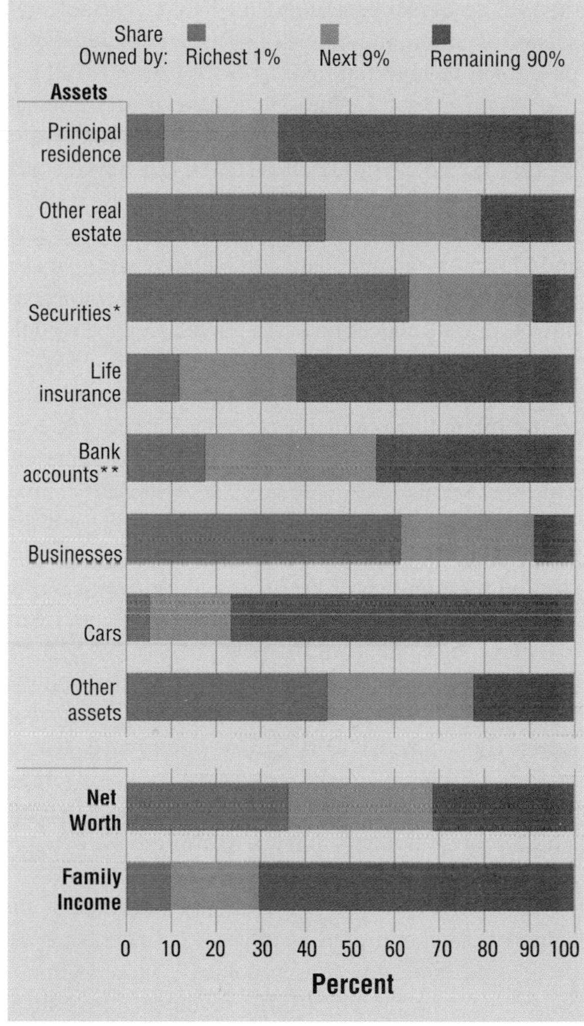

*Includes stocks, bonds, and trusts.
**Includes all deposits in checking, savings, and other accounts.
Source: Federal Reserve.

15% of African Americans have completed four years or more of college, compared to almost 26% of whites. And for Latinos the gap remains far wider. Approximately 56% of Latinos in the U.S. population (age 25 or older) have completed four years of high school or more, and about 11% have completed four years of college or more (*Statistical Abstract,* 2000).

Women, on the other hand, have nearly attained parity with men in terms of educational attainment. Eighty-three percent of adult women and 83% of men in the U.S. population had completed high school or beyond in 1999. Over twenty-seven percent of men and 23% of women had completed four years of higher education or more. We can see in Table 6.2, however, that educational attainment does not yield the same benefits for women as for men.

The last two decades of the twentieth century saw an important increase in the relationship between educational attainment and income. The advantage enjoyed by individuals with a college education over those with only a high school diploma has always been significant, but as shown in Table 6.2, it increased by almost 16% for men and by over 22% for women between 1975 and 1995. In 1975, a man with a college diploma could expect to earn $1,510 for every $1,000 earned by a male worker with only a high school diploma. These averages were somewhat lower for women. In 1995, by contrast, a female worker with a college degree earned, on the average, $1,658 for every $1,000 earned by a woman with only a high school diploma, and the figures are even higher for males.

In a rapidly changing economy like that of the United States, educated people are in increasing demand, and education thus becomes a more common route to upward mobility. Improvements in **educational achievement**—in basic reading, writing, and computational skills—are increasingly vital to individual careers. For the entire society, improvements in educational achievement represent improvements in the nation's "human capital," the wealth-producing capacity of its people (Harbison, 1973). But educational attainment and educational achievement are not always correlated. When students are moved through the school system without meeting certain minimum achievement criteria, attainment figures mask significant gaps in achievement. As a result, large segments of the population are unable to achieve nearly as much as they might, either for themselves or for their society.

These measures of wealth, income, and education indicate that not all Americans share equally in "the good life." But what are the larger consequences of these patterns of inequality, both for individuals and for society? Do they combine to form an identifiable system of social-class stratification in the United States? Do marked patterns of inequality contradict the American dream, the vision of the United States as a place where hard work and sacrifice will lead to success and material comfort?

TABLE 6.2
Payment by Degrees: 1975 Versus 1995 (average earnings of full-time workers, age 35 to 44 in the years' actual dollars)

	1975	
	Men	**Women**
High school diploma only	$14,007	$ 7,774
Bachelor's degree only	21,152	10,560
Bachelor's per $1,000 high school earnings	1,510	1,358
	1995	
	Men	**Women**
High school diploma only	$32,689	$22,257
Bachelor's degree only	57,196	36,901
Bachelor's per $1,000 high school earnings	1,750	1,658
Advantage for bachelor's degree (change from 1975 to 1995)	+15.8%	+22.1%

Source: Hacker, 1997.

Perceptions of Class Inequality

Sociologists often use opinion surveys when studying how people perceive their membership in a social class. They use more objective measures of income, wealth, or access to health services and education when they are interested in the effects of class membership on other aspects of a person's or a family's life.

The National Opinion Research Center measures subjective social class by asking thousands of respondents the following question: "People talk about social classes such as the lower class, the working class, the middle class, and the upper class. Which of these classes would you say you belong to?" Most people in the survey—at least 97%—assign themselves to a social class. The data in Table 6.3 show that 45% of men and over 46% of women classify themselves as members of the middle class, while far fewer African Americans (28%) assign themselves to this class. And even though the category "lower class" has a negative connotation for many people, we see in the table that almost 6% of women and almost 12% of African Americans rank themselves in this social class. These and many similar survey studies lead sociologists to conclude that instead of being clearly and rigidly defined, classes emerge in the public's awareness as clusters of people with similar socioeconomic standing (Jackman & Jackman, 1983).

The term *socioeconomic standing* (usually referred to as **socioeconomic status** or **SES**) requires further definition. When people think of social-class divisions in American society, they first assign various occupations to broad class ranks. When there is confusion about how an occupation is ranked, people tend to think of other aspects of social class, such as family prestige, education, and earned income. And as they think about these factors, they tend to reach a consensus about what social class they belong to and what classes others should be assigned to.

But that consensus is by no means perfect. Some blue-collar occupations—for example, many skilled trades, such as plumber and electrician—are relatively well paid; in fact, they are better paid than many office jobs. This leads people to classify the holders of such jobs as members of the middle class, while at the same time they may assign holders of jobs in offices and stores to the working class. These problems of classification make it extremely difficult to resolve the ambiguities in how people distinguish among social classes. Nevertheless, when we turn to how membership in social classes actually affects the lives of people in the United States or elsewhere, we see very important consequences of class stratification and inequality.

Social Class and Life Chances

The rich have far more money than the poor, and they tend to have more education and a great deal more wealth, as measured by the value of homes, cars, and

TABLE 6.3
Social Class, by Sex and Race

Social Class by Sex	Lower Class	Working Class	Middle Class	Upper Class	No Class	Total
Male						
Percent	45.0	47.4	45.0	3.1	0	100
N	607	6,423	6,087	425	0	13,542
Female						
Percent	5.7	45.0	46.2	3.1	0.01	100
N	977	7,681	7,899	524	1	17,082

Social Class by Sex	Lower Class	Working Class	Middle Class	Upper Class	No Class	Total
White						
Percent	4.1	44.2	48.7	3.0	0	100
N	1,049	11,380	12,537	783	1	25,750
Black						
Percent	11.6	58.6	28.2	3.6	0	100
N	482	2,350	1,170	150	0	4,152
Other						
Percent	7.3	51.8	38.6	2.2	0	100
N	53	374	279	16	0	722
Total number of cases	30,624					

Source: NORC, 2000.

investments. But what differences do these inequalities make in people's lives? Social scientists often answer this question by analyzing the life chances of people born into different social classes. As noted earlier, the term *life chances* refers to the likelihood that individuals will have access to the opportunities and benefits that their society values. Compare, for example, the life chances of a child born into a family in which the mother and father earn slightly more than the minimum wage by working in restaurants and supermarkets with the life chances of a child born into the home of a police officer and a teacher. Now compare both of these with the life chances of a child born into the home of a successful banker. Will these differences in the circumstances of birth affect the quality of education each child is likely to receive? Will the children's access to high-quality health care differ? Will the differences in the social class of their families influence whom they are likely to marry? Will it make a difference in the likely length of their lives? The answer to all of these questions is, emphatically, yes. But being born into a given social class does not determine everything about an individual's life chances. The more a society attempts to equalize differences in life chances—by improving health care for the poor, for example, or by creating high-quality institutions of public education—the more it reduces the impact of social class on life chances.

Numerous social-scientific studies have shown that one's social class tells a great deal about how one will behave and the kind of life one is likely to have. Following are some typical examples of the relationship between social class and daily life. These examples apply primarily to American society, but many of the same conditions can be found in other societies as well.

Class and Health. A child born into a rich upper-class family or a comfortable middle-class family is far less likely to be born premature or with a low birth weight than one born into a working-class or poor family. And a baby born into a family in which the parents are working at steady jobs is far less likely to be born with a drug addiction or AIDS or fetal alcohol syndrome than a child born to parents who are unemployed and homeless. These and many other disparities contribute to what is often called "the socioeconomic status (SES) gradient in health" (Sapolsky, 1998).

The SES gradient in health exists in all of the world's nations and is based on a complex combination of social class and culture. But the gradient is particularly marked in the United States:

> For example, in the United States the poorer you are, the more likely you are to contract and succumb to heart disease, respiratory disorders, ulcers, rheumatoid disorders, psychiatric diseases, or a number of types of cancer. And this is a whopper of an effect: In some cases disease or mortality risk increases more than fivefold as you go from the wealthiest to the poorest segments of our society, with things worsening each step of the way. (Sapolsky, 1998, p. 46)

Among adults, a salaried member of the upper class who directs the activities of other employees is less likely to be exposed to toxic chemicals or to experience occupational stress and peptic ulcers than wage workers at their machines and computer terminals. Those workers, in turn, are more likely to have adequate health insurance and medical care than the working poor—dishwashers, migrant laborers, temporary help, low-paid workers, and others whose wages for full-time work do not elevate them above the poverty level (Ellwood, 1988). The working poor are the largest category of poor Americans, and like those who lack steady jobs, they often depend on local emergency rooms for medical care and report that they have neither family doctors nor health insurance. The same poor and working-class population is also more likely to smoke, consume alcohol, and be exposed to homicide and accidents—while receiving less police protection—than members of the classes above them.

Education. Children of upper-class families are more likely to be educated in private schools than children from the middle or working classes. Sociologists have shown that education in elite private schools is a means of socializing the rich. Peter Cookson and Caroline Persell's (1985) study of socialization in elite American prep schools found that "preppies" develop close ties to their classmates, ties that often last throughout life and become part of a network they can draw upon as they rise to positions of power and wealth. The segregation of upper-class adolescents in prep schools also serves the purpose of limiting dating and marriage opportunities to members of the same class.

Although middle-class parents are more likely than rich parents to send their children to public schools, they tend to select suburban communities where the schools are known to produce successful college applicants. The public schools that serve the middle classes spend more per pupil than the schools attended by working-class and poor children, and they offer a wider array of special services in such areas as music, sports, and extracurricular activities. Children in the middle and upper classes also tend to have parents who insist that they perform well in school and who can help them with their schoolwork. Moreover, children from working-class and poor families are more likely to drop out of school than children from upper-class families.

Politics. The poor and members of the working class are more likely to vote for Democratic party candidates, whereas upper- and middle-class voters are more likely to choose Republican candidates. Throughout the industrialized world, voters with less wealth, prestige, and power tend to vote for candidates who promise to reduce inequalities, while voters with higher socioeconomic status tend to choose candidates who support the status quo. Thus, in their studies of social-class identification Robert and Mary Jackman (1983) found that 48.5% of poor respondents and 43% of respondents who assign themselves to the working class believe that the federal government should be doing more to achieve full employment and job guarantees, as opposed to only 24% of upper-middle-class respondents. They also found that about 48% of working-class respondents believe that "some difference" in levels of income (though less than currently exists) is desirable in order to sustain people's motives to achieve, whereas almost 50% of upper-middle-class respondents feel that a "great difference" is desirable. The most recent survey data (NORC, 2000) confirm the Jackmans' findings. Members of all classes tend to agree that whatever differences there are ought to be based on individual achievement rather than on advantages inherited at birth. Members of the working class and the poor, however, are more likely than members of other classes to vote for candidates who propose measures that would increase equality of opportunity.

Many other examples could be presented to illustrate the influence of social class on individuals in American society. But social-class divisions are so important in the United States and throughout the world that sociologists have devoted a great deal of thought and research to explaining why these divisions come about and persist in different types of societies and historical periods. Indeed, the effort to rid society of social classes is one of the great failed experiments of the twentieth century, as we will see in the next section.

THEORIES OF STRATIFICATION AND INEQUALITY

During the era of feudalism in the West, power and authority were a monopoly of the powerful landowning lords. All the strata below them—peasants, townspeople, and others—were subject to their rule and had no recourse to any other authority if the lords dealt with them cruelly or capriciously. As kings consolidated their power, they forced the feudal lords to recognize royal authority, often with the help of the growing middle class of artisans and merchants. This led the middle class to begin demanding rights and guarantees of protection under the law. Eventually a series of revolutions established the principle that rulers are elected by the public and are subject to laws enacted through legislative and judicial processes approved by the public. This enormous change in the conduct of human life has not occurred everywhere, however. There are still large areas of the world where aristocratic rule—usually in the form of dictatorship—continues, and the dictators usually represent and promote the interests of a rich and powerful ruling class. So when we speak of stratification in the modern era, and particularly of the rise of the middle classes, we are discussing a process that in many parts of the world lies in the future rather than in the past.

The feudal system was largely destroyed by a new social order known as capitalism. Capitalism is a form of economic organization based on private ownership and control of the means of production (land, machines, buildings, etc.). Whereas under feudalism land was viewed as belonging to a family as a matter of birthright, under capitalism land can be bought and sold like any other commodity. And whereas feudal labor systems were based on institutions like serfdom, in which peasants were bound to a particular lord, in capitalist systems workers are free to sell their labor to the highest bidder (and to endure unemployment when work is unavailable). Capitalism is also an ideology based on the value of individual rights; the ideology of feudalism, in contrast, stressed the reciprocal obligations of different strata of society.

In descriptions by observers of nineteenth-century society, capitalism seemed almost to explode onto the world scene. Capitalism made possible the dramatic change in production methods that we call the industrial revolution, which transformed England and Europe from a world of towns and villages, courts and cathedrals to a world centered on markets,

THE CENTRAL QUESTION: A CRITICAL VIEW

Can the social-class inequalities that affect our lives so dramatically ever be reduced?

Sociologists and other experts on world inequality believe that this central question in human affairs hinges more than ever on how societies address the problem of poverty among women and children. Contemporary social thought on this question is greatly influenced by feminism and direct social action by women activists. It emphasizes that almost everywhere in the world, women, as mothers and as citizens, face greater obstacles to their health and the ability to realize their full human potential than men do..This ubiquitous form of inequality is not due to biological factors or to any innate aspects of womanhood but to the dominance of men and the relegation of women to inferior status, a condition that adversely affects children in rich and poor nations alike.

In Africa and Asia, women and children are far more likely than adult men to be victims of starvation and disease. In affluent nations like the United States, women and children are far more likely to be living below the official poverty line than adult males. The accompanying graphs show these facts quite clearly. For instance, 22% of young people under the age of 18 were below the poverty level in 1994, almost twice the percentage in other age categories. And in every age category women are more likely to be living in poverty than men.

You will hear some people assert that these differences exist because too many young women are having babies outside of marriage and too many female single parents are having additional children in order to avoid going to work. These arguments blame the women and their children for being poor. We all know some people who are irresponsible, but does teenage pregnancy or a desire to avoid going to work explain why so many women and children are poor throughout the world? How can the study of social stratification help explain these social facts?

Children who are poor did not bring poverty on themselves. They were born into a poor household or one that became poor because the adults in that household became poor (O'Hare, 1996). Women are more likely to be poor than men because they have less control over their life situations and social class positions. In the developing regions of the world, women have far less access to education, jobs, health care, and political power than men do. In the United States and other industrial nations, women with children often become poor when they are divorced or separated and left with the full responsibility for their children. And contrary to what many people believe, the majority of female single parents are poor because although they are working, they are not earning enough to raise themselves and their children above the poverty line (Bianchi & Spain, 1996).

But these sociological explanations often seem to run counter to experience. People may point out to you that so-and-so "was loose and got pregnant and now is living an immoral life, having more children, . . ." Such individual stories may perhaps be true, but what about the sociological evidence? Is it based on individual stories or on larger trends and social facts? As you make up your mind on these issues, remember that both individual stories and sociological facts can be true. The sociological explanations, however, account for far more of the differences in rates of poverty among men, women, and children. Do sociological explanations absolve irresponsible people from blame? Hardly. But neither do individual examples of irresponsibility explain major social trends.

Poverty Rates by Gender and Age

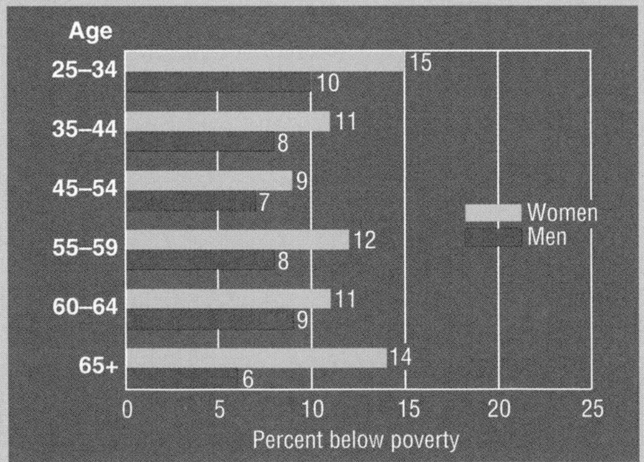

Source: Bianchi & Spain, 1996.

Poverty Rates of Persons by Selected Characteristics

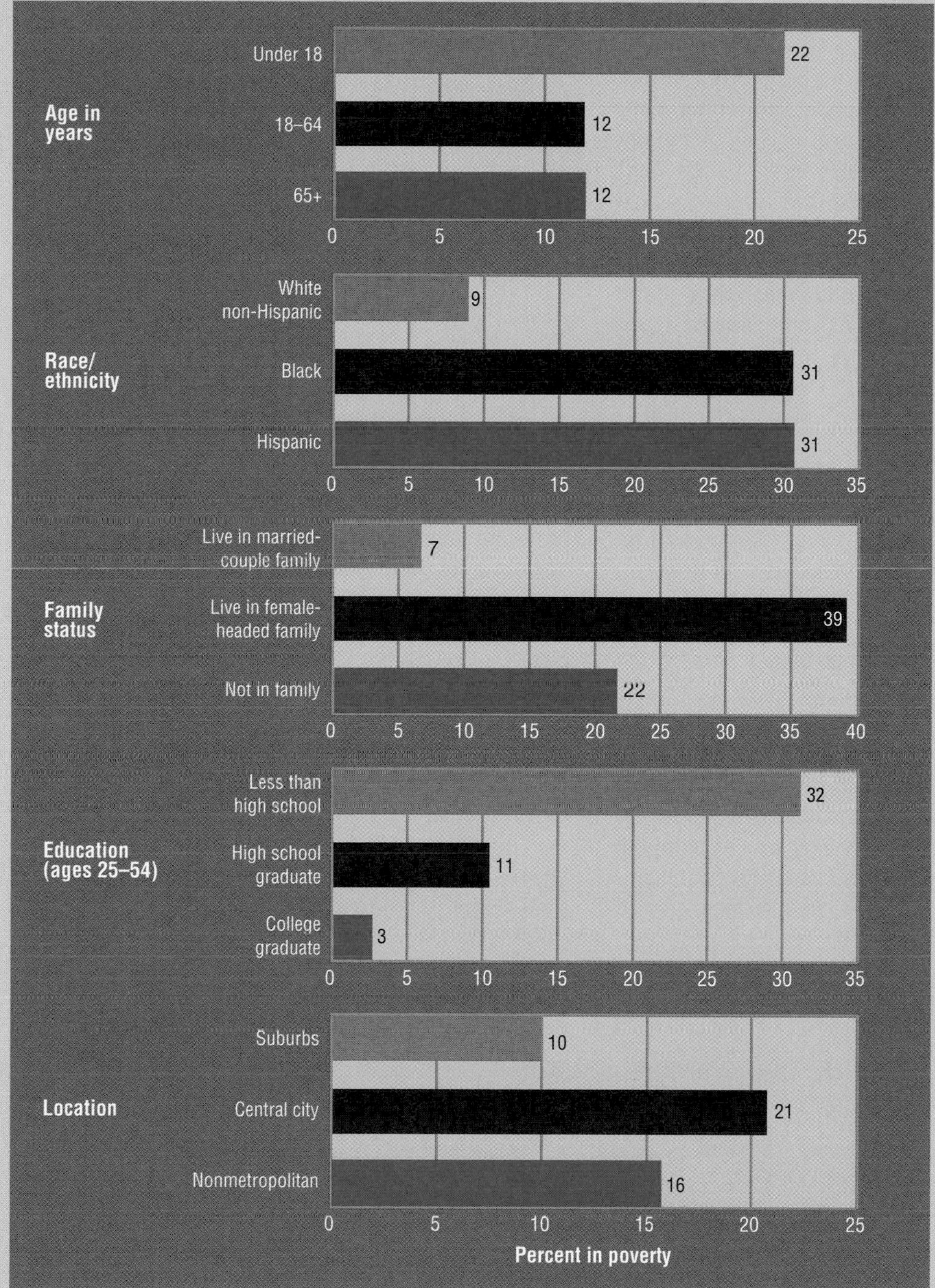

Source: O'Hare, 1996.

factories, and crowded cities. A few places where the physical structures of feudalism remain, such as Mont-Saint-Michel in France or Dubrovnik in Croatia, continue to illustrate the contrast between the old order and the new.

The industrial revolution began in England and other parts of Europe in the late seventeenth century. Today it is still occurring in other parts of the world, including China, India, and Africa. The former colonial outposts of England, France, and Germany are only now undergoing the transformation from rural agrarian societies to urban industrial states in which an ever-decreasing portion of the population is engaged in farming. Although not all of these industrializing societies have capitalist economies, they all are developing a class of managers, entrepreneurs, and political power brokers.

It is not enough, however, to describe the industrial revolution in terms of the spread of industrial technology and urban settlement. Technological innovations like the steam engine, the railroad, the mechanization of the textile industry, and new processes for making steel and mining coal were accompanied by equally important innovations in social institutions. In the words of sociologist and historian Karl Polanyi (1944), the industrial revolution was a "Great Transformation." For the first time in human history, the market became the dominant institution of society. By *market* we do not refer only to the places where villagers sold produce or traded handicrafts. Rather, the market created by the industrial revolution was a social network that gradually extended over the entire world and linked buyers and sellers in a system that governed the distribution of goods of every imaginable type, services of all kinds, human labor power, and new forms of energy like coal and fuel oil.

Among other key elements of the Great Transformation were these:

- Goods, land, and labor were transformed into commodities whose value could be calculated and translated into a specific amount of gold or its equivalent—that is, *money* (Marx, 1962/1867; Schumpeter, 1950; Weber, 1958/1922).
- Relationships that had been based on ascribed statuses were replaced with relationships based on *contracts*. A producer hired laborers, for example, rather than relying on kinship obligations or village loyalties to supply workers (Polanyi, 1944; Smith, 1910/1776).
- The business firm or *corporation* replaced the family, the manor, and the guild as the dominant economic institution (Weber, 1958/1922).
- Rural people, displaced from the land, began selling their labor for *wages* in factories and commercial firms in the cities (Davis, 1955).
- In the new industrial order, demands for *full political rights* and *equality of opportunity*, which originated with the middle class, slowly spread to the new class of wage workers, to the poor, and to women, especially in societies in which revolutions created more open stratification systems (Bendix, 1969; de Tocqueville, 1980/1835; Mannheim, 1941).

Conflict Theories

Marx's Theory: Class Consciousness and Class Conflict. The Great Transformation had a profound impact on the stratification systems of modern societies. It produced new and powerful social classes and thereby changed the way people thought about their life chances and the legitimacy of their society's institutions. As Karl Marx wrote in the last chapter of *Capital,* "The owners merely of labor-power, owners of capital, and landowners, whose respective sources of income are wages, profit, and ground rent . . . constitute the three big classes of modern society based upon the capitalist mode of production" (1962/1867, pp. 862–863). Here as elsewhere, Marx defined classes in terms of the "modes of production" that are characteristic of a society in a given historical period. The workers, by far the largest class in modern societies, must sell their labor to a capitalist or landowner in return for wages. The capitalists and landowners are far less numerous than the workers, but because they own and control the means of production, they command more of everything of value than the workers do.

Marx believed that the class of wage workers created by capitalism would inevitably rise up against the capitalist class and create a classless socialist society. But this prediction was proved wrong: Successful workers' revolutions never occurred in the most industrialized nations. Yet Marx's description of how members of the working class become conscious of their situation as a class remains an important subject in the study of stratification.

Sources of Class Conflict. "The history of all hitherto existing societies is the history of class struggles," wrote Marx and his collaborator Friedrich Engels in *The Communist Manifesto* (1969/1848, p. 11). In the modern era, "society as a whole is more and more splitting up into two great hostile camps, into two great classes directly facing each other" (p. 11). Those two classes were the owners of capital and the workers, which Marx termed the *bourgeoisie* and the *proletariat.* Why did Marx and other observers of the capitalist system believe that conflict between the bourgeoisie and the proletariat was inevitable? The answer can be found by taking a closer look at Marx's analysis of the evolution of capitalism.

To begin with, Marx observed the misery of industrial workers (many of whom were children) in the smoky factories of industrial England. He saw that however wretched the workers were, there would always be what he called a "reserve army" of unemployed people who would be willing to work for even lower wages. He noticed that the capitalists (and the intellectuals they paid to argue in their defense) always blamed the workers themselves for the miserable conditions in which they were forced to exist. If they were hungry, the capitalists claimed, it was because they did not work hard enough or because they spent their pay on too much alcohol or because they could not curb their sexual passions and bore too many children. Thus, the capitalists refused to accept blame for the misery of the working class.

Marx also argued that business competition would eliminate less successful firms and result in monopolies, which would control prices and wages and thereby contribute still more to the impoverishment of the workers. Moreover, the capitalists had the power to determine who ran the government and who controlled the police and the army. If the workers were to rebel, the armed forces and police would intervene as agents of the capitalists. Through these means, the workers and the unemployed would be forced to remain a huge, helpless population that could be manipulated by the capitalists. Over time, according to Marx and his followers, these masses of people would become increasingly conscious of their plight and would unite in a revolution that would destroy the power of the capitalists and their allies.

But why would workers become more conscious of their situation as a class, rather than merely remaining miserable in their impoverished families and communities? This problem of class consciousness became a central issue in Marxian thought. The study of changing patterns of class consciousness remains a major subject in sociology to this day.

Objective and Subjective Classes. In thinking about how a social class might be able to take collective action, revolutionary or otherwise, Marx distinguished between *objective* and *subjective* classes. All capitalist societies, according to Marx, have **objective classes**—social classes that an observer can identify by simply looking at people's visible, specific relationships to the means of production in their society. The workers, for example, are easily identified as an objective class that does not own the means of production and that sells its labor power in return for wages. The capitalists are the objective class that owns the means of production—the machines and property and railroads—and that buys the labor power of the working class.

These facts would be visible to an observer equipped with an understanding of what makes a social class, but they might not always be evident to the members of those classes themselves. Thus, Marx identified **subjective class** as the extent to which the people in a given stratum of society actually perceive their situation as a class. For example, if the workers in the chicken-processing industry in the American South are not aware that their low wages and miserable working conditions are similar to the conditions faced by millions of workers in other industries elsewhere in the world, and if they do not understand that they can improve their fortunes only by taking some of the power of another class (the capitalists), they are not yet a subjective class. Without this awareness of their situation, the workers are said to lack **class consciousness.** And without class consciousness, they cannot form the political associations that would enable them to fight effectively against the capitalists.

The Classless Society. Marx and other observers of early capitalism believed that the growing conflict between the working class or proletariat and the capitalist class or bourgeoisie would produce revolutions. In those revolutions, the proletariat and its allies would depose the bourgeoisie and establish a new social order known as socialism. Under socialism, the key institutions of capitalism—private ownership of the means of production, the market as the dominant economic institution, and the nation-state controlled by the bourgeoisie—would be abolished. The new society would be classless because the economic institutions that produced classes would have been eliminated and all the members of society would collectively own the means of production.

Class Conflict Theories Since Marx. As mentioned earlier, Marx's theory of stratification asserts that capitalist societies are divided into two opposing classes, wage workers and capitalists, and that conflict between these two classes will eventually lead to revolutions that will establish classless socialist societies. However, Marx's prediction was not borne out in any society that attempted to implement his ideas on a large scale. Deny it as they might, all of those societies developed well-defined systems of social stratification (Djilas, 1982; Parkin, 1971; Szelenyi, 1983). Each had an elite of high party officials; an upper stratum of higher professionals, scientists, managers of economic enterprises, local party officials, and high police officials; a middle level of well-educated technical workers and lower professionals; a proletariat of industrial and clerical workers and military personnel; and a bottom layer of people who were disabled, criminals, or political outcasts.

No persuasive evidence shows that class conflict is heightening the division between workers and owners of capital in capitalist societies. Conflict does exist, and we have seen that the gap between rich and poor is widening. But the industrial working class is shrinking, and the new occupational groups do not always share the concerns of the industrial workers. Moreover, reforms of capitalist institutions have greatly improved the workers' situation, thereby reducing the likelihood that the revolution predicted by Marx will occur in the urban industrial nations.

Contemporary theorists who draw upon the original theories of Marx and his followers do not deny the failures of efforts to create classless societies in the Soviet empire and China. On the other hand, they argue that class conflict continues to explain many aspects of class stratification in the modern world. These theorists argue that class inequalities often deprive people who are outside the upper and upper-middle classes of the ability to realize their full human potential (Tumin, 1967; Willis, 1983). Problems like wasted talent and poor self-image are among what Richard Sennett and Jonathan Cobb (1972) term "the hidden injuries of class"—meaning the ways in which a childhood of poverty or economic insecurity can leave its mark on people even after they have risen out of the lower classes.

Reproduction Theory. Contemporary theories of class conflict and its consequences often describe the ways in which the stratification systems of the United States and other industrial societies operate to reproduce class inequalities. Although American society is often claimed to have a highly open class structure in which people who work hard and strive for education both for themselves and for their children can gain upward mobility, reproduction theory suggests that in fact people from the lower classes are relegated to inferior schools and mediocre jobs, and that the media, controlled by the business elite, help convince them that this is how things should be (Egerton, 1999; Savage, 1997). Reproduction theory also attempts to explain why it is that among the poor and working classes women and members of racial or ethnic minorities are often doubly penalized by their class and status. These are issues to which we will return in detail in coming chapters.

Functionalist Theories of Class and Inequality

The functionalist view of stratification was originally stated by Talcott Parsons (1937, 1940) and Kingsley Davis and Wilbert Moore (1945). This theory holds that social classes emerge because an unequal distribution of rewards is essential in complex societies. Such societies need to reward talented people and channel them into roles that require advanced training, personal sacrifice, and extreme stress. Thus, the unequal distribution of rewards, which allows some people to accumulate wealth and deprives others of that chance, is necessary if the society is to match the most talented individuals with the most challenging positions.

Research by sociologists in the former Soviet Union often provided support for the functionalist idea that extreme "leveling" deprives people of the motivation to achieve more than a minimum of skills. On the basis of extensive surveys of workers in Russia and other former Soviet republics, sociologist Tatyana Zaslavskaya (cited in Aganbegyan, 1989) argues that unless there are incentives in the form of high wages and other advantages (such as better housing in areas where good housing is in short supply), engineers and scientists will resent their situation and will not work hard. In the United States, the same arguments are used to justify higher salaries for doctors, lawyers, and other professionals. Too much equality, it is said, reduces the incentive to master difficult skills, and as a result the entire society may suffer from a lack of professional expertise.

Critics of the functionalist view of inequality and stratification point to many situations in which people in positions of power or leadership receive what appear to be excessive benefits. In the early 1990s, during the most prolonged economic recession the United States had experienced since the Great Depression of the 1930s, the multimillion-dollar salaries of many corporate executives were frequently criticized in business magazines and news analyses. For example, Roberto Goizueta, former chief executive officer of the Coca-Cola Company, was criticized for taking home more than $10 million in salary and bonuses. Many other executives were earning equally high salaries even when their companies were not doing as well as they might in international competition. With the annual pay of top Japanese executives often running 5 to 10 times less than that of their American counterparts, it became difficult for the American corporate elite to argue that these enormous rates of executive compensation are "functional."

From the point of view of those who are critical of social inequality, the large sums paid to a few people seem wrong, especially when so many others are struggling just to survive. Indeed, the heads of large corporations in the United States often earn more than 50 times as much as the average employee of those corporations. Is such great disparity "functional" when it produces enormous gaps between the very rich and the working classes? Functionalist theory claims that it is, for in a capitalist system of free enterprise, top executives will seek the firms that are most willing to reward them for their talents. Those firms will benefit, and so will their workers.

The Interactionist Perspective

Conflict and functionalist theories explain stratification primarily in economic terms. Both trace the existence of certain classes to the central position of occupation, income, and wealth in modern life. But neither goes very far toward explaining the prestige stratification that occurs *within* social classes. Among the very rich in America, for example, people who have stables on their property tend to look down on people with somewhat smaller lots on which there is only a swimming pool. And rich families who own sailing yachts look down on equally rich people who own expensive but noisy powerboats. Within an economic class, people form status groups whose prestige or honor is measured not according to what they produce or how much wealth they own but according to what they buy and what they communicate about themselves through their purchases. Designer jeans and BMW cars are symbols of membership in the youthful upper class. Four-wheel-drive vehicles equipped with gun racks and fishing rods are symbols of the rugged and successful middle- or working-class male. Armani suits are symbols of urban professionalism; tweed suits and silk blouses are signs that a woman is a member of the "country club set." All of these symbols of prestige and group membership change as groups with less prestige mimic them, spurring a search for new and less "common" signs of belonging (Dowd, 1985).

Our tendency to divide ourselves up into social categories and then assert claims of greater prestige for one group or another is of major significance in our lives. The interactionist perspective on stratification therefore may not be very useful in explaining the emergence of economic classes, but it is essential to understanding the behaviors of the status groups that form within a given class. Those behaviors, in turn, often define, reinforce, or challenge class divisions. The stratification system, in this view, is not a fixed system but is created over and over again through the everyday behaviors of millions of people.

Stratification systems can change when new industries—for example, the production of computers and all their components—bring new jobs and new wealth to some regions of a nation. In other cases, new laws that affect the lives of people in specific social classes, particularly the poor, may affect people's access to the means of existence.

Interactionist views of social stratification tend to explain why people form status groups within classes. They also explain how, in their daily interactions with neighbors and associates, people come to take class inequalities for granted or even defend them as inevitable. But as noted earlier, an important test of the power of theories of class inequality hinges on the actual mobility of people and their children up and down (and especially up) the social-class hierarchy. We turn, therefore, to a brief examination of the issue of social mobility in modern societies.

Social Mobility in Modern Societies: The Weberian View

Max Weber's theory of social stratification cuts across all three of the perspectives just discussed. Well before the failure of Soviet communism, Weber took issue with Marx's view of stratification. Marx had defined social class in economic terms; classes are based on people's relationship to the means of production. Weber challenged this definition of social class.

People are stratified, Weber reasoned, not only by how they earn their living but also by how much honor or prestige they receive from others and how much power they command. A person could be a poor European aristocrat whose lands had been taken away during a revolution, yet his prestige could be such that he would be invited to the homes of wealthy families seeking to use his social status to raise their own. Another person could have little money compared with the wealthy capitalists and little prestige compared with the European aristocracy, yet she could command immense power. The late Mayor Richard J. Daley of Chicago (the father of the present mayor) was such a person. He was born into a working-class Irish American family, and although he was rather well-off, he was not rich. Yet his positions as mayor and chairman of the Cook County Democratic party organization made him a powerful man. Indeed, Mayor Daley's fame as a politician who led a powerful urban party organization but did not enrich himself in doing so helped his son continue the family tradition of political leadership: In 1989, Richard M. Daley was elected mayor of Chicago.

For Weber and many other sociologists, therefore, wealth or economic position is only one of three dimensions of stratification that need to be considered in defining social class. Prestige (or social status) and power are the others. Think of modern stratification systems as ranking people on all of these dimensions. A high ranking in terms of wealth does not always guarantee a high ranking in terms of prestige or power, although they often go together.

Other challenges to Marx's view of stratification focus on social mobility in industrial societies. Contrary to Marx's prediction, modern societies have not become polarized into two great classes, the rich and the poor. Instead, there is a large middle class of people who are neither industrial workers nor capitalists (Wright, 1989, 1997), and there is considerable social mobility, or movement from one class to another.

Forms of Social Mobility. Social mobility can be measured either within or between generations. These two kinds of mobility are termed *intragenerational* and *intergenerational* mobility. **Intragenerational mobility** refers to one's chances of rising to or falling from one social class to another within one's own lifetime. **Intergenerational mobility** is usually measured by comparing the social-class position of children with that of their parents. If there is a great deal of stability from one generation to the next, one can conclude that the stratification system is relatively rigid.

Unfortunately, not all mobility is upward. Downward mobility involves loss of economic and social standing. It is a problem for families and individuals at all levels of the class structure in the United States and elsewhere. In its broadest sense, downward mobility can be defined as "losing one's place in society" (Newman, 1988, p. 7). In fact, the term encompasses many different kinds of experiences. The married woman who works part-time and then is divorced, loses the family house, and must move with her children to a small apartment and work full-time is experiencing downward mobility. The couple who have been living with the wife's mother in public housing and are forced out during a check of official rosters and must seek refuge in a homeless shelter also experience downward mobility. The affluent young couple living in a downtown condominium experience downward mobility when he loses his job at a brokerage firm and she must go on maternity leave while he is looking for a new job. Often entire groups of people experience downward mobility, as happened to U.S. citizens of Japanese descent when their property was confiscated during World War II, and to people of Indian descent when they were expelled from Kenya and Uganda in the 1970s.

The General Social Survey, conducted annually by the National Opinion Research Center (NORC), charts changes in people's opinions about their prospects for mobility in either direction, using questions like these:

> Compared to your parents when they were the age you are now, do you think your own standard of living now is much better, somewhat better, about the same, somewhat worse, or much worse than theirs was?
>
> When your children are at the age you are now, do you think their standard of living will be much better, somewhat better, about the same, somewhat worse, or much worse than yours is now? (NORC, 1994, p. 199)

Fear of downward mobility weighs heavily on the minds of many Americans. A central feature of the American dream is that our standard of living, and that of our children, will continue to improve. But that vision is challenged by the realities of global competition and the "export" of jobs to lower-wage regions of the world (Block, 1990; Harrison, Tilly, & Bluestone, 1986). Indeed, the findings of the General Social Survey show

FIGURE 6.7
Perceptions of Mobility Opportunities (respondent's own and children's)

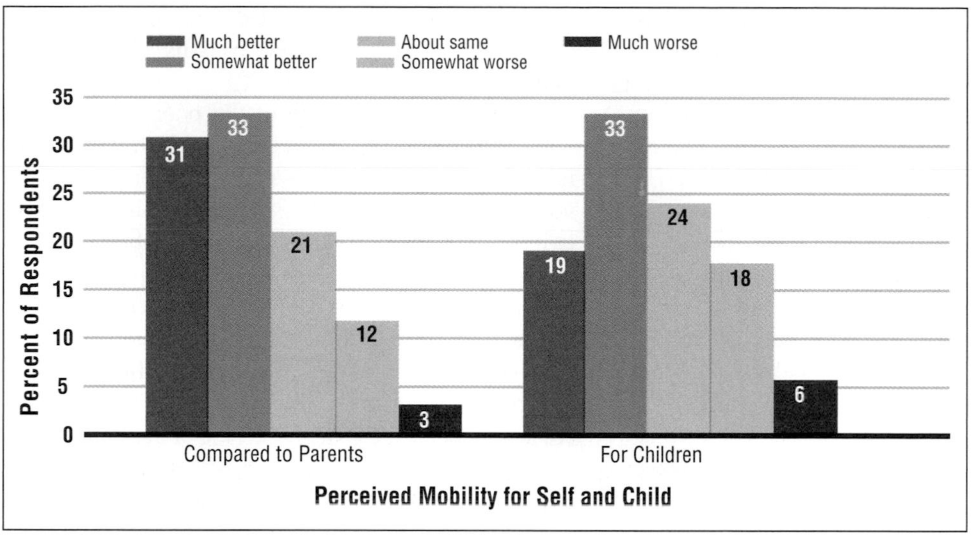

Source: NORC, 1994.

that more respondents claim to have experienced an improvement in their standard of living compared to their parents than believe their children will also enjoy such mobility. These results are presented in Figure 6.7. We will see in later chapters that this doubt about future opportunities for one's children is an extremely powerful force in contemporary politics.

MORE EQUALITY?

While people in the United States and other affluent nations worry about the economic and social opportunities open to their children, billions of people around the world are hoping that economic growth will lift them out of poverty. As we saw earlier in the chapter, a large proportion of the world's people are extremely poor. India, China, Bangladesh, the Philippines, Malaysia, most of sub-Saharan Africa, the mountain regions of Latin America, and many other parts of the world are inhabited by large numbers of people with drastically limited incomes and only the most rudimentary health care, education, and other resources. If poverty continues to increase and the hopes of impoverished people for a better life are frustrated, it will become increasingly difficult to maintain peace, achieve slower population growth, and prevent massive environmental damage in those regions.

Poverty is reduced through better health care, more schools, more food and shelter, and more jobs—all of which require economic and social development. The experience of the more affluent nations indicates that such development is likely to produce larger working and middle classes and reduce the rate of poverty. But the persistence of poverty and near-poverty in the United States and other industrialized nations shows how difficult it is to promote greater equality elsewhere in the world.

Poverty: A Relative Concept

Social scientists often point out that poverty is not an absolute concept but must be measured relative to the standards of well-being in particular societies. A family of four in the United States with an annual income of $16,400 is poor by official standards, yet compared with equivalent families in India or Africa, the U.S. family is doing quite well. The U.S. family probably has permanent shelter, even if it is in a slum or a rundown neighborhood. It probably receives some form of government income supplement, if only in the form of food stamps and federal medical insurance (Medicaid). The poor family in Africa or India, by contrast, may be living without permanent shelter, forced to beg and sift through scraps to obtain enough food to survive. Although there are poor families and individuals in the United States who exist under similar conditions, most of the U.S. poor are not yet in immediate danger of starvation.

Does this mean that the poor in the United States are better off? In absolute terms (that is, in comparison with the poor in very poor nations), they are not deprived of the necessities of life. In relative terms (that is, in comparison with the more affluent majority in the United States), they are severely deprived. This distinction between absolute and relative deprivation is important in explaining how people feel about being poor. It is no comfort to a child in a poor family in the United States whose parents cannot afford fancy sneakers or a color television that there are people starving in Africa. In short, people feel poor in comparison with others around them who have more.

Who Are the Poor?

One of the chief causes of poverty is that people who are working full-time are not being paid wages that give them enough income to raise them above the poverty line. Two-thirds of American children living in poverty have a parent who works full-time for wages that are too low to lift the household out of poverty (Tucker, 1998). Of course, many of these households are one-parent families, but in a study of two-parent families living in poverty, David Ellwood found that 44% had at least one member who was working full-time. Low wages, Ellwood states, are a major cause of poverty:

> Work does not always guarantee a route out of poverty. A full-time minimum-wage job (which pays $5.15 per hour) does not even come close to supporting a family of three at the poverty line. Even one full-time job and one half-time job at the minimum wage will not bring a family of four up to the poverty line. (1988, p. 88)

Since so many of the jobs created during the economic expansion of the 1990s were low-wage jobs, rates of employment and poverty have risen simultaneously.

Another reason for the increase in poverty is the increase in the number of single-parent, female-headed families. The breakup of marriages or long-term relationships generally leaves women alone with the responsibility for raising small children and earning the income to do so. Such families often become poor because it is more difficult for a woman to support a family alone than it is for a man. One result of this trend toward the "feminization" of poverty is that a rising proportion of the nation's children are growing up in poor, female-headed families and are deprived of the advantages enjoyed by children from more affluent homes.

A common stereotype of the poor in the United States is that they are heavily concentrated in inner-city minority ghettos. Ellwood's study of poverty shows this notion to be false. The poor in large central cities (those living in moderate- or high-poverty neighborhoods) account for just 19% of the total poor population, with only about 7% concentrated in high-poverty neighborhoods. Fully 29% of the poor reside in rural and small-town communities, and 19% live in the affluent suburbs of large cities. In short, the poor live everywhere and are not concentrated in a single type of community.

One reason for this pattern is that the poor are an extremely diverse population. A large portion of the poor are aged people living on fixed incomes (Social Security or very modest savings or pensions). Other categories of poor people include marginally employed rural workers and part-time miners in communities from Appalachia to Alaska, migrant farm workers in agricultural areas throughout the United States, chronically unemployed manual workers in the industrial cities, disabled workers and their families, and people who have been displaced by catastrophes like hurricanes, drought, or arson. For all of these groups, poverty brings enormous problems of insecurity and instability. (See the Mapping Social Change box on pages 178–179.)

Equality of Opportunity Versus Equality of Result

The presence of millions of poor people in a country as affluent as the United States is a major public-policy issue as well as a subject of extensive social research. However, policy debates on this issue are often clouded by problems of definition. Most Americans will say that they believe in equality, citing the claim of the nation's founders that "all men are created equal." But when pressed to define equality, they often fail to distinguish between **equality of opportunity** (equal opportunity to achieve material well-being and prestige) and **equality of result** (actual equality in levels of material well-being and prestige). Americans may believe that opportunities to succeed should be distributed equally and that the rules determining who succeeds and who fails should be fair. But their

TABLE 6.4
Which Americans Are Poor?

Number and Percentage in Each Group Who Are Poor		
Group	Number (in millions)	Percentage
In total population	**36.1**	**313.8**
Children under 18	14.7	20.8
Men 18–64	7.5	9.5
Women 18–64	10.9	13.3
Men 65 and over	0.8	6.2
Women 65 and over	2.5	13.6
White	16.3	8.5
Black	9.9	29.3
Hispanic	8.6	30.3
Asian	1.4	14.6

Americans Below the Poverty Line, 1960–1995			
Year	All Families	Persons Over 65	Children Under 18
1960	18.1%	35.2%	26.9%
1970	10.1	24.6	15.1
1980	10.3	15.7	18.3
1990	10.7	12.2	20.6
1995	10.8	10.5	20.8

Source: Hacker, 1997.

commitment to the ideal of equality falters in the face of their belief that hard work and competence should be rewarded and laziness and incompetence punished. Thus, many Americans believe that poverty is proof of personal failure. They do not stop to ask whether poor Americans have ever been given equality of opportunity, nor do they stop to notice how hard many poor people work (Jencks, 1994; Liebow, 1967).

Sociologists who study inequality in modern societies usually ask how equality of opportunity can be increased and the gaps between the haves and the have-nots decreased. Yet they are also highly aware of how difficult it is to narrow that gap in the United States or any other society. In response to those who believe that government can do nothing effective in its efforts to reduce poverty, sociologists point to the kinds of hard facts presented in Table 6.4. The table clearly shows that poverty among people over age 65 has been reduced dramatically, from 35.2% in 1960 to 10.5% in 1995, largely as a consequence of Social Security legislation and Medicare. Poverty among children, on the other hand, was greatly reduced between 1960 and 1980 but has been increasing rapidly since the 1980s.

How will new opportunities created by computers and information technology affect existing patterns of inequality? We are now living in what is often called the information revolution (Seiden, 2000). New technologies are rapidly changing the way people work throughout the world (a subject we discuss in more detail in Chapter 12). It is far too early to know the full implications of this technological and social revolution for world patterns of stratification or for inequality in the United States and other advanced societies, but some aspects of this "revolution" are summarized in the controversy over the so-called digital divide, to which we now turn.

MAPPING SOCIAL CHANGE

Poor Counties in the United States

The accompanying map shows all the counties of the United States, grouped by average income. The extremes of wealth and poverty in U.S. society are apparent. Or are they? Actually, such a map is more useful in revealing areas of poverty outside central cities than within them. This is because impoverished inner-city communities usually are only a portion of a densely urban county that may also include affluent and middle-class communities. The more affluent communities' income raises the average so that the pockets of inner-city poverty do not appear on the map.

This map still reveals a great deal about the places where poor people live. For example, along the Mississippi river, especially in Arkansas and Mississippi, we see groups of counties whose residents' average income is in the lowest range. The same is true in parts of the arid Southwest. Why are there such high concentrations of poor residents in these counties? The Mississippi delta region has rich agricultural land, but most of it is owned by a few white families and agricultural corporations. The land is farmed by African American and poor white workers who earn extremely little. Thus, the delta region has a pattern of agrarian stratification that results in high rates of poverty. In the impoverished counties of the Southwest, the land often contains mineral and water resources that are controlled by a small number of wealthy individuals and corporations; Chicanos, Native Americans, and poor whites do not share in the wealth created by these resources, and the counties in which they live have low average incomes.

The isolated poor counties of the mountain states in the West often suffer from a newer form of poverty. In these areas, economic growth is due to tourism, second-home developments, skiing, and other pursuits of the affluent upper and upper-middle classes. Farms, small-town businesses, and the trailer parks where the resort workers live may be located in a nearby valley, which may be part of a separate county with a lower average income. ■

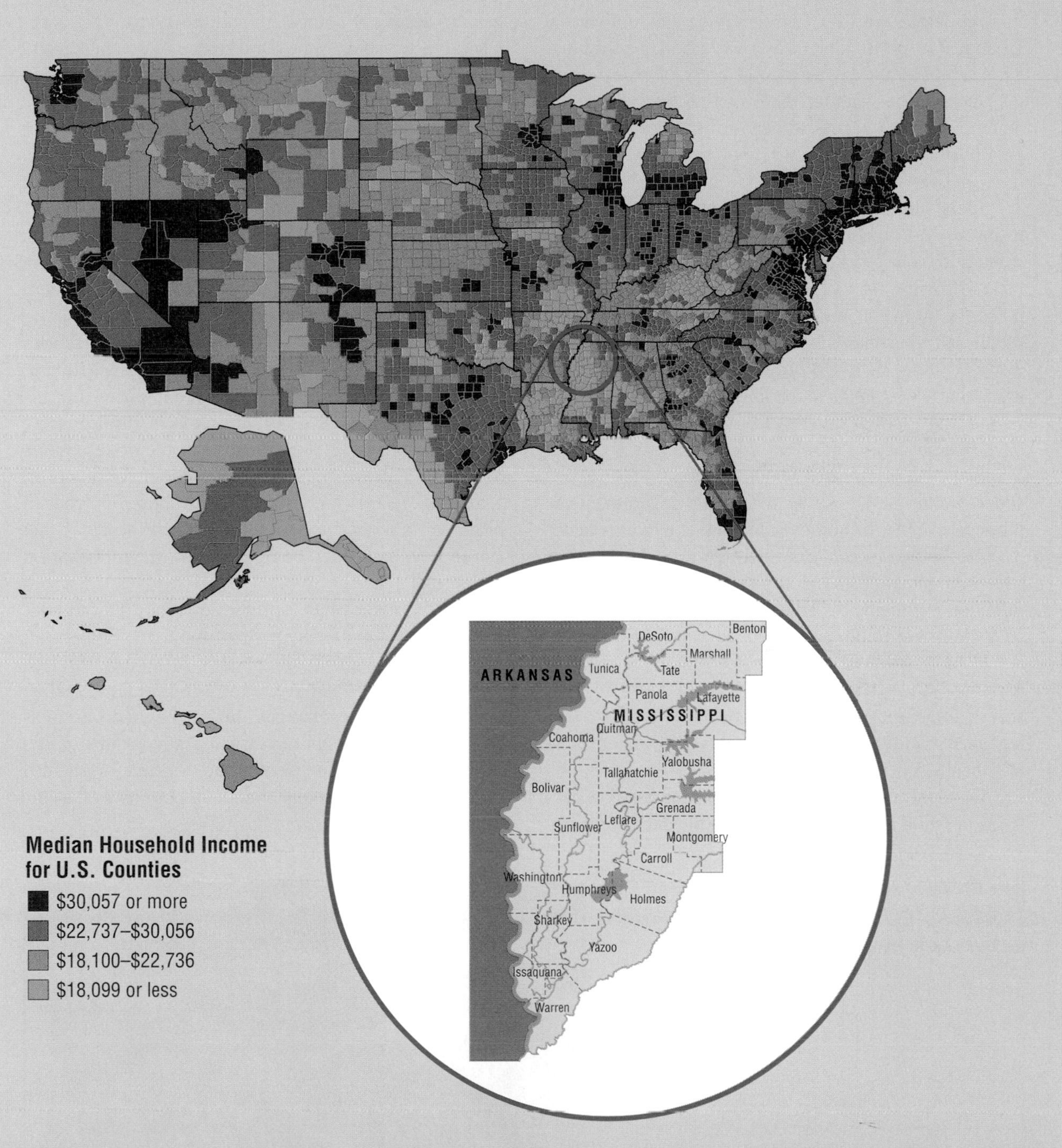

ARKANSAS
MISSISSIPPI
DeSoto
Benton
Marshall
Tunica
Tate
Panola
Lafayette
Quitman
Coahoma
Yalobusha
Tallahatchie
Bolivar
Grenada
Sunflower
Leflare
Montgomery
Carroll
Washington
Humphreys
Holmes
Sharkey
Yazoo
Issaquana
Warren
Median Household Income for U.S. Counties
$30,057 or more
$22,737–$30,056
$18,100–$22,736
$18,099 or less

TECHNOLOGY AND SOCIAL CHANGE

Is the rapid adoption of information technologies—computers, the Internet, e-mail, and much more—widening or closing the gap between rich and poor throughout the world? This is an issue that motivates a great deal of political activity and a good deal of bluster. Fortunately, there is a solid body of research on the question of whether there is a "digital divide," a widening gap in access to new information technologies by the affluent and the poor in the wealthy nations and elsewhere in the world.

In the United States, there is clearly a gap in access to information technologies between upper and lower strata of the class system, but recent data show that this "digital divide" is rapidly narrowing. A U.S. Department of Commerce study titled *Falling Through the Net* found that households with incomes over $75,000 are 20 times more likely to have access to the Internet at home than households at the lowest income level. Only 23% of high school students from low-income families have Internet access at home, compared to 7% of high school students from affluent neighborhoods. In response to this gap, however, the federal government has invested over $6 billion over the last eight years under the plan known as e-rate, which provides subsidies for telecommunications services in schools in poorer communities (Wilhelm & Thierer, 2000). As the cost of basic home computer equipment continues to decrease, and as more students receive higher-quality instruction in the use of these technologies, it is not surprising that access to the Internet and the World Wide Web is growing most rapidly among households with incomes of less than $25,000 (Ross, 2000).

On the world stage, there is a yawning gap in access to information technologies between the developing nations and the more affluent industrial nations (UNDP, 2000). But this gap is also a reflection of the larger patterns of inequality discussed earlier in the chapter. After all, with over 1.2 billion people living on one dollar a day or less, one would hardly expect the world's poor, the majority of whom have limited literacy skills, to be gaining access to computers and the Web. What is remarkable, however, is the rapid spread of computers and high-technology production outside the most advanced industrial and scientific centers of the world.

Globalization of production, made possible in large part by air travel and information technologies, stimulates the production of computers in Taiwan, China, and the Philippines, to name only a few important production centers outside the United States. While it remains true that there are more computers in the United States than in all other nations combined, production of hardware and software is stimulating economic growth in many parts of the Third World. The growth of a very important software production industry in Bangalore and other Indian cities, for example, does not address the problem of rural poverty in that nation, but it does encourage the governments of Indian states to make ever-greater investments in education, health, and urban infrastructure in order to promote the continued growth of a middle class that can supply the needed brain power for the information revolution.

E-commerce has similar effects. Throughout Latin America and many parts of Asia, one finds small businesses increasing their sales of goods through trade over the Internet. Again, it is important to realize that these positive changes do not begin to address long-standing patterns of class and gender inequality, but they do offer evidence that the information revolution is an engine of social change in poor as well as affluent nations (*UN Chronicle,* 2000).

India has one of the largest software production industries in the world. The income from these jobs is helping to accelerate the growth of the nation's middle class.

SOCIOLOGY AND SOCIAL JUSTICE

Roughly 7% of women in the United States work as Ann Berrios does, in the occupation of secretary (which includes stenographers and typists). This figure has declined steadily over the past generation as computers and other forms of office automation shift some kinds of work from the secretary to middle management. At the same time, the work secretaries do has become more stressful, with heavier workloads, more hassles with electronic equipment, increasing problems with repetitive strain injuries and the effects of chemical fumes, and eroding pay scales. Secretaries may be taken to lunch on International Secretaries Day, but what they are increasingly demanding is better pay and working conditions, and a chance, as Ann Berrios observes, to move up in the corporate hierarchy.

There are no easy answers to give secretaries or other mid- and lower-level office workers for improving their career paths or earnings. Career counseling, night school, and more technical training are possible individual solutions. But historically there is only one solution that addresses the occupation as a whole: organization of unions among office workers. Whenever clerical staff people join existing unions or form unions of their own, their wages and working conditions improve. In the antiunion U.S. business environment, this is an extremely difficult undertaking, but it pays off when hard-earned victories are won (Russo, 1999; Waldrop, 1994). ■

SUMMARY

Social strata are invisible strata, or layers, in a society. *Social stratification* refers to a society's system for ranking people hierarchically according to various attributes such as wealth, power, and prestige. Movement from one stratum (layer) of society to another is known as *social mobility.*

Castes are social strata into which people are born and in which they remain for life. *Classes* are social strata based primarily on economic criteria. Within any given class there are different status groups, defined by how much honor or prestige they receive from the society in general. The way people are grouped with respect to their access to scarce resources determines their *life chances*—the opportunities they will have or be denied throughout life.

Patterns of inequality around the world can be analyzed in terms of seven categories of insecurity: economic, food, health, personal, environmental, community and cultural, and political. These forms of insecurity are especially prevalent in the less-developed nations.

The principal forces that produce stratification are related to the ways in which people earn their living. For small farmers or peasants, social strata are based on land ownership and agrarian labor. Modern industrial societies are characterized by *structural mobility* (the elimination of entire classes as a result of changes in the means of existence) and *spatial mobility* (the movement of individuals and groups from one location to another).

Aspects of culture that justify the stratification system are learned through socialization. The system is often justified by an ideology and reinforced at the micro level by *deference* and *demeanor.* Changes in stratification systems may result from realignments of social *power. Authority* (legitimate power) is a major factor in maintaining existing relationships among castes or classes.

The basic measures of inequality in a society are income, wealth, and educational attainment. In the United States, the top fifth of all households receive almost half of total income, while the lowest fifth receive only 3.7% of the total. The distribution of wealth is even more skewed than that of income, but the distribution of educational attainment is more nearly equal.

When people think of social-class divisions, they first assign various occupations to broad class ranks. They then consider other aspects of social class, such as family prestige, education, and earned income. Together, these make up an individual's *socioeconomic standing* (SES). Differences in SES have a great impact on people's lives, especially in such areas as health, education, and politics.

The rise of industrial capitalism had far-reaching effects on stratification systems. According to Karl Marx, capitalism divided societies into classes based on ownership of the means of production. The largest of these classes, the workers, must sell their labor to capitalists or landowners in return for wages. In time, the workers would become conscious of their shared interests as a class and would rebel against the capitalist class. The outcome of the revolution would be a classless society.

Contemporary conflict theorists believe that conflict continues to explain many aspects of class stratification and that class inequalities often deprive people of the ability to realize their full human potential. Reproduction theory suggests that people from the lower classes are relegated to inferior schools and mediocre jobs and that the media help convince them that this is how things should be.

Functionalist theorists believe that classes emerge because an unequal distribution of rewards is necessary in order to channel talented people into important roles in society. This view has been criticized because it fails to account for the fact that social rewards in one generation tend to improve the life chances of the next generation; nor does it explain why talented people from lower-class families often are unable to obtain highly rewarded positions. From the interactionist perspective, the stratification system is not a fixed system but, rather, one that is created out of everyday behaviors.

Max Weber challenged Marx's definition of social class, pointing out that people are stratified not only by wealth but also by power and prestige. Studies of social mobility in industrial societies have shown that there is considerable movement between classes.

A significant proportion of the poor have jobs that do not pay enough to support their families. Another large percentage of poor families are single-parent families headed by women. Policy debates on the issue of poverty are often clouded by problems of definition. Although many Americans believe in *equality of opportunity,* they are less committed to the ideal of *equality of result.* Most sociologists agree that it is impossible to achieve a completely egalitarian society; instead, they concentrate on how much present levels of inequality can and should be reduced.

GLOSSARY

Social strata: Invisible strata, or layers, in a society that may be more or less closed to entry by people from outside any given layer. **(153)**

Social stratification: A society's system for ranking people hierarchically according to such attributes as wealth, power, and prestige. **(153)**

Closed stratification system: A stratification system in which there are rigid boundaries between social strata. **(153)**

Open stratification system: A stratification system in which the boundaries between social strata are easily crossed. **(153)**

Social mobility: Movement by an individual or group from one social stratum to another. **(153)**

Upward mobility: Movement by an individual or group to a higher social stratum. **(153)**

Downward mobility: Movement by an individual or group to a lower social stratum. **(153)**

Caste: A social stratum into which people are born and in which they remain for life. **(153)**

Ascribed status: A position or rank that is assigned to an individual at birth and cannot be changed. **(153)**

Achieved status: A position or rank that is earned through the efforts of the individual. **(153)**

Class: A social stratum that is defined primarily by economic criteria such as occupation, income, and wealth. **(153)**

Status group: A category of people within a social class, defined by how much honor or prestige they receive from the society in general. **(153)**

Life chances: The opportunities that an individual will have or be denied throughout life as a result of his or her social-class position. **(154)**

Structural mobility: Movement of an individual or group from one social stratum to another that is caused by the elimination of an entire class as a result of changes in the means of existence. **(158)**

Spatial mobility: Movement of an individual or group from one location or community to another. **(160)**

Status symbols: Material objects or behaviors that indicate social status or prestige. **(161)**

Deference: The respect and esteem shown to an individual. **(161)**

Demeanor: The way in which individuals present themselves to others through body language, dress, speech, and manners. **(161)**

Power: The ability to control the behavior of others, even against their will. **(162)**

Authority: Power that is considered legitimate both by those who exercise it and by those who are affected by it. **(162)**

Educational attainment: The number of years of school an individual has completed. **(163)**

Educational achievement: Mastery of basic reading, writing, and computational skills. **(164)**

Socioeconomic status (SES): A broad social-class ranking based on occupational status, family prestige, educational attainment, and earned income. **(165)**

Objective class: In Marxian theory, a social class that has a visible, specific relationship to the means of production. **(171)**

Subjective class: In Marxian theory, the way members of a given social class perceive their situation as a class. **(171)**

Class consciousness: A group's shared subjective awareness of its objective situation as a class. **(171)**

Intragenerational mobility: A change in the social class of an individual within his or her own lifetime. **(174)**

Intergenerational mobility: A change in the social class of family members from one generation to the next. **(174)**

Equality of opportunity: Equal opportunity to achieve desired levels of material well-being and prestige. **(176)**

Equality of result: Equality in the actual outcomes of people's attempts to improve their material well-being and prestige. **(176)**

QUESTIONS FOR THOUGHT AND DISCUSSION

1. In the United States, new students of sociology are sometimes upset when they study social inequality and stratification. Are there aspects of this subject that you find difficult or upsetting? What might the American values of individual achievement and mobility have to do with this feeling? Why is it that students from England or India are less likely to be perturbed by ideas of class and stratification?

2. Can you give some examples of how social stratification is related to how people earn their living in a society? In the contemporary world, how do technology and changes in world markets alter the fortunes of different social classes?

DIGGING DEEPER

Books

The Winner-Take-All Society. (Robert Frank; Free Press, 1996). Presents a compelling argument that more and more Americans are competing for even fewer and bigger prizes, encouraging economic waste, income inequality, and an impoverished cultural life.

Bringing Class Back in Contemporary and Historical Perspectives (Scott C. McNall, Rhonda F. Levine, & Rick Fantasia, eds.; Westview Press, 1991). A collection of essays dealing with theories and empirical studies of stratification in comparative and historical perspective.

The Structures of Everyday Life: Civilization and Capitalism, 15th–18th Century (vol. 1), ***The Wheels of Commerce*** (vol. 2), ***The Perspective of the World*** (vol. 3); (Fernand Braudel; Harper, 1981). A work of historical and sociological scholarship that is only beginning to influence the social sciences. The great French social historian shows in marvelous detail how social stratification has been largely explained by the way people earn their daily bread.

The Communist Manifesto (Karl Marx & Friedrich Engels; Penguin, 1969/1848). Marx and Engels's classic description of the role of class conflict in history, with a lucid analysis of the consequences of the rise of industrial capitalism and the inevitability of a proletarian revolution. Eclipsed by events in the modern world, ***The Communist Manifesto*** continues to inspire readers who are critical of human exploitation and imperialism.

The Debate on Classes (Erik Olin Wright; Verso, 1989). An excellent review of theories of social class in the United States and elsewhere by one of the nation's foremost researchers on social class.

The American Class Structure: A New Synthesis, 4th ed. (Dennis Gilbert; Wadsworth, 1993). A valuable summary of American sociological writing about class, together with clear explanations of modern studies, stratification, and mobility.

Class in Twentieth-Century American Sociology: An Analysis of Theories and Measurement Strategies (Michael D. Grimes; Praeger, 1991). A comprehensive review of the way social class has been studied in American sociology, with emphasis on why recent decades have brought renewed emphasis on the presence of classes and class conflict, in opposition to the functionalist theories and studies that were prevalent before the 1970s.

The War Against the Poor (Herbert Gans; Basic Books, 1995). A timely and thorough review of antipoverty policies and debates by one of the world's foremost experts on the subject.

Journals

Monthly Review. A monthly journal of critical social-scientific writing and research on class and inequality. Most articles are Marxian in orientation.

The Public Interest. A valuable interdisciplinary social science journal with highly readable articles on issues of class, race, poverty, and social policy, with special emphasis on policy debates.

Focus. The official publication of the Institute on Poverty of the University of Wisconsin. An invaluable source for current sociological and economic research on poverty and social policy (welfare reform, family policies, early education, and much more).

Other Sources

Current Population Reports. Published monthly by the Bureau of the Census, U.S. Department of Commerce. Presents data on income and employment patterns for American households.

Handbook of Economic Statistics. Published annually by the U.S. Department of Labor. Compares economic statistics for the European Community, the United States, and selected other nations, permitting measures of relative inequality and stratification.

Handbook of Labor Statistics. A compilation of statistical data published annually by the U.S. Department of Labor. Presents material on labor conditions and labor-force characteristics for the United States and selected foreign countries.

EXPLORING SOCIOLOGY ON THE INTERNET

International Sociological Association
www.ucm.es/OTROS/isa

Provides a fine introduction to global sociological work by professional researchers.

Rural Sociological Society
www.lapop.lsu.edu/rss

Balances the usual urban and industrial emphasis in sociology with access to research on rural topics and themes.

Population Reference Bureau
www.prb.org/prb/

A good summary; demographic data and links to other sources of useful demographic statistics.

Sociology Department, Trinity College
www.trinity.edu/departments/soc_anthro/index.html

A good place to seek help when writing research papers.

U.S. Census Bureau
www.census.gov/

Includes useful demographic data. Click on "Subjects A–Z" and then, for example, on "P" to look up Poverty or on "W" to look up Wealth.

The World Bank
www.worldbank.org/

An excellent source of comparative global statistics on indicators of social stratification in different nations and regions of the world, with particular emphasis on the developing nations.

The National Center for Children in Poverty
http:/cpmcnet.columbia.edu/dept/nccp/

Has updated information and ongoing research on the effects of changes in welfare policy on the lives of poor children and their communities.

Institute for Research on Poverty
www.ssc.wisc.edu/irp

Offers invaluable access to the most recent research on the causes and consequences of poverty and social inequality in the United States. Located at the University of Wisconsin–Madison.

Cato Institute
www.cato.org/

A conservative research institution devoted to criticism of the welfare state. Advocates market alternatives rather than government-driven policies to address problems of social stratification and inequality. Its home page has a search engine where, for example, you can type "Welfare" to access several hotly debated articles on the subject.

Chapter 7

Inequalities of Race and Ethnicity

Can knowledge of how racial and ethnic inequalities originate help societies improve race and ethnic relations?

Raul Garcia and two friends stop after work at a restaurant near their job in a factory on the outskirts of San Diego. On leaving the restaurant, Raul turns the key in the ignition, glances in the rear view mirror, and pulls away from the curb. Rounding the corner, his slightly worn tires slip on a fresh layer of light rain. He executes a perfect "California rolling stop" at an intersection and enters the freeway, accelerating to a comfortable 70 mph for the cruise into downtown. Seconds later, Raul and his friends are pulled over by a highway patrol car. The officer makes them exit the vehicle. After carefully going over Raul's license and registration, the officer searches the car, presumably for drugs or guns. Finding none, he issues Raul a ticket for going through the stop sign at the freeway entrance.

After the officer drives off, Raul and his friends feel abused and hurt. They are sure they have been victims of DWB, or Driving While Brown. Raul believes that his driving is no different from that of most other California drivers he sees every day, but because he is a Latino American, he has been singled out by the police.

According to the U.S. Supreme Court, any violation of the traffic code is a legitimate pretext for pulling a driver over, even if the officer is only on a fishing expedition for bigger offenses. If a police officer wants to target a particular driver, all he or she has to do is wait for the inevitable traffic violation to occur. African, Latino, and Asian Americans believe that the driver's skin color often determines whether or not the officer exercises his or her discretion to stop a vehicle. This behavior, known as "racial profiling," has been an issue in minority communities for many years.

Racial profiling, like all forms of prejudice and discrimination, has become a subject of civil rights demonstrations and political agitation in communities throughout North America. San Diego's police department has become the first in the nation to voluntarily record the race and ethnicity of every driver stopped. Preliminary results were recently released following a joint review by the police department and its consultants, the American Civil Liberties Union, the National Association for the Advancement of Colored People, the Urban League, the Latino Unity Coalition, and representatives from the Asian American community. The results of the research confirm what people in minority communities already believe: If you are black or brown or Asian American, you are more likely to be stopped by police. African and Latino Americans represent 28% of the driving population but account for 40% of those stopped and 60% of those searched. Two-thirds of all stops occurred during daylight hours, when the color of a driver's skin is more easily observed (Hills & Trapp, 2000).

These numbers need to be analyzed by age and proximity to the border before a definitive verdict of racial profiling can be rendered. But in the meantime, as elsewhere throughout the United States, the imbalance in treatment is inevitably perceived by most minorities as yet another symptom of their inequality and lower position in the society's system of social stratification.

In this chapter, we begin with a discussion of the different meanings of race and ethnicity in society and then present an overview of historical and contemporary patterns of intergroup relations. In later sections, we explore the cultural factors that shape these behaviors, as well as a range of theories that attempt to account for patterns of racial and ethnic inequality. The chapter ends with an analysis of the major dimensions of change in those patterns, particularly in the United States, and a discussion of why so many social problems are associated with race. ■

THE MEANING OF RACE AND ETHNICITY

Why is it that so many African Americans and other minorities have to struggle so hard to escape poverty and discrimination? How can we explain prejudice and racial discrimination, and how can we measure their effects on individuals and social institutions? Many social scientists focus their research on why some groups in a society have been subordinated and the consequences of that subordination for them and their children. The effects of various forms of subordination, such as slavery, expulsion, and discrimination, on a society's patterns of inequality are at the center of the study of race and ethnicity, not only in the United States but throughout the world.

The United States is not unique in the extent to which inequality and hostility among its ethnic and racial groups result in severe social problems. In Canada, conflict between Anglo and French Canadians in Quebec periodically threatens national unity. In Germany, there have been severe riots as native-born Germans have pressed for limits on the nation's acceptance of refugees from Eastern European nations that were formerly dominated by the Soviet Union. In what used to be Yugoslavia, hostility among Croats, Serbs, and Bosnians has produced the bloodiest ethnic war in recent history. In the former Soviet Union, there is great fear that ethnic nationalism unleashed by the demise of the communist dictatorship will produce even more terrible wars than the one in Chechnya because there are nuclear arms in the former Soviet republics.

Given its ongoing importance throughout the world, it is no wonder that the study of racial and ethnic relations has always been a major subfield of sociology. Most societies include minority groups, people who are defined as different according to the majority's perceptions of racial or cultural differences. And in many societies, as the ironic song from the musical *South Pacific* goes, "You've got to be taught to be afraid/of people whose eyes are differently made/or people whose skin is a different shade./You've got to be carefully taught." Sociologists try to get at the origins of these fears and groundless distinctions that categorize people as different and influence their life chances, often in dramatic ways.

Race: A Social Concept

Of the millions of species of animals on earth, ours, *Homo sapiens,* is the most widespread. For the past 10 millennia we have been spreading northward and southward and across the oceans to every corner of the globe. But we have not done so as a single people; rather, throughout our history we have been divided into innumerable societies, each of which maintains its own culture, thinks of itself as "we," and looks upon all others as "they." Through all those millennia of warfare, migration, and population growth, we have been colliding and competing and learning to cooperate. The realization that we are one people—the human race—despite our immense diversity has been slow to evolve. We persist in creating arbitrary divisions based on physical differences that are summed up in the term *race.*

In biology, **race** refers to an inbreeding population that develops distinctive physical characteristics that are hereditary. Such a population therefore has a shared genetic heritage (Coon, 1962; Marks, 1994). But the choice of which physical characteristics to use in classifying people into races is arbitrary. Skin color, hair form, blood type, and facial features such as nose shape and eyefolds have been used by biologists in such efforts. In fact, however, there is a great deal of overlap among the so-called races in the distribution of these traits. Human groups have exchanged their genes through mating to such an extent that any attempt to identify "pure" races is bound to be fruitless (Alland, 1973; Dobzhansky, 1962; Gould, 1981).

Yet doesn't common sense tell us that there are different races? Can't we see that there is a Negroid, or "black," race of people with dark skin, tightly curled hair, and broad facial features; a Caucasoid, or "white," race of people with pale skin and ample body hair; and a Mongoloid, or "Asian," race of people with yellowish or reddish skin and deep eyefolds that give their eyes a slanted look? Of course these races exist. But they are not a set of distinct populations based on biological differences. The definitions of race used in different societies emerged from the interaction of various populations over long periods of human history. The specific physical characteristics that we use to assign people to different races are arbitrary and meaningless—people from the Indian subcontinent tend to have dark skin and straight hair; Africans from Ethiopia have dark skin and narrow facial features; American blacks have skin colors ranging from extremely dark to extremely light; and whites have facial features and hair forms that include those of all the other supposed races. There is no scientifically valid typology of human races; what counts is what people in a society define as meaningful.

Unfortunately, once people in a society define a group as racially distinct or ethnically different—that is, as a people with a separate ancestry—unequal access to jobs, income, education, and power in the major political institutions of society may follow. This was certainly true in the case of slavery, but it can also be seen in contemporary societies. Conflict among groups labeled as racially or ethnically distinct often has to do with efforts to prevent one group or another from gaining access to decent housing and neighborhoods, better schools, or jobs and economic opportunity (Bobo & Hutchings, 1996; Jargowsky, 1996).

In short, race is a social concept that varies from one society to another, depending on how the people of that society feel about the importance of certain physical differences among human beings. Once made, racial distinctions (and ethnic ones as well) often become incorporated in a society's stratification system in ways that may severely diminish the life chances of minority groups. In reality, however, as Edward O. Wilson (1979) has written, human beings are "one great breeding system through which the genes flow and mix in each generation. Because of that flux, mankind viewed over many generations shares a single human nature within which relatively minor hereditary influences recycle through ever-changing patterns, between the sexes and across families and entire populations" (p. 52).

Racism

Throughout human history, many individuals and groups have rejected the idea of a single human nature. Tragic mistakes and incalculable suffering have been caused by the application of erroneous ideas about race and racial purity. They are among the most extreme consequences of the attitude known as *racism.*

Racism is an ideology based on the belief that an observable, supposedly inherited trait, such as skin color, is a mark of inferiority that justifies discriminatory treatment of people with that trait. In their classic text on racial and cultural minorities, Simpson and Yinger (1953) highlighted several beliefs that are at the heart of racism. The most common of these is the "doctrine of biologically superior and inferior races" (p. 55). Before World War I, for example, many of the foremost social thinkers in the Western world firmly believed that whites were genetically superior to blacks in intelligence. However, when the U.S. Army administered an IQ test to its recruits, the results showed that performance on such tests was linked to social-class background rather than to race. And when investigators controlled for differences in social class among the test takers, the racial differences in IQ disappeared (Kleinberg, 1935).

Since that time there have been other attempts to demonstrate innate differences in intelligence among people of different races. The most recent of these efforts is Herrnstein and Murray's controversial study, *The Bell Curve* (1994), which was described in Chapter 3. This study argues that because scores on intelligence tests are distributed along a "normal" bell curve, there will necessarily be a large number of people whose scores fall well below the mean and who are not capable of performing well in situations requiring reasoning ability or other academic skills. Because the bell curves for African Americans and whites are different, with that of blacks peaking at a somewhat lower mean score, Herrnstein and Murray argue that group differences in IQ establish biological limits on ability. This conclusion, in turn, leads them to argue that programs intended to correct differences in educational opportunities, such as Head Start, are doomed to failure.

Critics of these conclusions argue that studies like *The Bell Curve* merely dredge up old academic justifications for the status quo of racial and class inequality (Fischer et al., 1996; Gould, 1995). There is a great deal of evidence that IQ tests are biased against members of minority groups and that something as complex and elusive as intelligence cannot be summarized by a single score on a test. The history of efforts to address inequalities in education shows that when they have access to high-quality educational programs, minority students quickly begin to achieve at the same levels as white students (Sconing, 1999).

The notions that members of different races have different personalities, that there are identifiable "racial cultures," and that ethical standards differ from one race to another are among the other racist doctrines that have been debunked by social-scientific studies over the past 50 years. But even though these doctrines have been discredited, we will see shortly that they continue to play a major role in intergroup relations in many nations. Racist beliefs in the innate inferiority of populations that are erroneously thought of as separate races remain one of the major social problems of the modern world. And this tendency to denigrate socially defined racial groups extends to members of particular ethnic groups as well.

Ethnic Groups and Minorities

Ethnic groups are populations that have a sense of group identity based on a distinctive cultural pattern and, usually, shared ancestry, whether actual or assumed. Ethnic groups often have a sense of "peoplehood" that is maintained within a larger society (Dublin, 1996; Portes, 1996). Their members usually have migrated to a new nation or been conquered by an invading population. In the United States and Canada, a large proportion of the population consists of people who either immigrated themselves or are descended from people who immigrated to the New World or were brought there as slaves. If one traces history back far enough, it appears that everyone living in the Western Hemisphere can trace his or her origins to other continents. Even the Native Americans are believed to have crossed the Bering Strait as migrant peoples between 14,000 and 20,000 years ago.

European conquests of the Americas in the sixteenth and seventeenth centuries resulted in people of Iberian (Spanish and Portuguese) origins becoming the dominant population group in Central and South America. People of English, German, Dutch, and French origins became the dominant population groups throughout much of North America. By the late 1770s, almost 80% of the population of the 13 American colonies had been born in England or Ireland or were the children of people who had come from those countries. A century later, owing to the influence of slavery and large-scale immigration, the populations of the United States and Canada were far more diverse in terms of national origins, but people from England were still dominant.

Power and Dominance. According to sociological definitions of the term *dominance,* a dominant group is not necessarily numerically larger than other population groups in a nation. More important in establishing dominance is control or ownership of wealth (farms, banks, manufacturing concerns, and private property of all kinds) and political power—especially control of political institutions like legislatures, courts, the military, and the police. Thus, by the beginning of the twentieth century white Anglo-Saxon males (or WASPs, as they are sometimes called) had established dominance over U.S. society even though they no longer constituted a clear majority of the population. (See Figure 7.1 and the Global Social Change box on pages 192–193.)

FIGURE 7.1
Major Sources of Immigration to the United States

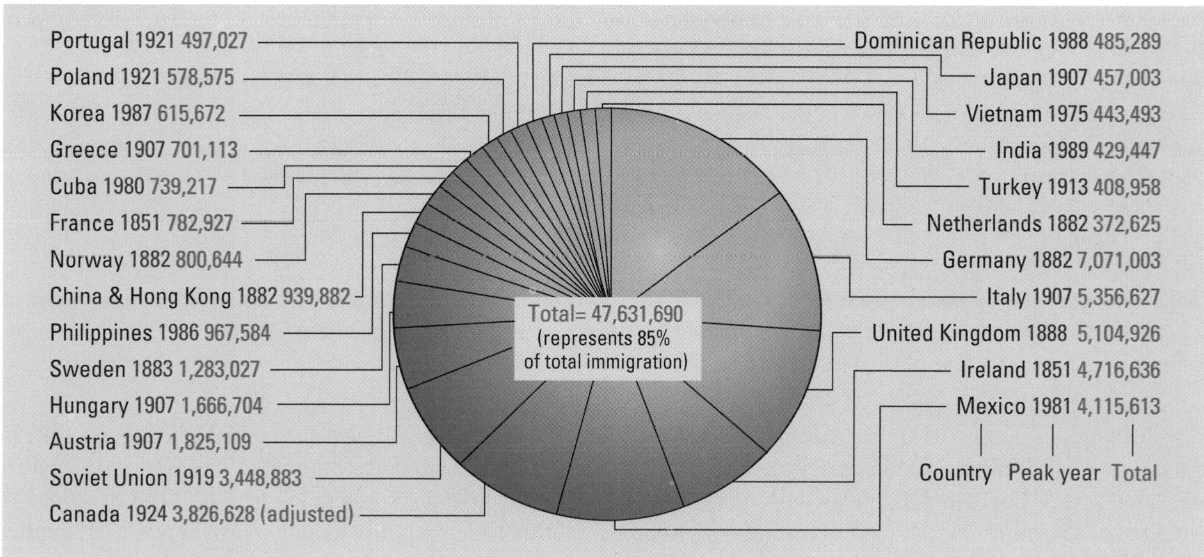

Source: Data from U.S. Immigration and Naturalization Service.
Note: Statistics are based on the last country of residence.

GLOBAL SOCIAL CHANGE

Periods of Migration and Settlement in the United States

In 1790, when the republic was formed, the vast majority of U.S. citizens and resident foreigners (77%) had come from Great Britain and the counties that are now included in Northern Ireland. People from Germany (7.4%), Ireland (4.4%), the Netherlands (3.3%), France (2%), Mexico (1%), and other countries were present in the population, but with the exception of the Irish, they were also white Protestants. Blacks—who were involuntary immigrants—and their children actually accounted for 19% of the total population, but neither they nor Native Americans were citizens.

1820–1885: The "Old" Northwest European and Asian Migration

By about 1830, the mass migration of different groups into the new nation was under way. The largest flows of immigrants came from Ireland and Germany, with well over a million people from each of these countries coming to the United States to escape political or economic troubles. Smaller waves of immigrants originated in the northern European nations of Sweden, Norway, and Denmark.

Although people from northern Europe made up the most important immigrant groups, far exceeding the continuing influx from Great Britain, 1885 marked the beginning of a flow of immigrants from China that continued at high rates for the next 30 years. Chinese immigrants settled primarily in the West and contributed immensely to the development of the western states, although they met with sporadic and often violent hostility.

1885–1940: The "Intermediate" Migration From Southern and Eastern Europe and the Beginning of Heavy Immigration From Mexico

From 1880 to 1900, the flow of immigrants from Italy, Poland, Russia, the Baltic states, and southern Europe (Serbs, Croats, Slovenes, Romanians, Bulgarians, Greeks, etc.) exceeded the flow of people from older immigrant groups by about four to one. In 1907 alone, more than 250,000 Italians and 338,000 Poles and other central Europeans were admitted through New York's Ellis Island. Many of these immigrants were Jewish, Roman Catholic, or Orthodox Christian. The newcomers' tendency to live in neighborhoods known as "Little Italy" or "Little Poland" convinced more established Americans that the immigrants would never learn English or become fully American. The mass arrival of physically distinctive groups such as Mexicans and Japanese during this period also aroused fear and hostility among those who disliked the newcomers.

During World War I, thousands of impoverished black southerners were attracted to the cities of the North and the Midwest. This began a trend that would continue throughout the century whenever wars or economic booms made it possible for blacks to overcome discrimination and get factory jobs.

1921–1959: Immigration by Quota and Refugee Status

Agitation against the "new" immigrants and demands for "Oriental exclusion" led Congress to establish a quota sys-

Immigrant groups that are distinct because of racial features or culture (language, religion, dress, etc.) are often treated badly in the new society and hence develop the consciousness of being a minority group. Louis Wirth (1945), a pioneer in the study of racial and ethnic relations, defined a **minority group** as a set of "people who, because of their physical or cultural characteristics, are singled out from the others in the society in which they live for differential and unequal treatment, and who therefore regard themselves as objects of collective discrimination" (p. 347). The existence of a minority group in a society, Wirth explained, "implies the existence of a corresponding dominant group with a higher social status and greater privileges. Minority status carries with it exclusion from full participation in the life of the society" (p. 347).

Over the past century, images of immigration to the United States have changed from portrayals of "huddled masses yearning to be free" to scenes at the customs counters of modern airports.

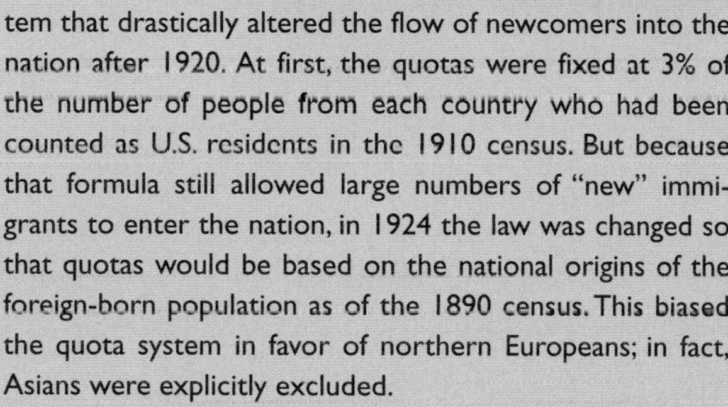

tem that drastically altered the flow of newcomers into the nation after 1920. At first, the quotas were fixed at 3% of the number of people from each country who had been counted as U.S. residents in the 1910 census. But because that formula still allowed large numbers of "new" immigrants to enter the nation, in 1924 the law was changed so that quotas would be based on the national origins of the foreign-born population as of the 1890 census. This biased the quota system in favor of northern Europeans; in fact, Asians were explicitly excluded.

World War II and the cold war produced millions of homeless and stateless refugees, especially from Germany, Poland, and the Soviet Union. Almost half a million of these displaced persons arrived in the United States between 1948 and 1950; another 2 million came between 1950 and 1957. Their arrival swelled the populations of ethnic enclaves in America's industrial cities.

1960 to the Present: Worldwide Immigration

In the 1960s, national quotas were replaced with a system based on preference for skilled workers and professionals, regardless of country of origin, as well as refugees and people with families already in the United States. Congress has gradually increased the number of immigrants who may enter the country legally—from 280,000 in 1980 to more than 450,000 in 1992. Today the United States is once again the foremost immigrant-receiving nation in the world. The largest streams of immigrants are from Mexico, other Central American and Caribbean nations, Asia (especially Korea, China, India, and Pakistan), and South America. Untold numbers of undocumented, or illegal, immigrants also arrive each year—especially from Mexico, Canada, and the Dominican Republic—but studies show that a high proportion of these newcomers eventually become legal residents or return to their country of origin (Bogue, 1985; Portes & Rumbaut, 1990). ■

In the United States, the term *minority* often suggests people of color, meaning African Americans, Native Americans, Latino Americans (many of whom have the darker coloring of Amerindian ancestry), and Asian Americans. But the term can be applied to any group of people who find themselves at the mercy of the society's dominant group. In Great Britain, for example, the Irish were a conquered people who were subjected to economic and social discrimination by the English. It is not surprising, therefore, that when the Irish began immigrating to the United States in the nineteenth century they were treated as an inferior minority by Americans of English ancestry. This attitude was especially prevalent in cities like Boston and New York, to which the Irish came in large numbers.

It should also be noted that the term *minority* does not always imply that the population is numerically inferior to the dominant group. There are counties in some states in the South, and entire cities in the North, in which African Americans constitute a numerical majority yet cannot be considered the dominant group because they lack the power, wealth, and prestige enjoyed by the white population.

PATTERNS OF INTERGROUP RELATIONS

Throughout history, when different racial and ethnic groups have met and mixed, the most usual outcome has been violence and warfare. In fact, the desire for peaceful and cooperative relations among diverse peoples has emerged relatively recently. In this section, we explore a continuum of relations between dominant and minority groups that extends from complete intolerance to complete tolerance, as shown in the study chart on page 198. At one end of the continuum is extermination or genocide; at the other is assimilation.

Genocide

In a study of a New Guinea tribe, the Siane, anthropologist Richard Salisbury (1962) found that the members of this isolated highland tribe believed that anyone from another tribe would want to kill them. Therefore, the Siane felt that they must kill any member of another tribe they might encounter. (Fortunately, they excluded the anthropologist from this norm.) We often think of such behavior as primitive, savage, or barbarous. Yet barbarities on a far greater scale have been carried out by supposedly advanced societies. The most extreme of these is **genocide,** the intentional extermination of a population, defined as a "race" or a "people," by a more dominant population.

There have been numerous instances of genocide in recent history. Those incidents have been characterized by a degree of severity and a level of efficiency unknown to earlier civilizations. Consider the following:

> The Native American populations of North, Central, and South America were decimated by European explorers and settlers between the sixteenth and twentieth centuries. Millions of Native Americans were killed in one-sided wars, intentional starvation, forced marches, and executions. The population of Native Americans in North America was reduced from more than 4 million in the eighteenth century to fewer than 600,000 in the early twentieth century (Thornton, 1987).
>
> When England, France, Germany, Portugal, and the Netherlands were engaged in fierce competition for colonial dominance of Africa during the nineteenth and early twentieth centuries, millions of native people were exterminated. The introduction of the Gatling machine gun made it possible for small numbers of troops to slaughter thousands of tribal warriors.
>
> Six million Jews, 400,000 gypsies, and about 2 million Russian civilians were killed by the Nazis during World War II; thousands of Pakistanis and Indians were slaughtered after the partition of India in 1947; and thousands of Tamils living in Sri Lanka were exterminated by Sinhalese in the 1980s. "Ethnic cleansing" by Serbian troops in the former Yugoslav republics of Bosnia and Herzegovina, and Kosovo, where entire Muslim-Bosnian villages have been wiped out, and the conflict between Hutu and Tutsi ethnic groups in Rwanda, which has resulted in hundreds of thousands of brutal deaths, are recent examples of genocidal action. (Block, 1994)

Mass executions and other forms of genocide are almost always "justified" by the belief that the people who are being slaughtered are less than human and in fact are dangerous parasites. Thus, the British and Dutch slaughtered members of African tribes like the Hottentots in the belief that they were a lower form of life, a nuisance species unfit even for enslavement. The Nazis rationalized the extermination of Jews and gypsies by the same twisted reasoning.

Expulsion

In many societies, extended conflicts between racial or ethnic groups have ended in **expulsion,** the forcible removal of one population from territory claimed by the other. Thus, on the earliest map of almost every major American city there appears a double line drawn at the edges where the streets end. This is the Indian Boundary Line, and it symbolizes the expulsion of Native Americans from lands that were taken from them in order to create a city in which they would be strangers.

Expulsion was the usual fate of North America's native peoples, who also experienced massacres, deadly forced marches, and other genocidal attacks. These actions were rationalized, in part, by the nineteenth-century doctrine of "Manifest Destiny," which asserted the inevitability of the westward expansion of Europeans across the territories of the Southwest and West (Davis, 1995). As white settlement expanded westward, Native Americans were continually expelled from their tribal lands. After the 1848 gold rush and the rapid settlement of the West Coast, the pressure of white settlement pushed the native tribes into the high plains of the West and Southwest. Between 1865 and the 1890 massacre of the Sioux at Wounded Knee, South Dakota, the remaining free tribes in the West were forced to settle on reservations. In the process, the Native Americans lost more than their ancestral lands. As the famous Sioux chief Black Elk put it, "A people's dream died" (quoted in Brown, 1970, p. 419).

The forced settlement of Native Americans on reservations is only one example of expulsion. Asian immigrants in the American West also suffered as a result of intermittent attempts at expulsion referred to as the *Oriental exclusion movement.* In an effort to prevent the large-scale importation of Chinese laborers into California and other states, Congress passed the Chinese Exclusion Act of 1882, which excluded Chinese laborers from entry into the United States for 10 years. But this legislation did little to relieve the hostility between whites and Asians. Riots directed against Chinese workers were common throughout the West during this period, and Chinese immigrants were actually expelled from a number of towns. In 1895, for example, a mob killed 28 Chinese immigrants in Rock Springs, Wyoming, and expelled the remaining Chinese population from the area (Lai, 1980).

The most severe example of expulsion directed against Asians in the United States occurred in 1942 after the Japanese attack on Pearl Harbor and the American entry into World War II. On orders from the U.S. government, more than 110,000 West Coast Japanese, 64% of whom were U.S. citizens, were ordered to leave their homes and their businesses and were transported to temporary assembly centers (Kitano, 1980). They were then assigned to detention camps in remote areas of California, Arizona, Idaho, Colorado, Utah, and Arkansas. When the U.S. Supreme Court declared unconstitutional the incarceration of an entire ethnic group without a hearing or formal charges (*Endo* v. *United States,* 1944), the Japanese were released, but by then many had lost their homes and all their possessions. In 1989 Congress finally voted to pay modest reparations to the families of Japanese Americans who had been imprisoned during the war.

Slavery

Somewhat farther along the continuum between genocide and assimilation is slavery. **Slavery** is the ownership of a population, defined by racial or ethnic or political criteria, by another population that not only can buy and sell members of the enslaved population but also has complete control over their lives. Slavery has been called "the peculiar institution" because, ironically, it has existed in some of the world's greatest civilizations. The socioeconomic systems of ancient Greece and Rome, for example, were based on the labor of slaves. And the great trading cities of late medieval Europe, such as Venice, Genoa, and Florence, developed plantation systems in their Mediterranean colonies that were based on slave labor. In fact, the foremost student of slave systems, Orlando Patterson, makes the following comment:

> There is nothing notably peculiar about the institution of slavery. It has existed from before the dawn of human history right down to the twentieth century, in the most primitive of human societies and in the most civilized. . . . Probably there is no group of people whose ancestors were not at one time slaves or slaveholders. (1982, p. vii)

Figure 7.2 indicates the magnitude of the transatlantic slave trade; the widths of the arrows represent the relative size of each portion of that terrible traffic in humanity. The arrows do not show, however, that for the Americas to acquire 11 million slaves who survived the voyage on the slave ships and the violence and diseases of the New World, approximately 24 million Africans had to be captured and enslaved (Fyfe, 1976; Patterson, 1982, 1991). It is evident from Figure 7.2 that the United States imported a proportionately small number of slaves. However, although somewhat less than 10% of all slaves were sold in the United States, by 1825 almost 30% of the black population in the Western Hemisphere was living in the United States (Patterson, 1982). This was due to the high rate of natural increase among the American slaves. In Brazil, by contrast, the proportion of slave imports was relatively high, but there was also a high mortality rate among the slaves owing to disease and frequent slave revolts.

At the time of the first U.S. census in 1790, there were 757,000 blacks counted in the overall population. By the outbreak of the Civil War, the number had

FIGURE 7.2
The Transatlantic Slave Trade

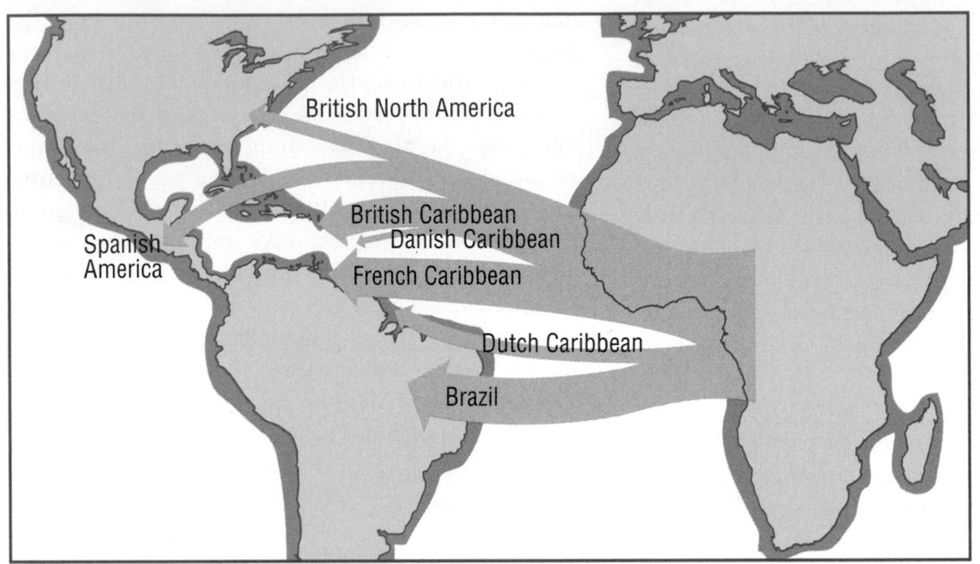

Source: Adapted from Curtin, 1969.
Note: From the end of the sixteenth century to the early decades of the nineteenth, approximately 11 to 12 million Africans were imported to the New World (Patterson, 1982). The thickness of the arrows shows the approximate volume of the slave trade to each region.

increased to 4.4 million, of whom all but about 10% were slaves. In the southern states that fought in the war, one third of white families owned slaves, the average being nine "chattels" per owner (Farley & Allen, 1987). (The term *chattels* refers to living beings that are considered property, including slaves and, in some cultures, women.)

It can be inferred from the great increase in the U.S. slave population that slaveholders in the United States treated their slaves less badly than did slaveholders in other parts of the Western Hemisphere. This does not mean, however, that the American slaves did not bitterly resent their condition, nor does it mean that they did not struggle against it. Patterson points out that the slave has always striven, against all odds, "for some measure of regularity and predictability in his social life. . . . Because he was considered degraded, he was all the more infused with the yearning for dignity" (1982, p. 337). Patterson concludes that one of the chief ironies of slavery throughout history has been that "without [it] there would have been no freedmen" (p. 342). In other words, the very idea of freedom developed in large part from the longing of slaves to be free.

Segregation

Although African American slaves gained their freedom during the Civil War and became citizens of the United States, this did not mean that they became fully integrated into American society. A long period of segregation followed. **Segregation** is the ecological and institutional separation of races or ethnic groups. The segregation of racially or ethnically distinct peoples within a society may be voluntary, resulting from the desire of a people to live separately and maintain its own culture and institutions. The Amish, the Hutterites, and the Hasidic Jews are examples of groups that voluntarily segregate themselves. But segregation is often involuntary, resulting from laws or other norms that force one people to be separate from others. Involuntary separation may be either **de jure**—created by laws that prohibit certain peoples from interacting with others or place limits on such interactions—or **de facto**—created by unwritten norms that result in segregation, just as if it were "in fact" legally required.

In the late 1980s, South Africa experienced a stunning reversal of its system of de jure segregation. Known as *apartheid,* South Africa's racial laws insisted that blacks, whites, and "coloreds" (people of mixed ancestry) remain separate. Intermarriage and integrated schools and communities were forbidden. In 1991, after years of struggle and violence, the white-dominated government agreed to repeal the apartheid laws. A constitutional convention—at which all segments of the population were represented—created a democratic regime in which all citizens' votes count equally. The African majority, led by Nelson Mandela and the African National Congress, assumed power in a relatively peaceful transition. Nevertheless, whites continue to hold most of the nation's wealth, while the majority of South African blacks are trapped in dire poverty and illiteracy, just as the former slaves were in the United States after the Civil War.

In the United States, legally sanctioned segregation no longer exists, but this has been true only in recent years. Before the civil rights movement of the 1960s, de jure segregation was common, especially in the southern states. The system that enforced segregation was supported by **Jim Crow** laws in many southern states. This term refers to laws that enforced or condoned segregation, barred blacks from the polls, and the like. (Jim Crow was a nineteenth-century white minstrel who performed in blackface and thereby reinforced black stereotypes.) This system was in effect for about a hundred years, from just after the Civil War until the early 1970s. During that period, the so-called color line was applied throughout the United States to limit the places where blacks could live, where they could work and what kinds of jobs they could hold, where they could go to school, and under what conditions they could vote.

At first, the color line was an unwritten set of norms that barred or restricted black participation in many social institutions. By the turn of the century, however, segregation had become officially sanctioned through legislation and court rulings. This official segregation was rationalized by the "separate but equal" doctrine set forth by the Supreme Court in *Plessy* v. *Ferguson* (1896). Under this doctrine, separate facilities for people of different races were legal as long as they were of equal quality. In addition, de facto segregation and the existence of a "job ceiling" based on race served to keep blacks in subordinate jobs and segregated ghettos.

Only after years of struggle by opponents of de jure segregation did the U.S. Supreme Court finally decide, in the landmark case of *Brown* v. *Board of Education* (1954), that "separate but equal" was inherently unequal. The *Brown* ruling put an end to legally sanctioned segregation of schools, hospitals, public accommodations, and the like. But it took frequent, often violent demonstrations and the mobilization of thousands of citizens in support of civil rights to achieve passage of the Civil Rights Act of 1964. That act mandated an end to segregation in private accommodations, made discrimination in the sale of housing illegal, and initiated a major attack on the job ceiling through the strategy known as affirmative action. Despite these judicial and legislative victories, however, de facto segregation remains a fact of life in the United States, especially in large cities (Massey & Denton, 1993).

Ecological studies of racial segregation show that despite some improvements in the past 30 years, residential segregation persists in almost all U.S. cities. Table 7.1 is based on a measure known as the segregation index. It shows "the minimum percentage of nonwhites who would have to change the block on which they live in order to produce an unsegregated distribution—one in which the percentage of nonwhites living on each block is the same throughout the city (0 on the index)" (Taeuber & Taeuber, 1965, p. 30). A value of 100 on the index would mean that all nonwhite people live on segregated blocks—that is, the city is 100% segregated. In 1970, for example, 81.2% of the nonwhite population in Boston would have had to move to another block to produce an unsegregated distribution of whites and nonwhites. By 1990, that proportion had declined to 68.2%, clearly an improvement.

Despite such improvements in many metropolitan regions, the table shows that most cities, especially in the North, continue to be highly segregated. Although segregation in some regions has fallen below the 70% level, high segregation indexes remain a fact of American life. Urban ecologists point out that most changes in segregation patterns are due to the movement of blacks into newer suburban communities and older white neighborhoods in the cities (Jaynes & Williams, 1989; Massey & Denton, 1993).

Hypersegregation. Douglas Massey, the leading researcher on residential segregation in the United States, makes the following comment:

> Although most white Americans now endorse open housing in principle, they remain reluctant to share neighborhoods with a high number of black residents. In metropolitan areas where blacks constitute a small share of the total population, they don't have to. For example, if blacks constitute 5% of the overall population, complete

desegregation can occur and each neighborhood will be only 5% black. On the other hand, in urban areas that are 20% black, desegregation will produce neighborhoods that are also, on average, 20% black—something most whites still won't tolerate.

Massey and his colleagues, Judith Denton and Mary Fischer, point out that these differences in the size of the black population in cities explain why black–white segregation levels have fallen in cities like Albuquerque, Tucson, and Phoenix while remaining high in Cincinnati, Detroit, Gary, New York, Newark, Chicago, and Philadelphia (see Table 7.1; note that "black" in this analysis can refer to Hispanic nonwhites as well as African Americans). In 20 U.S. metropolitan areas that together house 36% of all African Americans, Massey and his colleagues find that "racial segregation is so profound and pervades so many dimensions of life" that it can be considered **hypersegregation**—segregation of a racially or ethnically distinct population that results in segregation indexes of over 80% and profoundly affects the segregated population's life chances. Moreover, they show that "In no metropolitan area have Europeans, Latinos, or Asians ever experienced hypersegregation. Whereas 90–100% of Asians, whites, and Latinos experience no more than moderate segregation, three-quarters of African Americans are highly segregated."

The evidence Massey and his colleagues have collected over many years points to the conclusion that "segregation persists because of ongoing racial discrimination in the real estate and banking industries, the persistence of white prejudice against black neighbors, and the discriminatory impact of public policies. As a consequence, black ghettos have come to contain a disproportionate share of the nation's poor, creating an intensely disadvantaged environment that only blacks face" (Massey & Fisher, 1998, p. 25).

TABLE 7.1
Trends in Black–White Segregation in 20 Metropolitan Areas With Largest Black Populations, 1970–1990

Metropolitan Area	1970	1980	1990
Northern Areas			
Boston	81.2	77.6	68.2
Chicago	91.9	87.8	85.8
Cleveland	90.8	87.5	85.1
Detroit	88.4	86.7	87.6
Kansas City	87.4	78.9	72.6
Los Angeles–Long Beach	91.0	81.1	73.1
Milwaukee	90.5	83.9	82.8
New York	81.0	82.0	82.2
Philadelphia	79.5	78.8	77.2
St. Louis	84.7	81.3	77.0
San Francisco–Oakland	80.1	71.7	66.8
Northern Average	84.5	80.1	77.8
Southern Areas			
Atlanta	82.1	78.5	67.8
Baltimore	81.9	74.7	71.4
Birmingham	37.8	40.8	71.7
Dallas–Fort Worth	86.9	77.1	63.1
Houston	78.1	69.5	66.8
Memphis	75.9	71.6	69.3
New Orleans	73.1	68.3	50.3
Washington, D.C.	81.1	70.1	66.1
Southern Average	75.5	68.3	66.5

Source: Adapted from Massey & Denton, 1993.

STUDY CHART
A Continuum of Intergroup Relations

INTOLERANCE ⟷ TOLERANCE

Genocide	Expulsion	Slavery	Segregation	Assimulation
The intentional extermination of a population, defined as a "race" or a "people"	The forcible removal of a population from a territory claimed by another population.	Ownership of one population by another, which can buy and sell members of the enslaved population and controls every aspect of their lives.	Ecological and institutional separation of races or ethnic groups.	The process by which a minority group blends into the majority population and eventually disappears as a distinct people within the larger society.

Hypersegregation explains a great deal about how race and class combine to produce intense and durable barriers to the socioeconomic progress of African Americans. Elsewhere in the world, hypersegregation may present obstacles to efforts to improve relations among people of different races, ethnicities, or religions, such as in Northern Ireland or the Middle East. (See the Mapping Social Change box on pages 200–201.)

Other problems of inequality follow from these persistent patterns of racial segregation. It is extremely difficult, for example, to have integrated schools, especially in the primary grades, when children live in segregated neighborhoods. And it is difficult to establish tolerance and understanding among racial and ethnic groups whose members have grown up in segregated neighborhoods and schools.

Assimilation and Its Limits

One of the factors that led to segregation was the fear of racial intermarriage, since the ideology of white supremacy held that intermarriage would weaken the white race. As late as 1950, there were 30 states with laws prohibiting such marriages, and even after racist sentiments began to diminish in the 1950s, 19 states (17 of them in the South) maintained such laws until the Supreme Court declared them unconstitutional in 1967 (Holt, 1980).

Intermarriage between distinct racial and ethnic groups is an important indicator of **assimilation,** the pattern of intergroup relations in which a minority group is forced or encouraged or voluntarily seeks to blend into the majority population and eventually disappears as a distinct people within the larger society. Needless to say, it makes a great deal of difference whether an ethnic or racial group has been the victim of forced assimilation or has been allowed to absorb the majority culture at its own pace. Many Latin American societies offer examples of peaceful, long-term assimilation of various racial and ethnic groups. For example, as a result of generations of intermarriage, Brazilians distinguish among many shades of skin color and other physical features rather than relying on crude black–white distinctions (Fernandes, 1968).

Assimilation itself turns out to be a problematic process in most societies, even those with higher levels of racial tolerance than the United States. In the nineteenth century and the first half of the twentieth, many social thinkers believed in the ideal of a "melting pot" in which people from different ethnic heritages would become Americans by adopting many of the norms and roles of those who had arrived previously, while at the same time giving to the earlier Americans some of their own customs and behaviors (Gordon, 1964). But in the latter decades of the twentieth century it became clear that assimilation was never an entirely smooth or complete process for most groups. We now understand that while assimilation is a powerful process in diverse societies, so is the formation of distinct ethnic communities.

As new waves of immigrants from all over the world have streamed into the United States—Jews fleeing religious intolerance in Russia, Italians and Poles fleeing economic depression in their communities, Central and Latin Americans fleeing dictatorships and poverty, refugees from Asian countries torn by conflict—all have tended to join other members of their nationality group who had settled in the United States in earlier decades. Although the influx of new immigrants has led to conflicts between older and newer residents of some communities, in many cases it has resulted in new growth in those communities—in new ethnic businesses such as restaurants and grocery stores and in new social institutions such as churches and social clubs. It has also reinforced pride in ethnic identity as expressed in language and other aspects of ethnic subcultures. This infusion of new energy into ethnic communities confirms Glazer and Moynihan's (1970) thesis that the emergence of cultural pluralism is a significant aspect of life in American society.

The recognition that ethnic groups maintain their own communities and subcultures even while some of their members are assimilated into the larger society gave support to the concept of cultural pluralism. A **pluralistic society** is one in which different ethnic and racial groups are able to maintain their own cultures and lifestyles even as they gain equality in the institutions of the larger society. Michael Waltzer's (1980) comparative research on pluralism in the United States and other societies has shown that although white ethnic groups like the Italians, Jews, and Scandinavians may have the option of maintaining their own subcultures and still be accepted in the larger society, blacks and other racial minorities frequently experience attacks on their subcultures (e.g., opposition to bilingual education or African studies courses) and, at the same time, are discriminated against in social institutions. Waltzer concludes that "racism is the great barrier to a fully developed pluralism" (p. 787). In his study of Latinos in the United States, Earl Shorris (1992) essentially agrees with Waltzer. He finds that despite intense pressure to assimilate, many people from Spanish-speaking nations

MAPPING SOCIAL CHANGE

Hypersegregation in Israeli Cities

Although Judaism is the official religion of Israel, thousands of Arab Israelis practice their religion at mosques like this one.

Few places on earth are as tense with conflicted ethnic relations as Israel. At this writing, Arab Palestinians and Israeli Jews are once again confronting each other. Settlement by orthodox Jews in areas of the West Bank of the Jordan River continues to fuel Palestinian grievances. Palestinian claims over Israeli land and the presence of radical and terrorist segments in the Palestinian population embitter many Israelis. Will there ever be a solution to this chronic conflict? Although the answer is nowhere in sight, social scientists like Ghazi Falah (1996) point to some areas of hope.

Falah's analysis is parallel to that of Massey and Denton in the United States. It examines segregation and cross-ethnic interaction in the ancient cities inside Israel, where Arabs and Jews have been living peacefully for many decades, and in some cases for centuries. These are cities, like Acre and Jaffa, that go back to biblical times, as well as more modern, industrial cities like Haifa. Note that since the 1940s and the creation of Israel as a Jewish state, Jewish citizens have come to far outnumber Arab residents of these ancient cities (see table), but the two groups have continued to coexist. (As can be seen from the accompanying maps, these cities are not among those where violence has erupted recently.) Falah finds that although these cities contain both Arab and Jewish populations, they are generally characterized by patterns of hypersegregation similar to those found in many U.S. cities. On the other hand—and this is the hopeful aspect of the research—Arabs and Jews in these cities are not locked in violent conflict. Relations between them are often cool, but children often play together, adults socialize in the more mixed neighborhoods, and both sides voice hope for eventual peace. Are these old Arab-Israeli cities a blueprint for the future? Sociologists like Falah hope so, but they admit that it is far too soon to know.

Population of Mixed Cities, Various Years

	1945		1951		1990	
City	**Arabs**	**Jews**	**Arabs**	**Jews**	**Arabs**	**Jews**
Acre	12,310	50	4,220	11,480	9,290	31,010
Haifa	62,800	75,500	7,500	140,000	22,300	223,600
Jaffa	66,310	28,000	5,600	340,000	12,400	327,000*
Lydda	16,760	20	910	11,690	8,990	34,310
Ramia	15,160	0	1,960	15,240	7,980	39,920
Total	173,340	103,570	20,190	518,410	60,960	619,840

*Including Tel Aviv.

Source: Falah, 1996.

Sources: Palestinian Red Crescent Society; Israeli Defense Forces; Israeli Police

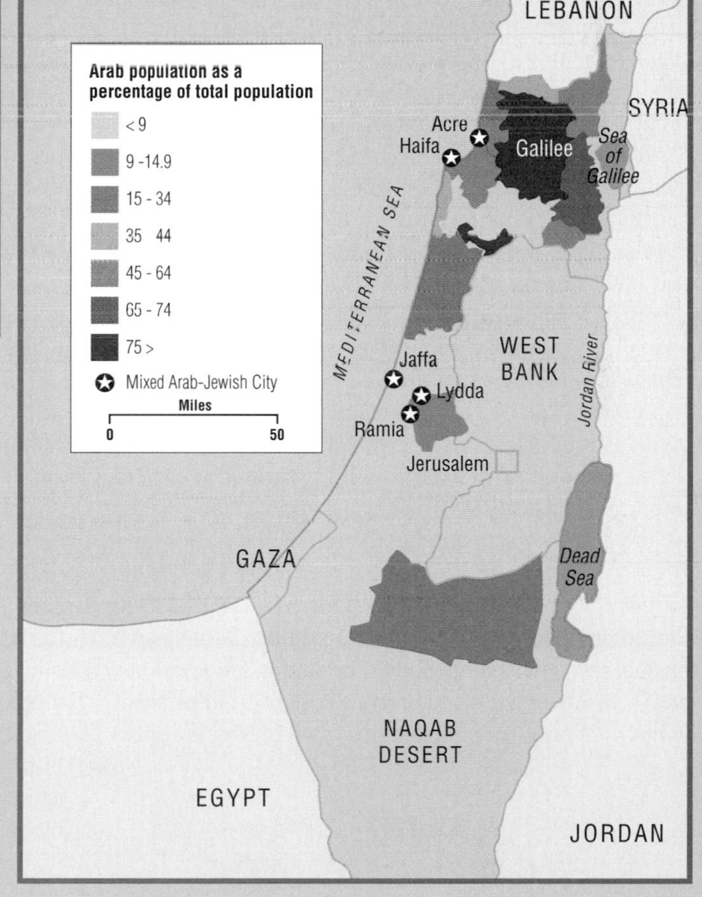

Sources: Falah, 1996.

continue to speak their native language and are using their growing influence in some parts of the country to promote bilingualism. Shorris also notes, however, that Latinos are embracing their own version of the American dream.

The problems of pluralism are illustrated by the case of the French-speaking Quebecois minority in Canada. The Quebecois account for almost 28% of the Canadian population and are the majority in the province of Quebec. This leads to a situation in which a group with its own culture and language seeks protection against pressure to assimilate into the dominant culture and yet demands equal access to the society's political, economic, and cultural institutions. The resulting tensions and hardships have led to the demand that Canada become a bilingual nation, with English and French given equal status. Although the Canadian government has taken firm steps to ensure that French is Quebec's first language, the memory of past discrimination has given rise to a social movement calling for the creation of an independent French-speaking nation. This movement is especially strong in small towns and rural areas whose entire population is French-speaking. This population strongly backed the separation referendum held in Quebec in 1995. However, the referendum was defeated by an extremely narrow margin, largely owing to voters in the more multicultural districts of Montreal, Quebec's largest and most diverse city. As this example shows, a truly pluralistic society is very difficult to achieve. The various ethnic and racial subgroups within a society often feel a strong sense of cultural identity, which they wish to preserve. They also demand equal access to the society's institutions—access to better schools, opportunities to obtain jobs in every field, opportunities to hold important positions—in short, a fair share of the wealth and power available in the society. These desires sometimes conflict, with the result that some groups may be tempted to "go it alone"—that is, form their own cultural and political institutions (their own businesses, newspapers and other media, labor unions, etc.)—or else give up their ethnic identity in order to gain greater access to the society's major institutions.

CULTURE AND INTERGROUP RELATIONS

Why do ethnic stratification and inequality persist in societies that are becoming increasingly pluralistic, like those of the United States and Canada? Sociologists have proposed a variety of theories to explain racial and ethnic inequality. Before we explore these theories, however, we need to understand the cultural basis of ethnic diversity and intergroup hostility—that is, the underlying values and attitudes that shape people's consciousness of other groups and, hence, their behavior toward members of those groups. Chief among these are the tendency to view members of other groups in terms of stereotypes and to use those stereotypes to justify differential attitudes (prejudice) and behaviors (discrimination) toward such individuals.

Stereotypes

People often express the opinion that specific traits of members of certain groups are responsible for their disadvantaged situation. Thus, in South Africa it was common for whites to assert that blacks were not ready for full citizenship because "they remain child-like and simple." In the United States, the fact that Hispanics are more likely to be found in low-paying jobs is explained by the assertion that "they don't want to learn English." And the fact that black unemployment rates are generally twice as high as white unemployment rates is explained by the statement that "they don't want to work; they like sports and music, but not hard work, especially in school." These explanations are **stereotypes,** inflexible images of a racial or cultural group that are held without regard to whether they are true.

Sociologist William Helmreich (1982) conducted a study of widely held stereotypes regarding America's major ethnic and racial groups. He found that "every single stereotype discussed turns out to have a reason, or reasons" (p. 242). Those reasons usually stem from earlier patterns of intergroup relations. For example, jokes about Poles stem from a period in the nineteenth century when uneducated Polish peasants immigrated to the United States. The idea that blacks are good at music or sports also has some basis, not because blacks are naturally superior in those areas but because when blacks were barred from other avenues to upward mobility they were able to succeed in entertainment and sports; as a result, many young blacks have developed their musical and athletic talents more fully than whites have. But even though stereotypes usually have some basis, they never take account of all the facts about a group. As the famous social commentator Walter Lippmann once quipped, "All Indians walk in single file, at least the one I saw did."

Prejudice and Discrimination

The fact that many people hold stereotypical ideas about other groups may be an indication that these individuals are ignorant or prejudiced, but it does not imply that they will actually discriminate against people whom they perceive as different. In a classic study of prejudice, social psychologist Richard LaPiere (1934) traveled throughout the United States with a Chinese couple, stopping at about 250 restaurants and hotels. Only one of the establishments refused them service. Six months later, LaPiere wrote to each establishment and requested reservations for a Chinese couple. More than 90% of the managers responded that they had a policy of "nonacceptance of Orientals." This field experiment was replicated for blacks in 1952, with very similar results (Kutner, Wilkins, & Yarrow, 1952; Shibutani & Kwan, 1965).

The purpose of such experiments is to demonstrate the difference between prejudice and discrimination. **Prejudice** is an attitude that prejudges a person, either positively or negatively, on the basis of real or imagined characteristics (stereotypes) of a group to which that person belongs. **Discrimination,** on the other hand, refers to actual unfair treatment of people on the basis of their group membership.

The distinction between attitude and behavior is important. Prejudice is an attitude; discrimination is a behavior. As Robert Merton (1948) pointed out, there are people who are prejudiced and who discriminate against members of particular groups. There are also people who are not prejudiced but who discriminate because it is expected of them. With these distinctions in mind, Merton constructed the typology shown in Table 7.2.

Merton's typology is valuable because it points to the variety of attitudes and behaviors that exist in multicultural and multiracial societies. However, it fails to account for situations in which certain groups are discriminated against regardless of the attitudes and behaviors of individuals. This form of discrimination is part of the "culture" of a social institution; it is practiced by people who are simply conforming to the norms of that institution and hence is known as *institutional discrimination.*

At its simplest, **institutional discrimination** is the systematic exclusion of people from equal access to and participation in a particular institution because of their race, religion, or ethnicity. But over time this intentional exclusion leads to another type of discrimination, which has been described as "the interaction of the various spheres of social life to maintain an overall pattern of oppression" (Blauner, 1972, p. 185). This form of institutional discrimination can be quite complex.

The conditions that led to the riot in South Central Los Angeles in 1992 after the acquittal of the police officers who beat a black motorist, Rodney King, conform very well to what sociologists have observed in similar communities where blacks or other minority groups are trapped in a self-perpetuating set of circumstances due largely to historical patterns of discrimination (Blauner, 1972, 1989). Blocked educational opportunities result in low skill levels, which together with job discrimination limit the

TABLE 7.2
Merton's Typology of Prejudice and Discrimination

DISCRIMINATION	PREJUDICE: Yes	PREJUDICE: No
Yes	TRUE BIGOT: does not believe in the American creed and acts accordingly	WEAK LIBERAL: not prejudiced, yet afraid to go against the bigoted crowd
No	CAUTIOUS BIGOT: does not believe in the American creed but is afraid to discriminate	STRONG LIBERAL: not prejudiced and refuses to discriminate

Source: Adapted from pp. 99–126 of *Discrimination and National Welfare,* edited by R. M. MacIver. Copyright © 1948 by the Institute for Religious and Social Studies. Reprinted by permission of Harper & Row Publishers, Inc.

incomes of minority group members. Low income and residential discrimination force them to become concentrated in ghettos. The ghettos never receive adequate public services such as transportation, thus making the search for work even more difficult. In those neighborhoods, also, the schools do not stimulate achievement, thereby repeating the pattern in the next generation. The young often grow up bitter and angry and may form violent gangs that engage in vandalism and other activities that the outside world sees as antisocial. At the same time, the police patrol ghettos to the point of harassment, with the result that young blacks are more likely to be arrested—and to be denied jobs because of their arrest records. Recent research shows that dark-skinned African American males with a criminal record have a jobless rate of 54%, compared with 41.7% for light-skinned African American males and 25% for white males with similar records (Johnson & Farrell, 1995). When these conditions are combined with a precipitating event, such as the savage beating of a black person and the acquittal of the police officers responsible for the beating, the results often take the form of violence like that which erupted in South Central Los Angeles.

All of the institutions involved—employers, local governments, schools, real estate agencies, agencies of social control—may claim that they apply consistent standards in making their decisions: They hire the most qualified applicants; they sell to the highest bidder; they apply the law evenhandedly; in short, they claim not to discriminate. Yet in adhering to its institutional norms each perpetuates a situation that was created by past discrimination. Many societies face the need to break this pattern and redress the conditions of inequality they have created. Unfortunately, it has proved extremely difficult to do so without also producing feelings of unfair competition among members of the majority population who do not consider themselves to blame for long-standing patterns of institutional discrimination and inequality.

Affirmative Action

Our society's foremost governmental and economic institutions are besieged with demands for *affirmative action*—that is, policies designed to correct persistent racial and ethnic inequalities in promotion, hiring, a nd access to other opportunities. Even more forceful in many parts of the country are the demands to do away with affirmative action policies. Conservatives strongly oppose affirmative action, whereas liberals feel that such policies are necessary if our society is to undo the effects of past discrimination (Buckley, 2000; Ezorsky, 1991).

At present, the conservative view is in the ascendance. Congress has done away with preferential treatment for minority-owned firms in awarding small government contracts. In California, controversial steps have been taken to eliminate affirmative action policies in the state's universities and colleges. It is likely that efforts to eliminate affirmative action will continue and will be increasingly successful. But it is also true, as Martin Kilson (1995) points out, that affirmative action has been essential to the advancement of African Americans and, increasingly, Latinos in job markets that were formerly out of bounds to minorities. In 1970, for example, "barely 2% of the officers in the armed forces were black; 20 years later, thanks (partly) to affirmative action, 12% of officers were African Americans" (p. 470).

We saw earlier that the Supreme Court struck down the doctrine that "separate but equal" facilities and institutions did not violate the constitutional rights of African Americans. It did so largely because social-scientific evidence showed that separate institutions are inherently unequal. Similar arguments have been advanced in affirmative action cases that have reached the Court. If a fire department in a city whose inhabitants are 30% black and Hispanic has no firefighters from those minority groups, it can be demonstrated that there is a pattern of discrimination that can be changed only if the department is required to hire a certain number of minority applicants—a quota—within a designated time. However, members of the majority may then feel that they are victims of "reverse discrimination," in which they are being penalized for the wrongs of earlier generations. Thus, difficult choices remain: Should employers mix or replace hiring decisions based on experience and merit with decisions based on race and ethnicity? In essence, the courts have said that they must. Opposition to affirmative action is increasing, however, and at this writing it is not clear whether the courts will continue to rule in favor of affirmative action policies.

It should be noted that affirmative action applies to women as well as to racial and ethnic minorities (Epstein, 1995). The effects of institutional discrimination against women are explored in the next chapter.

THEORIES OF RACIAL AND ETHNIC INEQUALITY

For large numbers of people, a dominant aspect of life in American society is racial or ethnic inequality and, often, hostility. How can we explain the persistence of these patterns of hostility and inequality? The first thought that comes to mind is that many people are prejudiced against anyone who is different from them in appearance or behavior. This may seem to explain phenomena like segregation and discrimination, but it fails to explain the variety of possible reactions to different groups. Social-psychological theories that focus on prejudice against members of out-groups find the origins of racism and ethnic inequality in individual psychological processes, but there are also more sociological theories that view prejudice as a symptom of other aspects of intergroup relations.

Social-Psychological Theories

The best-known social-psychological theories of ethnic and racial inequality are the frustration-aggression and projection theories. Both see the origins of prejudice in individual psychological orientations toward members of out-groups, but they differ in important ways.

Frustration-Aggression. The frustration-aggression hypothesis, which is associated with the research of John Dollard, Neil Miller, and Leonard Doob (1939), holds that the origin of prejudice is a buildup of frustration. When that frustration cannot be vented on the real cause, the individual feels a "free-floating" hostility that may be taken out on a convenient target, or **scapegoat.** For example, in the case of workers in eastern Germany who have lost jobs in antiquated factories as a result of unification with western Germany, with its far more modern industries, a convenient scapegoat may be found in gypsies, Turks, or even Jews, all groups that have historically been accused of causing negative conditions for native-born Germans. To justify the hostility directed at the out-group, the prejudiced individual often grasps at additional reasons (usually stereotypes) for hating the "others," such as "Gypsies steal children"; "Jews are usurers"; or "Turks will work for almost nothing."

Projection. The concept of projection is also used to explain hostility toward particular ethnic and racial groups. **Projection** is the process whereby we attribute to other people behaviors and feelings that we are unwilling to accept in ourselves. John Dollard (1937) and Margaret Halsey (1946) applied the concept of projection to white attitudes toward black sexuality. Observers had noted that southern whites frequently claimed that black males are characterized by an uncontrollable and even vicious sexuality. The theory of projection explains this claim as resulting from the white males' attraction to black females, an attraction that was forbidden by strong norms against interracial sexual contact. Thus, the white male projected his own forbidden sexuality onto blacks and developed an attitude that excused his own sexual involvement with black women.

Interactionist Theories

Not far removed from these social-psychological theories of intergroup hostility are theories derived from the interactionist perspective in sociology. But instead of locating the origins of intergroup conflict in individual psychological tendencies, interactionists tend to look at how hostility or sympathy toward other groups, or solidarity within a group, is produced through the norms of interaction and the definitions of the situation that evolve within and between groups. A few examples will serve to illustrate how the interactionist perspective is applied.

From his analyses of interaction in different groups, Georg Simmel concluded that groups often find it convenient to think of nonmembers or outsiders as somehow inferior to members of the group. But why does this familiar in-group–out-group distinction develop? Simmel explained it as arising out of the intensity of interactions within the group, which leads its members to feel that other groups are less important. Once they have identified another group as inferior, it is not a great leap to think of its members as enemies, especially because doing so increases their sense of solidarity (Coser, 1966; Simmel, 1904).

Conflict and hostility between racial or ethnic groups can be overcome by creating situations that require the groups to cooperate to achieve a common goal. Such situations occasionally occur—for example, when people from different ethnic and racial backgrounds compete in sports or when they work side by side in school and at jobs. Unfortunately, the integration and friendship found in such settings are not sufficient to overcome the more deep-seated prejudices and fears of many racist individuals. As film director Spike Lee observes, racists place all black people in one of two categories: "entertainers and niggers." By this he means that people with racist sentiments tend to exclude famous entertainers and sports figures from their negative feelings about blacks as a group, thereby accentuating their failure to see all blacks as individual human beings.

Functionalist Theories

The difficulty of reducing racism and discrimination through interaction leads to the question of whether racial and ethnic inequalities persist because they function to the advantage of certain groups. One answer was provided by the functionalist theorist Talcott Parsons (1968), who wrote that "the primary historic origin of the modern color problem lies in the relation of Europeans to African slavery" (p. 366). Parsons was not denying that racism is produced through interactions. Rather, he was pointing out that the specific form taken by those interactions—oppression, subordination, domination of blacks by whites—is directly related to the perceived need of white colonialists and traders to use blacks for their own purposes. The whites could abduct, enslave, and sell Africans because their societies had developed technologies (e.g., oceangoing ships and navigational instruments) and institutions (e.g., markets and trading corporations) that made them immensely more powerful than the Africans.

From the functionalist perspective, inequalities among ethnic or racial groups exist because they have served important functions for particular societies. Thus, in South Africa it was functional for the white government to insist on maintaining apartheid because to do otherwise would mean that whites would become a minority group in a black-dominated society. But as world opinion continued to condemn the white regime and blacks continued to build group solidarity and challenge that regime, it became less and less "functional" for the white government to insist on complete apartheid. Indeed, some have speculated that the nation's white rulers released Nelson Mandela and took steps to legalize the formerly banned African National Congress as a way of showing the world that they were willing to begin negotiations toward some form of shared power. Their move incurred the wrath of many South African whites, who feared the consequences of majority black rule under a new constitution. Similarly, the Israeli government's negotiations with the Palestine Liberation Organization enraged many Jewish settlers living in areas that would be placed under the control of Palestinians. This rage resulted in the shocking assassination of Israeli prime minister Yitzhak Rabin in November 1995 as peace negotiations were under way.

In both Israel and South Africa, it appears that the status quo of intergroup relations had become dysfunctional. Change had to come either through peaceful negotiations or through bloodshed. The processes whereby power is redistributed and new social relations develop are often best understood with the help of conflict theories.

Conflict Theories

Conflict theories do a better job than functionalist theories of explaining why groups with less power and privilege, like the South African blacks or the Palestinians in the Middle East, do not accept their place in the status quo. Conflict theories try to explain why these groups often mobilize to change existing intergroup relations (Blauner, 1989). Conflict changes societies, as we see clearly in the case of South Africa, where many years of conflict waged by the disenfranchised blacks led to the creation of a new constitution and majority black rule. But conflict can also destroy societies and nations, as we see all too clearly in the war-torn nations of the former Yugoslavia, where a long history of ethnic conflict has prevented the achievement of a stable, well-functioning society.

Conflict theories trace the origins of racial and ethnic inequality to the conflict between classes in capitalist societies. Marx believed, for example, that American wage earners were unlikely to become highly class conscious because ethnic and racial divisions continually set them against one another and the resulting strife could be manipulated by the capitalist class. Thus, in American history we see many examples of black and Mexican workers being brought in as strikebreakers by the owners of mines and mills, especially during the 1920s and 1930s. Strikebreakers from different racial and ethnic groups often absorbed the wrath of workers, anger that might otherwise have been directed at the dominant class. To forge class loyalties despite the divisions created by racial and ethnic differences, Marx believed, it would be necessary for workers to see that they were being manipulated by such strategies.

A PIECE OF THE PIE

Up to this point we have explored the patterns of intergroup relations, the cultural basis of those patterns, and some of the theories that have been proposed to explain them. But we have not discussed how groups actually win, or fail to win, a fair share of a society's valued statuses and other rewards—that is, a piece of the social pie. In particular, the persistence of racial inequality in the United States requires more attention. This is a complex problem that is a source of continuing controversy.

How can we explain the fact that so many social problems in the United States today are associated with race? As William Julius Wilson (1984) has stated, "Urban crime, drug addiction, out-of-wedlock births, female-headed families, and welfare dependency have risen dramatically in the last several years and the rates reflect a sharply uneven distribution by race" (p. 75). Wilson and others doubt that racial prejudice and discrimination adequately account for the severity of these problems, for the period since the early 1970s has seen more antidiscrimination efforts than any other period in American history. The answers are to be found, Wilson argues, in how older patterns of racism and discrimination affect the present situation.

In an important study of this issue, sociologist Stanley Lieberson (1980) asked why the European immigrants who arrived in American cities in the late nineteenth and early twentieth centuries have fared so much better, on the whole, than blacks. Sociologist Marta Tienda asks similar questions about people of Latino ancestry (Tienda & Singer, 1995; Tienda & Wilson, 1992). Both agree that the problem of lagging black mobility is a complex one. For African Americans, the situation can be summarized as follows:

1. African Americans have experienced far more prejudice and discrimination than any immigrant group, partly because they are more easily identified by their physical characteristics. As a result of the legacy of slavery, which labeled blacks as inferior, they have been excluded from full participation in American social institutions far longer than any other group.
2. Black families have higher rates of family breakup than white families. The problems of the black family are not part of the legacy of slavery, however. As Herbert Gutman (1976) has shown, slavery did not destroy black families to the extent that earlier scholars believed it did. Nor did the migration of blacks to the North during industrialization. A comprehensive review of the status of black Americans notes that "there was no significant increase in male-absent households even after the massive migration to the urban North" (Jaynes & Williams, 1989). Until the 1960s, three quarters of black households with children under age 18 included both husband and wife. "The dramatic change came only later, and in 1986, 49% of black families with children under age 18 were headed by women" (Jaynes & Williams, 1989, p. 528). The report reasons that if black two-parent families remained the norm through slavery, the Great Depression, migration, urban disorganization, and ghettoization, then it appears unlikely that there is a single cause for the dramatic decline in two-parent families during the past two decades.
3. Structural changes in the American economy—first the shift from work on farms to work in factories and then the shift away from factory work to high-technology and service occupations—have continually placed blacks at a disadvantage. No sooner had they begun to establish themselves as workers in these economic sectors than they began to suffer the consequences of structural changes in addition to job discrimination.

Recent research by social scientists points to the decrease in manufacturing jobs and the increase in service-sector jobs as contributing to the sharp decline in the fortunes of black males. (Black females also suffer, but they are more readily recruited into clerical employment.) Throughout the second half of the twentieth century, the unemployment rate for black males was twice that for white males. This remains true today, but the situation is even worse for black men who do not have high school diplomas. Their unemployment rate is more than three times higher than that for white males (Oliver & Shapiro, 1990). A great deal of research shows that difficulty in securing adequate employment contributes to the problems young couples experience in forming lasting relationships (Danziger & Gottschalk, 1993; Garfinkel & McLanahan, 1986).

It is significant that among Hispanic groups in the United States, Puerto Ricans have experienced northward migration and problems of discrimination that are very similar to those experienced by African

Americans in northern cities. Still a dependent territory of the United States, since World War II Puerto Rico has sent hundreds of thousands of working people to cities like New York, Philadelphia, and Chicago, which have been rapidly losing manufacturing jobs. The status of Puerto Ricans in the population is very similar to that of blacks. There is a growing Puerto Rican middle class and an even more rapidly growing number of Puerto Ricans among the poor. This situation is due largely to changes in the types of jobs available to unskilled blacks and Puerto Ricans (Sandefur & Tienda, 1988).

It is worth noting that social mobility is more available to blacks today than it was before the civil rights movement. Today there are millions of middle-class African Americans, and the earnings of black males who have completed college are about 85% to 90% of those of whites with comparable education. But although the black middle class is growing, the majority of black workers remain dependent on manual work in industry or lower-level service jobs—types of jobs with limited employment security and poor or nonexistent health benefits. Another important segment of black workers, male and female, are employed in the public sector. Cuts in public budgets drive their earnings down as well. Chronic unemployment is associated with family breakup, alcohol and drug addiction, and depression. These social problems, in turn, severely hamper the ability of individuals to learn the attitudes and skills they need for entry into available jobs (Oliver & Shapiro, 1990; W. J. Wilson, 1987).

Nor is the life of middle-class blacks free from the continuing consequences of racism and prejudice. In interviews with a sample of middle-class black Americans in 16 cities, sociologist Joe R. Feagin (1991) found that "they reported hundreds of instances of blatant and subtle bias in restaurants, stores, housing, workplaces, and on the street" (p. A44). And when the U.S. economy's ability to create good jobs declines, the number of crimes against minority individuals of all social classes increases.

TECHNOLOGY AND SOCIAL CHANGE

The historic presidential election of November 2000 raised issues of technology, racial and ethnic relations, and democratic processes that will be debated in the United States and throughout the world for many years. From a race relations perspective, the election ended on a sour note. Elderly voters in Palm Beach County, Florida, many of whom are Jewish Americans who originally lived in the Northeast, and African American voters in Broward County, Florida, claimed that their democratic rights had been infringed upon because faulty balloting technologies caused them to vote for the wrong candidate or to be denied ballots. In the confusion that followed the narrowest vote in U.S. history, many irregularities were revealed in voting processes in other states as well. But Florida was the pivotal state in the determination of electoral college votes. Naturally, therefore, the media's focus on protests in Florida made it appear that racial and ethnic considerations or biases were also at issue in the Florida vote.

In the aftermath of the election, many people raised questions about why the United States retains the electoral college system rather than deciding presidential elections on the basis of the popular vote. Others raised more technological issues, such as whether it would be feasible to vote over the Internet or to have a more heavily computerized system for recording and counting ballots. The first of these questions is about reform of political institutions (see Chapter 12). The second is about faith in the use of technology to solve social problems. Is there a "technological fix" for the problems of voter registration that barred African Americans in Broward Country from the polls? Would computer-based technologies have eliminated the problems in West Palm Beach? The answer to both questions is, of course, that yes, perhaps they would have. But as we know from our own use of computers and the Internet, these technologies are prone to problems of their own, and one could imagine that massive reliance on computer-based technologies rather than human administration at the polls could result in major disasters on Election Day as well. In fact, another technology, television, has had so much impact on the conduct of elections that many political experts feel that for elections to become more democratic it will be necessary first to reform the system of campaign finance and ensure greater equality of access of candidates to television time. A quick technological fix of polling systems is unlikely to solve these more fundamental problems in American democratic processes.

THE CENTRAL QUESTION: A CRITICAL VIEW

Can knowledge of how racial and ethnic inequalities originate help societies improve race and ethnic relations?

Here is how one important sociologist addresses this question. William Julius Wilson is the nation's most influential sociologist in the area of race and social policy: He believes strongly that "race blind" social policy is essential to improved race and ethnic relations. "I have in mind a vision," he writes,

> that promotes values of racial and intergroup harmony and unity and rejects the commonly held view that race is so divisive that whites, blacks, Latinos, and other ethnic groups cannot work together in a common cause. This vision recognizes that if a political message is tailored to a white audience, racial minorities draw back just as whites draw back when a message is tailored to minority audiences. (1996, p. xxi)

William J. Wilson, one of the most influential voices on race relations and social policy in the United States.

In studying the poorest and most segregated communities in the United States, Wilson comes to the conclusion that "the problems of ghetto joblessness are so severe," and so intertwined with other serious social problems, including crime, illegitimacy, ill health, and school failure, that we urgently need "a vision of interracial unity that acknowledges distinctively racial problems but nonetheless emphasizes common solutions to common problems" (1996, p. xxi). Here Wilson refers to efforts by government, corporations, and individuals to create economic opportunity in the nation's most impoverished and segregated neighborhoods. But he is quick to emphasize that whatever opportunity is extended to people of color or of different ethnic backgrounds needs to be equally available to poor people who are not members of minority groups. This is what he means by a "race-blind" social policy.

Why should we share this vision? Is it not true that many of the problems of the minority poor, and the nonminority poor as well, are due to immorality, lack of effort, drug abuse, and other individual failures? Wilson warns against falling into this trap. No one can deny that many people at the bottom of society are struggling with failures of many kinds, but can society afford to deny their humanity? Wilson and others who have studied the problems of the poor argue that recent changes in social policy that reduce government programs in poor neighborhoods and reduce aid to poor children will worsen, rather than improve, conditions in the ghettos. Thus, Wilson urges us to "endorse the idea that all groups, including those in the throes of ghetto joblessness, should be able to achieve full membership in society because the problems of economic and social marginality spring from the inequities in society at large and not from group deficiencies" (1996, p. xxii).

SOCIOLOGY AND SOCIAL JUSTICE

Racial profiling in police work raises enormous problems of social justice. This is why, for example, a federal district court in San Francisco decided in 2000 that Border Patrol agents may not consider "Hispanic appearance" as a factor in deciding whether to stop motorists for questioning near the U.S.–Mexico border. In a seven-to-four ruling, the court noted that "Stops based on race or ethnic appearance send the underlying message to all our citizens that those who are not white are judged by the color of their skin alone. Such stops also send a clear message that those who are not white enjoy a lesser degree of constitutional protection—that they are in effect assumed to be potential criminals first and individuals second" (Weinstein, 2000, p. 1).

Will decisions like this one stop the practice of racial profiling? That is quite unlikely. Although law enforcement officials theoretically make stops based on reasonable suspicion or actual driving infractions, in fact there are large areas of discretion that allow racial bias to creep in. One driver's rolling stop can be another's failure to respect a stop sign. For members of minority groups, however, the key lesson of the recent conflict over racial profiling is that eternal vigilance and protest are justified in the never-ending battle against racial and ethnic bias in all its manifestations. ■

Like most civil rights issues, racial profiling did not receive the attention of lawmakers until it became the subject of organized demonstrations and protests by minority citizens and their allies.

SUMMARY

Race is a social concept. It varies from one society to another, depending on how the people of that society feel about the importance of certain physical differences among human beings. *Racism* is an ideology based on the belief that an observable, supposedly inherited trait is a mark of inferiority that justifies discriminatory treatment of people with that trait.

Ethnic groups are populations that have a sense of group identity based on a distinctive cultural pattern and shared ancestry. Ethnic and racial populations are often treated as *minority groups;* that is, they are singled out for differential and unequal treatment because of their physical or cultural characteristics.

Genocide is the intentional extermination of one population by a more dominant population, usually justified by the belief that the people who are being slaughtered are less than human. *Expulsion* is the forcible removal of one population from territory claimed by another population. *Slavery* is the ownership of a population by another population that has complete control over the enslaved population.

Segregation is the ecological and institutional separation of races or ethnic groups. Involuntary segregation may be either *de jure* (created by laws) or *de facto* (enforced by unwritten norms). *Hypersegregation* refers to segregation of a racially or ethnically distinct population that results in segregation indexes of over 80% and profoundly affects the segregated population's life chances. It occurs when racial segregation is so profound that it pervades many dimensions of life.

Assimilation occurs when a minority group blends into the majority population and eventually disappears as a distinct people within the larger society.

Stereotypes are inflexible images of a racial or cultural group that are held without regard to whether or not they are true. They are often associated with *prejudice* (an attitude that prejudges a person on the basis of characteristics of a group of which that person is a member) and *discrimination* (behavior that treats people unfairly on the basis of their group membership). *Institutional discrimination* is the systematic exclusion of people from equal participation in a particular social institution because of their race, religion, or ethnicity. Policies designed to correct persistent racial and ethnic inequalities are referred to as affirmative action.

Social-psychological theories of ethnic and racial inequality argue that a society's patterns of discrimination stem from individual psychological orientations toward members of out-groups. Interactionist explanations focus on how hostility or sympathy toward other groups is produced by the norms of interactions that evolve within and between groups. The functionalist perspective generally seeks patterns of social integration that help maintain stability in a society, whereas conflict theories trace the origins of racial and ethnic inequality to the conflict between classes in capitalist societies.

The persistence of racial inequality in the United States is a result of a number of factors besides racial prejudice and discrimination. Among those factors are high rates of family breakup and the effects of structural changes in the American economy.

GLOSSARY

Race: An inbreeding population that develops distinctive physical characteristics that are hereditary. **(189)**

Racism: An ideology based on the belief that an observable, supposedly inherited trait is a mark of inferiority that justifies discriminatory treatment of people with that trait. **(190)**

Ethnic group: A population that has a sense of group identity based on shared ancestry and distinctive cultural patterns. **(191)**

Minority group: A population that, because of its members' physical or cultural characteristics, is singled out from others in the society for differential and unequal treatment. **(192)**

Genocide: The intentional extermination of one population by a more dominant population. **(194)**

Expulsion: The forcible removal of one population from a territory claimed by another population. **(194)**

Slavery: The ownership of one racial, ethnic, or politically determined group by another group that has complete control over the enslaved group. **(195)**

Segregation: The ecological and institutional separation of races or ethnic groups. **(196)**

De jure segregation: Segregation created by formal legal sanctions that prohibit certain groups from interacting with others or place limits on such interactions. **(196)**

De facto segregation: Segregation created and maintained by unwritten norms. **(196)**

Jim Crow: The system of formal and informal segregation that existed in the United States from the late 1860s to the early 1970s. **(197)**

Hypersegregation: Segregation of a racially or ethnically distinct population that results in segregation indexes of over 80% and profoundly affects the segregated population's life chances. **(198)**

Assimilation: A pattern of intergroup relations in which a minority group is absorbed into the majority population and eventually disappears as a distinct group. **(199)**

Pluralistic society: A society in which different ethnic and racial groups are able to maintain their own cultures and lifestyles while gaining equality in the institutions of the larger society. **(199)**

Stereotype: An inflexible image of the members of a particular group that is held without regard to whether it is true. **(202)**

Prejudice: An attitude that prejudges a person on the basis of a real or imagined characteristic of a group to which that person belongs. **(203)**

Discrimination: Behavior that treats people unfairly on the basis of their group membership. **(203)**

Institutional discrimination: The systematic exclusion of people from equal participation in a particular institution because of their group membership. **(203)**

Scapegoat: A convenient target for hostility. **(205)**

Projection: The psychological process whereby we attribute to other people behaviors and attitudes that we are unwilling to accept in ourselves. **(205)**

QUESTIONS FOR THOUGHT AND DISCUSSION

1. Is racial and ethnic understanding a particular problem in the United States, or is it a problem everywhere in the world? In framing your answer, think about the countries of the former Yugoslavia, the Middle East, and other countries and regions mentioned in the text. Can you cite examples from the United States and elsewhere that illustrate the continuum of race and ethnic relations, from genocide to assimilation or even beyond, to tolerance of multicultural diversity?

2. What was so important about the Supreme Court's decision in *Brown* v. *Board of Education?* What characteristics of the U.S. population make it difficult to achieve racial desegregation in the public schools? (Hint: Think about the measurement of residential segregation.)

DIGGING DEEPER

Books

Voices in Our Blood: America's Best on the Civil Rights Movement (John Meachan, ed.; Random House, 2001). A moving "oral history" of the civil rights movement, including contemporary voices.

Black Feminist Thought: Knowledge, Consciousness, and the Politics of Empowerment (Patricia Hill Collins; Routledge, 2000). An important contribution to the debate on how gender and racial minority status interact to give minority women special advantages and disadvantages in modern social life.

Native Americans in the News: Images of Indians in the Twentieth Century Press (Mary Ann Weston; Greenwood, 1996). An insightful study of how Native Americans are portrayed in the print media, and a reminder of the persistence of discrimination against the continent's original residents.

Threatened People, Threatened Borders: World Migration and U.S. Policy (Michael S. Teitlebaum & Myron Weiner; Norton, 1995). A review of world patterns of population movement and U.S. policy responses to the growing demand for entry into North America, by a panel of renowned experts on migration and immigration.

When Work Disappears: The World of the Urban Poor (William J. Wilson; Knopf, 1996). A trenchant analysis of how the drastic decline in manufacturing employment in a major U.S. city has devastated the social and economic prospects of an important African American community.

American Apartheid: Segregation and the Making of the Underclass (Douglas S. Massey & Nancy A. Denton; Harvard University Press, 1993). A review of the continuing persistence of residential segregation in American cities.

Latinos: A Biography of the People (Earl Shorris; Norton, 1992). A beautifully written and thoughtful book describing Spanish-speaking peoples' search for cultural unity in the face of historical, socioeconomic, and racial diversity.

U.S. Race Relations in the 1980s and 1990s: Challenges and Alternatives (Gail S. Thomas, ed.; Hemisphere, 1990). Covers the recent rise in racism, racially motivated incidents, and race-hatred groups, as well as efforts to address issues of education, poverty, and other problems that disproportionately affect racial minorities. Argues that improvement in race relations will be difficult in the years ahead and will require a far greater national commitment than exists at present.

Other Sources

Harvard Encyclopedia of American Ethnic Groups. A valuable collection of essays about ethnic and racial relations in American society. Covers almost every imaginable group in the United States and offers summary essays on such sociologically relevant subjects as slavery, assimilation, ethnic accommodation, and conflict.

EXPLORING SOCIOLOGY ON THE INTERNET

The Southern Institute at Tulane University

www.tulane.edu/%7Eso-inst

Offers information about antibias education and other research-oriented Web sites on civil rights, race, ethnicity, and "trouble spots around the world."

Interracial Voice

www.webcom.com/~intvoice/

An electronic publication that promotes the establishment of a multiracial category on the census.

The Movies, Race, and Ethnicity

www.lib.berkeley.edu/MRC/EthnicImages Vid.htm

Provides videographies, bibliographies, and full-text articles of a sample of Hollywood films emphasizing issues faced by African Americans, Asians, Latinos, Jews, and Native Americans.

The National Fair Housing Advocate

www.fairhousing.com/

Keeps track of legal cases concerning housing discrimination throughout the nation, links organizations that advocate fair housing practices, and provides updated information on federal guidelines, job openings, and other matters.

The Legal Information Institute at Cornell University

www.law.cornell.edu/topics/civil_rights.htm

Provides legal information on civil and constitutional rights, employment discrimination, human rights, and the latest judicial decisions.

International Sociological Organization

www.uem.es/

The Web site for the world organization of sociologists. Has links to many important researchers who are dealing with issues of ethnicity and race in the developed and developing world.

European Research Centre on Migration and Ethnic Relations (ERCOMER)

www.ercomer.org

A university-based research institute in the Netherlands that promotes "peaceful coexistence, justice, and harmony in inter-ethnic relations." Offers access to research papers and important research findings on ethnic and racial relations in global hot spots.

Chapter 8

Inequalities of Gender and Age

Will inequalities of gender and age diminish as nations become more industrialized and urbanized?

The brutal consequences of extreme sexism are common sights to Dr. Hani Jahshan. As a coroner and forensic medical specialist in the Jordanian capital of Amman, Dr. Jahshan sees case after case of women who are victims of "honor killings." "Women are brought to me to be examined after a sexual assault," the doctor explains. "I tell their parents they are innocent but a few weeks later I see them again, dead."

Asma Khader, a political activist and lawyer in Amman, tells of a woman who came to her because the woman's husband demanded that she turn over all her wages to him. When she refused, he told her that he would tell people he had seen her with another man. She would be killed. Defiant, the woman kept her money and joined a political movement known as Sisterhood Is Global. This organization dared to demand that the Jordanian government require that sentences for so-called honor killings be the same as those for any other form of murder. The Jordanian parliament almost unanimously rejected this measure.

The activists are not defeated. They are going back to parliament, this time with the support of several Christian and Muslim religious leaders as well as King Abdullah, Queen Rania, and the extremely popular Queen Noor, widow of the late King Hussein. "I think we may get it this time," Khader says. But she knows that changing the law is not enough. "We need a public-awareness campaign. We need to send a cultural message to the people that this isn't accepted anymore" (Armstrong, 2000).

Such messages will also have to convince many women in Jordan and other nations where such customs as honor killings are legitimate practices. Khader and many of her allies in the women's movement are younger women with their lives ahead of them. They are far more desirous of liberation from oppressive customs than women of their mothers' and grandmothers' generations, who have often, but by no means always, made their personal peace with the restrictions men place on their lives. ■

GENDER, AGE, AND SOCIAL CHANGE

All over the world, distinctions between men and women, boys and girls, the young and the elderly create boundaries that stratify people into different social categories. Whatever scientific evidence there may be about differences between the sexes or among various age groups, people's beliefs about such differences create divisions that separate men from women and young people from adults and the elderly. These social boundaries have enormous consequences for the way individuals are treated.

Two of the most significant examples of social change over the past century are the acceptance of women as full citizens in Western democracies like the United States and the end of widespread poverty among the elderly. Does it seem strange that women were once considered so different and inferior to men that they were not allowed to vote? Does it seem difficult to accept the notion that the elderly needed to demand the right to live out their lives in security and relative well-being? Inequalities based on a person's gender or age persist in the United States and other urban industrial nations, but they are not nearly as stark and significant as they are in the poorer nations of the world. (See the Global Social Change box on pages 222–223.) And those who feel comfortable with existing inequalities in the life chances of men and women or of people of different ages often continue to justify the status quo with arguments about biological differences or other kinds of seemingly immutable differences.

This chapter begins by defining the basic concepts of sexual and gender identity and discussing how people are socialized as boys and girls, men and women in a given society. We then move on to an examination of how gender roles and gender-based inequality are related, and then to issues of sexism and inequalities in the workplace. Inequalities of gender and age are closely related, especially because there are many more elderly women than men in the populations of economically advanced nations. But the developing nations have age structures that are distinct from those of older industrial nations like the United States; later sections of the chapter therefore consider how age structures divide a population into cohorts of people of roughly similar ages and the consequences of this structuring in terms of social stratification. The chapter ends with a discussion of ageism and social movements among the elderly.

GENDER AND INEQUALITY

Inequalities between men and women, even those that result in such extreme behaviors as "honor killings," are a fundamental aspect of social stratification. Every society has expectations for male and female behavior, expectations that prescribe a division of labor and responsibilities between women and men and grant them different rights and obligations (Mason, 1997). These differences produce various kinds of inequality that are not based on differences between the sexes or on any traits inherent in women and men, but instead emerge in "micro, middle, and macro-level processes that apply gender labels to jobs, skills, institutions, and organizations, as well as to people, and that use these labels to produce, express, and legitimate inequality" (Ferree & Hall, 2000, p. 476).

Knowledge of a society's system of gender stratification takes one away from mere concern with how men's attitudes toward women are formed, or why women submit to unequal treatment, to a broader understanding of the assumptions behind gender inequality. In the United States and other highly industrialized nations, for example, it is often assumed that a successful manager, usually a male, has a wife who has chosen to be a homemaker. These assumptions, although increasingly challenged, are part of what sociologists understand by gender stratification in operation. Another example is found in comparisons of the salaries received by men and women in similar occupations.

Table 8.1 shows that in the United States there is a continuing gap in income between men and women in the same occupations. We will see later in the chapter that even as increasing numbers of women join the labor force and compete for jobs and careers in ways that were almost unheard of early in the twentieth century, men still tend to move into the higher-paying jobs in each occupation. Figure 8.1 shows quite dramatically that in the less-developed regions women shoulder a far greater number of domestic burdens than men do. Much of the difference may seem to be due to the fact that women take primary responsibility for bearing and nurturing children. And one might argue that in a developed country like the United States, the differences in men's and women's incomes could be attributed to the fact that so many women take time out from work and careers to bear and raise children. These familiar (and flawed) arguments raise further questions about biological versus social causes of gender

TABLE 8.1
Median Weekly Earnings of Full-Time Wage and Salary Workers, 1998

Occupation	Men	Women
Managerial and professional	$952	$681
Technical, sales, and administrative support	626	431
Service	402	304
Precision production	606	428
Operators, fabricators, and laborers	472	337
Farming, forestry, and fishing	377	283

Source: *Statistical Abstract,* 2000.

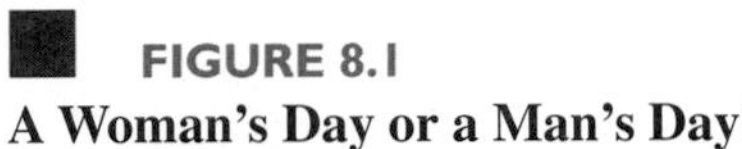

FIGURE 8.1
A Woman's Day or a Man's Day?

Which would you prefer if you could choose? The Swedish International Development Agency outlined a typical day for a man and a woman in a family that grows both cash crops and its own food supply, in its journal *Striking a Balance.* The family lives "somewhere in Africa."

A WOMAN'S DAY	A MAN'S DAY
rises first	
kindles the fire	
breast-feeds the baby	
fixes breakfast/eats	rises when breakfast is ready
washes and dresses children	eats
walks 1 km to fetch water	
walks 1 km home	walks 1 km to field
gives the livestock food and water	
washes cooking utensils, etc.	works in the field
walks 1 km to fetch water	
walks 1 km home	
washes clothing	
breast-feeds the baby	
walks 1km to field with food for husband	
walks 1 km back home	eats when wife arrives with food
walks 1 km to her field	
weeds field	works in the field
breast-feeds the baby	
gathers firewood on the way home	
walks 1 km home	walks 1 km home
pounds maize	rests
walks 1 km to fetch water	
walks 1 km home	
kindles the fire	eats
prepares meal/eats	walks to village to visit other men
breast-feeds the baby	goes to bed
puts house in order	
goes to bed last	

Source: Sadik, 1995.

inequality, which, in turn, may create confusion about the differences between sex and gender and the role of sexuality and sexual orientation in defining an individual's life chances. Before exploring the contours of gender inequality further, therefore, we turn to a discussion of the basic concepts of sex and gender.

SEX VERSUS GENDER

Masculine and *feminine* are among the most confusing words used both in everyday speech and in the social sciences. The confusion stems largely from the fact that our ideas of what is male and female in human individuals are based on overlapping influences from biology and culture. As in all such situations, it is difficult to sort out the relative contributions of each source of influence. But the underlying concepts are not hard to grasp: One's sex is primarily a biological quality, while one's gender is formed largely (but not entirely) by the cultural forces we experience through socialization from infancy on (Connell, 1995).

Sex refers to the biological differences between males and females, including the primary sex characteristics that are present at birth (i.e., the presence of specific male or female genitalia) and the secondary sex characteristics that develop later (facial and body hair, voice quality, and so on). Note that these biological qualities differ considerably among individuals, so that the differences between the sexes are not always as marked as the male–female distinction suggests. Biologist Helen H. Lambert notes that "biological sex differences are not necessarily universal, that is they do not obtain between any male/female pair chosen at random. Some women are taller than some men; the men of some racial groups have no beards" (1987, p. 125). She notes that the concept of a strict dichotomy between male and female is a social rather than a biological one. Differences between males and females are often as much a consequence of social learning and behavior (activities considered appropriate for males and females) as of innate biological differences.

Research on gender has benefited greatly from studies of people who were born with ambiguous sexual organs or whose external genitalia have been damaged. People born as **hermaphrodites** have primary sexual organs that exhibit both male and female features, making it difficult to categorize the person as either male or female. Hermaphroditism is uncommon, but it occurs in societies throughout the world. Research has shown that in some cultures hermaphroditic babies are allowed to have an uncertain sex. They may be treated as either girls or boys, or just as different individuals, with no sign of hostility. In most Western cultures, and especially in the United States, such infants are assigned to one sex or the other, a sign of the culture's concern with avoiding sexual ambiguity (Dreger, 1998).

Another ambiguous sexual category consists of people who feel very strongly that the sexual organs they were born with do not conform to their deep-seated sense of what their sex should be. These **transsexuals** sometimes undergo a course of endocrine hormonal treatments to change their secondary sex characteristics and may eventually have irreversible sex-change operations. Most such operations are performed to change the individual from a male to a female.

One of the most famous transsexuals in the contemporary world is the writer Jan Morris. As a male, James Morris reported on an expedition to Mount Everest and engaged in many activities that are associated with masculinity. But as Morris explains it, "I was born with the wrong body, being feminine by gender but male by sex, and I could achieve completeness only when the one was adjusted to the other" (quoted in Money & Tucker, 1975, p. 31). In making this distinction, Morris is calling attention to the difference between her biologically determined sex and her socially and emotionally influenced gender.

Gender refers to the culturally defined ways of acting as a male or a female that become part of an individual's personal sense of self. The vast majority of people develop a "gut-level" sense of themselves as male or female, boy or girl, early in life; however, as we will see shortly, some people's gender identities are more ambiguous. It is not entirely clear exactly how a person's gender identity is formed, but much of the research done to date suggests that the assignment of a gender at birth has a strong influence during the early years of life. In other words, children's feelings of being a boy or a girl are defined more by how they are treated by their parents than by their actual biological sex characteristics (Stockard & Johnson, 1992).

Gender and Sexuality

Sexuality refers to the manner in which a person engages in the intimate behaviors connected with genital stimulation, orgasm, and procreation. Like most areas of human behavior, sexuality is profoundly influenced by cultural norms and social institutions like the family

GLOBAL SOCIAL CHANGE

Women in the Developing World

Social scientists estimate that more than 80 million women in various parts of the world have undergone what is known as female circumcision. At most world population and development conferences this is a subject of hot debate, along with contraception and other reproductive issues. Female circumcision is common in about 40 nations, especially in East and West Africa and on the Arabian peninsula—but worldwide migrations have brought the practice to Europe and North America as well. Public health officials and critics note that female circumcision differs from male circumcision in that it involves the removal of all or part of a genital organ, the clitoris, and as such constitutes mutilation.

Female circumcision is done in order to diminish sexual desire and thus "protect" the woman from temptation and infidelity. Among the Maasai of Kenya, for example, a woman who is not circumcised is considered unsuitable for marriage. But the operation is performed with crude instruments and usually without anesthesia or antibiotics. It can result in death from infection, shock, and psychological trauma (Fathalla, 1992).

The International Conference on Population and Development, held in Cairo in 1994, was the first such gathering to debate the practice and to call for its eradication on the ground that it violates basic human rights and constitutes a lifelong risk to women's health. A number of African nations, including Ghana and Ivory Coast, have passed laws forbidding the practice, and many others are debating such measures at this writing. But as the proceedings of the Cairo conference note, traditions are deeply rooted. To eradicate female circumcision, more national and community education programs need to be created (Sadik, 1995).

Some participants in international conferences take an extremely relativistic position, insisting that the practices of a culture must be understood from that culture's point of view and should not be subject to Western assumptions and values. A far larger number of voices are raised in criticism of the Western legalistic approach, which calls for prohibitions and bans. According to this line of thought, while prohibitions based on theories of human rights are desirable, they will have little effect unless they are accompanied by social action, especially action that promotes social and economic development (Murti, Guio, & Dreze, 1995).

In this regard, female genital mutilation is only one of many severe threats to women's equality and human rights. Reproductive diseases, death during childbirth, AIDS, and other sexually transmitted diseases claim the lives of millions of women each year. They are indicators of the low

Iranian President Mohammad Khatami, right, fills out his ballot before voting in the elections for the Iranian parliament in February 2000 as women wait in line to get their ballots. More women are using their power at the ballot box throughout the world, not just in developing nations.

Women's Share of National Legislatures in Selected Countries

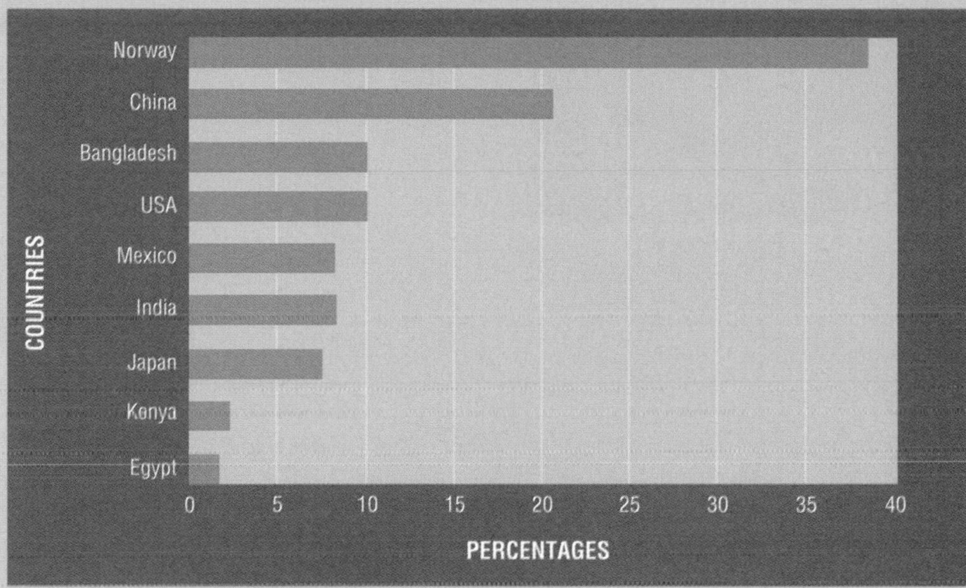

Source: United Nations, 1995.

status and condition of women in poor regions of the world, and their cost to their societies and the world is immense. So rather than merely attacking the symptoms of women's underdevelopment, world gatherings like the Cairo conference are working to reach a consensus around action programs based on the concept of women's empowerment. The conference proceedings stated that "power to make decisions within the family is the basis for the power of decision in other areas. Empowerment begins with winning equality, autonomy, and respect for women within the household" (Sadik, 1995, p. 19). The empowerment approach insists that "social and economic development cannot be secured in a sustainable way without the full participation of women" (United Nations, 1995a, p. 1).

Women's full participation in improving their lives cannot proceed without far greater attention to female literacy. The developed nations have established parity between women and men in basic literacy. Many nations in Latin America and East Asia have made significant strides toward universal female literacy as well, although they lag somewhat, with about 80% female literacy rates as opposed to almost 100% in the developed nations. In South Asia, the Arab nations, and sub-Saharan Africa, however, while female literacy rates are increasing, only about 30% to 40% of women are minimally literate (UNDP, 1997).

The accompanying chart presents data on women's share of political representation. It was among those used at the Cairo conference in discussions of women's empowerment and development. The participants agreed overwhelmingly that in the poor and underdeveloped regions of the world there must be more investment in basic education and literacy for women so that they can participate in decisions in the family and in the larger community. Eventually, with more education and local empowerment would come increased participation in national political institutions. ■

As women increasingly question traditional definitions of their gender, they have also opened up new possibilities for redefining the biological potential of their sex.

and the school, as well as by social structures like the class system of a society. Sociologists and historians believe that there have been major changes in sexuality as a result of changes in the economy, politics, and the family. Over the past 300 years, for example, the meaning and place of sexuality in American life have changed: from a family-centered system for ensuring reproduction and social stability during the colonial period to a romantic and intimate sexuality in nineteenth-century marriage, with many underlying conflicts, and to a commercialized sexuality in the modern period, when sexual relations are expected to provide personal identity and individual happiness apart from reproduction (D'Emilio & Freedman, 1988).

While these changes are important and controversial, they hardly begin to exhaust the range of sexual norms in cultures throughout the world. Polygyny, which is illegal (bigamy) in North America and Europe, is condoned throughout much of the Islamic world. Marriage between adult men and preadolescent girls is practiced in some Asian cultures but would be considered a form of child abuse in the West. Women bare their breasts on the beaches of Europe but may be issued summonses for the same behavior on the beaches of North America. On American television screens, women and men expose their bodies in a fashion that is horrifying to people in many parts of Asia. Catholic priests and nuns are expected to remain celibate, while Protestant ministers are encouraged to have families, as are Jewish rabbis. With all these contrasting cultural norms governing sexuality, can it be said that there are any universal sexual norms?

Universal cultural norms exerting social control over sexuality include the *incest taboo, marriage,* and *heterosexuality,* but even these norms include variations and differing degrees of sanction. As noted in Chapter 2, the incest taboo is known in every existing and historical society and serves to protect the integrity of the family. Incest norms prohibit sexual intimacy between brothers and sisters and between parents and children, and usually specify what other family relatives are excluded. However, the specific relatives involved vary in different cultures; for example, sex between first cousins is permissible in some societies and strictly taboo in others.

Every society also has marriage norms that specify the relationships within which sexual intimacy is condoned. Marriage norms protect the institution of the family, confer legitimacy on children, and specify parental rights and obligations. But while marriage norms are found in all cultures, there are vast differences in how they operate. In addition to variations having to do with polygyny, there are some important variations with regard to the strength of sanctions on adultery (sexual intimacy outside marriage). The majority of cultures prohibit adultery, but the sanctions vary widely. Former French President François Mitterand, who died in 1996, specified in his funeral instructions that both his wife and his mistress were to be present at his funeral. Although adultery is not openly condoned in France, it is far less strongly sanctioned than in the United States, where evidence that a presidential candidate may have had an extramarital affair can terminate a candidacy (Stockard & Johnson, 1992).

Sexual Orientation

Heterosexuality refers to sexual orientation toward the opposite sex, in contrast to **homosexuality,** or sexual orientation toward the same sex, and **bisexuality,** or sexual orientation toward either sex. Heterosexuality is a norm in every society, but it too is subject to wide variations. Norms of heterosexuality function to ensure that there is genital sexual intercourse between men and women in the interest of population replacement and growth. The Shakers of colonial North America were an example of a society that attempted to curtail or eliminate heterosexuality in favor of celibacy in the interest of religious piety. Because Shaker society could continue only through recruitment of new members, it eventually declined and essentially disappeared.

While heterosexuality is practiced in some form in all societies, about one third of them actively ban homosexual practices, while many other societies tolerate them in some form (Murdock, 1983). Historically, the best-known examples of heterosexual societies that condoned some forms of homosexuality and bisexuality were ancient Greece, Confucian China, and Hawaii before it was colonized by Europeans (Connell, 1995). In most Western cultures, while the norm of heterosexuality is quite strong, older taboos against homosexuality and bisexuality are changing.

Homosexuality and Bisexuality. The intensity of debate over homosexuality in the United States is an indication, on the one hand, of how controversial the behavior remains and, on the other, of the growing, if grudging, acceptance of alternatives to traditional norms of heterosexuality. In the mid-twentieth century, homosexuality was a taboo subject. Homosexual men and women remained secretive or "in the closet" about their true sexual feelings. Today there is at least one homosexual member of Congress and many other openly homosexual elected officials and public figures. At the same time, fear and loathing of homosexuals—gay men and lesbian women—is a major social problem that continues to cause immense human suffering.

Estimates of the rate of incidence of homosexuality—that is, the proportion or number of women and men in a population who are sexually attracted to people of the same sex—raise a number of social-scientific and political problems. In his pathbreaking and courageous studies of male and female sexual behavior, Alfred Kinsey interviewed thousands of Americans about their sexual preferences and practices. Kinsey described sexuality as a continuum extending from exclusive heterosexuality to equal attraction for the same and the opposite sex (bisexuality) to exclusive homosexuality. By his estimates, about 4% of men and 2% of women were exclusively homosexual, while far more, perhaps as many as one third of men and about 12% of women, said that they had had a homosexual experience leading to orgasm at least once in their lives (Kinsey, Pomeroy, & Martin, 1948, 1953). Subsequent reinterpretations of Kinsey's data led social scientists to estimate that 10% of the population was homosexual or had strong homosexual tendencies. This estimate became a commonly cited statistic in the politics of the gay rights movement. However, the most recent major study of sexuality in the United States, conducted by the National Opinion Research Center (NORC), finds this estimate to be too high. For instance, in the NORC sample about 4% of men and about 1.4% of women said that they are exclusively homosexual, but only about 10% of men (not the 30% to 35% cited by Kinsey) admitted to having had prior homosexual experiences even if they did not claim to be homosexual in their daily lives (see Figure 8.2). Data from the NORC study also indicate that there are wide differences among men and between men and women in how sexual pleasure is defined and derived (see Table 8.2; Laumann, Gagnon, Michaels, & Michaels, 1994).

Given the anxiety that many people feel about discussing their sexual orientation and behavior, however, there are some who feel that the NORC study may underestimate the incidence of homosexuality in the population (Gould, 1995). It should also be noted that this ambitious study might have included a far larger sample and somewhat greater scientific accuracy if the federal funds that had been scheduled for its budget had not been withdrawn by members of Congress opposed to research on sexuality.

Only about 0.8% of men and 0.9% of women in the NORC sample ($n = 3{,}432$) said that they were bisexual. But when they were asked more indirectly whether they were sexually attracted to both sexes, the proportions who said yes increased to 4.1% of women and 3.9% of men. Martin S. Weinberg observes that many people "would say that the person who feels the sexual attraction to both sexes but never acts on it is not bisexual, but in my definition they are" (quoted in Gabriel, 1995b, p. A12). The findings of his recent study of bisexuality refute the common assumption

TABLE 8.2
The Appeal of Various Sexual Practices (percentages of respondents)

	Appealing to Men Ages 18–44				Appealing to Women Ages 18–44			
	Very	Somewhat	Not Really	Not at All	Very	Somewhat	Not Really	Not at All
Vaginal intercourse	83%	12%	1%	4%	78%	18%	1%	3%
Watching partner undress	50	43	3	4	30	51	11	9
Receiving oral sex	50	33	5	12	33	35	11	21
Giving oral sex	37	39	9	15	19	38	15	28
Active anal intercourse	5	9	13	73	—	—	—	—
Passive anal intercourse	3	8	15	75	1	4	9	87
Group sex	14	32	20	33	1	8	14	78
Same-sex partner	4	2	5	89	3	3	9	85
Sex with a stranger	5	29	25	42	1	9	11	80
Forcing someone to do something sexual	0	2	14	84	0	2	7	91
Being forced to do something sexual	0	3	13	84	0	2	6	92

Source: Laumann et al., 1994.

that bisexuality is a stage leading to eventual homosexual orientation. His sample included women and men who had come to their bisexuality from both heterosexual and homosexual orientations and for whom bisexuality appeared to be a stable lifestyle.

Many homosexuals, along with an increasing number of biologists and social scientists, believe that homosexuality will be shown to have genetic origins (Tobach & Rosoff, 1994; E. O. Wilson, 1979). In accounts of their earliest sexual feelings, homosexual men, and to a lesser degree lesbian women, recount their experiences of sexual attraction to members of the same sex. These experiences convince many gay people that their sexual orientation is not merely a "lifestyle choice," as critics often claim. However, as yet there is insufficient evidence from genetic studies to resolve this issue (Kemper, 1990).

Homophobia. Many young boys use negative images of homosexuals well before they ever encounter a gay person (Thorne, 1993). Fear of homosexuals and same-sex attraction is known as *homophobia.* Although it is quite common in the United States, its causes are not entirely clear. Most social-scientific (as opposed to religious or ideological) explanations of homophobia hinge on an analysis of the problems of masculinity and the male role in Western societies.

Identification with the male gender and the ability to assume male roles, as these are defined for young boys and later for men, are two different aspects of what it means to "be a man." Every society has its notions of what distinguishes men from women and seeks to teach boys and girls how to perform the roles assigned to their gender. In Western societies, the male role is often depicted in movies, on television, and in advertising as distinct from that of the female and imbued with more strength, power, and rationality—a subject to which we return later in the chapter. Because as children they lack the power and strength they admire in images of masculinity, young boys typically become concerned about being seen as lacking in masculinity and the ability to take on male roles. Thus, their vocabulary is rich with terms of abuse for anyone they believe to be lacking in masculinity: wimp, nerd, turkey, sissy, lily liver, yellowbelly, candy ass, ladyfinger, cream puff, mother's boy, dweeb, geek, and so on (Connell, 1995). As they grow older, a small but

FIGURE 8.2
Defining "Gay"

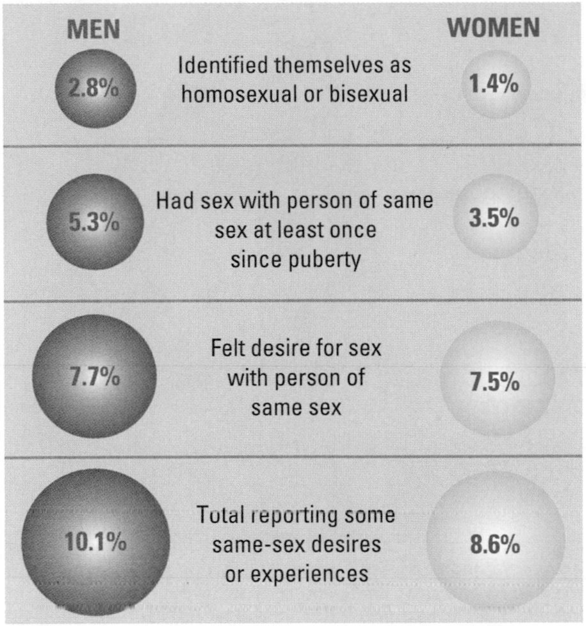

Source: Data from Laumann et al., 1994.

socially significant proportion of boys may become violently homophobic and go out of their way to abuse males whom they consider effeminate or "queer."

Girls may also develop homophobic attitudes early in life, often through identification with their male siblings, but homophobia among females often occurs during the teenage years. Female socialization stresses the ability to appear feminine and attract men, and anxiety over these aspects of the female role may lead young women to reject other women whom they do not regard as feminine enough.

These early patterns can have lasting consequences. Throughout their lives, the early habit of avoiding emotions for fear of seeming effeminate tends to make men in the United States and other Western societies wary of expressing their feelings and of admitting vulnerability: "Men are thus denied an important part of their . . . well-being when they cannot touch and cannot express their tender feelings for other men. Needless to say, gay men are damaged by the negative implications of homophobia" (Blumenfeld, 1992, p. 37; Richmond-Abbott, 1992).

Fundamentalist religious beliefs are another important source of negative attitudes and strong disapproval of homosexuality, including vehement homophobia. For example, 77% of fundamentalist Baptists disagreed or strongly disagreed with the statement, "Even if homosexuality is wrong, the civil rights of gays should be protected." They believe that homosexuality is immoral and inimical to the propagation of the species. In the United States, it is common for prominent fundamentalist Christian leaders to state that AIDS is God's punishment for the homosexual lifestyle (Ammerman, 1990). Despite the vehemence of such religiously motivated attacks, gay men and women in smaller towns and communities with fundamentalist congregations are increasingly asserting their right to live as they choose as long as they do not violate the rights of others (Miller, 1989).

Controversies over male and female roles, homosexuality, norms of sexual conduct, and the origins of sexual orientation all point to the extent to which these central areas of social life are subject to change and reactions to change. In the study of gender inequalities, the origins and patterns of change in the condition of women and the continuing inequalities between the sexes are a central focus, to which we now turn.

GENDER STRATIFICATION

After class and race, the most important dimensions of inequality in modern societies are gender and age. As stated earlier, *gender* refers to a set of culturally conditioned traits associated with maleness or femaleness. There are two sexes, male and female; these are biologically determined statuses. There are also two genders, masculine and feminine; these are socially constructed ways of being a man or a woman. **Gender roles** are the sets of behaviors considered appropriate for individuals of a particular gender. Controversies over whether women in the armed forces should serve in combat or whether men with children ought to be eligible for family leave from work are examples of issues arising out of the definition of gender roles.

All human societies are stratified by gender, meaning that males and females are channeled into specific statuses and roles. "Be a man"; "She's a real lady"—with these familiar expressions, we let each other know that our behavior is or is not conforming to the role expectations associated with our particular gender. When women's roles are thought to require male direction, as is the case in many households and organizations, the unequal treatment of men and women is directly related to gender roles. The roles assigned to men and women are accorded differing amounts of income, power, or prestige, and these patterns of inequality contribute to the society's system of stratification.

Under the Soviet regime, the dominant image of women was that of a factory hand or street sweeper in a lumpy gray overcoat and babushka. In fact, Russian women were always style conscious and wished to be romantic and glamorous. As in the United States, feminists in Russia often criticize the beauty ideal portrayed in *Cosmopolitan*, yet it has undeniable appeal to Russian women.

The accompanying photos, which portray two vastly different images of women in Russia, illustrate the social construction of gender roles. The older photo is from the pre-1989 Soviet era. Women wearing scarves on their heads, known as babushkas, were often assigned heavy work as laborers and street cleaners. The woman in the foreground was likely to be more educated and better able to compete for a scientific or technical career. Her clothes, however, only begin to suggest the dowdiness of Soviet-era fashions. No wonder that after the fall of the Soviet empire, magazines like *Cosmopolitan* became great hits among Russian women. The growing middle class wanted to participate in Western fashions and styles, even if that meant accepting stereotypical Western conceptions of gender roles.

Feminists around the world object to depictions of women like this one in *Cosmo*. Despite the popularity of gender roles in which a woman is valued primarily as a sexual object, feminist sociological research demonstrates that women are oppressed by the beauty ideal, which often leads them to become obsessed with weight, looks, and youth, to the detriment of their intellectual and political growth (Moore, 1998).

Gender Roles and Gendered Organizations

Until quite recently, it was assumed that there were two spheres of life, one for women and the other for men:

> Church, school and family—these chief agents of a child's socialization—worked together to transmit the notion that boys should grow up to be the doers, thinkers, and movers in the world at large. Girls, on the other hand, were expected to grow up to be wives and mothers. Any involvement in the world outside the home would be indirect rather than direct. (Scanzoni & Scanzoni, 1976, pp. 18–19)

Out of this gender-based division of labor, which defined the activities that were appropriate for men and women, grew the notion of differences in men's and women's abilities and personalities. These differences were thought to be natural—an outgrowth of biological and psychological differences between males and females (Epstein, 1985). Behaviors that did not fit these patterns were viewed as deviant and in some cases as requiring severe punishment.

In the twentieth century, evidence from the social sciences has called into question the assumption that there are innate biological or psychological reasons for the different roles and temperaments of men and women. Margaret Mead's (1950) famous research in New Guinea directly challenged this assumption. Mead was one of the first social scientists to gather evidence to show that gender-specific behavior is learned rather than innate. In her study of gender roles in three tribes, Mead found that different tribes had different ways of defining male and female behavior. In one tribe, the Mundugumor, men and women were equally aggressive and warlike, traits that Westerners usually associate only with men. In a second tribe, the Tchambuli, the men spent their time gossiping about women and worrying about their hairdos, while the women shaved their heads and made rude jokes among themselves. In the third tribe, the Arapesh, both men and women behaved in sympathetic, cooperative ways and spent a great deal of time worrying about how the children were getting along, all behaviors that Westerners traditionally associate with women.

The lesson of cross-cultural research is that gender roles are heavily influenced by culture. Although the relationship of earlier societies to their natural environment often required that women tend the hearth and home while men went out to hunt for big game, women were also hunting for small game around the encampment and experimenting with new seeds and agricultural techniques. The division of labor by gender was never fixed; it could always be adapted to new conditions.

Gendered Organizations. Recent research, especially that of feminist sociologists, has expanded our understanding of gender in society by demonstrating that gender stratification is a feature of organizations of all kinds, and not merely an attribute of individuals and their roles (Lorber, 1994; Staggenborg, 1998). The concept of "gendered organizations" encourages researchers to study how assumptions about gender are ingrained in the culture of organizations ranging from the family to corporations and universities. Many law firms, for example, strongly suggest that success is based on long hours of time spent in the office. This assumption distinctly favors male workers whose wives have chosen to be homemakers. And as we will see in the next chapter, the family is often a highly gendered organization as well.

Especially in the second half of the twentieth century, women have organized in social movements to bring down barriers to their employment based on the gendered nature of organizations and the assumption that they could not do certain types of work (e.g., policing, fire fighting, or construction). Tables 8.3 and 8.4 show that women have increased their share of employment in many occupational groups that were formerly male "turf." The proportions of female executives and managers have increased dramatically, as have those of technicians and professionals. But older barriers and assumptions continue to stand in the way of equal access to male-dominated occupations such as precision production. Conversely, women continue to be disproportionately represented in "pink-collar" occupational sectors such as clerical and administrative support and domestic service (Bianchi & Spain, 1986).

TABLE 8.3

Percentage of Women in Major Occupational Groups, 1970–1999

Major Occupational Group	1970	1999
Executive, administrative, and managerial	18.5%	45.1%
Professional specialty	44.3	53.5
Technicians and related support	34.4	51.9
Sales occupations	41.3	50.1
Administrative support, including clerical	73.2	78.7
Private household	96.3	95.2
Protective service	6.6	18.9
Other service	61.2	65.4
Farming, forestry, fishing	9.1	19.7
Precision production, craft, and repair	7.3	9.0
Machine operators, assemblers, and inspectors	39.7	37.2
Transportation and material-moving occupations	4.1	9.9
Handlers, equipment cleaners, helpers, and laborers	17.4	20.5
Total	38.0	46.5

Sources: Data from Bianchi & Spain, 1986; and *Statistical Abstract,* 2000.

TABLE 8.4
Proportion of Jobs in Certain Occupations Filled by Women, 1970–1990

Occupation	1970	1990
Bartender	27%	57%
Chemist	17	29
Doctor	11	22
Economist	14	44
Farmer	7	17
Industrial engineer	3	27
Lawyer, judge	6	27
Librarian	84	85
Nurse	91	94
Police detective	5	13
Psychologist	43	59
Public official	24	59
Secretary	98	98
Teacher	74	74

Source: Program for Applied Research, Queens College.

Gender Inequality in Industrial Societies

In modern industrial societies, age and gender interact to shape people's views of what role behavior is appropriate at any given time. Before puberty, boys and girls in the United States tend to associate in sex-segregated peer groups. Because they model their behavior on what they see in the home and on television, girls spend more of their time playing at domestic roles than boys do; boys meanwhile play at team sports more than girls do. These patterns are changing at different rates in different social classes, but they remain generally accepted norms of behavior. And they have important consequences: Women are more likely to be socialized into the "feminine" roles of mother, teacher, secretary, and so on, while men are more likely to be socialized into roles that are considered "masculine," such as those of corporate manager or military leader. It is expected that men will concern themselves with earning and investing while women occupy themselves with human relationships (Baron & Bielby, 1980; Chodorow, 1978; Rossi, 1980).

Childhood socialization explains some of the inequalities and differences between the roles of men and women, but we also need to recognize the impact of social structures. In the United States, for example, it was assumed until fairly recently that boys and girls needed to be segregated in their games. Boys were thought to be much stronger and rougher than girls, and girls were thought to need protection from unfair competition with boys. This widespread belief translated into school rules that did not permit coeducational sports. Those rules, in turn, reinforced the more general belief that girls' roles needed to be segregated from those of boys. Such patterns have significant long-term effects. As sociologist Cynthia Epstein points out, human beings have an immense capacity "to be guided, manipulated, and coerced into assuming social roles, demonstrating behavior, and expressing thoughts that conform to socially accepted values" (1988, p. 240). Through such means, gender roles become so deeply ingrained in many people's consciousness that they feel threatened when women assert their similarities with men and demand equal opportunity and equal treatment in social institutions.

In their adult years, men enjoy more wealth, prestige, and leisure than women do. Working women earn less than men do, and they are frequently channeled into the less prestigious strata of large organizations. Even as executives, they are often shunted into middle-level positions in which they must do the bidding of men in more powerful positions. Similar patterns are found in all advanced industrial nations.

Sexism

Gender stratification is reflected in attitudes that reinforce the subordinated status of women. The term **sexism** is used to refer to an ideology that justifies prejudice or discrimination based on sex. It results in the channeling of women into statuses considered appropriate for women and their exclusion from statuses considered appropriate for men. Sexist attitudes also tend to "objectify" women, meaning that they treat women as objects for adornment or sex rather than as individuals worthy of a full measure of respect and equal treatment in social institutions. This can be seen in the case of beautiful women. Such women receive special treatment from both men and women, but their beauty is a mixed blessing. The beautiful woman is often viewed as nothing more than an object for admiration. Being a woman is a master status (see Chapter 4) in that gender tends to outweigh the person's achieved statuses. This is even more painfully true for beautiful women. Someone like Marilyn Monroe is

THE CENTRAL QUESTION: A CRITICAL VIEW

Will inequalities of gender and age diminish as nations become more industrialized and urbanized?

The answer to this question is clearly no. Gender and age inequalities will not be reduced automatically by macro-level social processes. Without continuing vigilance and more effective social movement organization, industrialization and urbanization are no guarantee of greater equality.

In the United States, for example, there is a significant lag in men's sympathy for gender issues. In the 2000 presidential elections, men voted far more heavily for George W. Bush, while women tended to vote for Al Gore. The magnitude of this gender gap was at least 10 percentage points, showing that there are clear differences in how women and men view their interests. Men do not care nearly as much about what they view as "women's issues" like abortion and reproductive health, or even education. They tended to support Bush because he was more clearly interested in professional sports and other stereotypical male activities.

Feminist sociologists have shown that men move away from such patriarchal views as they become active in support of feminist causes, but first they need to become active (Staggenborg, 1998). Pioneering research by feminist sociologist Robert W. Connell (1998) shows that men who do become active in movements for greater gender equality do so through involvement with women in other social movements. On the basis of interviews with men in many different social groups, Connell finds that men who are active in the peace or civil rights movements, or in the environmental movement, at first tend to hold stereotypical male attitudes about gender. Over time, however, their active participation with women in these movements makes them realize the need for greater sharing of domestic obligations and greater concern for the consequences of gender inequality.

Men are often compelled by gender norms to behave in stereotypically "masculine" ways in public.

FIGURE 8.3
Labor Force Participation of Women

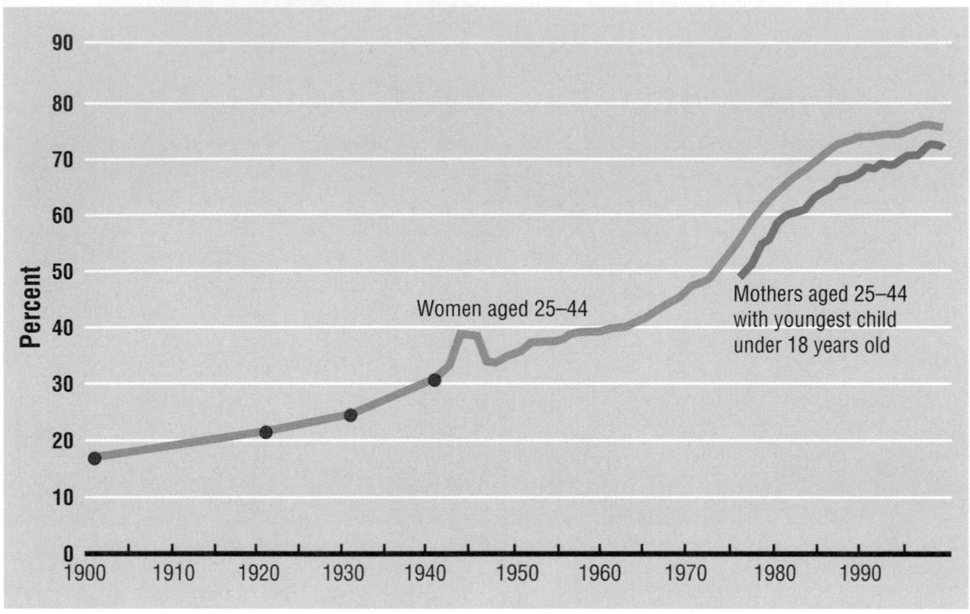

Note: Annual data are available only since 1942. Dots indicate decennial census data.
Source: Council of Economic Advisers, 2000.

thought of only in terms of her beauty; the person beneath the surface is ignored.

The objectification of women can be seen in the beauty contest, which came into being in the United States in the summer of 1921 when the first Costume and Beauty Show was held at a bathing beach on the Potomac River. There the women wore tunic bathing suits and hats, but later that year a similar contest was held at Atlantic City, New Jersey, that eventually developed into the Miss America Pageant. In that contest, women wore one-piece bathing suits that showed their calves and thighs and created a sensation in the tabloid newspapers. These contests and the publicity they generated made beauty a way for women to gain celebrity and wealth, but they also reflected the dominant male view that the most extraordinary women are those with the most stunning faces and the shapeliest figures (Allen, 1931). Women have been struggling against this view throughout the modern era, a struggle that has frequently been opposed not only by men but also by women who feel threatened by changes in their traditional statuses.

Sexism is also expressed in violence against women. Two million American women are severely beaten in their homes every year, and 20% of visits by women to hospital emergency rooms are caused by battering. (Thousands of men are battered by their wives each year as well, but they are far outnumbered by women victims.) In popular American culture, especially movies and television, although violence against women is not condoned, it is presented as a form of thrilling entertainment. Examples like these hardly exhaust the overt types of sexism that exist in employment and other aspects of social life in the United States, but they are aspects of American culture that too often go unnoticed. Despite advances in women's rights and changes in women's access to careers, sexism remains commonplace in many areas of American life (Benokraitis & Feagin, 1986; Connell, 1995).

WOMEN AT WORK

In the industrial nations of North America, Europe, and Oceania, the increasing proportion of women who are in the labor force as paid employees is one of the most important aspects of social change in those societies. In the mid-twentieth century, 34% of U.S. women were in the labor force, but that proportion has increased steadily since then; in the 1990s, over 70% of women considered themselves to be employees, either at work or looking for work (*Statistical Abstract,* 2000). These important changes are illustrated in Figure 8.3, which also shows that labor force participation rates for women with young children are almost equally high.

In fact, women assumed an important economic role well before industrialization created a sharp distinction between work in the paid labor force and work at home or in the fields. When the United States was an agrarian society, women planted and harvested crops, including extensive household gardens, and were also expected to take responsibility for housework and child rearing, both of which were highly labor intensive in the pre-electricity era. And since women have higher life expectancies than men, widows routinely ran farms or businesses after their husbands died, and many middle-class women gained additional income by taking in lodgers or selling handicrafts or other products of their labor. With the industrial revolution came growing demand for factory workers. Although male workers were often preferred, thousands of women and children also swelled the ranks of the new industrial working class. A recent Census Bureau study showed that many of the working women were immigrants or poor single women from rural backgrounds. Large numbers of African American women were also working for wages in fields and factories and as domestic servants (Richmond-Abbott, 1992).

World War II marked a significant turning point in women's labor force participation. Although returning servicemen "bumped" women (and minority men) from jobs in factories and offices, there were thousands of war widows and single women who had to continue earning wages whether they wanted to or not. As a result, although in the 1950s women's labor force participation was still far lower than it is today, in fact female employment was increasing at a far faster rate than male employment.

At the same time, the 1950s were a time of relative economic prosperity. The middle classes were expanding, and a suburban home became a central feature of the American dream. Popular culture, including the powerful new medium of television, emphasized the ideal norms of the middle-class nuclear family in which the mother was a homemaker and the father worked in the labor force. This "feminine mystique" asserted that women would find fulfillment as wives and mothers. In reality, thousands of women were entering the labor force in a trend that would continue throughout the second half of the century (Epstein, 1988).

Whether they worked outside the home because they wanted to or because necessity forced them to do so, women in the labor force encountered problems of gender segregation and discrimination. In a survey of 150,000 working women conducted during the mid-1980s by the National Commission on Working Women, the most frequently cited problems were low wages, differentials in fringe benefits, dead-end jobs with little opportunity for training or advancement, sexual harassment on the job, lack of child care or difficulty obtaining it, stress over multiple roles, and lack of leisure time (Richmond-Abbott, 1992).

An enduring problem for women in the labor force is segregation into what are known as "pink-collar ghettos." Secretarial and clerical work especially remain heavily gender segregated, but child care, nursing, and dental assistants also remain largely female occupations despite some increases in the proportions of males in these fields in recent decades. Largely as a result of antidiscrimination laws and women's efforts to break into occupations that previously were more or less closed to them, women have made considerable gains in some areas. Occupations like bus driver, psychologist, and others show significant gains in female employment, but clerical occupations remain a pink-collar ghetto: Clerical work employs one in five working women, a figure that has not changed since the 1950s (Roberts, 1995).

Even when women successfully enter a formerly male-dominated field, their wages are often lower even though they do the same work and have the same length of service. Among psychologists, for example, the median income for men was $31,000 in 1990 while that of women was $22,000. For psychologists under age 35, the men's median was $19,000 while that of women was $17,000. According to sociologists and activists in the movement for pay equity, these differences reflect a more general trend: As women hold an increasing proportion of the jobs in an occupation, its average pay decreases, as does its status. This is particularly true of fields in which the proportion of minority women is increasing (Wright & Jacobs, 1994).

Figure 8.4 shows the ratio of women's to men's earnings for major developed nations. Clearly, the northern European nations—especially the social democracies of Scandinavia, where affirmative action programs are well established—do a far better job of arriving at wage parity for men and women than the United States or the United Kingdom, which are more reluctant to create gender-neutral hiring policies.

The Second Shift

Sociologist Arlie Hochschild (1989) coined the term *the second shift* to describe the extra time working women spend doing household chores after working at a job outside the home. This term emphasizes the

FIGURE 8.4
Ratio of Women's Earnings to Men's, Selected Countries

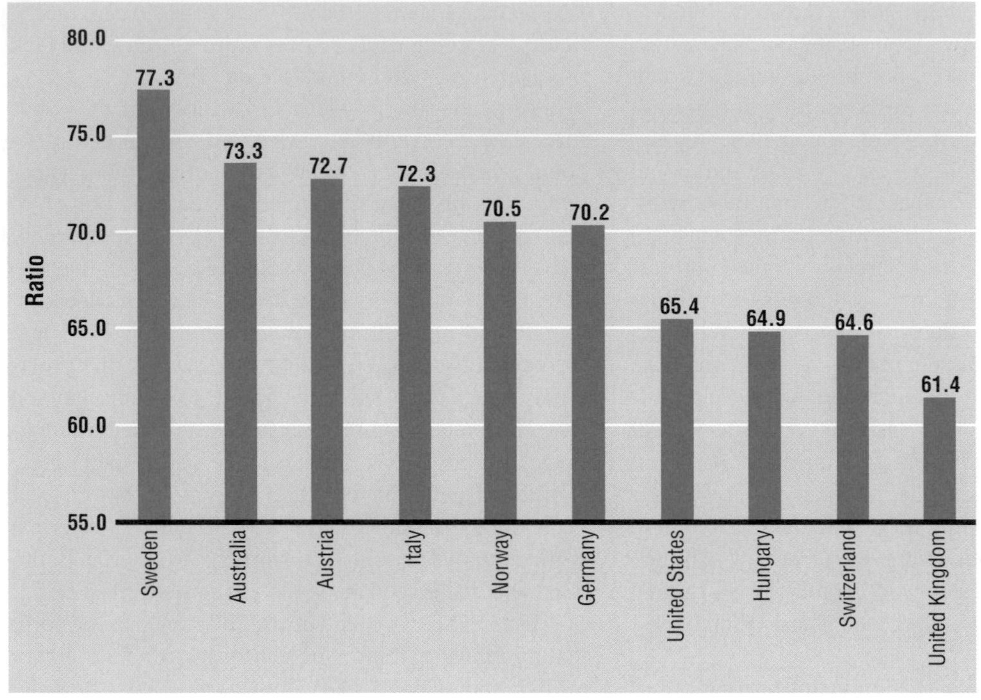

Source: Blau & Kahn, 1999.

expectation that women who work will also perform the bulk of domestic and child-care work. This is an example of the persistence of **patriarchy,** the dominance of men over women. Although women often attempt to influence their male partners to share household chores that were formerly considered "women's work," the men continue to resist despite slow changes—especially in two-career families in which both partners are highly educated.

Some sociologists have proposed that as women gain greater parity with men in occupations, the disparities between men's and women's roles will begin to diminish and gender roles in the household will become more symmetrical (Bernard, 1982; Willmott & Young, 1971). But as Hochschild (1989) and others report, this remains a speculative hypothesis despite the exceptions one might cite here and there. On the average, data on time budgets show that working women in the United States have at least 10 hours a week less leisure time than their husbands or partners because they shoulder far more of the domestic responsibilities. How much extra responsibility they take on at home is clearly indicated in Figure 8.5, which shows in detail the gap between the amount of time married women spend on domestic chores and the amount spent by their husbands.

Do large increases in married women's labor force participation mean that men are helping more at home? Trend data do show some increases in men's involvement in domestic chores, but Figure 8.5 shows that major gaps remain, with women taking on a far larger share of domestic work even when they also work full-time outside the home. The same disparities do not exist in all industrial nations. France and Holland, for example, incorporate gender equality into national social policies. These nations provide excellent universal child care, family leave, and longer vacations than do lagging nations like the United States, and these policies ease the burdens of dual-earner families in coping with the demands of work and family life (Kammerman & Kahn, 1981).

As women pass through the life cycle from childhood to old age, the nature of the inequalities they face changes. College women, for example, may be

FIGURE 8.5
Average Hours per Week Wives and Husbands Spend Doing Household Tasks

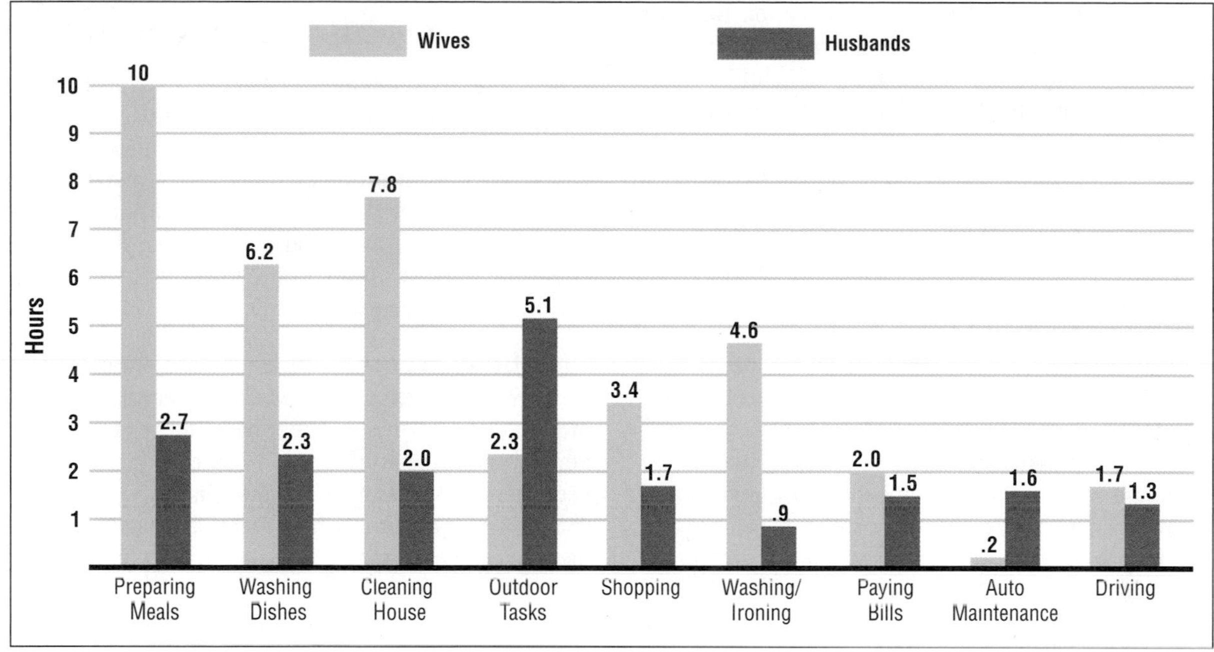

Source: Chadwick & Heaton, 1999.

concerned about unequal allocation of resources to their education and to their participation in sports. Elderly women, who are often widows, may be concerned that their declining strength interferes with their responsibilities as grandparents. And although the elderly are far less likely to be poor than they were two or three generations ago, poverty rates among elderly women are second only to those of young children. These facts point to age as another important dimension of stratification and inequality.

SOURCES OF AGE STRATIFICATION

In many societies, age determines a great deal about the opportunities open to a person and what kind of life that person leads. Only a little over half a century ago, for example, becoming old in the United States almost automatically meant becoming poor. Today children are the most impoverished and vulnerable population group in many nations, including the United States. This section examines how age stratification—the roles assigned to young people, adults, and the elderly—contributes to inequality.

The Life Course—Society's Age Structure

All societies divide the human life span into "seasons of life" (Hagestad & Neugarten, 1985; Zerubavel, 1986). This is done through cultural norms that define periods of life, such as adulthood and old age, and channel people into **age grades**—sets of statuses and roles based on age. These systems of age grades "create predictable, socially recognized turning points that provide roadmaps for human lives" (Hagestad & Neugarten, 1985, p. 35). Graduations, communions, weddings, retirements, and funerals are among the ceremonies that are used to mark these turning points.

Age strata are rough divisions of people into layers according to age-related social roles. We speak of infants, preschoolers, elementary school children, teenagers, young adults, and so on; these categories form a series of younger-to-older layers, or strata, in the population. People in different age strata command different amounts of scarce resources like wealth, power, and prestige (Riley, Foner, & Waring, 1988). Numerous laws establish inequalities between youth and adults; they include laws governing the rights to vote, to purchase alcoholic beverages, to incur debt, and the like. In theory, a person who lacks the rights of adult citizenship will be protected by adults, who are responsible for providing him or her with adequate food, shelter, and education (and are presumed to have the resources to do so). In practice, however, hundreds of thousands of children and teenagers do not receive the care that is intended to offset their unequal status under the law.

Social scientists often refer to the **life course,** which may be defined as a "pathway along an age-differentiated, socially created sequence of transitions" (Hagestad & Neugarten, 1985, p. 36; see also Cain, 1964; Clausen, 1968; Elder, 1981). The cultural norms that specify the life course and its important transitions create what is thought of as "the normal predictable life cycle" (Neugarten, 1969, p. 121). We expect that we will go to school, find a job, get married, have children, and so on at certain times in our lives, and we consider it somewhat abnormal not to follow this pattern. Social scientists often refer to ceremonies that mark the transition from one phase of life to another as **rites of passage** (Van Gennep, 1960/1908). The confirmation, the bar mitzvah, the graduation, and the retirement party are examples of rites of passage in modern societies.

In the United States and other Western cultures, the life course is constructed from categories like childhood, adolescence, young adulthood, adulthood, mature adulthood, and old age. But our definitions of these categories lack the stability and uniformity of the age grades found in many traditional societies. For example, the French historian Philippe Ariès (1962) showed that in Western civilization the concept of childhood as a phase of life with distinct characteristics and needs did not develop until the late seventeenth century. Before that time, children were treated as small adults. They were expected to perform chores and to conform to adult norms to the extent possible. When they reached puberty, they were usually married, often to spouses to whom they had been promised in infancy.

Norms regarding gender are closely linked to the life course established by a society. Thus, Ariès's study of the emergence of childhood revealed that ideas about the appropriate forms of play and education for boys, and indeed the very concept of boyhood, developed at least a century before the concept of girlhood emerged. In eighteenth-century European societies, boyhood was conceived of as a time when male children could play among themselves and receive education in the skills they would need as adults. Girls, in contrast, were treated as miniature women who were expected to work alongside their mothers and sisters.

Cohorts and Age Structures

When we think about age, we tend to think in terms of **age cohorts,** or people of about the same age who are passing through the life course together (Bogue, 1969). We measure our own successes and failures against the standards and experiences of our own cohorts—our schoolmates, our workmates, our senior circle—as we pass through life. (See the Sociological Methods box on pages 238–239.)

The Baby Boom. When a population experiences marked fluctuations in fertility, there are bulges in its population pyramid that have important effects. Perhaps nowhere in the world has this phenomenon been better studied than in the United States. The "baby boom" cohorts, which were produced by rapid increases in the birthrate from about 1946 through the early 1960s, have profoundly influenced American society and will continue to do so for the next three decades.

Throughout Europe and North America, the baby boom generations did not have nearly as many children as their parents had. A mean family size of 2.1 children per couple is required for a population to remain constant over time (rapid growth requires a mean number of children closer to 3.0 per family). But since the 1970s the mean number of children per family in industrial societies has been about 1.85 (and much lower in nations like Japan and Germany); as a result, the baby boom has been followed by a relative shortage of children known as the "baby bust" (Keyfitz, 1986). One way to visualize the impact of the baby boom and baby bust is to compare the population pyramids in Figure 8.6.

As these cohorts mature, they make new demands on the society's institutions. During the 1960s and 1970s, for example, when the baby boom cohorts passed through their college years, the nation's universities and colleges expanded; the slogan of the day was "Never trust anyone over 30." Now the baby boom cohorts are moving into the dominant age groups of the population. Their succession to national leadership in the 1990s represents a transfer of power from the generation that fought in World War II and is now in its late sixties and early seventies, to the generation that experienced the Vietnam War and the social movements of the 1960s.

FIGURE 8.6

Impact of the Baby Boom on the U.S. Population, 1950–2020

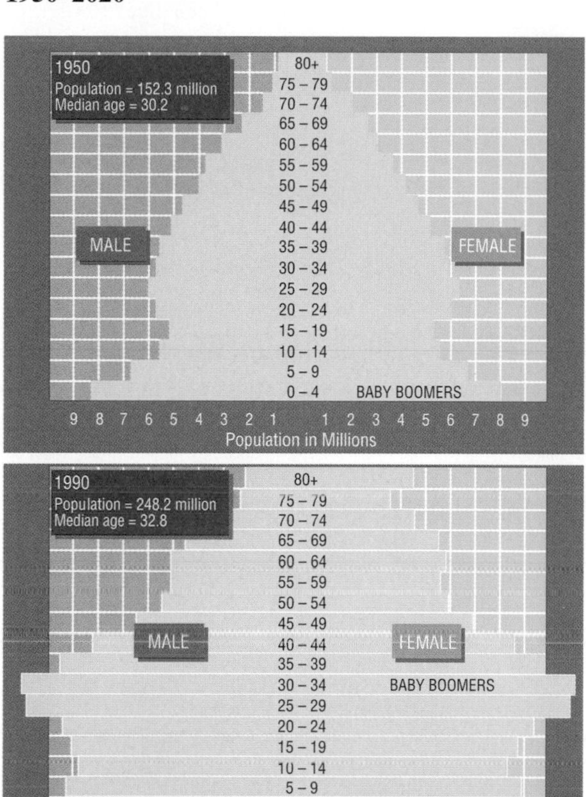

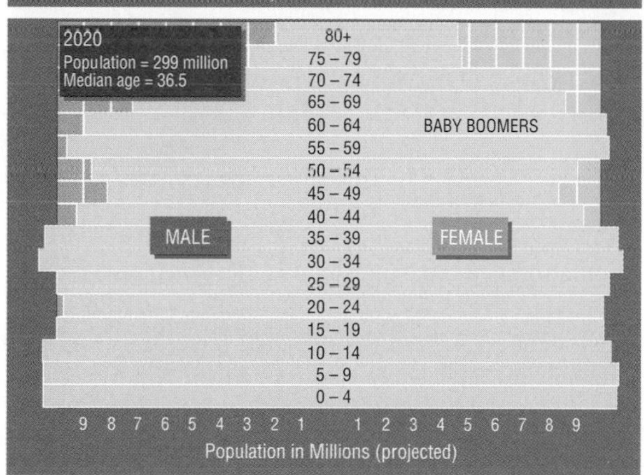

Source: Data from Census Bureau, 1985; and United Nations, 1992.

FIGURE 8.7

Racial and Ethnic Composition of the Baby Boom and Boomlet Generations

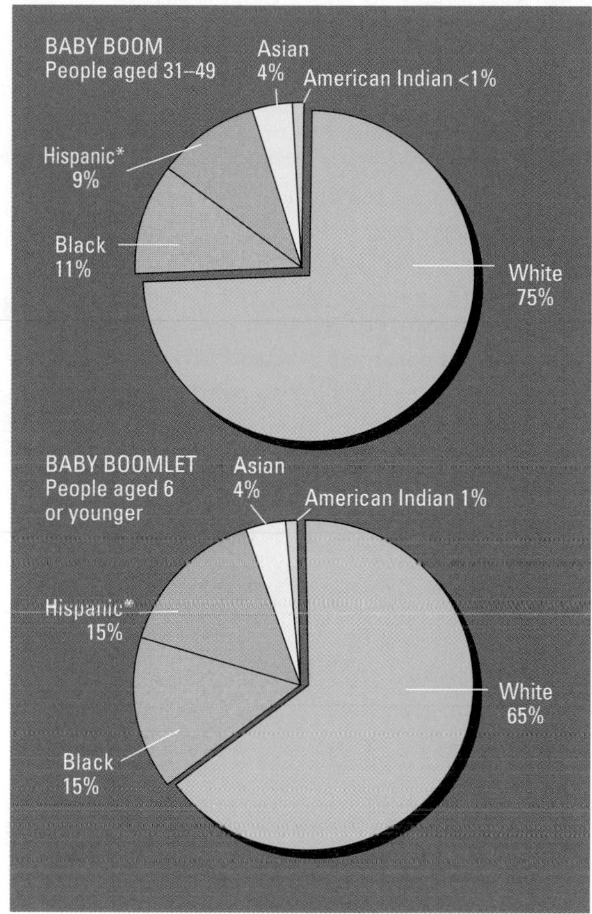

*Hispanic people can be of any race.
Note: Because of rounding, figures do not add to 100%.
Source: Data from Census Bureau.

The Baby Boom Echo. Sometimes referred to as the "baby boomlet" or the "baby boom echo," the children of parents who were part of the original baby boom are themselves an important demographic phenomenon. In the late 1980s, the number of births exceeded 4 million per year, a threshold reached over a longer period during the original baby boom, which extended from the 1940s to the 1960s. Like any "bulge" in a society's age structure, the wave of children born during the 1980s and early 1990s promises to have significant effects on social institutions like schools and businesses.

Figure 8.7 shows that a sizable proportion of the "second wave" or "boomlet" children are members of minority groups. This trend, combined with high rates of immigration, will increase the proportion of minority children moving through the nation's schools and

SOCIOLOGICAL METHODS

Population Pyramids

Demographers use the cohort concept in studying how populations change. If we divide populations into five-year cohorts, grouped vertically from age 0 to 100+ and divided into males and females, we can form a population pyramid, a useful way of looking at the influence of age on a society. The accompanying charts show population pyramids for an advanced industrial nation (Japan) and a Third World nation (Mexico). Note that Mexico's high birthrate expands the base of its pyramid, but its high rate of infant mortality causes the base to decrease rather dramatically within the first 10 or 15 years. High mortality rates in later cohorts bring the pyramid to a sharp point. In Japan, by contrast, the birthrate is far lower

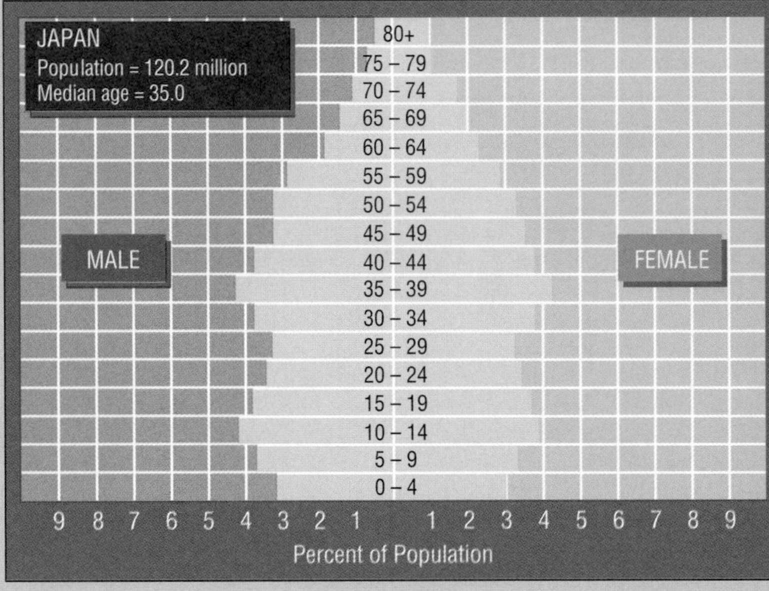

Source: Data from United Nations.

colleges in coming decades. Between 1988 and 1993, for example, kindergarten enrollments rose by 7.2% and preschool enrollments by 14%, both trends that resulted in increasing enrollments in many school districts. This demographic effect, in turn, spurred increased debate over the investment of public funds in schools versus other uses, such as medical care for the poor and elderly (Gabriel, 1995a).

A Graying Population. As the parents of these children, the baby boom cohorts, move into middle age, the problems of their own parents, the aged, also take on increasing importance. Table 8.5 shows that the proportion of elderly people in the American population is increasing steadily. This increase—often referred to as the "graying" of America—will result in greater concern about the needs of the elderly and will augment the influence of the aged on American culture and social institutions.

Demographers and biologists disagree over estimates of how many very elderly people there will be in the U.S. population during the twenty-first century. Most Census Bureau demographers estimate that the population over age 85 will increase from about 3.3 million persons (mainly women) in 1990 to about 18.7 million in 2080. But demographers working with the U.S. Institute on Aging argue that there may be no biological limits on life expectancy. With advances in and much more constant. There are bulges in the pyramid for particular cohorts, but for the most part the cohorts pass in regular fashion through the stages of life.

Note that the way sociologists construct a nation's population pyramid allows for easy comparisons between cohorts of men and women at each age. For example, the population pyramids for Japan and Mexico both show that in early childhood there are slightly more boys than girls in each cohort because slightly more boys than girls are born each year. But near the tops of the pyramids there are slightly more women than men who have reached old age, a situation that is far more common in industrialized nations than in developing ones.

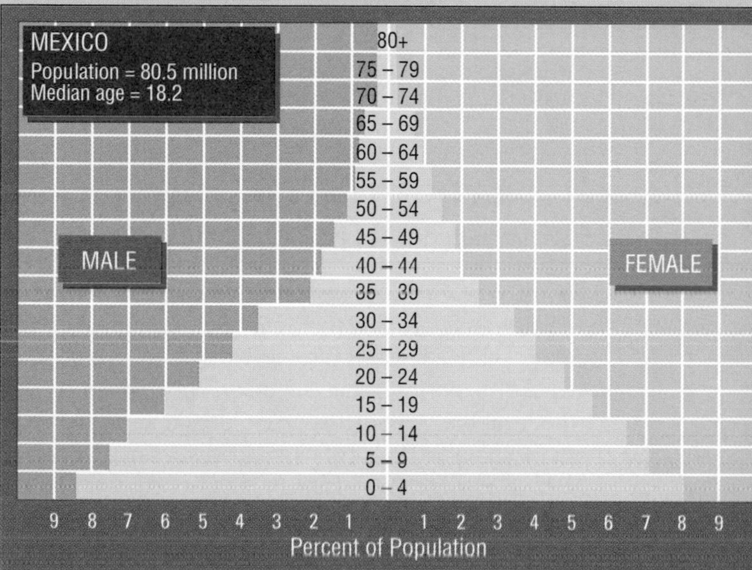

Source: Data from United Nations, 1992.

medicine and better prevention of disease, more people may live into their nineties. If that happens, there could be as many as 70 million people over age 85 in the U.S. population late in the twenty-first century, a change that would have enormous effects on all aspects of social life (Kolata, 1992).

Life Expectancy

At age 65 and beyond, more than 50% of North American women are widowed, whereas only 13.6% of men have lost their wives. This is because the life expectancy of females is at least 7 years longer than that of males. (By **life expectancy** we mean the average number of years a member of a given population can expect to live beyond his or her present age.) Typically, husbands die before their wives do. And they tend to marry younger women, making widows the single largest category in the elderly population.

In the future, if male and female roles become more similar and women experience the stresses and risks that are thought to cause earlier death in men, the gap between the life expectancies of the sexes may narrow. Research on sex differences in the causes of death indicates that women may be less vulnerable to death for genetic reasons, but it is extremely difficult to prove this scientifically. The research of Lois M. Verbrugge (1985) on health trends among males and

TABLE 8.5
Projected Proportions of Population in Various Age Categories: United States, 1980–2080

	Percent of Population by Age*		
Year	Under 18 Years	18–64 Years	65+ Years
1980	28.0%	60.7%	11.3%
1985	26.3	61.7	12.0
1990	25.8	61.5	12.7
1995	25.9	61.1	13.1
2000	25.1	61.8	13.0
2010	22.9	63.3	13.8
2020	22.3	60.4	17.3
2030	21.6	57.2	21.2
2050	21.0	57.3	21.8
2080	20.3	56.2	23.5

*Totals may not add to 100% because of rounding.

Source: Reprinted with permission of The Free Press, a Division of Macmillan, Inc., from *The Population of the United States: Historical Trends and Projections,* by Donald J. Bogue. Copyright © 1985 by The Free Press.

females indicates that the lifestyles of men and women in the United States have become similar; in particular, women's lives are more like men's. This change has unfortunate implications for women's health and longevity. For example, women are now smoking almost as much as men. Smoking among men declined from 50.2% in 1965 to 25.9% in 1998, but the rate for women decreased much less, from 31.9% to 22.1% (*Statistical Abstract,* 2000). Among older smokers, men were twice as likely to quit as women. It is no surprise that lung cancer has surpassed breast cancer as a leading cause of death for women.

Age and Dependency. People in the working adult cohorts—that is, those between the ages of 18 and 64 (although many continue to work well after age 64 and may start before age 18)—contribute disproportionately to the well-being of the young and the elderly. Of course, societies justify this pattern of dependency by recognizing that adults are merely doing in their turn what was done for them as children or will be done for them when they are elderly. Institutions of modern societies, such as public education and social security, ensure that a share of wealth passes to the dependent cohorts.

When there are very large numbers of children in a society, as is the case in the developing nations, or increasing numbers of elderly people, as is the case in the older nations of Western Europe and North America, working adults may shoulder an increased burden. Table 8.5 shows, for example, that children being born now will be adults in the early decades of the next century. If population projections are accurate, working adults will constitute a decreasing proportion of the total population, and larger proportions of young people and the elderly will be dependent on their productivity. Will this burden be too great? It may, but on the other hand improvements in productivity and changes in social policy, such as improved health care policies, may actually ease the burden compared with that of contemporary adults.

AGE STRATIFICATION AND INEQUALITY

In urban industrial societies, there are distinct patterns of stratification in which age defines the roles one plays and the rewards one can expect. We speak of the "age of majority"—the age at which a person crosses the legal boundary between childhood and adulthood. In fact, this age is not always clearly defined. A person can vote at age 18 but cannot legally consume alcohol until age 21. In addition to the age of majority, there are other distinct ages that mark the passage toward the full rights of adulthood. At age 18, one can join the armed forces without parental permission. At age 16, in many states one is no longer obliged by law to attend school, even though parents still share responsibility for the behavior of the school dropout. Teenagers can drive at age 16 in many U.S. counties, but their parents remain responsible, through insurance systems, for their actions and can be held liable for the consequences. We also make clear distinctions between children in the primary grades and teenagers in high school, and in most homes a child's passage through these age grades is accompanied by various privileges and responsibilities. So although the passage to full adult status may be somewhat vague, it is clear that to be young is to be less equal.

Later in life, as people become elderly, they may yield some of their autonomy to their grown children—either willingly through trusts and living wills or unwillingly as they are committed to nursing homes because it has become impossible or impractical for their children to manage their care. Thus, it is at the early and late extremes of youth and age that the relationship between age and inequalities of power and material resources are most evident.

There are also age-based inequalities that may affect the life chances of nonelderly adults. As corporations downsize—that is, lay off employees in order to increase profit margins—higher-paid people in their fifties may find that their jobs are combined with other work and offered to younger, less well-paid replacements. Mature adult status and years of accumulated seniority in the corporation may be of no avail. Once on the job market, these former employees—still in their productive adult years and well below the normal age of retirement—may find that they are less likely to be hired than younger workers; often they must reduce their expectations and experience downward mobility (Newman, 1988).

Youth and Inequality

The effects of age stratification can also be seen in the youth of most nations, rich or poor. There is a growing consensus among social scientists on the need for a quantitative measure of well-being among youth. In recent years, as ideological debates over social problems like teenage fertility have become more rancorous, this need has become even more pressing (Zill, 1995). Toward this end, a group of eminent sociologists has been working to create a set of data books entitled *Kids Count.* These data books provide an invaluable and up-to-date array of indicators of trends in child well-being in the United States. They are available for use by local officials, scholars, and those who work with the youth of their communities.

Although in many communities the large majority of children are healthy and well cared for, the *Kids Count* data indicate that increasing numbers of young people are experiencing poverty and near-poverty conditions. Table 8.6 presents some indicators of well-being among youth. Measures of children in poverty, rates of violent death, infant mortality, teenagers not in school, and births to single teenagers are among the most important ways of looking at the proportion of children and young people whose well-being is questionable and whose daily lives are fraught with risk of further trouble. Perhaps the most controversial of these indicators are those dealing with births to teenagers. In the United States and other urban industrial societies, there is a great deal of debate about increasing rates of teenage childbearing. But the facts show that, contrary to what many people believe, rates of births to teenagers decreased from 1960 through 1980, when they began increasing again but never reached the same levels as those prevailing at the end of the 1950s (see Figure 8.8). But by separating out births to teenagers aged 15 to 17 and 18 to 19, one sees that births to older teenagers have been decreasing since 1960, while births to younger teenagers have not decreased significantly, and in fact have risen in recent years.

The data on births outside marriage are most revealing. The most significant change is not so much the overall rate of fertility among teenage women as the rate of marriage among teenagers, which has declined drastically. This fact is at the center of debates

TABLE 8.6
Indicators of the Well-Being of U.S. Children

Indicator	1985	1999
Percentage of low–birth weight babies	6.8	7.4
Infant mortality rate (per 1,000 live births)	10.6	7.3
Child death rate, ages 1–14 (per 100,000 children)	33.8	26.0
Percentage of all births that are to single teens	7.5	9.0
Juvenile violent crimes arrest rate, ages 10–17 (per 100,000 youths)	305	457
Percentage of teens not in school and not in labor force, ages 16–19	5.3	9.0
Teen violent death rate, ages 15–19 (per 100,000 teens)	62.8	62.0
Percentage of children in poverty	20.8	20.0
Percentage of children in single-parent families	22.7	27.0

Source: *Focus,* Spring 1995; Annie E. Casey Foundation, 1999.

FIGURE 8.8
Births to Teenage Mothers in the United States, 1960–1997

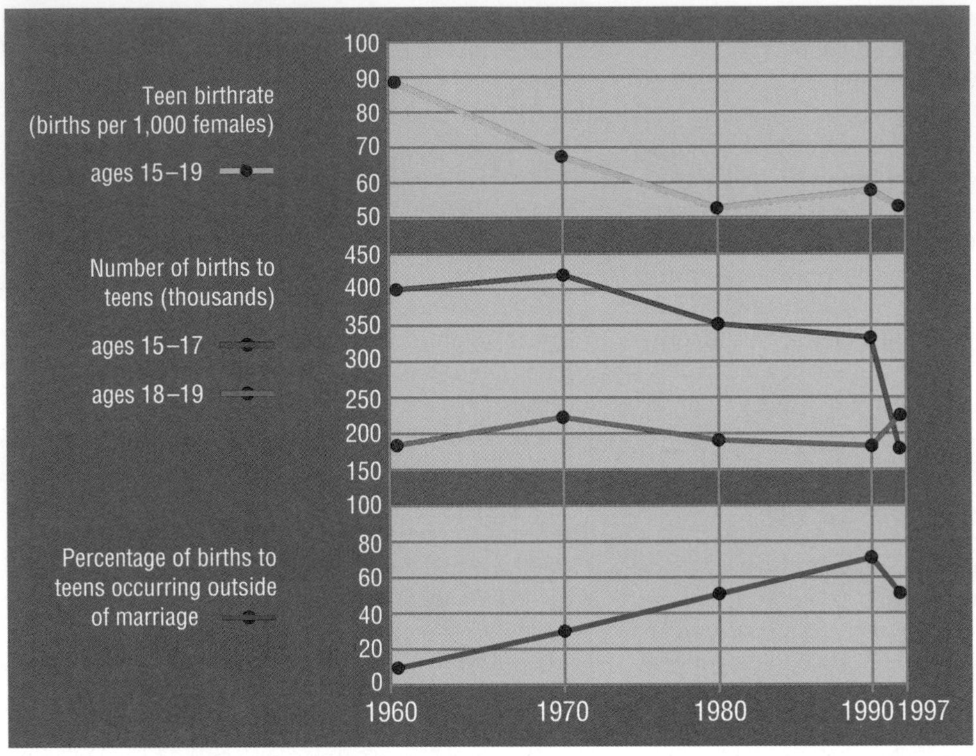

Source: Data from *Focus,* Spring 1995; *Statistical Abstract,* 1999.

over "family values." While there is no question that children have a better chance of avoiding poverty if there are two parents present in the household, it is also true that separation and divorce are among the leading causes of poverty for young mothers and children. In consequence, efforts to influence and oblige young fathers, whether married or not, to take responsibility for their children are an important social policy trend in American communities and throughout the world (United Nations, 1995c).

It should be noted that some of the indicators presented in Table 8.6, such as the rate of violent death among teenagers, have actually improved in the past few years. Others, however, such as the percentage of low–birth weight babies, have worsened. These indicators are sensitive to changes in society (declining use of crack-cocaine, efforts to reduce truancy, dropout prevention programs, etc.). The low–birth weight indicator is especially sensitive to changes in the economic situation of low-income households.

Throughout the world, the infant mortality rate is considered the single most important indicator of the relative well-being of a nation's people. It measures the extent and efficacy of a nation's investment in the health of its people. The infant mortality rate in the United States—6.3 per 1,000 live births—ranks among the world's best, but it is higher than those in other advanced nations such as Japan (3.9), the Netherlands (4.4), and Australia (5.0), while it is much lower than those in less affluent nations like Russia (20.3) and far lower than those in relatively impoverished nations like India (64.9) (*Statistical Abstract,* 2000). Thus, the United States, one of the world's most affluent nations, actually ranks on the lower end of the scale of infant mortality among the developed nations with which it is often compared. Much of the difference is due to the large number of underweight babies born in the United States.

The infant mortality rate for African Americans in the United States (14.1 in 1998) is more than twice the rate for whites (6.0). Much of this disparity is due to the high percentage of babies with low birth weights among members of minority groups, especially those in poverty. Birth weights below 5.5 pounds are a cause of infant death and are also associated with higher rates of illness in later years as well as a higher incidence of neurological and developmental handicaps and poor academic performance (Partin & Palloni, 1995). The incidence of low–birth weight babies among African Americans declined from more than 14% in 1970 to a little over 12% in the mid-1980s, but began climbing again after 1984.

In explaining these trends, sociologists Melissa Partin and Alberto Palloni (1995) point out that while poverty is always associated with low birth weight, a woman's access to prenatal care in her community combined with her decision not to have an abortion are important factors. As more poor minority women choose not to have abortions and find it more difficult to obtain adequate care during pregnancy, the risk of low birth weight increases. Neither side in the abortion debate wishes to see more abortions among poor women. The central implication of these findings is that the rate of low–birth weight babies would be reduced by improved access to health care for young mothers and more education about the need to decrease behaviors such as smoking during pregnancy.

Age and Inequality

At the beginning of the twentieth century, the largest segment of the U.S. population living in poverty or near-poverty conditions was the elderly (Preston, 1984). This was also true throughout much of Europe. Until the development of modern systems of employee pensions, Social Security, and basic health insurance (both private insurance systems and public systems like Medicare), people over age 60 were highly likely to be poor. Although poor and working-class elderly people without substantial wealth or savings might receive assistance from adult children and other family members, even those relatively fortunate individuals often felt deep resentment about having worked all their lives only to end up powerless and dependent (Atchley, 1991).

The Social Security Act of 1935 established federal old-age pensions based on employee and employer contributions (Social Security) as well as unemployment insurance and other antipoverty programs. The results of these programs, and the Medicare system instituted in 1965, have been dramatic. Rates of poverty among the elderly, which were close to 40% in the early decades of the twentieth century, declined to about 28% in the early 1960s and then to about 14% after Medicare had gone into effect. At present, 10.5% of people age 65 and over in the United States live at or below the poverty line, compared to 13.3% of the population as a whole.

Among elderly members of minority groups, however, the situation is not nearly so positive. Twenty-six percent of African Americans and 20.5% of Hispanics over age 65 are living in poverty (*Statistical Abstract,* 2000). These differences are often due to the fact that elderly people who previously worked at lower-paying jobs, which contributed little or nothing to private pension plans, may be forced to live almost exclusively on their federal Social Security payments, which by themselves are not enough to raise them above the official poverty line. And elderly blacks and Hispanics are especially likely to have had jobs that did not provide pensions.

As more people join the ranks of the elderly, concern about their economic situation becomes an ever more powerful theme in American life. Many Americans in their early sixties are subject to mandatory retirement. Once they are forced into retirement, they are limited in what they can earn without experiencing cuts in their Social Security payments and decreases in Medicare support. Matilda White Riley, a leading authority on the sociology of aging, asserts that there is a growing gap between the number of skilled and energetic elderly people in the population and the availability of "meaningful opportunities in work, family, and leisure" (Riley, Kahn, & Foner, 1994). We return to this theme shortly.

Age and Disability. As life expectancies have increased throughout the world, so have the numbers of people with major disabilities that seriously impair their ability to function effectively in their daily lives. In the United States, the average life expectancy for people age 60 is 19 years for women and 15 years for men. On the average, women who live that many years after their sixtieth birthday can expect to experience major disabilities for five of those years (four years for men). In Germany, another affluent industrial nation, life expectancies for women and men over age 60 are 22 years and 18 years, respectively, and the chances of experiencing major disabilities are the same as those in the United States. In Egypt, a developing nation, the comparable figures are 13 years of life expectancy for women after age 60 and 12 years for men, but because of less advanced health care, the average Egyptian woman can expect to experience six years of disability (four years for men; Haber & Dowd, 1994). People who live longer do not automatically suffer major disabilities. As they age and become more frail, however, the experience of living with disability inevitably affects higher proportions of women because of their greater life expectancy.

The problems of health and disability among the elderly are compounded by inequalities of socioeconomic status. The more affluent and educated people are as they enter their sixties, the healthier they tend to be (House et al., 1994). People who have worked for decades in factories tending machines, often in polluted and stressful conditions, are more likely to have health problems and to have difficulty obtaining regular medical care than people who have worked in more physically and emotionally favorable conditions. Data on aging and health also show, however, that people from poor and working-class backgrounds are more likely to engage in **psychosocial risk behaviors** like smoking and heavy drinking (90 or more drinks a month). They also tend to experience a higher than average number of **lifetime negative experiences** (e.g., the death of a child or spouse, divorce, physical assault), which cause long-term stress. Even when risk behaviors and sources of chronic stress are considered, however, quantitative studies show a strong positive relationship between higher socioeconomic status and better health in elderly people (House et al., 1994).

Loss of Social Functions With Age. As people age, they experience more medical problems and disabilities—but does this mean that they must inevitably withdraw from social life? Can anyone deny the biological facts of aging and death? Isn't it only natural for the elderly to perform less important roles as their physical capacities diminish? The answer, of course, is that it is not obvious at all. The increasing participation of elderly widows in securities investment clubs throughout the nation is a good example of how elderly people may perform socially important roles even as their physical strength diminishes. It is not uncommon for members of these clubs to impress far younger stockbrokers with their skill as players in the volatile securities market.

Sexual behavior among the elderly is another area in which popular notions about the influence of biology on age roles have been disproved. In the late nineteenth century and well into the twentieth, it was widely believed that people lose their sexual desire and potency after middle age. In fact, however, a number of studies have shown that the image of the elderly as lacking sexual desire and the ability to enjoy sex is an ageist stereotype. For example, Eric Pfeiffer, Adrian Verwoerdt, and Glenn Davis (1972) gathered data from several samples of elderly people—including individuals as old as 94—that showed conclusively that although sexual interest and activity tended to decline with age, sex remained an important aspect of the subjects' lives. The researchers also found, however, that elderly men are more interested in sex than women of the same age. The explanations for this difference are cultural rather than biological. Elderly men are in short supply. They tend to be married to women with whom they have had a long-standing relationship that includes an active sex life. If they are not married, they are in such great demand that they have less difficulty than women of the same age in finding a sexually compatible partner (Greeley, Michael, & Smith, 1990; Kornblum & Julian, 2001; Pfeiffer, Verwoerdt, & Davis, 1972).

Ageism

The attitude known as **ageism** is similar to sexism. The term refers to an ideology that justifies prejudice or discrimination based on age. Ageism limits people's lives in many ways, both subtle and direct. It may label the young as incapable of learning. It labels the elderly as mentally incapable or asexual or too frail to get around. But people of all ages increasingly reject these notions. In their everyday lives in families and communities, for example, older people continually struggle against the debilitating effects of ageism. "Just because I need help crossing the street doesn't mean I don't know where I'm going," an elderly woman said to community researcher Jennie Keith (1982, p. 198).

Gerontologist Robert Butler observes that "ageism allows the younger generation to see older people as different from themselves; thus they subtly cease to identify with their elders as human beings" (1989, p. 139). Butler, a physician and social scientist, has found that as the proportion of older people in a society increases (as is occurring in the United States and

Europe), the prevalence of ageism also increases. The younger generations, he notes, tend to fear that the older, increasingly frail and dependent generations will deprive them of opportunities for advancement. This fear is expressed in demands for reduced spending on Medicare and other programs that assist the elderly, as well as in the belief that the elderly are affluent and do not need social supports.

Changing Views of the Elderly

Although they have been less far-reaching than the women's movement, social movements among the elderly—led by organizations like the Gray Panthers and the American Association of Retired Persons—have had a significant impact on American society. And as the population continues to age, we can expect to see more evidence of the growing power of the elderly. Consider, for instance, how changes in the consciousness of elderly people themselves are altering the way sociologists formulate questions about old age.

Until the social movements of the 1960s prompted the elderly to form movements to oppose ageism and fight for their rights as citizens, the most popular social-scientific view of aging was *disengagement theory.* Numerous empirical studies had shown that old people gradually disengage from involvement in the lives of younger people and from economic and political roles that require responsibility and leadership. In a well-known study of aging people in Kansas City, Elaine Cumming and William Henry (1971) presented evidence that as people grow older they often gradually withdraw from their earlier roles, and that this process is a mutual one rather than a result of rejection or discrimination by younger people. From a functionalist viewpoint, disengagement is a positive process both for society as a whole (because it opens up roles for younger people) and for the elderly themselves (because it frees them from stressful roles in their waning years).

The trouble with disengagement theory is that, on the one hand, it appears to excuse policy makers' lack of interest in the elderly and, on the other hand, it is only a partial explanation of what occurs in the social lives of elderly people. An alternative view of the elderly is that they need to be reengaged in new activities. Known as *activity theory,* this view states that the elderly suffer a sense of loneliness and loss when they give up their former roles. They need activities that will serve as outlets for their creativity and energy (Palmore, 1981).

Theories of aging are not simply abstract ideas that are taught in schools and universities. The disengagement and activity theories lead to different approaches that often impose definitions of appropriate behavior on people who do not wish to conform to those definitions. Today gerontologists tend to reject both theories. They see older people as needing opportunities to lead their lives in a variety of ways based on individual habits and preferences developed earlier in life. Elderly people themselves express doubt that activity alone results in successful adjustment to aging or happiness in old age. For example, in her study of a French retirement community, Jennie Keith (1982) made this observation:

> The residents . . . seem to offer support to the gerontologists who have tried to mediate the extreme positions, disengagement vs. activity, by introducing the idea of styles of aging. . . . Some people are happy when they are very active, others are happy when they are relatively inactive. From this point of view, life-long patterns of social participation explain the kinds and levels of activity that are satisfying to different individuals. (p. 59)

In sum, for the elderly as well as for women, there is a growing tendency among social scientists to emphasize individual needs and capabilities. The social movements for gender and age equality also advance the needs of individuals, but in a collective manner, by asserting the needs of entire populations and rejecting preconceived notions of what is best for all women, all youth, all men, or all the elderly.

TECHNOLOGY AND SOCIAL CHANGE

Is telecommuting in your future? About 6.3 million Americans are telecommuters, meaning that they have a formal arrangement with their supervisors at work that allows them to spend part of the week working at home. They use computers and modems to receive and send work-related materials, and, of course, they make frequent use of the telephone. Most of these employees are in professional or service occupations. It is rare for blue-collar workers to telecommute because

SOCIOLOGY AND SOCIAL JUSTICE

The brutal murder of women because they assert their independence is not a feature of the Islamic religion. Such killings occur in some Islamic nations but not in others. They are unknown in the Islamic regions of black Africa, for example. Where they do occur, especially in the Middle East and Pakistan, they are based on tribal traditions of extreme patriarchy and male dominance. These traditions predate the rise of Islam. Some extremists, such as the Taliban regime in Afghanistan, may link the subjugation of women to Islamic laws, but honor killings are not condoned by the Kuran, the holy book of Islam (Armstrong, 2000).

The existence of so-called honor killings calls our attention to all the ways, violent and nonviolent, in which social stratification systems keep women "in their place." How many families do you know in which parents' expectations of success in education or earnings or professional status are higher for their sons than for their daughters? How many of these parents would admit to this? Would they not be more likely to reason that since women are expected to spend more time on domestic matters, their lower aspirations for their daughters are justified? Certainly you also know families in which parents have equal expectations for their daughters and sons. And you may also know families in which the women encounter more violent or abusive forms of dominance and gender stratification.

most manufacturing jobs must be done in shops and factories.

Telecommuters report that the arrangement allows them to avoid the stress and expense of commuting and gives them more flexibility in balancing domestic and work responsibilities. They can schedule deliveries to their home, for example, without worrying about taking time away from work. Potentially, telecommuting can make it easier for older workers to remain in the labor force and for couples to balance the demands of parenting and work (Mariani, 2000).

Like all social innovations, telecommuting entails some complications and unintended consequences. For women, a common problem is establishing reasonable boundaries between work and child rearing. The same is true for many men, but since more women choose telecommuting in order to work at home while raising children, they face this issue more frequently than men. Just because a mother is working at home does not mean that she can be available whenever her children demand attention. Most mothers who telecommute report that it takes some patience and creativity to

It is important to realize that throughout the world women are struggling against low expectations as well as against the violence directed at keeping them in their place. If feeling sympathy for or somehow joining that struggle makes you a "women's liberationist," you might decide that the label is a badge to be worn proudly. ■

An Iranian woman covered with the traditional head-to-toe "chador" looks at paintings by the American pop artist Roy Lichtenstein at the Museum of Contemporary Art in Tehran.

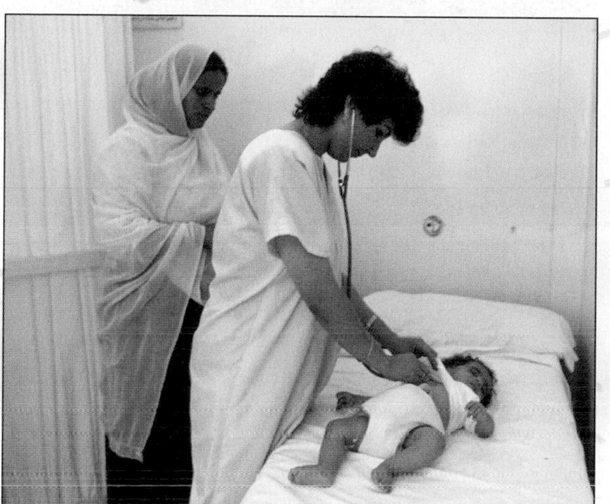

Women in the Islamic world often choose traditional roles, recognizing that their access to careers like medicine may be barred by males in positions of power.

balance work and family demands when both responsibilities are fulfilled in the home.

Another issue that telecommuters face, also an especially difficult one for women, is a sense of isolation from fellow workers. It may be even more difficult for female telecommuters to advance in the organization than for their counterparts who go to the office every day. When it comes to promotion decisions, it does not appear that absence makes the heart grow fonder.

Despite these problems, the option to telecommute can be a lifesaver for working parents, especially mothers. For elderly workers with the necessary skills, telecommuting can be a way to solve the problems of declining energy levels and satisfy the desire to feel fulfilled in the world of work. In a labor market in which technologically sophisticated workers are in increasing demand, telecommuting can be an attractive option for managers seeking such workers. As such, it is likely to involve an increasing proportion of the labor force in the world's most technologically advanced nations (Bray, 1999).

SUMMARY

Sex refers to the biological differences between males and females, including the primary sex characteristics that are present at birth and the secondary sex characteristics that develop later. *Gender* refers to the culturally defined ways of acting as a male or a female that become part of an individual's personal sense of self.

Sexuality refers to the manner in which a person engages in the intimate behaviors connected with genital stimulation, orgasm, and procreation. *Heterosexuality* refers to sexual orientation toward the opposite sex, in contrast to *homosexuality,* or sexual orientation toward the same sex, and *bisexuality,* or sexual orientation toward either sex.

All human societies are stratified by gender, meaning that males and females are channeled into specific statuses and roles. Until quite recently, it was assumed that there were two separate spheres of life for men and women. Out of this gender-based division of labor grew the notion of differences in men's and women's abilities and personalities. These differences were thought to be based on biological and psychological differences between males and females. Evidence from the social sciences has established that gender roles are not innate but are strongly influenced by culture.

In modern industrial societies, age and gender interact to shape people's views of what is deemed appropriate role behavior. The resulting gender stratification is reflected in attitudes that reinforce the subordinated status of women. *Sexism* refers to an ideology that justifies prejudice or discrimination based on sex.

The most frequently cited problems of women in the workplace are low wages, differentials in fringe benefits, dead-end jobs, sexual harassment, lack of child care, stress over multiple roles, and lack of leisure time. Another problem is the segregation of women into the pink-collar ghettos of secretarial and clerical work. Women who work outside the home are also expected to perform the bulk of domestic and child-care work. This "second shift" is an example of the persistence of *patriarchy,* the dominance of men over women.

In many societies, age determines a great deal about the opportunities open to a person and what kind of life that person leads. All societies channel people into *age grades,* or sets of statuses and roles based on age. The transitions among these age grades create a *life course* and are often marked by ceremonies known as *rites of passage.*

Age cohorts are people of about the same age who are passing through life's stages together. The baby boom cohorts, which were produced by rapid increases in the birthrate from about 1946 through the early 1960s, have profoundly influenced American society. A sizable proportion of the children of the baby boom generation, the "baby boom echo," are members of minority groups.

By *life expectancy* we mean the average number of years a member of a given population can expect to live beyond his or her present age. As life expectancy in a population increases, the proportion of the population that is dependent on the working adult cohorts also increases.

As people age, they experience more medical problems and disabilities, but this does not mean that they must inevitably withdraw from social life. *Ageism* is an ideology that justifies prejudice or discrimination based on age.

The most popular social-scientific view of aging before the 1970s was disengagement theory, the belief that as people grow older they gradually disengage from their earlier roles. An alternative view of the elderly, known as activity theory, states that the elderly need activities that will serve as outlets for their creativity and energy. Today gerontologists tend to reject both of these theories, seeing older people as needing opportunities to lead their lives in a variety of ways based on individual habits and preferences.

GLOSSARY

Sex: The biological differences between males and females, including the primary sex characteristics that are present at birth (i.e., the presence of specific male or female genitalia) and the secondary sex characteristics that develop later (facial and body hair, voice quality, etc.). **(221)**

Hermaphrodite: A person whose primary sexual organs have features of both male and female organs, making it difficult to categorize the individual as either male or female. **(221)**

Transsexuals: People who feel very strongly that the sexual organs they were born with do not conform to their deep-seated sense of what their sex should be. **(221)**

Gender: The culturally defined ways of acting as a male or a female that become part of an individual's personal sense of self. **(221)**

Sexuality: The manner in which a person engages in the intimate behaviors connected with genital stimulation, orgasm, and procreation. **(221)**

Heterosexuality: Sexual orientation toward the opposite sex. **(225)**

Homosexuality: Sexual orientation toward the same sex. **(225)**

Bisexuality: Sexual orientation toward either sex. **(225)**

Gender role: A set of behaviors considered appropriate for an individual of a particular gender. **(227)**

Sexism: An ideology that justifies prejudice and discrimination based on sex. **(230)**

Patriarchy: The dominance of men over women. **(234)**

Age grade: A set of statuses and roles based on age. **(235)**

Life course: A pathway along an age-differentiated, socially created sequence of transitions. **(236)**

Rite of passage: A ceremony marking the transition to a new stage of a culturally defined life course. **(236)**

Age cohort: A set of people of about the same age who are passing through the life course together. **(236)**

Life expectancy: The average number of years a member of a given population can expect to live beyond his or her present age. **(239)**

Psychosocial risk behaviors: Behaviors that are detrimental to health, such as smoking and heavy drinking. **(244)**

Lifetime negative experiences: Experiences that cause long-term stress, such as the death of a child or spouse. **(244)**

Ageism: An ideology that justifies prejudice and discrimination based on age. **(244)**

QUESTIONS FOR THOUGHT AND DISCUSSION

1. Is it possible to believe that men and women are different without resorting to biological determinism? What kinds of differences would you identify? What sociological explanations would you offer?

2. When you hear statements that begin with phrases like "Young people nowadays just don't understand" or "Old people nowadays are fortunate that . . . ," do your sociological antennae send warning signals? What kinds of ideas give rise to these notions? From a sociological perspective, what is wrong with these observations?

DIGGING DEEPER

Books

Gender, Family and Social Movements (Suzanne Staggenborg; Pine Forge Press, 1998). An extremely useful analysis of the importance of social movements to address gender and age inequalities.

Manhood in America: A Cultural History (Michael S. Kimmel; Free Press, 1996). A sociological analysis of changing norms of masculinity and ideas of manhood in North America.

Women and Children First: Environment, Poverty, and Sustainable Development (Filomena Cioma Steady; Schenkman Books, 1993). A recent review of international issues of gender, inequality, and development.

Women in Law, 2nd ed. (Cynthia Fuchs Epstein; University of Illinois Press, 1993). A fine empirical study of the structural and cultural barriers women face in the legal profession.

Gender, Family, and Economy: The Triple Overlap (Rae Lesser Blumberg, ed.; Sage Publications, 1991). A collection of original essays based on recent research on gender stratification. Includes some excellent material on sharing, and lack of sharing, of household chores among heterosexual couples, and valuable historical material on the stratification of minority women in the United States and elsewhere.

Age and Structural Lag (Matilda White Riley, Robert L. Kahn, & Anne Foner; Wiley, 1994). A comprehensive collection of articles covering many recent research initiatives. Riley is one of the world's foremost authorities on the demography of aging and the comparative situation of elderly people throughout the world.

Number Our Days (Barbara Myerhoff; Simon & Schuster, 1978). A haunting study of how elderly people struggle to survive and die with dignity. The setting is a senior-citizens center in Venice Beach, California, and the people are elderly Jewish immigrants and refugees from Europe.

Journals

Signs. A quarterly journal published by the University of Chicago Press that presents new research and theory in gender studies and feminist social science.

Gender & Society. The official publication of Sociologists for Women in Society.

Other Sources

Kids Count Data Book. A timely and scientifically accurate source of comparative demographic data on the situation of children in the United States and elsewhere.

Age in America: Trends and Projections. A chartbook that analyzes demographic and socioeconomic trends affecting the United States and its elderly population.

Handbook of Aging and the Social Sciences. A valuable source of interdisciplinary research and policy studies dealing with the condition of the aged in the United States and other Western societies.

EXPLORING SOCIOLOGY ON THE INTERNET

Population Profile, U.S. Census Bureau
www.census.gov/population/www/pop-profile/

Provides excellent comparisons of women and men in the labor force, education, health care, and much more. The profile is divided into several short essays with colorful graphs on a number of critical topics.

National Organization for Women (NOW)
www.now.org/

Features calls for action, press releases, and the NOW newspaper, all of which exemplify the activities of a mainstream feminist organization.

The Eagle Forum
www.eagleforum.org/

A conservative political organization that speaks out on issues of gender and inequality.

Women's Studies Program at the University of Maryland
www.inform.umd.edu:8080/EdRes/Topic/WomensStudies/

Offers reports on issues like women's status in higher education, women in the workforce, sex discrimination, and sexual harassment. Also offers links to international organizations for comparative data.

Socionet
www.socio.com

The Web site for the journal *Sociometrics*. Devoted to the study of social groups and networks and how they are related to social issues of all kinds (for example, adolescence, health, disability).

The Ethical Spectacle
www.spectacle.org/995/sd.html

Presents discussions about many sociologically relevant topics, especially those involving ethical dilemmas or challenges, such as assisted suicide or protection versus development of public lands.

U.S. Department of Labor
www.dol.gov

The official site of the United States Department of Labor; includes a source for recent labor relations data. Current information about gender unemployment and salary rates are updated regularly.

Chapter 9

The Family

Is the family breaking down or simply adapting to new social conditions?

The Nature of Families

Defining the Family
Variations in Family Structure
Family Futures
Family Life Cycles
Parenting, Stepparenting, and Social Change

Dynamics of Mate Selection and Marriage

Marriage as Exchange
Norms of Mate Selection
Marriage and Divorce

Perspectives on the Family

The Interactionist Perspective
The Conflict Perspective
The Functionalist Perspective

Technology and Social Change

Kerry and Deborah Henry of Fayetteville, North Carolina, face the challenges of blending two families into one. When the couple married two years ago, Kerry's two teenagers and Deborah's young son took their place among the more than 1,300 blended families that are formed in the United States every day (Davis, 2000). "The odds are always against blended families. I knew it wasn't going to be like *The Brady Bunch,"* says Kerry.

What's right about this picture? Clearly, the happy bride and groom are celebrating not only their marriage but the blending of two families created by earlier marriages.

All three children had been used to the way things were before their parents divorced. That did not stop Deborah and Kerry from making a commitment to each other and to making their new family work. "Our first priority was us," Deborah says. "We wanted to be a family."

As they planned to marry, the couple conferred with their children, who gave their approval and joined the wedding party. "I asked their permission, and they told me that they wanted to see me happy," the new husband says of his two children. "I was content with putting my personal happiness after theirs." S. Malone-Hawkins, an expert on blended families, notes that "For stepfamilies to work, the married couple has to show loving, patient, and tolerant behavior. They also have to recognize the rights and feelings of their children and stepchildren. No matter how long their preparation, however, there will probably be a period of adjustment," (quoted in Davis, 2000).

All families face the same challenges in dealing with the feelings, thoughts, and emotions of their children. But for blended families there may be additional challenges, Malone-Hawkins observes: "No matter how good the relationship is between the biological parents [after the separation], there are going to be those instances when children feel guilty about the failure of [their biological parents'] relationship." Parents and members of the extended family must intervene at this point to make sure children understand that they are loved and that the failure of their biological parents' marriage is not their fault. An effective strategy for doing this is to maintain a good relationship with the biological parent. This demonstrates good family behavior for all members of the extended families, including grandparents, aunts, uncles, and cousins. "People need to be grown-up about these things," says Malone-Hawkins, herself a stepchild and a stepmother. "They need to be models for the children."

Blended families are just one of several types of families in North America and other parts of the world. This chapter begins, therefore, by introducing some of the basic concepts in the study of the family, as well as important variations in family structure that occur as families undergo major changes like divorce and remarriage. We then move on to the study of mate selection, marriage and divorce, and the special needs of gay and lesbian couples. This section of the chapter closes with an analysis of the long-term effects of divorce on individuals and on the family as an institution. The final section explores some theoretical perspectives on the family. ■

THE NATURE OF FAMILIES

Defining the Family

The family is a central institution in all human societies, although it may take many different forms. A **family** is a group of people related by blood, marriage, or adoption. Blood relations are often called *consanguineous attachments* (from the Latin *sanguis,* meaning "blood"). Relations between adult persons living together according to the norms of marriage or other intimate relationships are called *conjugal relations.* The role relations among people who consider themselves to be related in these ways are termed **kinship.**

The familiar kinship terms—*father, mother, brother, sister, grandfather, grandmother, uncle, aunt, niece, nephew, cousin*—refer to specific sets of role relations that may vary greatly from one culture to another. In many African societies, for example, "mother's brother" is someone to whom the male child becomes closer than he does to his father and from whom he receives more of his day-to-day socialization than he may from his father. It must be noted that biological or "blood" ties are not necessarily stronger than ties of adoption. Adopted children are usually loved with the same intensity as children raised by their biological parents. And many family units in the United States and other societies include "fictive kin"—people who are so close to members of the family that they are considered kin and may carry out the roles of uncle or aunt, or even parent, despite the absence of blood ties (Council of Economic Advisors, 2000; Liebow, 1967; Stack, 1974). Finally, neither blood ties nor marriage nor adoption adequately describes the increasingly common relationship between unmarried people who consider themselves a couple or a family.

The smallest units of family structure are usually called **nuclear families.** This term is usually used to refer to a wife and husband and their children, if any. Nowadays one frequently hears the phrase "the traditional nuclear family" used to refer to a married mother and father and their children living together. But as we will see throughout this chapter, there is no longer a "typical" nuclear family structure. Increasingly, therefore, sociologists use the term *nuclear family* to refer to two or more people related by consanguineous or conjugal ties or by adoption who share a household; it does not require that both husband and wife be present in the household or that there be any specific set of role relations among the members of the household (Laslett & Wachter, 1978).

The nuclear family in which a person is born and socialized is termed the **family of orientation.** The nuclear family a person forms through marriage or cohabitation is known as the **family of procreation.** The relationship between the two types of families is shown in the Sociological Methods box on page 256, which presents the social scientist's method of diagramming family relationships to create a graphic depiction of family structure.

The vast majority of people not only live in a nuclear family but also have an **extended family** that includes all the nuclear families of their blood relatives—that is, all of their uncles, aunts, cousins, and grandparents. The chart in the box shows Ego's extended family. Ego's spouse also has an extended family (not indicated in the chart), which is not defined as part of Ego's extended family. But relationships with the spouse's extended family are likely to occupy plenty of Ego's time. Indeed, as the chart shows, the marriage bond brings together far more than two individuals. Most married couples have extensive networks of kin to which they must relate in many varied ways throughout life.

Variations in Family Structure

Changes in human societies over the past 300 years or more have brought enormous changes to the family. Industrialization, urbanization, and contemporary technological advances have changed the nature of family roles and the relationships between families and the communities and societies in which they exist. As we outline some of these changes, it is important to remember that whereas the industrial revolution occurred in England, Europe, and the United States primarily during the 1800s and early 1900s, elsewhere in the world it is still occurring. And in the advanced industrial nations technological advances are accelerating global social change and having major impacts on the family (Goode, 1963, 1993).

Industrial Revolution and Changing Families. Some of the most important changes brought on by industrialization are:

- A decline in the number of families that live and work on farms and an increase in the movement of individuals and families to cities in search of work in manufacturing and related industries.

SOCIOLOGICAL METHODS

Kinship Diagrams

Kinship diagrams are used in both anthropology and sociology to denote lines of descent among people who are related by blood (children and their parents and siblings) and by marriage. Kinship terms are often confusing because families, especially large ones, can be rather complex social structures. It may help to devote a little time to the kinship diagram shown here.

To understand the chart, one must know the meanings of the symbols used; these are explained in the key to the chart. "Ego" is the person who is taken as the point of reference. You can readily see that Ego has both a family of orientation and a family of procreation. So does Ego's spouse. Ego's parents become the in-laws of the spouse, and the spouse's parents are Ego's in-laws.

Kinship diagrams like this one provide a visual model of family statuses extending over more than one generation. They are a very useful way of analyzing the social structure of families. We can quickly see the generations in the family, from Ego's grandparents to Ego's children. We can also compare the nuclear and extended families of Ego's parents and cousins.

Nuclear and Extended Family Relationships

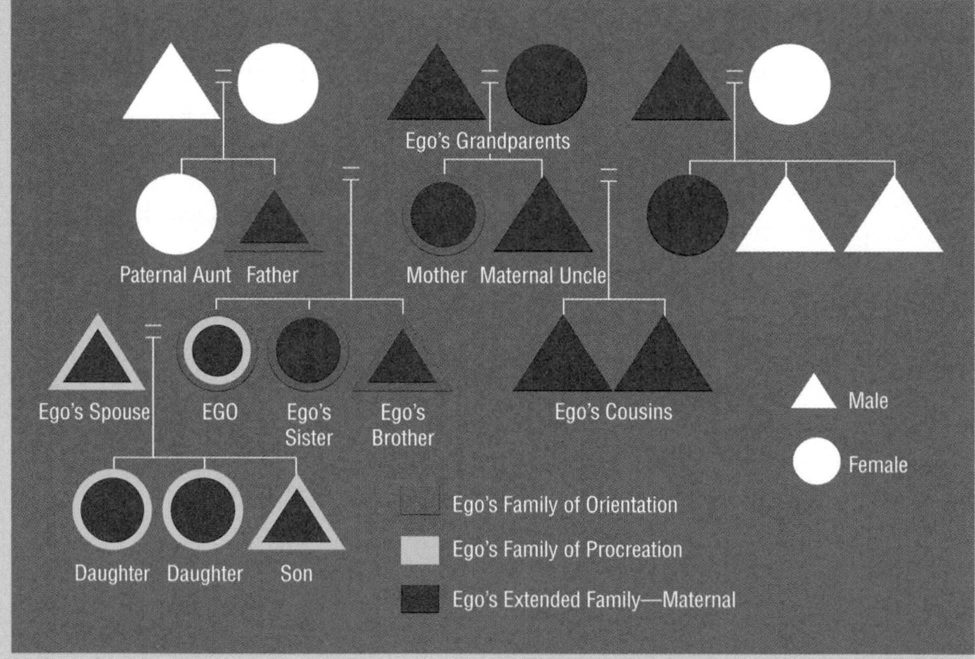

In the mid-twentieth century, the ideal image of the family had the mother caring for children and creating a comfortable domestic world for her working husband. Today's images of the family are much more likely to reflect the realities of women in the labor force who are trying to juggle home and family responsibilities. In fact, ads in magazines often refer to young mothers as "jugglers" to emphasize their economic and domestic responsibilities.

- Increases in the proportion of women who enter the paid labor force.
- Increases in the demand for educated workers and the rise of public schooling for women as well as for men.
- Greater investments in public health, especially separation of drinking water from sewage systems, which lowers rates of infant mortality dramatically and also contributes to longer life spans for adults, which in turn makes it possible for more men and women to become grandparents.

These and many other far-reaching changes brought on by industrialization and the massive movement of people from country farms to city tenements place great pressure on families to adapt. Agrarian families are typically quite large, often including extended family members in the same household or nearby households. Moreover, the family is the primary economic unit and as such allocates productive roles to its members (e.g., work assignments on the farms and nearby village markets). During industrialization, the family's economic function is shifted to other institutions, such as the corporation. In addition, when they migrate to cities, women typically find work at jobs outside the home. The demands of the workplace make it necessary for parents to rely on educational and other institutions (day-care centers, summer camps, etc.) to take over important aspects of their children's socialization.

Geographic mobility from country to city disrupts traditional family patterns but also creates opportunities that were unheard of in traditional agrarian communities. Young men and women migrate alone to cities and often must wait long years to find mates and marry. When they do marry, their choice of mates is often based on considerations that differ from the values and norms of their parents and their extended families. They may marry people of different religions or races or from widely different cultural or social-class backgrounds. They may not marry at all, or they may choose not to have children. All these are new (and sometimes very threatening) ways of carrying out family life that are made possible by the greater choices available to people in urban environments where they are no longer living "in the bosom of their families." The rise of divorce and remarriage, the

TABLE 9.1
Change and Continuity in Family Structure and Roles

	1900	1950	1998
Average household size (persons)	4.8	3.4	2.6
Households with seven or more people (percent)	20.4	4.9	1.2
Living arrangements of children by family status (percent)			
Two-parent farm family	41	17	—
Two-parent nonfarm family			
Father breadwinner, mother homemaker	43	56	24
Dual earner	2	13	44
Single-parent	9	8	28
Not living with parent	5	6	4
Males and females by marital status (percent)			
Males aged 15 and over			
Married	54.6	68.9	58.0
Divorced	.3	2.0	8.2
Widowed	4.6	4.2	2.5
Never married	40.3	24.9	31.2
Females aged 15 and over			
Married	57.0	67.0	54.9
Divorced	.5	2.4	10.3
Widowed	11.2	12.0	10.2
Never married	31.2	18.5	24.7
Median age at first marriage			
Men	25.9	22.8	26.7
Women	21.9	20.3	25.0
Life expectancy at birth (years)			
Men	46.3	65.6	73.9
Women	48.3	71.1	79.4
Infant mortality rate (deaths per 1,000 live births)	99.9	29.2	7.2
Labor force participation rate of women (percent)	20.0	33.9	60.0
Women in labor force by marital status (percent)			
Single	66.2	31.9	26.8
Married	15.4	52.2	53.1
Widowed, divorced, or separated	18.4	16.0	20.0

Source: Council of Economic Advisors, 2000.

formation of blended families, the emergence of same-sex marriages, reliance on child care outside the family, the growing role of grandparents in rearing grandchildren, and the dilemmas parents face in caring for elderly and infirm parents are also among the changes in family structure and behavior brought on by industrialization and urbanization.

Changes in American Families. Table 9.1 offers a concise statistical summary of the changes and continuities in family structure and family roles in the United States during the twentieth century. Note, for example, that in 1900, 41% of children lived in a two-parent farm family and 43% lived in a nonfarm family in which the father was in the labor force as "breadwinner" and the mother was a "homemaker." In total, over 80% of families had a "traditional" two-parent structure and "traditional" roles in which the mother performed unpaid household work. Only .5% of women were divorced, but over 31% were never married. Life expectancy had not yet reached the fifties; most women who were in the labor force were single (66.2%); and the median age at which women married was just under 22 (21.9 years). By the 1950s, only 17% of two-parent families lived on farms, but the families that Americans often still think of as "traditional" (father breadwinner, mother homemaker) had increased to 56% of all two-parent nonfarm families. The proportion of never-married men and women was declining, but already the proportion of divorced individuals in the population was beginning to rise. The period after World War II was one in which many millions of Americans were marrying and having children, often after having delayed marriage due to the impact of the Great Depression and then the war, so the marriage rates were extremely high, producing the famous "baby boom" of the 1950s and 1960s.

Today two-parent farm families account for less than 2% of all family households. Overall, two-parent families in which the mother works as a homemaker account for only 24% of two-parent families, and "dual-earner" and single-parent families are an overwhelming majority. The female labor force is composed predominantly of married women (53.1%), and divorced or never-married men and women are much larger segments of the population than they were around 1950.

Family Futures

These trends point to a growing diversity of family types, including far more people living alone and far more women raising children alone. There are also more households composed of unrelated single people who live together not just because they are friends but because only by doing so can they afford to live away from their families of orientation. Thus, when they discuss family norms and roles, sociologists must be careful not to represent the traditional nuclear family, or even the married couple, as typical. As an institution, the contemporary family comprises a far greater array of household types than ever before. For those who worry that the family is an endangered institution, however, it is important to look at the continuing high rates of marriage and remarriage in the U.S. population and those of other urban industrial nations. While family forms are increasingly diverse, the norms of marriage and family formation appear to be as strong, measured by the proportion of men and women who marry, as they were before the revolution in family structure that took place during the twentieth century.

Sociologists are divided on the meaning and implications of these facts. For some, the trend toward greater diversity of family forms means that the family as an institution is adapting to change (Skolnick, 1991; Stacey, 1990). For others, the decline of the family composed of two married parents living with their biological children poses a serious threat to the well-being of children (Popenoe, 1994).

Ideal Versus Actual Family Patterns. When politicians and government officials talk about "preserving family values," they often have in mind the traditional two-parent nuclear family. To many people, this remains the ideal family form. In reality, as shown in Figure 9.1, such families are less prevalent today than they were in earlier decades. Increasing numbers of families are headed by a single parent, usually the mother but sometimes the father. Moreover, two-parent families may actually be blended—that is, created by remarriage after divorce or the death of a spouse. Thus, what appears to be a conventional nuclear family is really a far more complicated family form with extended ties among parents and siblings in different nuclear families (Richie, 2000).

Among lower-income people, these patterns are especially prevalent—so much so that some researchers have concluded that families without fathers are a direct consequence of poverty (Oppenheimer, 1994). Although they tend to share the same norms of family life as other members of society, poor people bear added burdens that make it difficult, if not impossible, for them to meet these norms. In particular, men in such families may not be able to provide their expected share of family income and hence may become demoralized and unable to maintain family relationships. This is especially true when teenagers father children before they are adequately prepared to take economic responsibility for them. Thus, in poor families the father often is not a central figure, and female-headed families are an increasingly common family form (Furstenberg & Cherlin, 1991; Osmond, 1985).

Contemporary social-scientific research on family structure and interaction increasingly focuses on issues like those just described. At the same time, there is growing recognition that the actual form the family takes may vary greatly at different stages of the life course. In his research on homeless women, for

FIGURE 9.1
U.S. Family Households in 1950 and 1999

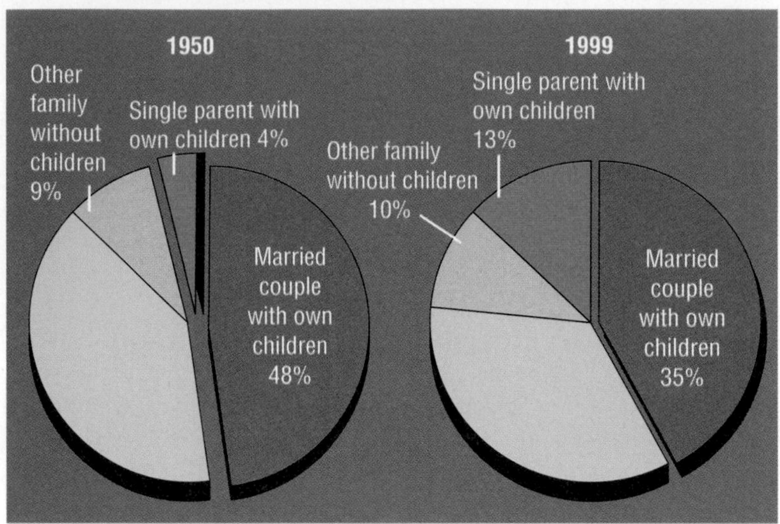

Source: Richie, 2000.

example, Elliot Liebow (1995) observed that as they grow older, women who have lived lives marked by poverty and abuse are often left without real families and must create surrogate family relationships in caregiving institutions. And as the proportion of elderly people in the population increases, more families must decide whether to care for an aging parent at home. Let us therefore take a closer look at how families change throughout the family life cycle.

Family Life Cycles

Sociologist Paul Glick, an innovator in the field of family demography and ecology, developed the concept of the family life cycle, the idea that families pass through a sequence of five stages:

1. Family formation: first marriage
2. Start of childbearing: birth of first child
3. End of childbearing: birth of last child
4. "Empty nest": marriage of last child
5. Family dissolution: death of one spouse (Glick & Parke, 1965)

These are typical stages in the life cycle of conventional families. Although there are other stages that could be identified within each of these—such as retirement (the years between the time of retirement and the death of one spouse) or the "baby" stage (during which the couple is rearing preschool-age children)—the ones that Glick listed are most useful for comparative purposes.

As the "typical" family structure becomes ever more difficult to identify owing to changes in family norms, the stages of family development also vary. In fact, the stages of the family life cycle have become increasingly useful as indicators of change rather than as stages that all or most families can be expected to experience. The Census Bureau estimates, for example, that there are almost 5 million heterosexual couples in the United States who are cohabiting. This number represents a sevenfold increase since 1970, when an estimated 500,000 cohabiting couples were counted (*Statistical Abstract,* 2000). (Note also that the census figures include only heterosexual couples and therefore vastly underestimate the total number of cohabiting couples.) Although the majority of cohabiting couples will eventually marry, many others will break up or continue to live together without marrying (Bumpass & Sweet, 1989). A study based on data from the 1990 U.S. census concluded that 55% of white women and 42% of black women in their first cohabiting union eventually marry their male partner (Condon, 1991).

THE CENTRAL QUESTION: A CRITICAL VIEW

Is the family breaking down or simply adapting to new social conditions?

It would be understandable to fear that the family is breaking down and that it is a declining institution. After all, as we have seen in this chapter, the family form that was once so dominant in American society and many others—the family with mother, father, and children living in the same home—is now a minority of family households. Approximately 50% of children in the United States live in a household with their two parents, compared with 73% in 1972. The percentage of children living with single parents rose to 18.2% in 1998, compared with 4.7% in 1972 (NORC, 1998). Especially among the poor, the proportion of single-parent families is high. Since disproportionate numbers of African Americans and Latinos are poor, the proportion of minority single-parent families is higher as well. These changes are alarming because there is little doubt that overall it is more advantageous for children to grow up in two-parent families than in single-parent ones, although there may be many exceptions to this generality (Popenoe, 1999).

On the more positive side, we have seen that high rates of remarriage testify to the desire of adults to have families, even if they fail the first time. Another indication of the continuing strength of the family as an institution, especially in minority families, is the importance of the extended family as a support network, particularly for its younger members. Although it is true that the nuclear family is becoming an increasingly diverse institution, and geographic mobility makes it difficult to hold large extended families together, family reunions are a thriving institution. Among African Americans, Latino Americans, and people from the rural South and the Midwest, large family reunions like those shown in these photos are a common scene in public parks and picnic grounds during good weather. Do members of these large families lend each other support? Do they help in times of need? Are there schisms and conflicts within the larger family groups? Almost certainly the answer to all of these questions is yes, but the fact that these families continue to get together to celebrate their kinship and to honor their elders is perhaps the most important sign that the extended family can survive in modern society.

Large family reunions are extremely popular among African Americans, Mexican Americans, and many other groups in the United States. Evidence of the continuing strength of extended-family networks does not contradict statistics on divorce or family breakup, but it does show that as a social institution the family retains great strength even under the pressures of a highly individualistic culture.

When sociologists look at the median age at which people experience the various stages of the family life cycle, significant trends emerge. Consider age at first marriage. In 1890, the median age of Americans at first marriage was 26.1 years for men and 22 years for women; in the 1950s and 1960s, it reached the historic lows of 22.8 and 20.3. Then it began rising again slowly, until by 1990 it was 25.9 for men and 24.0 for women (*Statistical Abstract,* 2000). Age at first marriage tends to be somewhat higher for males than for females and is associated with educational attainment; more highly educated people are more likely to delay marriage. There is also a tendency for members of minority groups to marry later owing to the proportionately lower income levels of these populations. People with lower incomes tend to delay marriage or never marry because they lack the material means to sustain a marital relationship (Bumpass & Sweet, 1989; Cherlin, 1981; Oppenheimer, 1994). (See the Global Social Change box on page 263.)

As it passes through the family life cycle, every family experiences changes in its system of role relations. In analyzing these changes, social scientists often modify Glick's stages so as to focus more sharply on interactions within the family. An example is the set of developmental stages shown in Table 9.2 on page 264. There are major emotional challenges at each of these stages. For example, families with adolescent children must adapt to the children's growing independence. This may involve going through a stage of negotiation over such issues as money, cars, and dating. But researchers who study family life note that parents are often confused about how to interact with their adolescent children. They may assume that it is normal for adolescents to leave the family circle and to become enmeshed in their own peer groups, which often get into trouble. However, as Carol Gilligan (1987) has pointed out, adolescents also want the continuing guidance and involvement of adults.

Social changes that create shortages of jobs for young people—especially declines in less-skilled manufacturing jobs and increases in low-wage service-sector jobs—have a major impact on the transition from dependence to independence for many young adults in the United States and elsewhere in the world. The Population Reference Bureau reports that "compared with the 1970s, more young adults (ages 18 to 24) are living at home with their parents. . . . Over half (54%) of all 18- to 24-year-olds lived with their parents in 1991" (Ahlburg & De Vita, 1992, p. 9). Shortages in low-rent housing and continuing difficulties in finding work will oblige many young adults to live with their family of orientation, a situation that often frustrates family members' expectations about "normal" movement through the life course (Elder, Modell, & Parke, 1992; Rossi & Rossi, 1990).

In later stages of family life, the parents must be willing to watch their grown children take on the challenges of family formation while they themselves worry about maintaining their marital roles or caring for their own parents. The latter issue is taking on increasing importance in aging societies like the United States. And since women are still expected to be more nurturing and emotionally caring than men, it often falls to women to worry about the question of "where can Mom live?" (Hochschild, 1989). According to Elaine Brody, a leading researcher on this issue, "It's going to be primarily women for a long time. Women can go to work as much as they want, but they still see nurturing as their job." Moreover, "with many more very old people, and fewer children per family, almost every woman is going to have to take care of an aging parent or parent-in-law" (quoted in Lewin, 1989, p. A1).

As if these stages were not stressful enough, consider the complications resulting from divorce, remarriage, and the combining of children of different marriages in a new family. It is increasingly common, for example, for teenagers and young adults to have parents who are dating and marrying and for children to have four parental figures in their lives instead of two. We therefore turn to a discussion of role relations in reconstituted families.

Parenting, Stepparenting, and Social Change

It has never been easy to be a parent at any stage of the family life cycle. Keeping a family intact and raising children who are confident and capable and will become loving, capable parents themselves is a domestic miracle. For better or (sometimes) worse, most people learn to be parents by following the examples set by their own parents when they were children. Much as we may swear that when our turn comes we will improve on the way our parents performed their roles, we often find ourselves acting toward our children much as our parents acted toward us. In fact, many parents seek help from family counselors because they are shocked to find themselves doing and saying the same things to their children that their parents did and said to them (Scarf, 1995).

GLOBAL SOCIAL CHANGE

The Significance of Delayed Marriage

In rural India, it is not unusual for a girl to be married at the age of 12 or 14. If she is not married, it is likely that she has been promised in marriage to a young man from another family in the village or one nearby; such arrangements are made when the bride is extremely young. This pattern of arranged marriages is found throughout agricultural societies where the desire for children is high. It is common in many parts of Africa, Latin America, and Asia. Indeed, demographers note that an upward trend in age at first marriage in these regions would indicate that women are becoming better educated and more empowered to resist the pressure to marry at a very young age (United Nations, 1995b).

In the United States, sociologists note that changes in the median age at first marriage, and lower marriage rates overall, are indicators of global social change as well (Oppenheimer, 1994). The median age at first marriage for American women has changed dramatically from the 1950s (20.3) to the present (24.0). This has occurred because when couples find it difficult to achieve job and income security they tend to delay marriage, as they did in earlier periods of American history. However, the median figures mask even more striking differences among population age cohorts—groups of people born during different five-year periods. For example, among white women between the ages of 20 and 24 in 1985, about 50% were single; 30% of them were still single between the ages of 25 and 29. In contrast, only about 28% of white women who were in their early twenties in 1965 were single, and only about 10% of those women were still single in their late twenties. For black women, the changes are even more dramatic. Of the 1985 cohort, over 70% were single in their early twenties, and over 50% were still single between the ages of 25 and 29.

These cohort differences reveal a major trend toward delay of marriage, with many more women, especially minority women, remaining single far longer and perhaps never marrying. What influence does global social change have on these marriage patterns? Although they do not explain all of the change, global trends such as the "export" of manufacturing jobs from developed to less-developed nations mean that in the United States young men with less education find it increasingly difficult to gain enough income and job security to be attractive marriage partners. To compete well in the new global economy, men and women in older industrial nations seek to increase their education and skills, which means more years spent in universities and training programs. For those who gain less knowledge and skills, it becomes ever more difficult to find decent jobs that will help support a family. This situation is worse for young minority men than for whites because of past patterns of discrimination and the effects of growing up in lower-income families and neighborhoods (Oppenheimer, 1994; W. J. Wilson, 1996).

Throughout the world, women crave opportunities to become literate or, better yet, to become more broadly educated. Education may delay their marriage in some cases, but it also makes them better parents and empowers them to participate more fully in the development of their societies.

TABLE 9.2
Stages of the Family Life Cycle

Stage	Emotional Process	Required Changes in Family Status
1. Between families: The unattached young adult	Accepting parent–offspring separation	a. Differentiation of self in relation to family of origin b. Development of intimate peer relationships c. Establishment of self in work
2. The joining of families through marriage: The newly married couple	Commitment to new system	a. Formation of marital system b. Realignment of relationships with extended families and friends to include spouse
3. The family with young children	Accepting new generation of members into the system	a. Adjusting marital system to make space for children b. Taking on parenting roles c. Realignment of relationships with extended family to include parenting and grandparenting roles
4. The family with adolescents	Increasing flexibility of family boundaries to include children's independence	a. Shifting of parent–child relationships to permit adolescents to move in and out of system b. Refocus on midlife marital and career issues c. Beginning shift toward concerns for older generation
5. Launching children and moving on	Accepting a multitude of exits from and entries into the family system	a. Renegotiation of marital system as a dyad b. Development of adult-to-adult relationships between grown children and their parents c. Realignment of relationships to include in-laws and grandchildren d. Dealing with disabilities and death of parents (grandparents)
6. The family in later life	Accepting the shifting of generational roles	a. Maintaining own and/or couple functioning and interests in face of physiological decline; exploration of new familial and social role options b. Support for a more central role for middle generation c. Making room in the system for the wisdom and experience of the elderly; supporting the older generation without overfunctioning for them d. Dealing with loss of spouse, siblings, and other peers, and preparation for own death; experiencing life review and integration

Source: Adapted from McGoldrick & Carter, 1989.

Many other adults and children will experience major reorganizations of their families as parents divorce and remarry. When blended or stepfamilies are formed, family members must adapt to new role relations as stepchildren and stepparents, often while maintaining many of their obligations in the original family. Sharon L. Hanna, an expert on stepfamilies, notes that there are more than 20.6 million stepparents in the United States; it is estimated that almost 50% of all Americans are part of a stepfamily (Hanna, 1994).

"Stepfamilies," sociologists Frank Furstenberg and Andrew Cherlin point out, "are a curious example of an organizational merger; they join two family cultures into a single household" (1991, p. 83). This joining of family cultures also occurs in first marriages, but in those marriages the couple usually has time to work out differences before children are born. In the case of stepfamilies, the task of working out a mutually acceptable concept of the family may be more stressful than the partners anticipated.

Members of all kinds of families experience particular stress in times of economic hardship or when other social changes influence the family. Stress and change in the larger society may cause parents to feel unable to perform their roles as providers of material and emotional support. Early in the twentieth century, for example, the combination of frequent severe recessions, high rates of industrial accidents, and lack of adequate health care had a significant impact on families. Fathers were often forced to leave their families in search of work; infection often claimed the lives of mothers during childbirth. Demographers estimate that at the beginning of the twentieth century, when divorce rates were still extremely low, almost 25% of all children lost at least one parent through death before reaching the age of 15 (Uhlenberg, 1980).

By the middle of the twentieth century, improvements in health care had reduced this proportion to 5%, but divorce rates had increased to 11%, and another 6% of children were born to unmarried parents. Thus, even in the 1950s—the most stable decade for families and children—demographers estimate that 22% of children were being raised in families that had only one parent as a result of death, separation, divorce, or single parenthood (Bumpass & Sweet, 1989). Even this relative stability lasted only a decade. By the end of the 1950s, divorce rates had begun their precipitous rise, and so had rates of births to unmarried parents. By 1998, almost 33% of all births occurred out of wedlock (*Statistical Abstract,* 2000). When individuals fail to maintain their marital commitments or do not marry despite pregnancy, the results are cause for concern in any society. The difficulties faced by children in "intact" families as well as in troubled ones raise the age-old questions of how people decide whom to marry and what makes marriages last. In the next section, therefore, we examine the sociology of mate selection and marriage.

DYNAMICS OF MATE SELECTION AND MARRIAGE

"In the animal kingdom, mating involves only the two partners. For mankind, it is the joining of two enormously complex [social] systems" (McGoldrick & Carter, 1982, p. 179). We may think of mate selection and marriage as matters that affect only the partners themselves, but in reality the concerns of parents and other family members are never very far from either person's consciousness. And as we will see shortly, the values of each partner's extended family often have a significant impact on the mate-selection process.

Marriage as Exchange

People in Western cultures like to think that interpersonal attraction and love are the primary factors in explaining why a couple forms a "serious" relationship and eventually marries. But while attraction and love are clearly important factors in many marriages, social scientists point out that in all cultures the process of mate selection is carried out according to basic rules of bargaining and exchange (G. S. Becker, 1973; W. J. Goode, 1964; Levitan & Belous, 1981). Sociologists and economists who study mate selection and marriage from this exchange perspective ask who controls the marriage contract, what values each family is attempting to maximize in the contract, and how the exchange process is shaped by the society's stratification system.

Among the upper classes of China and Japan before the twentieth century, marriage transactions were controlled by the male elders of the community—with the older women often making the real decisions behind the scenes. In many societies in the Middle East, Asia, and preindustrial Europe, the man's family negotiated a "bride price" with the woman's family. This price usually consisted of valuable goods like jewelry and clothes, but in some cultures it took the form of land and cattle. Throughout much of Hindu India, in contrast, an upper-class bride's family paid a "groom price" to the man's family.

Although such norms appear to be weakening throughout the world, arranged marriages remain the customary pattern of mate selection in many societies. The following account describes factors that are often considered in arranging a marriage in modern India:

> Every Sunday one can peruse the wedding ads in the classifieds. Many people still arrange an alliance in the traditional manner—through family and friend connections. Caste is becoming less important a factor in the selection of a spouse. Replacing caste are income and type of job. The educational level of the bride-to-be is also a consideration, an asset always worth mentioning in the ad. A faculty member at a college for girls has estimated that 80–90% of the students there will enter an arranged marriage upon receiving the B.A. Perhaps 10% will continue their studies. In this way, educating a daughter is parental investment toward securing an attractive, prosperous groom. (D. Smith, 1989)

In all of these transactions, the families base their bargaining on considerations of family prestige within the community, the wealth of the two families and their ability to afford or command a given price, the beauty of the bride and the attractiveness of the groom, and so on. Different cultures may evaluate these qualities differently, but in each case the parties involved think of the coming marriage as an exchange between the two families (W. J. Goode, 1964). But do not get the idea that only selfish motives are involved in such marriages. Both families are also committing themselves to a long-standing relationship because they are exchanging their most precious products, their beloved young people. Naturally, they want the best for their children (as this is defined in their culture), and they also want a climate of mutual respect and cooperation in their future interfamily relationships.

Norms of Mate Selection

Endogamy/Exogamy. All cultures have norms that specify whether a person brought up in that culture may marry within or outside the cultural group. Marriage within the group is termed **endogamy;** marriage outside the group and its culture is termed **exogamy.** In the United States, ethnic and religious groups normally put pressure on their members to remain endogamous—that is, to choose mates from their own group. These rules tend to be especially strong for women. Among Orthodox Jews, for example, an infant is considered to have been born into the religion only if the mother is Jewish; children of mixed marriages in which the mother was not born a Jew are not considered Jewish. This conflict between Orthodox and Reform Jews over the status of children born to non-Jewish mothers—even when the mothers have converted to Judaism and the children have been raised as Jews from birth—is an example of conflict over endogamy/exogamy norms. Many African tribes have developed norms of exogamy that encourage young men to find brides in specific villages outside the village of their birth. Such marriage systems tend to promote strong bonds of kinship among villages and serve to strengthen the social cohesion of the tribe while breaking down the animosity that sometimes arises between villages within a tribe.

Homogamy. Another norm of mate selection is **homogamy,** the tendency to marry a person from a similar social background. The parents of a young woman from a wealthy family, for example, attempt to increase the chances that she will associate with young men of the same or higher social-class standing. She is encouraged to date boys from "good" families. After graduating from high school, she will be sent to an elite college or university, where the pool of eligible men is likely to include many who share her social-class background. She may surprise her parents, however, and fall in love with someone whose social-class, religious, or ethnic background is considerably different from hers. But when this happens she will invariably have based her choice on other values that are considered important in the dating and marriage market—values like outstanding talent, good looks, popularity, or sense of humor. She will argue that these values outweigh social class, especially if it seems apparent that the young man will gain upward mobility through his career. Often the couple will marry and not worry about his lower social-class background. On the other hand, "the untalented, homely, poor man may aspire to a bride with highly desirable qualities, but he cannot offer enough to induce either her or her family to choose him, for they can find a groom with more highly valued qualities" (W. J. Goode, 1964, p. 33).

Homogamy in mate selection generally serves to maintain the separateness of religious groups. Because the Census Bureau does not collect systematic data on religious preferences, it is extremely difficult to obtain accurate data on religious intermarriage. Yet sociologists and religious leaders agree that although parents continue to encourage their children to marry within their religion, there is a trend away from religious homogamy, particularly among Protestants and Catholics (Scanzoni & Scanzoni, 1976). This trend may affect rates of divorce and separation. Recent research shows that families that maintain an active religious life tend to have lower divorce rates than those that do not participate in religious worship together (National Commission on America's Urban Families, 1993). Similarly, a study by Howard Weinberg (1994) found that shared religion has the strongest effect on the likelihood that couples will be able to overcome a period of separation and achieve a successful reconciliation.

The norm of homogamy also applies to interracial marriage. Many states had laws prohibiting such marriages before 1967, when the U.S. Supreme Court struck them down as unconstitutional. After that decision, marriages between blacks and whites began to increase rapidly, although they remain a very small proportion of all marriages. In 1960, there were about 51,000 black–white couples; by 1999, that number had risen to 307,000, a sixfold increase. Intermarriage of Hispanic and non-Hispanic whites also increased during the same period, from about half a million marriages in 1960 to over 1.6 million in 1998 (*Statistical Abstract,* 2000). Both of these trends are signs that more people are choosing to marry across racial and ethnic lines, but since the vast majority of marriages are within racial categories, it is clear that the norm of racial homogamy remains strong in the United States.

Marriage and Divorce

More than any other ritual signifying a major change in status, a wedding is a joyous occasion. Two people are legally and symbolically joined before their kin and friends. It is expected that their honeymoon will be pleasant and that they will live happily ever after. But about 20% of first marriages end in annulment or divorce within three years (Bogue, 1985; Furstenberg & Cherlin, 1991). Of course, divorce can occur at any time in the family life cycle, but the early years of family formation are the most difficult for the couple because each partner experiences new stresses that arise from the need to adjust to a complex set of new relationships. As Monica McGoldrick and Elizabeth A. Carter (1982) point out, "Marriage requires that a couple renegotiate a myriad of personal issues that they have previously defined for themselves or that were defined by their parents, from when to sleep, have sex, or fight, to how to celebrate holidays and where and how to live, work, and spend vacations" (p. 178). For people who were married before, these negotiations can involve former spouses and shared children, resulting in added stress for the new couple.

In the United States and other Western societies, the rate of divorce rose sharply after World War II, accelerated even more dramatically during the 1960s and 1970s, and has decreased only slightly since then. These statistics often lead sociologists to proclaim that there is an "epidemic" of divorce in the United States. But demographer Donald Bogue has concluded that "the divorce epidemic is not being created by today's younger generation. It has been created by today's population aged 30 or more, who married in the 1960s and before" (1985, p. 190). This generation was noted for its search for self-realization, often at the expense of intimate family relationships. It is not yet clear whether subsequent generations, who appear to be somewhat more pragmatic, will continue this trend. If they do, we can expect high divorce rates to continue.

Most states in the United States now have some form of no-fault divorce that reduces the stigma of divorce by making moral issues like infidelity less relevant than issues of child custody and division of property. While the growing acceptance of divorce helps account for why divorce rates are so much higher in the United States than in other nations with highly educated populations, Americans today also place a higher value on successful marriage than their parents may have. This means that they often divorce in the expectation of forming another, more satisfying and mutually sustaining relationship (Ahlburg & De Vita, 1992).

Trial Marriage. In the 1980s, it was widely believed that the practice of "trial marriage," or cohabitation before marriage, would result in greater marital stability: Couples who lived together before marriage would gain greater mutual understanding and a realistic view of marital commitment, and this would result in a lower divorce rate among such couples after they actually married. However, by the 1990s it had become evident that these expectations were unfounded; in fact, the divorce rate among couples who had lived together before marriage was actually higher than the rate for couples who had not done so. Within 10 years of the wedding, 38% of those who had lived together before marriage had divorced, compared to 27% of those who had married without cohabiting beforehand.

On the basis of an analysis of data from a federal government survey of more than 13,000 individuals, Larry Bumpass and James Sweet (cited in Barringer, 1989) concluded that couples who cohabit before marriage are generally more willing to accept divorce as a solution to marital problems. They also found that such couples are less likely to be subject to family pressure to continue a marriage that is unhappy or unsatisfactory. In addition, cohabitation has become a predictable part of the family life cycle, not only before marriage but in the interval between divorce and remarriage. Once again, these changes in families mean that family life has become ever more variable and uncertain for children. Thus, while some advocate a return to traditional family norms and values, others urge greater tolerance for a variety of family types and accommodation to their needs. This conflict is most evident in discussions of same-sex relationships.

Gay and Lesbian Relationships. In 1995, the Disney Corporation, long viewed as a defender of traditional family values, shocked many conservative

Americans by announcing that it would provide spousal benefits (health insurance, pension rights, etc.) for its employees who were part of a same-sex relationship or marriage. In taking this action, Disney was simply acknowledging that norms are changing and it is becoming less acceptable for companies to impose their definitions of the family on their employees.

Margaret L. Andersen, a noted expert on issues of sex and gender, makes this observation:

> Sociological and popular understanding of gay and lesbian relationships has been greatly distorted by the false presumption that only heterosexual relationships are normal ways of expressing sexual intimacy and love. We live in a culture that tends to categorize people into polar opposites: men and women, black and white, gay and straight. (1993, p. 57)

However, an unknown but significant number of people do not conform to conventional heterosexual behaviors. They may have felt a preference for intimacy with members of the same sex very early in life without knowing how to identify their feelings. Often they went through long periods of confusion and self-doubt before learning to understand and accept their homosexuality.

Our culture tends to make homosexual love into a form of deviance. Homosexuals, in consequence, are victims of homophobia of all descriptions (from individual acts of insensitivity or aggression to antigay legislation), which places additional burdens on homosexual relationships (Meyer, 1990). Because homosexual couples are not welcome or do not feel comfortable in many social situations, they often feel a need to keep their relationship secret. This deprives homosexual men and women of the social approval that heterosexual couples enjoy—and take for granted—in their peer and work relations.

In large measure owing to the stigma attached to homosexual relationships, there are no accurate census counts of homosexual families, nor is there much research about life in gay and lesbian households. The research that has been done, however, shows that fewer children live in male homosexual families than in female homosexual ones, mainly because women tend to win custody of their children in divorce cases. Recent research on gay fathers shows that compared with heterosexual fathers they tend to be better at setting limits on their children's behavior, are somewhat more nurturing, and place more emphasis on verbal communication with their children—features of family life that are often the province of the mother in a heterosexual family (Bigner & Bozett, 1989).

The Impact of Divorce. In the mid-twentieth century, it was still considered almost impossible for a divorced man to run for the presidency. Ronald Reagan was the first president who had previously been divorced. Senators Robert Dole and Phil Gramm, candidates in the 1996 presidential election, had each been divorced. Divorce has become much more prevalent than it was earlier in the century. Nevertheless, it remains a significant event in people's lives—as significant as marriage itself. But unlike a wedding, a divorce is not a happy event. Although some divorces turn out well for both partners, the majority do not.

Research on divorce has shown that many of its most disruptive consequences are due to its economic impact. Women suffer an average decline of about 30% in their income in the year following separation, while men experience a 15% increase. In fact, the majority of women who apply for various forms of public support do so because they have recently experienced a drastic decline in income due to divorce, separation, or abandonment. In addition, almost 40% of divorced mothers (and the children in their custody) move within the first year after divorce, and another 20% move after a year, a rate far higher than that for married couples (Furstenberg & Cherlin, 1991; McLanahan, 1984). And as if the breakup of their families were not stressful enough, many children also experience the loss of friends and familiar neighborhoods.

Beyond the material effects of divorce, there are the longer-term effects on family roles and the feelings of family members. In an important longitudinal study, Wallerstein and Blakeslee (1989) tracked 60 families with a total of 131 children for a period of 10 to 15 years after divorce. Both parents and children were interviewed at regular intervals. The data from those interviews show that the turmoil and stress of divorce may continue for a year or more. Many divorced adults continue to feel angry, humiliated, and rejected as much as 18 months later, and the children of divorced parents tend to exhibit a variety of psychological problems. Moreover, both men and women have a diminished capacity for parenting after divorce. They spend less time with their children, provide less discipline, and are less sensitive to their needs. Even a decade after the divorce, the parents may be chronically disorganized and unable to meet the challenges of parenting. Instead, they come to depend on their children to help them cope with the demands of their own lives, thereby producing an "overburdened child"—one who, in addition to handling the normal stresses of childhood, also must help his or her parent ward off depression (Wallerstein & Blakeslee, 1989).

There is a silver lining to the dark cloud of divorce, however. Since so many adults who are now marrying for the first time come from families that have experienced divorce, they are likely to take more time in selecting their mates in an effort to make sure that their choice is best for both partners; in addition, they try to become economically secure before marrying, thereby eliminating a major source of stress in a new marriage (Blumstein & Schwartz, 1983). If this is true, the recent modest downturn in the divorce rate may be expected to continue in the future. In any case, the data on the effects of divorce on adults and children suggest that societies need to do more to ease the stress experienced by young families—for example, by providing more day-care facilities, establishing more flexible work schedules, and offering opportunities for family leave.

PERSPECTIVES ON THE FAMILY

High divorce rates do not indicate that the family is about to disappear as an institution in modern societies. Even though many marriages end in divorce and increasing numbers of young adults postpone marriage or decide not to marry at all, the large majority do marry, and the majority of those who divorce will eventually remarry. Far from disappearing, the family is adapting to new social values and to changes in other institutions, especially economic ones. In this section, we review some current research on family roles and relationships from the standpoint of the three basic sociological perspectives.

The Interactionist Perspective

Interactions within the family cover a wide range of emotions and may take very different forms in different families. Families laugh and play together, work together, argue and bicker, and so on. All of these aspects of family interaction are important, but frequently it is the arguing and bickering that drives family members apart. Studies of family interaction therefore often focus on the sources of tension and conflict within the family.

Problems of family interaction can stem from a variety of sources. Often problems arise in connection with critical life stages or events, such as the loss of a job or the time when adolescent children begin to assert their independence in ways that threaten established family roles and arrangements. And conflicts often occur because of the particular ways in which the family's experiences are shaped by larger social structures. In the armed forces, for example, families often experience severe stress because of frequent moves from one base to another (J. A. Shaw, 1979). In many cases, such moves draw family members closer together, but in other cases family interactions are marked by tension because children resent their inability to maintain stable friendships.

The context within which family life occurs can affect family interactions in other ways as well. At the lower levels of a society's stratification system, for example, money (or the lack of it) is often a source of conflict between parents or between parents and children. But the rich are by no means immune to problems of family interaction. Because they do not have to be concerned about the need to earn a living as adults and because their parents can satisfy any desires they may have, the children of the very rich often develop a sadness that resembles anomie (Wixen, 1979). Their lack of clear goals, which sometimes expresses itself in a compulsion to make extravagant purchases, may give rise to conflict between them and their parents.

All families need to resolve a contradiction inherent in the institution of the family: the need to maintain the individuality of each member while providing love and support for him or her within a set of interdependent relationships. Many families never succeed in developing ways of encouraging each member to realize his or her full potential within the context of family life. And research shows that the core problem is usually the failure of the adult couple, even in intact families, to understand and develop their own relationship. Such a couple may become what John F. Cuber and Peggy B. Haroff (1980) term "conflict-habituated" or "devitalized." Couples of the first type have evolved ways of expressing their hostility toward each other through elaborate patterns of conflict that persist over many years. In contrast, the devitalized or "empty-shell" marriage may have begun with love and shared interests, but the partners have not grown as a couple and have drifted apart emotionally. Although each has the habit of being with the other—a habit that may be strongly

SOCIOLOGY AND SOCIAL JUSTICE

More than for blended families or even single-parent families, gay couples' desire to have families, adopt children, and live peacefully in communities of their choice is a fundamental issue of social justice. The status of gay and lesbian families is a "wedge" issue in American life. People with different opinions regarding the morality of homosexuality are driven apart by the political question of whether gay and lesbian marriages should be recognized as legal and thus qualify for full rights under marital and family laws. Ideological conservatives like William Bennett (1996), an outspoken critic of homosexual relationships, argue that broadening the definition of marriage to include same-sex unions would stretch it almost beyond recognition. Marriage, says Bennett, "is not an arbitrary construct which can be redefined simply by those who lay claim to it." On the other hand, defenders of homosexual marriage, such as Andrew Sullivan (1996), former editor of *The New Republic,* believe that permitting same-sex marriages would have a positive moral influence—"It would promote monogamy and the disciplines of family life for those cast on the margins."

You may have strong opinions on this issue, or you may be confused about what to think and how society should deal with the issue. How can a sociological perspective on the controversy help you make up your mind?

First we must look at the facts. There are many gay and lesbian people in this and other societies. It seems increasingly likely that homosexuality is not a mere lifestyle choice but a deeply felt aspect of a person's social being that cannot be changed easily, if at all. Moreover, it is clear that many gay and lesbian relationships are in fact marriages, judging from the couple's love and commitment to each other. How, then, can one limit membership in such fundamental institutions as marriage and the family only to heterosexuals?

But there are other sociological facts to consider. Clearly, there are parts of the nation where the vast majority of the population is bitterly opposed to homosexual marriage. Must norms that are morally

supported by the norms of a particular ethnic or religious community—neither partner is satisfied by the relationship. At the same time, neither feels that he or she can do anything to change the situation. Thus, the conflicts that might have produced change are reduced to indifference.

Today social scientists who study family interaction must deal with family structures that are more complex than the traditional nuclear family. Divorce and remarriage create many situations in which children have numerous sets of parental figures—parents and stepparents, grandparents and surrogate grandparents, and so on. These changes in family form result in new patterns of family interaction. For example, in a study of 2,000 children conducted over a five-year period, sociologist Frank Furstenberg found that 52% of children raised by their mothers do not see their fathers at all, not only because the fathers are absent by choice but because the opportunities for visits decrease as parents remarry or move

offensive to antihomosexual majorities be forced upon them by the federal government? A sociologist would argue that such an action would only engender further hatred and perhaps violence. On the other hand, there are also parts of the nation where homosexual unions are more common, and they have gained support in the name of tolerance and fairness. Many states and municipalities are likely to vote in favor of legal recognition of homosexual unions if given the opportunity. Most sociologists would argue, therefore, that as states and localities decide on the issue through democratic means, the norms will shift toward greater recognition of the rights of gay and lesbian marriages. ■

away (cited in Aspaklaria, 1985). This produces situations in which parents strive to maintain long-distance relationships with their children and occasionally have brief, intense visits with them. Although this specific type of relationship may not be the most desirable, it appears that if parents and children can express love and affection even when the parents are divorced, the children's ability to feel good about themselves and to love others in their turn may not be impaired.

The Conflict Perspective

The conflict perspective on families and family interactions assumes that "social conflict is a basic element of human social life." Conflict exists "within all types of social interaction, and at all levels of social organization. This is as true of the family as it is of any other type of social entity" (Farrington & Chertok, 1993, p. 368). From this perspective, one can observe actual family interactions (the micro level) and ask why conflict occurs, what the issues are, and how they are resolved. At a more macro level, one can ask how conditions of inequality and class conflict influence actual families or the laws and policies governing the family as a central institution in society.

Sociologists who study poverty argue that most families are poor for the obvious reason that they do not earn or receive enough income. But critics of liberal social-welfare policies argue that self-destructive behaviors of family members themselves, such as dropping out of school, drug use, and out-of-wedlock childbearing, are causes of poverty. Social-welfare policies that are too generous or take the form of entitlements that do not have to be earned are said to encourage behaviors that cause poverty.

Social scientists who study class conflict and poverty policies, notably Herbert Gans (1995), Christopher Jencks (1993), and Michael Katz (1983), point out that early in the twentieth century a large proportion of impoverished families were composed of elderly couples who did not have Social Security or Medicaid, two programs that have drastically reduced poverty among elderly Americans. Yet at the time that these policies were being debated, many members of the upper classes opposed them, viewing them as undemocratic transfers of wealth to less-deserving members of society. The situation today is similar, as members of the upper classes oppose adequate funding for day care and other policies that can help move adult family members from dependence on public support to employment at decent jobs.

The Functionalist Perspective

From a functionalist perspective, the family evolves in both form and function in response to changes in the larger social environment. As societies undergo such major changes as industrialization and urbanization, the family must adapt to the effects of those changes. Functionalist theorists have called attention to the loss of family functions that occurs as other social institutions like schools, corporations, and social-welfare agencies perform functions that were previously reserved for the family. We have discussed numerous examples of this trend in earlier chapters—the tendency of families to have fewer children as the demand for agricultural labor decreased, the changing composition of households as people were required to seek work away from their families of orientation, the increasing number of dual-income households, and so on. The functionalist explanation of these changes is that as the division of labor becomes more complex and as new, more specialized institutions arise, the family too must become a more specialized institution. Thus, modern families no longer perform certain functions that used to be within their domain, but they do play an increasingly vital part in early-childhood socialization, in the emotional lives of their members, and in preparing older children for adult roles in the economic institutions of industrial societies (Parsons & Bales, 1955).

TECHNOLOGY AND SOCIAL CHANGE

The family as a social institution is immensely influenced by technological change. Think of the impact of electricity and all the labor-saving devices made possible by electrification. Washing machines, sewing machines, vacuum cleaners, and refrigerators, to name only a few, revolutionized the way home labor was carried out. Much of the drudgery was taken out of housework. Women had more disposable time to seek education and enter the labor force. Men had fewer excuses for shirking housework (although most continued to do so). The family car made the family vacation a significant experience of childhood. The telephone made it far easier for families to stay in touch over great distances, just as today many families use the cellular telephone to coordinate pick-ups and gatherings even when they are out of the home (McGaw, 1987).

But technology has problematic aspects as well. Television, one of the most powerful innovations of the second half of the twentieth century, continues to have vast impacts on family life. The medium is often used as an "electronic baby-sitter," which means that children spend time immersed in television programs at the expense of family interactions that might better develop their emotional and verbal skills. The Internet can be a valuable learning tool, but it can also be a source of family conflict when children download information their parents disapprove of. Overall, technologies tend to increase the flexibility and adaptability of family life so that individual members can participate in many activities outside the family and still fulfill family expectations. But within individual families it is also true that particular technologies, like the family car or the Internet, may become contentious issues and areas of family conflict.

SUMMARY

In all known societies, almost everyone is socialized within a network of family rights and obligations that are known as family role relations. A *family* is a group of people related by blood, marriage, or adoption; the role relations among family members are known as *kinship* relations. A *nuclear family* consists of two or more people, related by consanguineous ties or by adoption, who share a household. The nuclear family in which one is born and socialized is one's *family of orientation,* and the nuclear family one forms through marriage or cohabitation is one's *family of procreation.* Changes in human societies over the past 300 years have brought enormous changes to family structure and roles.

The typical stages of the family life cycle are family formation, start of childbearing, end of childbearing, "empty nest," and family dissolution. As it passes through this cycle, every family experiences changes in its system of role relations.

In all cultures, the process of mate selection is carried out according to basic rules of bargaining and exchange. All cultures also have norms that specify whether a person brought up in that culture may marry within the cultural group *(endogamy)* or outside it *(exogamy).* In societies in which marriages are based on attraction and love, individuals tend to marry people who are similar to themselves in social background. This tendency, known as *homogamy,* generally serves to reproduce the society's system of social-class stratification in the next generation and to maintain the separateness of religious and racial groups.

In the United States and other Western societies, the rate of divorce has risen sharply since World War II. Both men and women have a diminished capacity for parenting after divorce and may come to depend on their children to help them cope with the demands of their own lives.

From the interactionist perspective, a basic contradiction inherent in the family is the need to maintain the individuality of each member while providing love and support for him or her within a set of interdependent relationships. From a conflict perspective, changes in the family as an institution cannot occur without conflict both within the family and between the family and other institutions. Functionalist theorists call attention to the loss of family functions that occurs as other social institutions assume functions that were previously reserved for the family. However, modern families play a vital part in early-childhood socialization, in the emotional lives of their members, and in preparing older children for adult roles.

GLOSSARY

Family: A group of people related by blood, marriage, or adoption. **(255)**

Kinship: The role relations among people who consider themselves to be related by blood, marriage, or adoption. **(255)**

Nuclear family: Two or more people related by blood, marriage, or adoption who share a household. **(255)**

Family of orientation: The nuclear family in which a person is born and raised. **(255)**

Family of procreation: The nuclear family a person forms through marriage or cohabitation. **(255)**

Extended family: An individual's nuclear family plus the nuclear families of his or her blood relatives. **(255)**

Endogamy: A norm specifying that a person brought up in a particular culture may marry within the cultural group. **(266)**

Exogamy: A norm specifying that a person brought up in a particular culture may marry outside the cultural group. **(266)**

Homogamy: The tendency to marry a person who is similar to oneself in social background. **(266)**

QUESTIONS FOR THOUGHT AND DISCUSSION

1. Does recognition of the greater diversity of family types in the contemporary world imply acceptance of the single-parent family? What is the difference between empirical research on family structure and the politics of "family values"?

2. Does research on the consequences of divorce make you feel that divorce is more or less problematic as a social phenomenon than it has been in the past? What are some ways in which the divorce rate might be decreased if certain kinds of social change were to occur (e.g., in the economy or in the socialization of children)?

DIGGING DEEPER

Books

Divided Families: What Happens to Children When Families Part (Frank F. Furstenberg, Jr., & Andrew J. Cherlin; Harvard University Press, 1991). An assessment of the impact of divorce on the emotional, educational, and economic lives of children who experience the divorce of their parents.

The Family Crucible (Augustus Y. Napier & Carl Whitaker; Bantam, 1980). An introduction to the dynamics of family interaction and family therapy.

The Reconstruction of Family Policy (Elaine A. Anderson & Richard C. Hula, eds.; Greenwood Press, 1991). A series of original essays on the problems of developing family policies, with special emphasis on conflicting interests of parents, children, and society. Includes good material on latchkey children and after-school care, welfare policies, and problems of the elderly in families.

The Changing American Family (United States Council of Economic Advisors; Government Printing Office, 2000). Contains the latest facts about changes and trends in family composition, earnings, education, and a host of other important indicators of the health of the family as a social institution.

Journals

Journal of Marriage and the Family. Published quarterly by the National Council on Family Relations, Minneapolis, Minnesota. The best journal on family research in the United States.

Other Sources

Vital Statistics of the United States, Vol. III: Marriage and Divorce. An annual compilation that presents a complete count of marriages and divorces in the preceding year. Data are broken down by age, race, previous marital status, and other characteristics.

Statistical Handbook on the American Family (Bruce A. Chadwick & Tim B. Heaton; Oryx Press, 1999). Includes excellent survey material from various sources on family roles, equality and inequality, and family conflict.

EXPLORING SOCIOLOGY ON THE INTERNET

American Sociological Association, Section on the Family
www.asanet.org/family.htm

Contains references to sociologists doing original research on the family.

Administration for Children and Families of the U.S. Government
www.rcp.dhhs.gov

A rich mine of official information about family stability and change; suggests many other useful sites to browse for research on the family in the United States and elsewhere.

Civic Practices Network (CPN)
www.cpn.org/

A collaborative and nonpartisan project dedicated to building a movement for "new citizenship" and "civic revitalization." Its Web site features research and discussion on how to strengthen the two-parent family.

Anthropology Department at the University of Manitoba
www.umanitoba.ca/anthropology/kintitle.html

Offers "Kinship and Social Organization," an online interactive tutorial that is engaging and informative and will reinforce or greatly expand your knowledge of family systems.

Children's Defense Fund
www.tmn.com/cdf/index.html

An organization that pays "particular attention to the needs of poor and minority children and those with disabilities." Includes a section of "Facts & Figures" that presents updated information on the costs of raising children as well as ongoing demographic data for different ethnic and racial groups.

Legal Net
www.legal.net/family.htm

Offers an overview of family law as it is changing in the United States.

Trinity College
www.trinity.edu/departments/soc_anthro/index.html

A good place to get help with kinship diagrams and related subjects dealing with family social structure.

U.S. Census Bureau
www.census.gov/population/socdemo/

Offers an array of statistical information, including information about the estimated median age at first marriage, the number of interracial couples, the number of young adults living at home, and language preferences within homes. Much of this data is gathered from the U.S. Census Bureau during the data collection cycles.

Chapter 10

Religion

Why is it difficult for people of different religions to realize their shared quest for peace and understanding?

Religion in Society

Defining Religion

The Power of Faith

Classification of Religious Beliefs

Simple Supernaturalism

Animism

Theism

Abstract Ideals

Civil Religion

Religion and Social Change

Forms of Religious Organization

Churches and Sects

Denominations

Cults

Trends in Religion

Religiosity

Fundamentalism

Technology and Social Change

Urvashi Butalia is one of India's foremost intellectuals and a founder of the nation's leading feminist publishing company. Her most recent book, *The Other Side of Silence* (2000), is about one of the darkest episodes in the tormented history of world religions and world politics. The book is a sociological history of Partition, the sudden and bloody separation of Muslim Pakistan from Hindu India in 1947, just as British colonial rule of the Indian subcontinent was ending. Partition resulted in two separate nations, Pakistan and India. A million people were killed and 12 million lost their homes in the violence that accompanied the political separation of the two nations. Partition also caused the largest mass migration in history—and the most rapid, since it took only a few months (Kapur, 2000).

As violent as the "ethnic cleansing" in Bosnia and Kosovo were more recently, and as bloody as the interminable conflict is between Palestinian Muslims and Israeli Jews, Partition was far worse. Nor was it a complete separation. There remain millions of Muslims in India, and entire provinces, such as Kashmir, remain divided and perennially on the brink of precipitating war between the world's two newest nuclear powers, India and Pakistan.

These Muslim women in Kashmir, a mountain province claimed by both India and Pakistan, are protesting against human rights abuses. Strife between Hindu and Muslim populations in Kashmir and elsewhere in India is a bitter reminder that Partition did not solve the problems of religious animosity.

Butalia applies her sociological imagination to coaxing stories out of Muslim and Hindu women who experienced Partition, witnessed the deaths of loved ones in the religious warfare, and struggled for the rest of their lives to raise families despite the hatreds surrounding them. She spent over a decade speaking to the "small actors and the bit players," focusing especially on women of the lower castes and people who were children during Partition. They did not foment the religious and nationalistic fervor that resulted in Partition, but they were swept up in the events and acted according to what they thought their religions demanded of them.

Throughout the world people are trying to come to terms with conflicts that often have their origins in religious differences within large populations. Urvashi Butalia's work is one among many examples of efforts to look squarely at the beauty of religion and to understand how even such a beautiful and spiritual human experience can stimulate the kind of "us versus them" distinctions that result in rancorous conflict and even brutal events like Partition.

In this chapter, we begin with definitions of religion and religious experience. We also introduce some of the central sociological questions about religion and review some data that reveal how people feel about their religious beliefs and how they practice them. We then turn to a discussion of religion and social change and end with a discussion of some important aspects of contemporary religion as practiced in the United States and elsewhere in the world. ■

RELIGION IN SOCIETY

Religion, one of the oldest human institutions, is also among the most changeable and complex. On the one hand, religion expresses our deepest yearnings for spiritual enlightenment and understanding; on the other, conflicts over religious beliefs and practices have given rise to persecution, wars, and much human suffering, as can be seen, for example, in the violence between Muslim and Serbian Orthodox Christians in the war-torn Balkans or the interminable conflict between Jews and Palestinian Muslims in the Middle East. Little wonder, therefore, that the founders of sociology—including Émile Durkheim, Karl Marx, and Max Weber—all wrote extensively about the power of religion and the great changes religion has undergone as societies have evolved.

Defining Religion

Religion is not easy to define. One could begin with a definition that has a concept of God as its core, but many religions do not have a clear concept of God. One could define religion in terms of the emotions of spirituality, oneness with nature, awe, mystery, and many other feelings, but that would not be a very helpful definition because emotions are extremely difficult to capture in words. Taking another tack, one might think in terms of organized religion—churches, congregations, ministers, rabbis, and so on—but clearly the organizational aspect of religion is just one of its many dimensions. It is frustrating to have to work so hard to define something that seems so commonplace, yet without a good working definition of religion it is impossible to compare different religions or refer to particular aspects of religion.

We can approach a working definition of religion by saying that **religion** is any set of coherent answers to the dilemmas of human existence that makes the world meaningful. From this point of view, religion is how human beings express their feelings about such ultimate concerns as sickness or death or the meaning of human life. Almost all religions involve their adherents in a system of beliefs and practices that express devotion to the supernatural and foster deep feelings of spirituality. In this sense, we say that religion functions to meet the spiritual needs of individuals.

But religion has also been defined in terms of its social function: It is a system of beliefs and rituals that serves to bind people together through shared worship, thereby creating a social group. **Rituals** are formal patterns of activity that symbolically express a set of shared meanings; in the case of religious rituals such as baptism or communion, the shared meanings are sacred. The term **sacred** refers to phenomena that are regarded as extraordinary, transcendent, and outside the everyday course of events—that is, supernatural. That which is sacred may be represented by a wide variety of symbols, such as a god or set of gods; a holy person such as the Buddha; various revered writings like the Bible, the Torah, or the Koran; holy objects such as the cross or the star of David; holy cities like Jerusalem or Mecca; and much else. The term **profane** refers to all phenomena that are not sacred.

The Power of Faith

Until comparatively recent times, religion dominated the cultural life of human societies. Activities that are now performed by other cultural institutions, particularly education, art, and the media, used to be the province of religious leaders and organizations. In hunting-and-gathering bands and in many tribal societies, the holy person, or shaman, was also the teacher and communicator of the society's beliefs and values. In early agrarian societies, the priesthood was a powerful force; only the priests were literate and, hence, able to interpret and preserve the society's sacred texts, which represented the culture's most strongly held values and norms. For example, in ancient Egypt, where the pharaoh was worshipped as a god, his organization of regional and local priests controlled the entire society.

Today religion continues to play an important part in the lives of people throughout the world even though the influence of organized religions is diminishing in many societies. In the United States, for example, polls routinely find that over 90% of Americans believe in God. The strength of religious attitudes and the influence of some religions can also be seen in the conflict over abortion, which plays such a prominent role in American politics, and in the controversies generated by Christian fundamentalists who believe in the literal interpretation of the Bible and may therefore deny the validity of evolutionary theory. Outside the United States, the Islamic world is torn by religious strife between liberals and fundamentalists, and Lebanon and Northern Ireland remain deeply divided owing largely to conflict between Protestants and Catholics or between Christians and Muslims.

At the same time that religion is a source of division and conflict, it can also be a force for healing social problems and moving masses of people toward greater insight into their common humanity. This occurs at the micro level of interaction, for example, in groups like Alcoholics Anonymous, in which spirituality is an essential part of the recovery program. At the macro level, the power of faith can be seen in impoverished rural and urban communities throughout Latin America. In those communities, Catholic church leaders, parish priests, and lay parishioners have embraced the ideals of a "social gospel" that seeks the liberation of believers from poverty and oppression. These "base Christian communities," as they are often called, with their "liberation theology," have become a powerful force in the movement for social justice and other far-reaching changes in their societies (Tabb, 1986).

Secularization and Its Limits. Despite the great power of spirituality, since medieval times the traditional dominance of religion in many spheres of life has been greatly reduced. The process by which this has occurred is termed **secularization.** This process, according to Robert A. Nisbet (1970), "results in . . . respect for values of utility rather than of sacredness alone, control of the environment rather than passive submission to it, and, in some ways most importantly, concern with man's present welfare on this earth rather than his supposed immortal relation to the gods" (p. 388). Secularization usually accompanies the increasing differentiation of cultural institutions—that is, the separation of other institutions from religion. In Europe during the Middle Ages, for example, there were no schools separate from the church. The state, too, was thought of as encompassed by the church or at least as legitimated by the official state religion and church organization. Laws and courts were guided by religious doctrine, and clerical law could often be as important as civil law—indeed, to be tried as a heretic often meant torture and death. Churches engaged in large-scale economic activity, owned much land and property, and often mounted their own armed forces.

The Renaissance, the Enlightenment, and the revolutions of the eighteenth and nineteenth centuries all speeded the process of differentiation in which schools, science, laws, courts, and other institutions gained independence from religious control. However, this process has not occurred at the same pace throughout the world. For example, the removal of education from the control of religious institutions has occurred more slowly in some societies than in others. In Eastern Europe, all education was controlled by the state until very recently. In most Western European and American countries, there are religious schools, but these are separate from and overshadowed by the state-run educational system (van den Berghe, 1975). The emergence of new cultural institutions and the weakening of the influence of religion does not result in complete secularization. People who are free to determine their own religious beliefs and practices may attend church less often or not at all, but total secularization does not occur (Finke & Stark, 1992). Moreover, in almost every society that has experienced secularization one can find examples of religious revival. Indeed, modern communication technologies, especially television, have contributed immensely to the revival of interest in religion, as witnessed by the popularity of "televangelists" like Billy Graham and Pat Robertson (Gutwirth, 1999).

CLASSIFICATION OF RELIGIOUS BELIEFS

The religions practiced throughout the world today vary from belief in magic and supernatural spirits to complicated ideas of God and saints, as well as secular religions in which there is faith but not God. With such a wide range of religious beliefs and practices to consider, it is useful to classify them in a systematic way. One often-used system classifies religions according to their central belief. In this scheme, the multiplicity of religious forms is reduced to a more manageable list consisting of five major types: simple supernaturalism, animism, theism, abstract ideals, and civil religion. (See the study chart on page 281.) In this section, we describe each type briefly. Be warned, though, that not all religions fit neatly into these basic categories.

Simple Supernaturalism

In less complex and rather isolated societies, people may believe in a great force or spirit, but they may not have a well-defined concept of God or a set of rituals involving God. Studies by anthropologists have found that some isolated peoples—for example, South Pacific island cultures and Eskimo tribes—believe strongly in the power of a supernatural force but do not attempt to embody that force in a visualized

STUDY CHART

Forms of Religion

Form	Description	Example
Simple Supernaturalism	A form of religion in which there is no discontinuity between the world of the senses and the supernatural; all natural phenomena are part of a single force	Some Inuit Eskimo cultures
Animism	A form of religion in which all forms of life and all aspects of the earth are inhabited by gods or supernatural powers	Native American culture; some African tribal cultures
Theism	A form of religion in which gods are conceived of as separate from humans and from other living things on the earth, although the gods are in some way responsible for the creation of humans and for their fate	Many religions of the ancient empires
Polytheism	A form of theism in which there are numerous gods, all of whom occupy themselves with some aspect of the universe and of human life	The pantheon of gods of the ancient Greeks and Romans
Monotheism	A form of theism that is centered on belief in a single, all-powerful God who determines human fate and can be addressed through prayer	Christianity, Islam, Judaism
Abstract Ideals	A form of religion that is centered on an abstract ideal of spirituality and human behavior	Buddhism, Confucianism
Civil Religion	A collection of beliefs, and rituals for communicating those beliefs, that exists outside religious institutions	Marxism-Leninism; some versions of humanism

conception of God. In this form of religion, called **simple supernaturalism,** there is no discontinuity between the world of the senses and the supernatural; all natural phenomena are part of a single force. Consider these remarks by an Inuit Eskimo:

> When I was small I knew a man who came from the polar bears. He had a low voice and was big. That man knew when he was a cub and his bear mother was bringing him to the land from the ocean. He remembered it. (quoted in Steltzer, 1982, p. 111)

Animism

More common among hunting-and-gathering societies is a form of religion termed **animism,** in which all forms of life and all aspects of the earth are inhabited by gods or supernatural powers. Most of the indigenous peoples of the Western Hemisphere were animists, and so were many of the tribal peoples of Africa before the European conquests. Europeans almost invariably branded American Indians "heathens and barbarians" because, among other things, the Indians believed that "people journeyed into supernatural realms and returned, animals conversed with each other and humans, and the spirits of rocks and trees had to be placated" (Jennings, 1975, p. 48). The same can be said of European attitudes toward African religions. Determined to subjugate nature and make the earth yield more wealth for new populations, the Europeans could not appreciate the meanings of animism for people who lived more closely in touch with nature.

Yet if one takes some time to read about the perceptions of animistic religions, it becomes clear that they contain much wisdom for our beleaguered planet. In one beautifully written account, an Oglala Sioux medicine man, Black Elk, speaks "the story of all life that is holy and is good to tell, and of us two-leggeds sharing in it with the four-leggeds and the wings of the air and all green things; for these are children of one mother and their father is one Spirit" (Neihardt, 1959/1932, p. 1). Black Elk's prayer continues:

Muslim prayer occurs at designated times of the day and with essentially the same rituals throughout the Islamic world.

> Grandfather, Great Spirit, lean close to the earth that you may hear the voice I send. You towards where the sun goes down, behold me; thunder Beings, behold me! You where the White Giant lives in power, behold me! You where the sun shines continually, whence come the day-break star and the day, behold me! You in the depths of the heavens, an eagle of power, behold! And you, Mother Earth, the only Mother, you who have shown mercy to your children! (p. 5)

Traces of animism can also be seen in the religious beliefs of the ancient Egyptians, Greeks, and Romans. The Greeks, for example, spoke of naiads inhabiting rivers and springs and of dryads inhabiting forests. These varieties of nymphs were believed to be part of the natural environment in which they dwelled, but sometimes they took on semihuman qualities. They thus bridged the gap between a quasi-animistic religion and the more familiar theistic systems that evolved in Greece and Rome.

Theism

Religions whose central belief is **theism** usually conceive of gods as separate from humans and from other living things on the earth—although these gods are in some way responsible for the creation of humans and for their fate. Many ancient religions were **polytheistic,** meaning that they included numerous gods, all of whom occupied themselves with some aspect of the universe and of human life. In the religion of the ancient Greeks, warfare was the concern of Ares; music, healing, and prophecy were the domain of Apollo; his sister Artemis was concerned with hunting; Poseidon was the god of seafaring; Athena was the goddess of handicrafts and intellectual pursuits; and so on. A similar division of concerns and attributes could be found among the gods of the Romans and, later, among the gods of the Celtic tribes of Gaul and Britain.

The ancient Hebrews were among the first of the world's peoples to evolve a **monotheistic** religion—one centered on belief in a single all-powerful God who determines human fate and can be addressed through prayer. This belief is expressed in the central creed of the Jews: "Hear O Israel, the Lord our God, the Lord is One." Jewish monotheism, based on the central idea of a covenant between God and the Jewish people (as represented in the written laws of the Ten Commandments, for example), helped stimulate the codification of religious law and ritual, so that the Jews became known as "the people of the book." As they traveled and settled throughout the Middle East, the Jews were able to take their religion with them and hold on to the purity of their beliefs and practices (P. Johnson, 1987; H. Smith, 1952).

Christianity and Islam are also monotheistic religions. The Roman Catholic version of Christianity envisions God as embodied in a Holy Trinity consisting of God the Father, Christ the Son, and the Holy Spirit of God, which has the ability to inspire the human spirit. The fundamental beliefs of Islam are similar in many respects to those of Judaism and Christianity. Islam is a monotheistic religion centering on the worship of one God, Allah, according to the teachings of the Koran as given by Allah to Mohammed, the great prophet of the Muslim faith. In his early preachings, Mohammed appears to have believed that the followers of Jesus and the believers in Judaism would recognize him as God's messenger and realize that Allah was the same as the God they worshipped. The fundamental aim of Islam is to serve God as he demands to be served in the Koran.

Another basically monotheistic religion, Hinduism, is difficult to categorize. On the one hand, it incorporates the strong idea of an all-powerful God who is everywhere yet is "unsearchable"; on the other hand, it conceives of a God who can be represented variously as the Creator (Brahma), the Preserver

Dance, song, and ceremonial costume in traditional African societies almost always have religious and social meanings, which usually are known only to members of that culture and do not assert universal significance.

(Vishnu), and the Destroyer (Shiva). Each of these personifications takes a number of forms in Hindu ritual and art. Of all the great world religions, Hinduism teaches most forcefully that all religions are roughly equal "paths to the same summit."

Abstract Ideals

In China, Japan, and other societies of the Far East, religions predominate that are centered not on devotion to a god or gods but on an abstract ideal of spirituality and human behavior. The central belief of Buddhism, perhaps the most important of these religions, is embodied in these thoughts of Siddhartha Gautama, The Buddha:

> Life is a Journey
> Death is a return to the Earth
> The universe is like an inn
> The passing years are like dust

Like all of the world's great religions, Buddhism has many branches. The ideal that unifies them all, however, is the teaching that worship is not a matter of prayer to God but a quest for the experience of godliness within oneself through meditation and awareness.

Another important religion based on abstract ideals is Confucianism, which is derived from the teachings of the philosopher Confucius (551–479 B.C.). The sayings of Confucius are still revered throughout much of the Far East, especially among the Chinese, although the formal study of Confucius's thought has been banned since the communist revolution of the early 1950s. The central belief of Confucianism is that one must learn and practice the wisdom of the ancients. "He that is really good," Confucius taught, "can never be unhappy. He that is really wise can never be perplexed. He that is really brave can never be afraid" (quoted in McNeill, 1963, p. 231).

In Confucianism, the central goal of the individual is to become a good ruler or a good and loyal follower and thus to carry out the *tao* of his or her position. *Tao* is an untranslatable word that refers to the practice of virtues that make a person excellent at his or her discipline. As is evident even in this brief description, Confucianism is a set of ideals and sayings that tend toward conservatism and acceptance of the status quo, although the wise ruler should be able to improve society for those in lesser positions. Little wonder that under communism this ancient and highly popular set of moral principles and teachings was banned in favor of what sociologists call a civil religion.

Civil Religion

Some social scientists, notably Robert Bellah (1970), have expanded the definition of religion to include so-called **civil religions.** These are collections of beliefs, and rituals for communicating those beliefs, that exist outside religious institutions. Often, as in the former Soviet communist societies, they are attached to the institutions of the state. Marxism-Leninism can be thought of as a civil religion, symbolized by the reverence once paid to Lenin's tomb. Central to communism as a civil religion is the idea that private property is evil while property held in common by all members of the society (be it the work group, the community, or the entire nation) is good. The struggle against private property results in the creation of the socialist personality, which values all human lives and devalues excessive emphasis on individual success, especially success measured by the accumulation of property. Although the communist regimes of the Soviet Union and Eastern Europe have fallen, there are millions of people in those nations and in China who were socialized to believe in these principles.

In the United States, certain aspects of patriotic feeling are sometimes said to amount to a civil religion: Reverence for the flag, the Constitution, the Declaration of Independence, and other symbols of America is cited as an example. Thus, most major public events, be they commencements, political rallies, or Super Bowl games, begin with civil-religious rituals like the singing of the national anthem or the recitation

THE CENTRAL QUESTION: A CRITICAL VIEW

Why is it difficult for people of different religions to realize their shared quest for peace and understanding?

In Northern Ireland, a fragile peace agreement is continually threatened by the deep mistrust between Protestants and Roman Catholics. As the photo below demonstrates, children and young people there have grown up in an environment in which the cycle of violence and revenge has overwhelmed the population's shared Christian ideals. But Northern Ireland is hardly unique. In India, Christian missionaries face persecution from Hindus. Muslim Pakistanis and Hindu Indians threaten war over religiously divided Kashmir. At this writing, peace between Jews and Muslims in the Middle East seems a remote dream. And in the United States, deep divisions over religious beliefs polarize the electorate and challenge our ability to get along in a civil society (Gardner, 2000).

All the conflicting religious groups believe in a God of love and claim to seek peace and understanding. But if that is true, why are so many of the world's conflicts based on differences in religion? There are no simple answers to these questions, but one can begin by understanding that religious conflict is often especially violent because people believe that the most basic aspects of their beliefs are being threatened by outsiders. Before the holocaust, Christians in Germany were told that Jews were a direct threat to their lives and to their faith. Hindus in India are told that Christians are bribing poor families to renounce their faith and convert to Christianity. When overlaid by religious fervor, the human tendency to make in-group–out-group distinctions can become social dynamite.

In his classic study of social conflict, Lewis Coser (1966) pointed out that religious conflicts are often a way for groups to strengthen their internal solidarity. By constructing an external threat, enemy, or bogeyman, religious leaders (or any other type of group leader) thereby unify the ranks of their followers. Indeed, much of the religious conflict in the world today actually masks the motives of smaller groups of leaders within these groups who have political rather than religious purposes in mind. The trained sociological observer quickly learns to go beyond the irony of the fact that both sides ostensibly believe in peace and love. The best conflict mediators have sociological insights that enable them to convince religious leaders to meet to resolve conflicts, to establish the basis of trust, and to reduce the hysteria of partisan religious crowds. These are difficult steps to accomplish. They require extremely hard work and great patience, and a clear understanding of the basic differences in beliefs that separate the contestants. ■

Until recently, violence was a fact of everyday life for people in Northern Ireland.

The Wailing Wall and the mosque on the Dome of the Rock in Jerusalem, sacred places for Jews and Muslims.

The pope, shown here mingling with a crowd of devout followers, often takes a conservative position on social issues like abortion and women's roles. In many parts of Latin America, however, the Catholic Church is a major force for social change, and its leaders play an active part in the struggle against severe economic inequality.

of the Pledge of Allegiance, in which a nonsectarian God is invoked to protect the nation's unity ("One nation under God"; Demerath, 1998).

Although there is no doubt that Lenin's image and the American flag may be viewed as sacred in some contexts, neither communism nor American patriotism can compete with the major world religions in the power of their central ideals and their spirituality. In consequence, sociologists tend to concentrate on religions in the traditional sense—that is, on the enactment of rituals that represent the place of sacred beliefs in human life. In the remainder of this chapter, we discuss the structure of religious institutions and the processes by which new ones arise.

RELIGION AND SOCIAL CHANGE

Now that the nations behind the former "iron curtain" are enjoying new freedoms, the role of religion in bringing about social change is an important aspect of life in those societies. Indeed, in many parts of the world, religion is one of the primary forces opposing or supporting change. However, it is not always a simple matter to predict whether religion will encourage change or hinder it. In Israel, for example, highly orthodox Jews, though they account for a small minority of the electorate, hold the votes necessary to keep the ruling party in power. The orthodox political parties favor continued settlement on the West Bank of the Jordan River and oppose the creation of a Palestinian state near Israel's borders.

In the United States, the Catholic church plays an active role in seeking social change. The church strongly opposes women's right to obtain abortions legally. Instead, it supports a return to the traditional view of abortion as a crime, which is based on the belief that humans must submit to the will of God and not use their technological skills to achieve power over life and death. In this instance, therefore, although the church is promoting social change, the change represents a return to an earlier and more absolute moral standard. In contrast, in much of Latin America as well as in the United States, the Catholic church is fighting in support of the urban and rural poor, who seek social justice and equitable economic development. In Brazil, for example, the typical Catholic priest or nun favors the political left, which seeks a more egalitarian distribution of wealth and income and is highly critical of the rich (Allen, 2000).

We could add many other examples of the role played by religion and religious organizations in social change. But in attempting to generalize about the relationship between religion and social change, it is useful to return briefly to classic sociological theories. The pioneering European sociologists, particularly Karl Marx and Max Weber, noted the prominent role of religion in social change. But they wondered whether the influence of religious faith is a determining force in social change or whether religious sentiments and the activities of religious organizations are an outgrowth of changes in more basic economic and political institutions.

Marx, as we have seen, believed that economic institutions are fundamental to all societies and that they are the source of social change. In his view, religion and other cultural institutions are shaped by economic and political institutions; they are a "superstructure" that simply reflects the values of those institutions—of markets, firms, the government, the military, and so on. The function of cultural institutions, especially religion, is to instill in the masses the values of the dominant class. In this sense, they can be said to shape the consciousness of a people, but they do so in such a way as to justify existing patterns of economic exploitation and the existing class structure. Religion, in Marx's words, is "the opium of the people" because it eases suffering through prayer and ritual and deludes the masses into accepting their situation as divinely ordained rather than organizing to change the social system.

For Weber, on the other hand, religion can be the cause of major social change rather than the outcome or reflection of changes in economic or other institutions. Weber set forth this thesis in one of his most famous works, *The Protestant Ethic and the Spirit of Capitalism* (1974/1904). Noting that the rise of Protestantism in Europe had coincided with the emergence of capitalism, Weber hypothesized that the Protestant Reformation had brought about a significant change in cultural values and that this was responsible for the more successful development of capitalist economic systems in Protestant regions. As Weber explained it, Protestantism instilled in its followers certain values that were conducive to business enterprise, resulting in the accumulation of wealth. Because the early Protestants believed that wealth was not supposed to be spent on luxuries or "the pleasures of the flesh," the only alternative was to invest it in new or existing business enterprises—in other words, to contribute to the rapid economic growth that was characteristic of capitalist systems (R. Brown, 1965). This view was reinforced by the belief—also part of the Protestant ethic—that a person who worked hard was likely to be among those predestined for salvation.

Some have questioned the validity of Weber's thesis regarding Protestantism and early capitalism, but there can be no doubt that religious institutions are capable of assuming a major role in shaping modern societies. Throughout the Islamic world, there are currents of orthodoxy and reform that threaten to cause both civil and international wars. And as noted earlier, in many Latin American communities the Catholic church leads the movement for large-scale social change. So Marx was wrong in his claim that religion functions largely to maintain the existing values of more basic social institutions. On the contrary, religion can often lead to new ways of organizing societies—to new political and economic institutions as well as whole new lifestyles.

The Marxian view of religion is still relevant to those who are critical of the influence of religious institutions. These critics assert that religion, along with other cultural institutions such as education, serves to reaffirm and perpetuate inequalities of wealth, prestige, and power. When the poor are encouraged to pray for a better life, for example, they are further oppressed by a religion that prevents them from realizing that they need to marshal their own power to challenge the status quo. Thus, there is still some question as to whether (and how much) religious institutions change society in any fundamental way.

To make informed judgments about this and related issues, we need to have a better understanding of the nature of religious organizations and how they function in modern societies. In the next section, therefore, we describe the main types of religious organizations. We then focus on trends in religious belief and practice in the United States.

FORMS OF RELIGIOUS ORGANIZATION

Religion as a fully differentiated institution developed in agrarian societies, and it was in such societies that formal religious organizations first appeared. Agrarian societies produce enough surplus food to support a class of priests and other specialists in religious rituals. In those less complex societies, religion was incorporated into village and family life; it had not yet become differentiated into a recognized, separate institution with its own statuses and roles (Parsons, 1966). Over time, however, the development of religious institutions resulted in a wide variety of organizations devoted to religious practice. Today those organizations include the church, the sect, and the denomination.

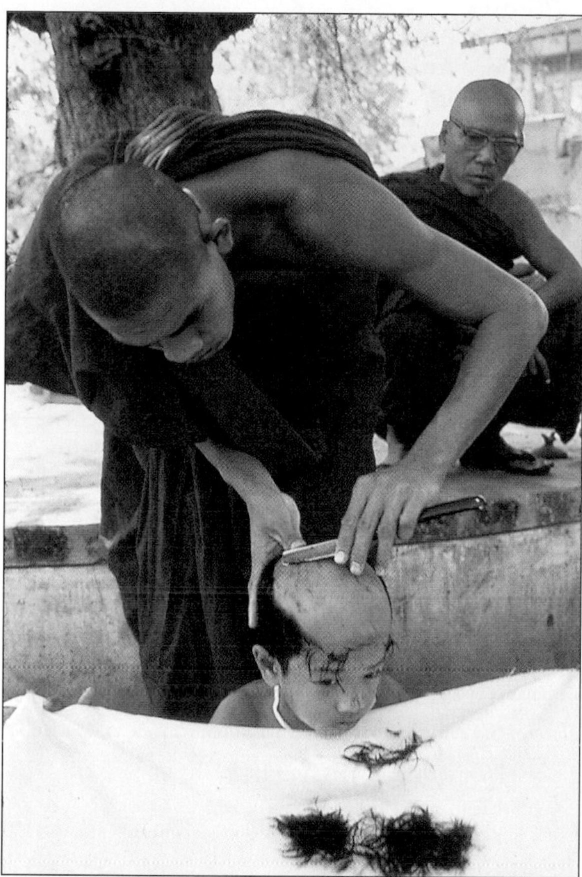

In the agrarian villages of Asia, where Buddhism is a popular religion, children who are destined to become monks are initiated at an early age.

Churches and Sects

A **church** may be defined as a religious organization that has strong ties to the larger society. Often in its history it has enjoyed the loyalty of most of the society's members; indeed, it may have been linked with the state itself (Weber, 1963/1922). An example is the Church of England, or Anglican church. A **sect,** by contrast, is an organization that rejects the religious beliefs or practices of established churches. Whereas the church distributes the benefits of religious participation to anyone who enters the sanctuary and stays to follow the service, the sect limits the benefits of membership (i.e., salvation, fellowship, common prayer) to those who qualify on narrower grounds of membership and belief (McGuire, 1987; Weber, 1963/1922).

Sects require strong commitment on the part of their members and usually are formed when a small group of church members splits off to form a rival organization. The sect may not completely reject the beliefs and rituals of the church from which it arose, but it changes them enough to be considered a separate organization. Most "storefront churches" are actually sects that have developed their own particular interpretations of religious ritual. An important difference between churches and sects is that churches draw their adherents from a large social environment—that is, from a large pool of possible members—whereas the size of the population from which a sect draws its members tends to be small (Stark & Bainbridge, 1979, 1997). Also, churches make relatively limited demands on their members, while sects make heavy claims on their members' time, money, and emotional commitment. As Robert Wuthnow (1988) writes, "Churches attempt to regulate or fulfill a few of the activities or needs of large numbers of people; sects attempt to regulate or fulfill many of the activities or needs of small numbers of people" (p. 495).

Denominations

A third type of religious organization is the **denomination.** Unlike a sect, a denomination is on good terms with the religious institution from which it developed, but it must compete with other denominations for members. An example of a denomination is the United Methodist Church, a Protestant denomination that must compete for members with other Protestant denominations such as Presbyterians, Episcopalians, and Baptists. Denominations sometimes evolve from sects. This occurs when the sect is successful in recruiting new members and grows in both size and organizational complexity. Sociologists who study religion have found that the bureaucratic growth of a sect and its increasing influence over nonreligious matters in the community is associated with a decline in the fervor with which the sect pursues its spiritual ideals and a decrease in its efforts to remain faithful to the claims that originally made it so different. This aspect of religious change is an active area of research (H. R. Niebuhr, 1929; Pope, 1942; Stark & Bainbridge, 1997; Troeltsch, 1931).

Cults

Still another type of religious body, the **cult,** differs in significant ways from the organizations just described. Cults are usually entirely new religions whose members hold beliefs and engage in rituals that differ from those of existing religions. Some cults have developed out of existing religions. This occurred in the case of early Christianity, which began as a cult of Jews who believed that Jesus of Nazareth was the Messiah and who practiced rituals that were often quite different from those of Judaism. Cults may also be developed by people who were not previously involved in a church or sect, such as those who become active in pagan cults like those based on ancient forms of witchcraft (Adler, 1979). Most major religions began as seemingly insignificant cults, but new cults are formed every day throughout the world, and very few of them last long enough to become recognized religions. The Mormon church, which began as a cult in 1830 with six members, grew into a large religious organization with more than 60,000 members by 1850 and today is one of the fastest growing Christian churches in the world (Finke & Stark, 1992).

Sects and cults are a major source of change in religious organizations. People who are not satisfied with more established churches and denominations, or are otherwise alienated from society, often form or join a cult or sect (Nelson, 1984; Smelser, 1984). One of the most convincing explanations of the emergence of sects was suggested by H. Richard Niebuhr (1929), borrowing from Max Weber's (1922) pioneering analysis of churches and sects. According to Weber, churches tend to justify the presence of inequality and stratification because they must appeal to people of all classes. Sects, on the other hand, may be led by charismatic individuals who appeal to people who have felt the sting of inequality. Niebuhr agreed with Weber that class conflict is a primary cause of sect formation. But he observed that as a sect becomes more successful and better organized, it becomes more like a church and begins to justify existing systems of stratification. This creates the conditions in which new sects may emerge.

TRENDS IN RELIGION

Contemporary religious beliefs and practices show two somewhat contradictory trends. In the advanced industrial nations, there is a decline in attendance at formal religious services, on one hand, but a rise in spirituality and religious beliefs, on the other (Inglehart & Baker, 2000). These trends are occurring even in the United States, which of all the industrial nations has had the highest proportions of people who attend church regularly and who express extremely strong

TABLE 10.1
Percentage Attending Religious Services at Least Once a Month, by Country and Year

Country	1981	1990–1991	1995–1998	Net Change
Advanced Industrial Democracies[a]				
Australia	40	—	25	–15
Belgium	38	35	—	–3
Canada	45	40	—	–5
Finland	13	13	11	–2
France	17	17	—	0
East Germany	—	20	9	–11
West Germany	35	33	25	–10
Great Britain	23	25	—	+2
Iceland	10	9	—	–1
Ireland	88	88	—	0
Northern Ireland	67	69	—	+2
South Korea	29	60	27	–2
Italy	48	47	—	–1
Japan	12	14	11	–1
Netherlands	40	31	—	–9
Norway	14	13	13	–1
Spain	53	40	38	–15
Sweden	14	10	11	–3
Switzerland	—	43	25	–18
United States	60	59	55	–5
Ex-Communist Societies[b]				
Belarus	—	6	14	+8
Bulgaria	—	9	15	+6
Hungary	16	34	—	+18
Latvia	—	9	16	+7
Poland	—	85	74	–11
Russia	—	6	8	+2
Slovenia	—	35	33	–2
Developing and Low-Income Societies[c]				
Argentina	56	55	41	–15
Brazil	—	50	54	+4
Chile	—	47	44	–3
India	—	71	54	–17
Mexico	74	63	65	–9
Nigeria	—	88	87	–1
South Africa	61	—	70	+9
Turkey	—	38	44	+6

[a]Sixteen of 20 advanced industrial democracies declined; mean change = –5.
[b]Of ex-Communist societies, five of seven increased; mean change = +4.
[c]Of developing and low-income societies, five of eight declined; mean change = –4.

Source: Inglehart & Baker, 2000.

faith in the existence of God (Bishop, 1999). Despite declining church attendance, people in the United States tend to be more active in their religious participation and more firm in their belief in God. These facts are shown quite clearly in Tables 10.1 and 10.2, which appeared in a recent worldwide survey of values, religion, and cultural change (Inglehart & Baker, 2000). Note that among industrial nations, the United States, with monthly religious attendance at 55% of Americans, is surpassed only by highly religious Ireland and Poland, where under the Soviet empire religious practice was a form of protest against communism.

Developing nations like India and Mexico have experienced significant decreases in attendance at religious services. However, these changes are not accompanied by equal declines in religious beliefs. Most sociologists believe that as societies undergo the transition to greater reliance on industrial production and their populations become more urban and educated, religious practice seems less connected to people's direct survival and more a matter of personal choice. In contrast, in the former communist nations, where religion was suppressed for much of the last century, there is a resurgence of religious practice. Much of this change may be related to the greater social and economic insecurity of these nations, but it is also due to pent-up demand for religious freedom. In Russia, for example, the tables show that strong belief in God has increased, as has religious attendance, but both remain at extremely low levels compared to rates of belief and attendance in nations like Poland. Sociologists cannot yet determine how important the return to religion will be in nations like Russia after many decades during which people were socialized in highly secular schools and other public institutions.

In studying religion in the United States, sociologists are unable to use census data because the Census Bureau does not collect data on religious affiliation. However, some statistics are available from smaller sample surveys such as those conducted by the National Opinion Research Center (NORC). The available data indicate that more than half of Americans over age 18 identify themselves as Protestant, about one quarter are Roman Catholic, and about 2% identify themselves as Jewish. All other religions combined account for about 5% of the population. The remaining 7% say that they have no religious affiliation; they constitute the third-largest category of religious preference in the United States. (See Figure 10.1.)

TABLE 10.2

Percentage Rating the "Importance of God in Their Lives" as "10" on a 10-Point Scale, by Country and Year

Country	1981	1990–1991	1995–1998	Net Change
Advanced Industrial Democracies[a]				
Australia	25	—	21	–4
Belgium	9	13	—	+4
Canada	36	28	—	–8
Finland	14	12	—	–2
France	10	10	—	0
East Germany	—	13	6	–7
West Germany	16	14	16	0
Great Britain	20	16	—	–4
Iceland	22	17	—	–5
Ireland	29	40	—	+11
Northern Ireland	38	41	—	+3
Italy	31	29	—	–2
Japan	6	6	5	–1
Netherlands	11	11	—	0
Norway	19	15	12	7
Spain	18	18	26	+8
Sweden	9	8	8	–1
Switzerland	—	26	17	–9
United States	50	48	50	0
Ex-Communist Societies[b]				
Belarus	—	8	20	+12
Bulgaria	—	7	10	+3
Hungary	21	22	—	+1
Latvia	—	9	17	+8
Russia	—	10	19	+9
Slovenia	—	14	15	+1
Developing and Low-Income Societies[c]				
Argentina	32	49	57	+25
Brazil	—	83	87	+4
Chile	—	61	58	–3
India	—	44	56	+12
Mexico	60	44	50	–10
Nigeria	—	87	87	0
South Africa	50	74	71	+21
Turkey	—	71	81	+10

[a]Eleven of 19 advanced industrial democracies declined; mean change = –1.
[b]Of ex-Communist societies, six of six increased; mean change = +6.
[c]Of developing and low-income societies, five of eight increased; mean change = +6.

Source: Inglehart & Baker, 2000.

FIGURE 10.1
Percentage Distribution of Religious Preference in the United States (persons 18 years of age and older)

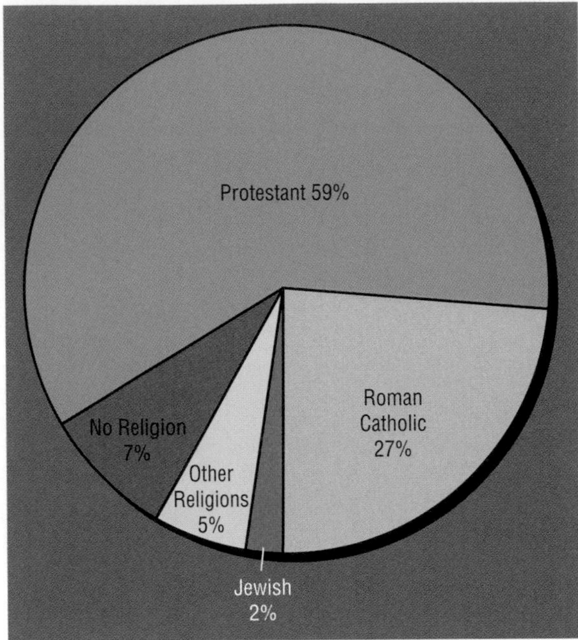

Source: Data from *Statistical Abstract,* 2000.

The religious preferences of Americans have been changing slowly but steadily over the last two generations. The Protestant majority has gradually declined, owing partly to upward mobility and the tendency of more affluent and educated people to be less active in religious institutions. Meanwhile, the proportion of the population who identify themselves as Catholics has been growing, primarily as a result of the large numbers of Hispanic immigrants who have entered the United States in recent years. (See the Global Social Change box on page 291.) Also, since the mid-1960s there has been an increase in the percentage of people who express no religious preference (NORC, 1998).

Membership in a religious organization is quite different from identification with a religious faith, and this is reflected in statistics on Christian church membership in the United States. Because data on membership are obtained from the organizations themselves, there are significant differences in who is included. (Some churches, for example, count all baptized infants, whereas others count only people above a certain age who are enrolled as members.) Different regions of the country are quite different with respect to the church membership of the majority of their residents. Protestants outnumber members of all other religions in the South, for example, and in the West and Southwest the Roman Catholic church is dominant. There are also significant differences in church membership by size of place: Protestants are most likely to live in rural places, small cities, and suburbs of smaller cities, while native-born Catholics are found in medium-sized cities and suburbs of larger cities. Much of the rapid growth of Catholic congregations in the United States is accounted for by immigrants to major cities like Miami, Los Angeles, and New York (G. Niebuhr, 1995). Jews, who originally migrated to the nation's largest cities, are most numerous in major metropolitan regions like New York, Los Angeles, Chicago, and Miami (Bogue, 1985).

Religiosity

Neither church membership figures nor self-reports of religious identification are accurate indicators of the aspect of religious behavior known as **religiosity.** This term refers to the depth of a person's religious feelings and how those feelings are translated into religious behavior. Responses to questions about whether one believes in God and how strongly, whether one believes in a life after death, whether one's religious beliefs provide guidance in making major decisions, and the frequency of one's church or temple attendance all can be used to measure religiosity.

When sociologists study religiosity as opposed to church membership, some important results emerge. For example, Rodney Stark and William S. Bainbridge (1985) found that about 62% of Americans are church members. But church membership varies rather sharply by region. The southwestern states have the highest rate of membership—about 65% of adults. Other regions have rates ranging between 53% and 62%, but the Pacific states have by far the lowest proportion of adult church members, only 36%. Quite different results are obtained on measures of religiosity: About 90% of Americans believe in the existence of God, and 86% express certainty in a life after death. Adults in the Pacific states do not differ very much on these measures from those in other regions. These figures highlight the difference between church membership and religiosity. The lower rates of church membership on the West Coast are correlated with the greater spatial mobility and lower median age of the

GLOBAL SOCIAL CHANGE

The Impact of Immigration

The wave of immigration to the United States over the past 20 years has dramatically changed the nation's religious profile. New immigrants are also posing new opportunities and challenges for established churches and denominations, not only in major metropolitan centers, where religious and ethnic diversity is nothing new, but also in formerly more homogeneous areas, where traditions of religious tolerance are somewhat weaker (Keely, 2000). Cambodian fishing families in small towns along the Gulf coast, Hindu merchants in midwestern towns, turbaned Sikhs, Muslim Syrians, Presbyterian Koreans, all living or working in neighborhoods populated by Latino Catholics and African American Protestants—these are only a few examples of how immigrants are bringing a diversity of religious beliefs and practices to American communities.

For the newcomers, the first goals are to get an economic foothold in the region and then, as soon as possible, build a place of worship. For the established religions, the influx of newcomers allows local religious leaders to call attention to the universality of human spirituality and offers lessons in tolerance and religious freedom. Invariably, however, the newcomers experience some prejudice and even direct hostility, especially when there is direct economic competition between the newcomers and older populations (such as in the shrimp fishing regions of the Gulf states).

The Catholic church in the United States is feeling the effects of immigration most directly. Catholics make up about 25% of the U.S. population, but more than 41% of recent immigrants are Catholic. This is by far the single largest faith category cited by immigrants, whose national origins are extremely diverse. Most of the Catholic immigrants are arriving from Mexico, the Philippines, Poland, the Dominican Republic, and Vietnam. Far fewer come from the Western European countries traditionally associated with Catholicism in the United States. After Catholics, the next largest category of immigrants—14.8%—list "no religion," followed by Protestant, 8.8%; Muslim, 7.4%; Buddhist, 4%; and Hindu, 3.4%.

Although these last percentages are rather small, the numbers of legal immigrants the United States receives in a year is large (between 600,000 and 800,000). This means that over a decade the number of Hindu residents of a community that is receiving immigrants from India, for example, can quickly grow to many thousands, as is occurring in some communities in the New York New Jersey metropolitan region. The religious beliefs and ceremonies these and other immigrants bring to their new communities will at first seem strange and even threatening to some. Eventually, however, their presence will expand American ideas of tolerance and religious understanding.

populations of those states. People who move frequently tend to sever their attachments not only to churches but to all organizations in the community. Yet as we see in Tables 10.1 and 10.2, people who do not belong to organized churches can nevertheless hold deeply cherished religious beliefs.

Fundamentalism

Religious *fundamentalists* are believers (and their leaders) who are devoted to the strict observance of ritual and doctrine. They hold deep convictions about right and wrong in matters of faith and lifestyle, with little tolerance for differences in belief and practice, and they are fiercely opposed to astrology, magic, unorthodox conceptions of religion, and any form of civil religion. In all the world religions, there are fundamentalist movements or divisions between liberal and conservative approaches to religious norms that guide daily life. In the United States, fundamentalist believers may be found in every major religion, but fundamentalism is especially strong among some Protestant churches and sects.

Historically, the influence of religious fundamentalism on the culture and politics of American life reached its height in 1919, when the Eighteenth Amendment to the U.S. Constitution prohibited the manufacture, sale, or transportation (and hence the consumption) of "intoxicating liquors," regardless of whether one supported the fundamentalist view of alcohol consumption. The repeal of Prohibition represented a defeat for religious fundamentalism in American culture and politics.

Another famous episode involving fundamentalism was the so-called monkey trial of 1925, in which John T. Scopes, a biology teacher in Tennessee, was charged with teaching the then forbidden theory of evolution, which many religious fundamentalists oppose because it contradicts the account of creation presented in the Bible. The prominent lecturer and religious fundamentalist (and former presidential candidate) William Jennings Bryan prosecuted the controversial case, which received worldwide attention. Despite a strong defense by Scopes's lawyer, Clarence Darrow, Scopes was convicted and fined $100. (Although courts later reversed the conviction on a technicality, it would be more than 30 years before the Tennessee legislature repealed the law that prohibited teaching evolution theory in state-supported schools.) Despite the conviction, the Scopes trial is seen as a turning point for the intellectual opponents of fundamentalism. Although fundamentalism continued to pose a challenge to mainline Baptist, Methodist, and Presbyterian denominations, after the Scopes trial it lost influence—until recently—as a religious ideology (Simpson, 1983).

In the 1980s, the Southern Baptist denomination became embroiled in conflicts between fundamentalist and more moderate leaders. Sociologist Nancy Tatom Ammerman used her professional training and her extensive personal knowledge of Baptist congregations to try to understand the causes and consequences of these divisions. She views issues of pluralism and social change as central to understanding the split:

> The disruption against which fundamentalists struggle is often labeled "modernity." Whatever else that label may mean, the "modern" world is one in which change is a fact of life, in which people of multiple cultures live side by side, and in which religious rules have been largely relegated to a private sphere of influence. (1990, p. 150)

To assess differences among Baptists in the way they view social change and diversity, Ammerman surveyed hundreds of Baptist leaders in congregations throughout the United States. Her survey sample (n = 944) included men who were pastors and church deacons and women who were presidents of their local Baptist women's organizations. After examining their responses to key questions about their beliefs, Ammerman developed categories (or "theological parties") with labels corresponding to the terms respondents often used to describe themselves (e.g., *fundamentalist, conservative, moderate*). She also developed a scale for measuring individuals' degree of opposition to modernity. The questions on which the scale is based and the method of scoring responses are discussed in the Sociological Methods box on page 293. Table 10.3 shows that "among those who disapprove of pluralism, fundamentalists outnumber moderates nearly ten to one" (Ammerman, 1990, p. 151). But the data in the table and in the Sociological Methods box also show that Baptists classified as "conservative," whose views fall somewhere between those of the fundamentalists and moderates, are the most numerous category and are quite divided in their views of pluralism.

When she examined how respondents felt about modernity in relation to such variables as education, occupation, and rural versus urban residence, Ammerman found some surprising results. Although people often think of individuals who hold strong fundamentalist beliefs and reject modernity as uneducated

SOCIOLOGICAL METHODS

Scales and Composite Scores

Sociologists who are analyzing questionnaire responses often create scales in order to arrive at a composite score for a particular variable. Nancy Tatom Ammerman created scales to measure the strength of a person's fundamentalist beliefs and opposition to modernity. (See the accompanying tables.) A *scale* is a set of statements that fit together conceptually in that they deal with aspects or examples of the same attitude or behavior. For each statement, the respondent may be asked to choose among responses ranging, for instance, from 1 ("Strongly disagree") to 5 ("Strongly agree"), with "Unsure" in the middle. The statements in a particular scale (e.g., fundamentalist beliefs) are scattered throughout the questionnaire to avoid *response set*, the tendency to answer all such items similarly without giving them much thought; response set is a frequent problem when similar items are grouped together. To ensure that the statements in a scale actually measure the behavior or attitude the researcher is studying, great care must be taken in their selection and wording.

Once the questionnaires have been completed and the data entered into a computer, the researcher can arrive at a simple composite score by adding up the scores on all the items in a scale. Then the researcher can either use each respondent's composite score as a continuous variable or establish cutoff points that will create categories for analysis, as Ammerman did. In her measurement of opposition to modernity, for example (see table), Ammerman decided that those who agreed or strongly agreed with all the items (i.e., whose scores ranged from 5 to 9) could be classified as "strongly agreeing with pluralism." Those who agreed with most of the items but not all of them (i.e., scored from 10 to 14) were classified as showing "moderate approval." Those who scored 15 or above—in other words, those who disagreed as often as they agreed—were classified as "disapproving."

Fundamental Beliefs

Items	Scoring
The Scriptures are the inerrant Word of God, accurate in every detail.	Strongly agree = 5
God recorded in the Bible everything He wants us to know.	Strongly agree = 5
The Genesis creation stories are there more to tell us about God's involvement than to give us a precise "how and when."	Strongly disagree = 5
The Bible clearly teaches a premillennial view of history and the future.	Strongly agree = 5
It is important that Christians avoid worldly practices such as drinking and dancing.	Strongly agree = 5

Distribution

Range	Percent of Respondents
5–10	5
11–15	14
16–20	47
21–25	34

Opposition to Modernity

Items	Scoring
I like living in a community with lots of different kinds of people.	Very untrue = 5
Public schools are needed to teach children to get along with lots of different kinds of people.	Strongly disagree = 5
I sometimes learn about God from friends in other faiths.	Very untrue = 5
One of the most important things children can learn is how to deal creatively with change.	Strongly disagree = 5
Children today need to be exposed to a variety of educational and cultural offerings so they can make informed choices.	Strongly disagree = 5

Distribution

Range	Percent of Respondents
5–9	29
10–14	52
15–19	17
20–25	2

TABLE 10.3
Percent in Different Theological Parties, by Responses to Pluralism

Theological Party	Strong Approval of Pluralism	Moderate Approval of Pluralism	Disapproval of Pluralism	Total
Self-Identified Moderate	21%	5%	3%	9%
Moderate Conservative	16	8	3	9
Conservative	38	55	43	48
Fundamentalist Conservative	18	24	27	23
Self-Identified Fundamentalist	8	9	23	11
Total	101%	101%	99%	100%
(Number of Cases)	(256)	(500)	(187)	(944)

Note: Difference statistically significant at $p < .001$. Some percentages do not total 100 due to rounding.

Source: Ammerman, 1990.

people living in remote farming hamlets, and of individuals who approve of modernity as educated middle-class urban dwellers, Ammerman's results challenge this conventional wisdom:

> Those among Southern Baptists who were the most skeptical of change, choice, and diversity were blue-collar workers, those with middle incomes, people who had moved from farm to city, and those who had been to college, but did not have a degree. And combinations were important: people who had moved from farm to city and had some college or people with middle income and blue-collar households, for instance. These were people who knew exactly what the modern world was all about, and they were less enthusiastic about embracing it than were any other Southern Baptists. And the less enthusiasm they had for modern attitudes, the more likely they were to adopt fundamentalist beliefs and identity. (1990, p. 155)

Eventually, Ammerman believes, these differences in outlook may lead to an irremediable schism in which moderates may form a new Baptist denomination.

Sociologists of religion tend to view the rise of fundamentalist religious movements such as that occurring among Southern Baptists as part of a larger effort by people in modern societies to make moral and spiritual sense of their lives. Clearly, religion continues to play an important role in the contemporary world. This fact is reflected in the remarks of Robert Bellah, a well-known commentator on American culture. For several centuries, he writes, Americans "have been embarked on a great effort to increase our freedom, wealth, and power. For over a hundred years, a large part of the American people, the middle class, has imagined that the virtual meaning of life lies in the acquisition of ever-increasing status, income, and authority, from which genuine freedom is supposed to come" (Bellah et al., 1985, p. 284). Yet many Americans seem uneasy about their lives despite their material comfort. They seem to yearn for spiritual values without necessarily wanting to return to traditional religious practices. They adhere to the values of individualism, but at the same time they long for the stronger sense of community and commitment that one finds in religious congregations. Bellah predicts that Americans will continue to seek self-actualization as individuals, but that increasingly they will express their desire for community attachments and higher values, either in traditional religions or in civil-religious practice.

TECHNOLOGY AND SOCIAL CHANGE

For those who seek fulfillment of their spiritual needs in new communities of faith, there is a lively marketplace for new forms of religious affiliation. In fact, some of the most sociologically important innovations

Contemporary technologies are essential to the working of any large church congregation. Imagine, for example, how members of this congregation would arrive at church without automobiles and highways, or how the minister could project his voice to such a gathering without the benefit of amplified sound.

in religious affiliation are taking place in the parts of the United States that are often thought of as highly conservative when it comes to religious life. And creative adaptations of major technologies are making much of this innovation possible, as the following case from Texas suggests.

President George W. Bush calls him a "social entrepreneur" and recognizes him as a religious and social force in the Dallas metropolitan region. He is Bishop T. D. Jakes, one of the newest sensations in the volatile world of high-profile evangelical Christianity. Jakes "is a multitasking, multiplatform marketing whiz pitching a single message for a hurting world: 'By God's grace you can make it'" (Grossman, 2000).

Bishop Jakes is the founder and spiritual leader of The Potter's House, "a 26,000-member church in a new facility the size of two football fields, so state-of-the-art there's computer technology built into the 8,000-seat sanctuary." Jakes is also the author of numerous popular books, and his records have been nominated for Grammy awards. His ministries include Bible classes in hundreds of U.S. prisons, preaching podiums in Africa, and weekly broadcasts on American television. Clearly, Jakes is a religious innovator who knows how to harness a variety of technologies in the interest of building his congregations and spreading his interpretation of the gospels.

The interstate highway system and the almost universal ownership of automobiles in the United States make "mega" churches like those of Bishop Jakes possible. Television and the Internet allow potential congregants to become aware of the new church, its leader, and his charismatic appeal. Homebound adherents can participate in church services and exchange e-mail with other members of the congregation. They can order the bishop's books and records online.

In the early seventeenth century, the astronomer Galileo, using a primitive version of the telescope, made observations that convinced him that, contrary to church teachings and received wisdom, the earth was not the center of the solar system but instead moved in regular orbit around the sun. When he published his observations, he was brought before the Inquisition of the Church of Rome and forced to renounce his scientific findings. It is said that as he was leaving the trial the astronomer muttered, "E pur si muove" (Nevertheless it moves). Today, as modern evangelists like Bishop Jakes selectively adapt the latest products and techniques of modern science to building their churches, perhaps somewhere in the universe Galileo continues to enjoy his insight. For "move it does." The most successful religious organizations in the world, including the Roman Catholic church, depend heavily on all the technologies that science produces for their organization and their continual growth.

SOCIOLOGY AND SOCIAL JUSTICE

Born in Bangladesh, Taslima Nasreen has devoted her career to helping women throughout the world balance the demands of conservative religions against their desire to realize their potential as human beings. Few people are as well qualified as she is to speak about the issues of religion and social justice for women.

Nasreen has led a twin career as a doctor and writer. She is the author of six novels, several collections of poetry and essays, and an autobiography. Her works have been translated into more than a dozen languages. Two of her novels, *Shame* and *My Girlhood,* are banned in her country as blasphemy, and Islamic fundamentalists have issued a *fatwa,* or death decree, against her. In consequence, Nasreen has been living in exile since 1994. She has received numerous international awards, including India's Ananda Award, the European Parliament's Sakharov Prize, and the International Humanist Award from the United States.

In the following remarks, Nasreen applies her sociological imagination to the issue of women's rights in religiously orthodox societies.

> Every day, women continue to be victims of rape, trafficking, acid-throwing, dowry deaths, and other kinds of torture. At the opening of this new century, women are still not considered as equal human beings in many parts of the world. Religion and patriarchy continue to have an all-encroaching hold on their lives, maintaining and justifying their age-old oppression. In some South Asian societies, this hold is even increasing.
>
> I do not believe that there can be real equality in a society dominated by religion. Western countries speak repeatedly about the necessity of economic development to alleviate poverty. But this is not enough. Societies such as Saudi Arabia may be economically developed, but women are deprived of all rights. The supremacy of religion is incompatible with freedom of expression, women's rights, and democracy. This is why I see religion as the main enemy of women's development.
>
> We have to act on several fronts at once. First of all, improving access to education. In a society like Bangladesh, 80% of women are illiterate. For centuries women have been taught they are the slaves of men. It is very hard to change their minds, to make them aware of their oppression, to give them a sense of their independence. This educational effort has to go hand in hand with a secular feminist movement in society. Such movements have to start within the country and they cannot take hold when people are uneducated and unaware of their oppression. I'm not sure you can accomplish much from the outside, except to expose in the media the atrocities women in all too many countries face in their day to day lives. . . .
>
> For women's status to change, we also need enlightened leaders who believe in equality. In countries such as mine, women with a strong voice do not have the support of political leaders, whether they be men or women. Look at

"Education, a secular feminist movement, and leaders—both men and women—committed to equality and justice. This is what it will take to change the dire conditions which too many women still face today. It will take a very long time, but we are here to work towards that end" (Nasreen, 2000).

the countries in which women are in politics, or even heads of state. Does it follow that women in those countries are emancipated? Because of long-standing vested interests, such leaders continue to back measures that oppress women. They are not ideologically committed to changing these conditions. In South Asia, most of the women who become heads of state are religious, and like men, they adhere to the religious objectives of the Establishment. I am the victim of a country where the prime minister is a woman. Because I went one step too far in denouncing religion and the oppression that it keeps women under, I had to leave my country.

I have seen women oppose me when I talked about women's rights. They said straight out that God did not believe that women should have so many rights. And I have met men in my country who are against what is said in the religious scriptures and believe in equality between men and women. It does not depend on gender. It depends on one's conscience. Muslim women who are wearing the veil and glorifying their subservience are obviously not going to better the lives of the oppressed....

Education, a secular feminist movement, and leaders—both men and women—committed to equality and justice. This is what it will take to change the dire conditions which too many women still face today. It will take a very long time, but we are here to work towards that end. (Nasreen, 2000) ■

SUMMARY

Religion has been defined as any set of coherent answers to the dilemmas of human existence that makes the world meaningful. It has also been defined as a system of beliefs and rituals that serves to bind people together into a social group. *Rituals* are formal patterns of activity that express a set of shared meanings; in the case of religious rituals, the shared meanings are *sacred,* pertaining to phenomena that are regarded as extraordinary, transcendent, and outside the everyday course of events. Until comparatively recent times, religion dominated the cultural life of human societies. Since medieval times, however, the traditional dominance of religion over other institutions has been reduced by a process termed *secularization.*

In simpler and rather isolated societies, people may believe in a great force or spirit, but they do not have a well-defined concept of God or a set of rituals involving God. This form of religion is called *simple supernaturalism.* More common among hunting-and-gathering societies is *animism,* in which all forms of life and all aspects of the earth are inhabited by gods or supernatural powers. *Theism* comprises belief systems that conceive of a god or gods as separate from humans and from other living things on the earth. Many ancient religions were *polytheistic,* meaning that they included numerous gods. The ancient Hebrews were among the first of the world's peoples to evolve a *monotheistic* religion, one centered on belief in a single, all-powerful God.

In China, Japan, and other societies of the Far East, religions predominate that are centered not on devotion to a god or gods but on an abstract ideal of spirituality and human behavior. Some social scientists have expanded the definition of religion to include *civil religions,* or collections of beliefs and rituals that exist outside religious institutions.

Karl Marx believed that cultural institutions like religion are shaped by economic and political institutions and function to instill in the masses the values of the dominant class. Max Weber argued that religion can cause major social change by instilling certain values in the members of a society, in turn causing changes in other institutions.

A *church* is a religious organization that has strong ties to the larger society and has at one time or another enjoyed the loyalty of most of the society's members. A *sect* rejects the religious beliefs or practices of an established church and usually is formed when a group of church members splits off to form a rival organization. A *denomination* is on good terms with the religious institution from which it developed but must compete with other denominations for members.

A *cult* is an entirely new religion. People who are not satisfied with more established churches and denominations may form or join a cult or sect.

Religiosity refers to the depth of a person's religious feelings and how those feelings are translated into religious behavior. Studies of religiosity find high percentages of Americans believing in the existence of God and in a life after death.

Religious fundamentalists are believers who are devoted to the strict observance of ritual and doctrine and lack tolerance for differences in belief and practice. In the United States, fundamentalism experienced a resurgence during the 1980s, encouraged by conservative social movements.

GLOSSARY

Religion: Any set of coherent answers to the dilemmas of human existence that makes the world meaningful; a system of beliefs and rituals that serves to bind people together into a social group. **(279)**

Ritual: A formal pattern of activity that symbolically expresses a set of shared meanings. **(279)**

Sacred: A term used to describe phenomena that are regarded as extraordinary, transcendent, and outside the everyday course of events. **(279)**

Profane: A term used to describe phenomena that are not considered sacred. **(279)**

Secularization: A process in which the dominance of religion over other institutions is reduced. **(280)**

Simple supernaturalism: A form of religion in which people may believe in a great force or spirit but do not have a well-defined concept of God or a set of rituals involving God. **(281)**

Animism: A form of religion in which all forms of life and all aspects of the earth are inhabited by gods or supernatural powers. **(281)**

Theism: A belief system that conceives of a god or gods as separate from humans and from other living things on the earth. **(282)**

Polytheistic: A term used to describe a theistic belief system that includes numerous gods. **(282)**

Monotheistic: A term used to describe a theistic belief system centered on belief in a single all-powerful God. **(282)**

Civil religion: A collection of beliefs and rituals that exist outside religious institutions. **(283)**

Church: A religious organization that has strong ties to the larger society. **(287)**

Sect: A religious organization that rejects the beliefs and practices of existing churches; usually formed when a group leaves the church to form a rival organization. **(287)**

Denomination: A religious organization that is on good terms with the institution from which it developed but must compete with other denominations for members. **(287)**

Cult: A new religion. **(287)**

Religiosity: The depth of a person's religious feelings. **(290)**

QUESTIONS FOR THOUGHT AND DISCUSSION

1. Some people think religions oppose social change, but history shows clearly that this is not always the case. Can you give some examples from both the present and the past in support of this idea?

2. What are some differences between a church and a cult that would affect an individual's experiences in each of these types of religious structures?

DIGGING DEEPER

Books

Religion, Deviance, and Social Control (Rodney Stark & William Sims Bainbridge, Routledge, 1997). An excellent source of material on religion, religious cults, and religious leadership.

One God: Peoples of the Book (Edith S. Engel & Henry W. Engel, eds.; Pilgrim Press, 1990). An introduction to the major monotheistic religions with a message of peace, commonality, and openness.

"The Sociology of Religion" (Robert Wuthnow; in Neil Smelser, ed., *Handbook of Sociology;* Sage Publications, 1988). An extremely useful review of current trends in the sociology of religion by one of the field's leading scholars. Includes an excellent bibliography of classic and recent sociological research on religion.

Base Communities and Social Change in Brazil (W. E. Hewitt; University of Nebraska Press, 1991). A fine case study of the influential Catholic ecclesiastical base communities movement in Latin America, based primarily on research in São Paulo, Brazil. A good example of empirical research on religion and social change at the community level.

Religion: The Social Context, 3rd ed. (Meredith B. McGuire; Wadsworth, 1992). A comprehensive text on the sociology of religion with excellent material on religion and social change.

Baptist Battles (Nancy Tatom Ammerman; Rutgers University Press, 1990). A seminal study of the conflicts that arise within a major Protestant denomination over issues of morality and religious practice.

Journals

Journal for the Scientific Study of Religion. Available in most college and university libraries; publishes recent research on religious practices, religiosity, and changes in religious institutions.

Other Sources

The Encyclopedia of American Religions, 4th ed. (Gale Research). Contains useful descriptions of religions in America; covers beliefs, organization, distribution in the population, and other aspects.

EXPLORING SOCIOLOGY ON THE INTERNET

Boston College Center for International Higher Education
www.bc.edu

Provides links to research centers and international news and library information dealing with the Catholic religion throughout the world.

The Anti-Defamation League
www.adl.org

Monitors hate activities in the United States and throughout the world; includes addresses of other useful research sites.

Ontario Consultants on Religious Tolerance
www.religioustolerance.org

Explores many religions and states; does not promote or denounce a specific belief. The site does list religious beliefs and news topics for over 20 denominations.

Andrew Greeley: Author, Priest, Sociologist
www.agreeley.com

Offers insights into Andrew Greeley's recent writings. The site includes previews of his recently published works and articles addressing recent issues in the Catholic religion and others.

Chapter 11

EDUCATION

Can education in contemporary societies provide equality of opportunity for new generations of students?

Education for a Changing World

The Nature of Schools
Who Goes to School?
Education and Citizenship

Attainment, Achievement, and Equality

Educational Attainment
Barriers to Educational Attainment
Educational Achievement
Education and Social Mobility
Inequality in Higher Education
Education for Equality

The Structure of Educational Institutions

Schools as Bureaucracies

Technology and Social Change

In 2001, Dr. Roderick Raynor Paige, formerly the superintendent of schools in Houston, Texas, became Secretary of Education in the cabinet of President George W. Bush. Known as an educational leader who raised standards in a large urban school district, Paige is also known for his support for using public funds, raised through local, state, and federal taxes, to fund private educational initiatives. Since this is a policy favored by the president, educators look for Paige to stimulate more innovations like charter schools and the use of vouchers to pay for private schooling. Critics fear that these policies will weaken public education, but even the critics are encouraged by Paige's appointment. He is widely regarded as a person of great integrity and someone whose own biography is a testament to the need for good schools in a democratic society (Steinberg, 2000).

Paige, the oldest of five siblings, was born in 1933 in Monticello, Mississippi, at a time when racial segregation was the norm in the American South. Elsewhere in the nation, segregation was often just as strict, but it was less openly supported by explicit norms of conduct. Paige worked his way up through segregated schools, displaying great talent both in the classroom and on the sports field. After graduating from Jackson State University in Mississippi, he went on to study for his doctorate in physical education at the University of Indiana. A successful football coach at several colleges, including Texas Southern in Houston, he eventually was named dean of the school of education there. As dean, he became active in debates over reforming the Houston schools, which led to his appointment as the city's educational leader. As superintendent in Houston, he is noted for having pushed teachers and students to perform more successfully on standardized tests; the proportion of Hispanic students passing the math exam increased from 44.2% to 80.3% in 1994. Successes like that one soon put Paige in the national educational spotlight.

Dr. Roderick Paige being sworn in as Secretary of Education on January 24, 2001, with President Bush and Vice President Cheney and Paige's son and brother in the background.

It will not be easy for Paige to achieve the same success on a national level that he achieved in Houston. Given the diversity of the nation's population and the deep ideological divisions that exist over educational issues, educational reform is fraught with great political difficulty. But in an advanced industrial society education is essential, both for the success of individuals and for the health of the economy and society as a whole. Indeed, so much of the hope for progress in a democracy is pinned on education that, despite all the controversy, national educational leaders like Paige have a major opportunity to shape the debates and guide the course of change in educational institutions.

This chapter provides a sociological view of many critical issues related to education in our changing world. The first sections deal with the place of education in the United States and other societies. We then turn to the concepts of educational achievement and attainment and the relationship between education and citizenship. Later sections address problems such as dropping out and degree inflation, as well as the link between education and social mobility. The final sections deal with the school as a formal organization with its own characteristic bureaucratic structure, and as a social institution subject to controversial demands for reform and change. ■

EDUCATION FOR A CHANGING WORLD

Controversies over the goals and methods of education are not new. One of the most famous trials in history took place in ancient Greece when Socrates was accused of corrupting the morals of Athenian youth with his innovative ideas and educational methods that encouraged critical thinking. Today there is great concern about the need to improve education in order to train new generations of workers. Throughout the world, efforts are also being made to extend the benefits of education to more people than a narrow elite, and especially to more women. These efforts range from literacy and health education to vocational training, school reform, and increasing opportunities in higher education. The diversity of these efforts calls our attention to the distinction between education and schooling, for they are not the same.

Education may be defined as the process by which a society transmits knowledge, values, norms, and ideologies and in so doing prepares young people for adult roles and adults for new roles; in other words, it transmits the society's culture to the next generation. Education occurs in many different social groups, in religious ceremonies, at work, in peer groups, at home, and, in most contemporary societies, in schools. Schooling thus is just one form of education. In modern societies, schools are specialized places where education takes place, especially for younger people. As sociologist Steven Brint (1998) notes, an indicator of the importance we attach to education is the amount of time it occupies in young people's lives. Typically, a student spends six hours in school five days a week, nine months a year, for twelve years, or a total of about 13,000 hours between ages 6 and 18. And for those who attend and complete college or university, that total increases to over 17,000 hours of schooling.

Education is a form of socialization that is carried out by institutions outside the family, such as schools, colleges, preschools, and adult education centers. Each of these is an educational institution because it encompasses a set of statuses and roles designed to carry out specific educational functions—it is devoted to transmitting a specific body of knowledge, values, and norms of behavior. In everyday language, a particular school or college may be referred to as an *institution,* but in sociological terms it is an *organization* that exemplifies an educational institution. Thus, El Centro College in Dallas is an organization that exemplifies the institution of higher education. The high school you attended was also an organization, but its curriculum and norms of conduct were those of the institution known as secondary education.

Educational institutions have a huge effect on communities in the United States and other modern societies. Upwardly mobile couples often base their choice of a place to live on the quality of the public schools in the neighborhood. Every neighborhood has at least an elementary school, and every large city has one or more high schools and at least one community college or four-year college. Cities usually also have a variety of school administrations—public and private—and some owe their existence, growth, and development to the presence of a college or university (Ballantine, 1993).

The Nature of Schools

Educational institutions affect not only the surroundings but also the daily lives of millions of Americans: children and their parents, college and university students, teachers and professors. Hence, education is a major focus of social-scientific research. To the sociologist, the most common educational institution, the school, is a specialized structure with a special function: preparing children for active participation in adult activities (F. E. Katz, 1964). Schools are sometimes compared with total institutions (see Chapter 3), in which a large group of involuntary "clients" is serviced by a smaller group of staff members (Boocock, 1980). The staffs of such institutions tend to emphasize the maintenance of order and control, and this often leads to the development of elaborate sets of rules and monitoring systems. This comparison cannot be taken too literally (schools are not prisons, although some of their "inmates" may think of themselves as such), but the typical school does tend to be characterized by a clearly defined authority system and set of rules. In fact, sociologists often cite schools as examples of bureaucratic organizations (Mulkey, 1993; Parelius & Parelius, 1987).

A more interactionist viewpoint sees the school as a set of behaviors; that is, the central feature of the school is not its bureaucratic structure but the kinds of interactions and patterns of socialization that occur in schools. In the words of Frederick Bates and Virginia Murray (1975), the basic feature of schools is "the behavior of a large number of actors organized into groups that are joined together by an authority structure, and by a network of relationships through which information, resources, and partially finished projects flow from one group to another" (p. 26). In other words, "school-related behavior (e.g., doing homework or grading papers) is part of the school as a social system, whether or not it takes place in the school building" (Boocock, 1980, p. 129). So, too, is the involvement of parents in schools and in the school careers of their children. Students with parents who feel comfortable dealing with teachers, who understand what activities to encourage their children to join, who know where and how to challenge and when to defer to the principal's authority—such children, according to interactionists, have a significant advantage over others.

Conflict theorists, by contrast, view education in modern societies as serving to justify and maintain the status quo (Aronowitz & Giroux, 1985; Bowles & Gintis, 1977). For example, on the basis of his field research on secondary school students in a working-class community, the British sociologist Paul Willis (1983) concluded that "education was not about equality, but inequality. . . . Education's main purpose . . . could be achieved only by preparing most kids for an unequal future, and by insuring their personal underdevelopment" (p. 110). The group of boys from low-income homes whom Willis observed thought of themselves as "the lads" and delighted in making fun of higher achievers, whom they called "earholes." But in many different ways "the lads" indicated that they believed their teachers were pushing them into low-prestige futures. Willis interpreted the hostility and alienation of these students as a way of resisting the social forces channeling them into working-class careers.

As we will see shortly, this critical perspective challenges the more popular view that education is the main route to social mobility and that it can offset inequalities in family background (Bell, 1973). When sociologists analyze the impact of educational institutions on society, they generally conclude that the benefits of education are unequally distributed and tend to reproduce the existing stratification system (Fullan, 1993; Jencks et al., 1972).

Who Goes to School?

The idea that all children should be educated is a product of the American and French revolutions of the late eighteenth century. In the European monarchies, the suggestion that the children of peasants and workers should be educated would have been considered laughable. In those societies, children went to work with adults at an early age, and adolescence was not recognized as a distinct stage of development. Formal schooling, generally reserved for the children of the elite, typically lasted three or four years, after which the young person entered a profession.

Even after the creation of republics in France and the United States and the beginning of a movement for universal education, the development of a comprehensive system of schools took many generations. In the early history of the United States, the children of slaves, Native Americans, the poor, and many immigrant groups, as well as almost all female children, were excluded from educational institutions. The norm of segregated education for racial minorities persisted into the twentieth century and was not overturned until 1954 in the Supreme Court's famous ruling in *Brown v. Board of Education of Topeka*. Even after that decision, it took years of civil rights activism to ensure that African Americans could attend public schools with whites. Thus, although the idea of universal education in a democracy arose early, it took many generations of conflict and struggle to transform that idea into a strong social norm (Ariès, 1962; Cremin, 1980).

The idea of mass education based on the model created in the United States and other Western nations has spread throughout the world. Mass education differs from elite education, which is designed to prepare a small number of privileged individuals (generally sons of upper-class families) to run the institutions of society (the military, the clergy, the law, etc.). Mass education focuses instead on the socialization of all young people for membership in the society (Trow, 1966). Mass education is seen as a way for young people to become citizens of a modern nation-state. It also establishes an increasingly standardized curriculum and tries to link mastery of that curriculum with personal and national development (Benavot, Cha, & Kamen, 1991; Meyer, Ramirez, & Soysal, 1992).

Figure 11.1 indicates that by 1980 all the nations of the world had adopted the basic model of a mass educational system (although many had not extended educational opportunities to the majority of young citizens). Note the sharp upsurge in the 1950s. This was the decade when large numbers of colonial nations in Africa and Asia became independent. Commitment to a mass educational system became a hallmark of modernity for these and other new nations. It was also a time when agencies like the World Bank, UNESCO (United Nations Educational, Social, and Cultural Organization), and the U.S. Agency for International Development began to actively encourage and provide financial support for mass educational institutions.

FIGURE 11.1
Percentage of Nations Developing Systems of Mass Education, by Decade

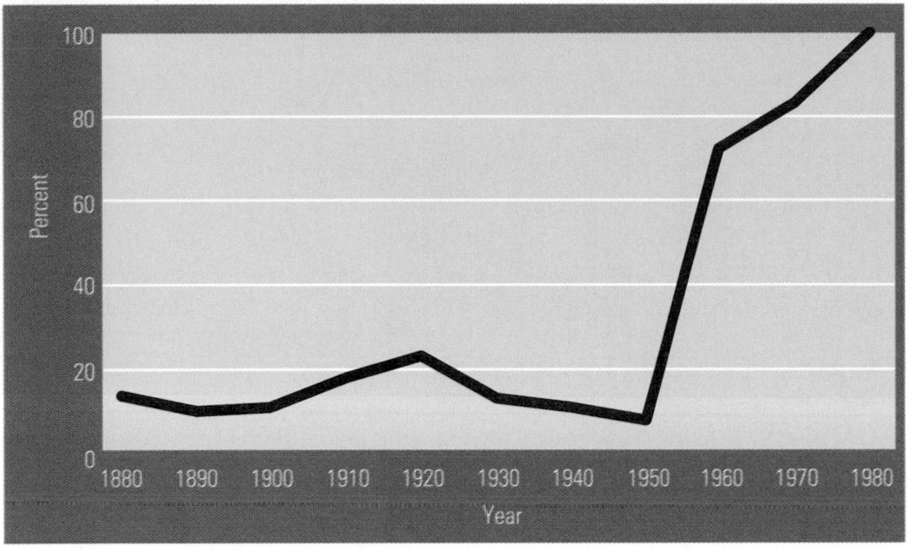

Source: Adapted from Meyer, Ramirez, & Soysal, 1992.

An extension of mass education in the United States was the expansion of public institutions of postsecondary education (and the parallel expansion of private colleges and universities, often with federal aid and research funding). This expansion was fueled by the post–World War II baby boom and the massive increase in the college-age population in the 1960s. During those years, there was a parallel boom in employment for elementary school teachers and then for high school teachers and college professors. But after the boom came the bust: The birthrate fell sharply beginning in the late 1950s, and by the mid-1970s college enrollments also began to decline. These changes had dramatic effects on primary schools, but at the college level they were partially offset by an unprecedented countertrend: the immense increase in the number of older students seeking higher education. Today large numbers of adults are returning to college. Unlike the typical student of earlier years, they are in the labor force, are married and living with their spouses, are going to school part-time, and are seeking skills and knowledge to enhance their careers (Bogue, 1985; Mulkey, 1993). The return of so many adults to educational institutions has led many social scientists to describe a future in which education will be a lifelong process in which people of all ages will move in and out of educational institutions (Parelius & Parelius, 1987; Reich, 1992).

Education in the Developed World. Although all urban industrial nations accept the principle of mass education, there are profound differences in their approaches to schooling, especially in regard to ideas of who gets educated and for what purposes. When sociologists compare different educational systems, they often employ a set of major comparative dimensions like the following:

- the age at which students are sorted into higher- and lower-track schools
- the proportion of secondary-school students who are in vocational versus academic schools, and the extent to which the vocational schools are directly tied in to future employment
- the proportion of students who attend universities, and the degree to which elite universities are linked to top-level corporate and government careers (Brint, 1998)

TABLE 11.1
Secondary Schooling, Selected Nations

	Percentage completing at typical age	Percentage of graduates in academic programs	Secondary school-leaving examinations
United States	75	Over 75	No
Japan	90	75	Yes
France	80	45	Yes
Sweden	75	35	No
England	60	40	Yes
Germany	90	25	Yes

Source: Adapted from Brint, 1998.

The United States has by far the broadest and most egalitarian approach to education; Germany has one of the narrowest. In the United States, children are largely untracked (although we will see that there are many ways in which inequalities are incorporated into schooling in this country). In Germany, tracking begins very early and is linked to vocational training and apprenticeships that prepare young people for employment at a relatively early age. Entry into universities is restricted in Germany, as it is in France and Sweden, although tracking begins somewhat later in those nations than it does in Germany. Japan is similar to the United States in that it provides largely untracked schooling for its primary- and secondary-school students, but it offers far more vocational schooling in the secondary years and restricts access to universities far more than does the United States. However, in the United States there are elite private universities that have more direct links to the best careers in some professions and public agencies than the state universities do. England, with its system of private secondary schools and elite universities (Cambridge and Oxford), is more restrictive at the top levels of education than the United States.

Table 11.1 summarizes some of these differences among the industrial nations. We see that the United States has the highest proportion of students who attend four-year colleges or universities, and more college graduates, but that the percentage of students who graduate "on time"—that is, with their age cohort—is lower than in Japan (75% as opposed to 90%). The table also shows that the United States does not require students to pass a national exam to qualify for secondary-school diplomas, although that is changing in many states at this writing.

Education in the Developing World. Despite the rise of mass education all over the world, there are persistent problems in the educational systems of many of the developing nations. Illiteracy, especially among women, remains high in some regions, particularly in the Muslim world and most parts of Africa below the Sahara. About 98% of people in the industrial nations are considered literate by United Nations standards, while 70% of people in the developing world, and only about 50% in the least-developed nations, are literate. In the past 20 years, the least-developed nations have managed to raise their literacy rates by more than 20 percentage points, due in part to funding specifically for that purpose from world development agencies. But in some parts of the world, particularly in Africa, mounting debt, corruption, the ravages of AIDS, and political instability have reversed progress toward literacy and mass education (IBRD, 2000). (See the Global Social Change box on pages 310–311.)

Another persistent educational issue in the developing world is overproduction of university graduates and the "brain drain." India and Latin America have had great success in improving access to universities but less success in producing university graduates with the qualifications and skills needed for economic development. Graduates with law or humanities degrees quickly become too numerous for existing job openings. Those with technical expertise in engineering and medicine are often offered jobs in the industrial nations. This brain drain leaves some nations with a skills deficit after paying the costs of educating university graduates who have emigrated (Brint, 1998). In consequence, the World Bank, the leading funding agency for education in the developing world, recommends that developing nations place more emphasis on investment in base education—that is, the first years of schooling—a strategy that the industrialized nations are applying to the educational needs of their disadvantaged populations.

Education and Citizenship

An important aspect of education in the United States is the relationship between education and citizenship. Throughout its history, this nation has emphasized public education as a means of transmitting democratic values, creating equality of opportunity, and preparing new generations of citizens to function in society. In addition, the schools have been expected to help shape society itself. During the 1950s, for example, efforts to combat racial segregation focused on the schools. Later, when the Soviet Union launched the first orbiting satellite, American schools and colleges came under intense pressure and were offered many incentives to improve their science and mathematics programs so that the nation would not fall behind the Soviet Union in scientific and technological capabilities.

Education is often viewed as a tool for solving social problems, especially social inequality. The schools, it is thought, can transform young people from vastly different backgrounds into competent, upwardly mobile adults. Yet these goals seem almost impossible to attain (Cahill, 1992). In recent years, in fact, public education has been at the center of numerous controversies arising from the gap between the ideal and the reality. Part of the problem is that different groups in society have different expectations. Some feel that students need better preparation for careers in a technologically advanced society; others believe children should be taught basic job-related skills; still others believe education should not only prepare children to compete in society but also help them maintain their cultural identity (and, in the case of Hispanic children, their language). On the other hand, policy makers concerned with education emphasize the need to increase the level of student achievement and to involve parents in their children's education (B. L. Wilson, 1993).

Some reformers and critics have called attention to the need to link formal schooling with programs designed to address social problems. Sociologist Charles Moscos, for example, is a leader in the movement to expand programs like the Peace Corps, VISTA, and Outward Bound into a system of voluntary national service. National service, as Moscos defines it, would entail "the full-time undertaking of public duties by young people—whether as citizen soldiers or civilian servers—who are paid subsistence wages" and serve for at least a year (1988, p. 1). In return for this period of service, the volunteers would receive assistance in paying for college or other educational expenses.

Advocates of national service and school-to-work programs believe that education does not have to be confined to formal schooling. In devising strategies to provide opportunities for young people to serve their society, they emphasize the educational value of citizenship experiences gained outside the classroom. Early in his administration, President Clinton, a believer in national service, implemented a modest program known as Americorps, a volunteer community service program targeted at young people who normally would not apply to existing volunteer programs like the Peace Corps. During the present administration, it is likely that the emphasis in developing youth service programs will shift from government agencies to churches and other "faith-based" organizations, as well as private companies, in keeping with a more conservative approach to social action.

ATTAINMENT, ACHIEVEMENT, AND EQUALITY

While some educational reformers focus on the need to expand learning opportunities through nonschool service experiences, by far the majority of scholars and administrators seek improvements in educational institutions themselves. Their efforts often focus on *educational attainment,* or the number of years of schooling that students receive, and *educational achievement,* or the amount of learning that actually takes place. Both aspects of education are closely linked to economic inequality and social mobility.

GLOBAL SOCIAL CHANGE

Literacy in a Changing World

There is a strong correlation between illiteracy and poverty throughout the world. The poor nations of the Sahel, such as Mali, Chad, Niger, Ethiopia, and the Sudan, suffer from some of the highest rates of illiteracy in the world. High rates of illiteracy are also evident in much of South Asia, especially on the Indian subcontinent. On the other hand, illiteracy rates are quite low in poor nations like Mexico and Cuba, where the ideology of social development places strong emphasis on educating the mass of citizens to the fullest extent possible.

Sociological theories of modernization stress the need for populations to become literate so that their members will be better-informed voters, more highly skilled workers, more careful parents, and generally better able to realize their human potential. Literacy, it is argued, significantly increases a society's human capital. Reductions in the level of ignorance yield improvements in every aspect of a nation's social and civic life.

This line of argument is clearly supported by the correlation between poverty and illiteracy in India (see chart). Overall, the chart shows that as illiteracy decreases, the amount of money a household is able to spend each month increases. The chart also shows that illiteracy and the poverty associated with it are much more prevalent among women than among men, with rural women showing the highest rates of illiteracy and poverty (about 90% illiterate in the high-poverty category). That men are far more likely than women to become literate in India is a reflection of the immense gap in prestige between the sexes in that society; men are considered far more worthy of education than women.

Recent research on the effects of literacy on vital measures of social change, such as reduced fertility, clearly shows the importance of educating women. Data from research conducted in Thailand and other Southeast Asian nations suggest that until women gain access to at least minimal educational

These women have chosen to stay in the fields where they work as agricultural laborers to attend a literacy class.

opportunities, fertility rates in those nations will remain high. This research confirms the hypothesis that "demographic change is unlikely if the movement towards mass schooling is confined largely to males" (London, 1992, p. 306).

These relationships among gender, illiteracy, and social indicators like poverty and fertility offer a warning that investments in literacy alone are necessary but not sufficient to accelerate social change in a population. Investments in literacy must be accompanied by strategies to reach the most impoverished and discriminated-against segments of the population (such as rural women in many societies). Such stratagems, in turn, are difficult to develop in a society in which the powerful may fear their effects.

In nations where poor rural women are offered more education, for example, the women often begin to demand greater equality. It takes farsighted leadership to actively promote such strategies. But if the full benefits of literacy and education are to be applied to national development, such policies are necessary.

Poverty and Illiteracy in India

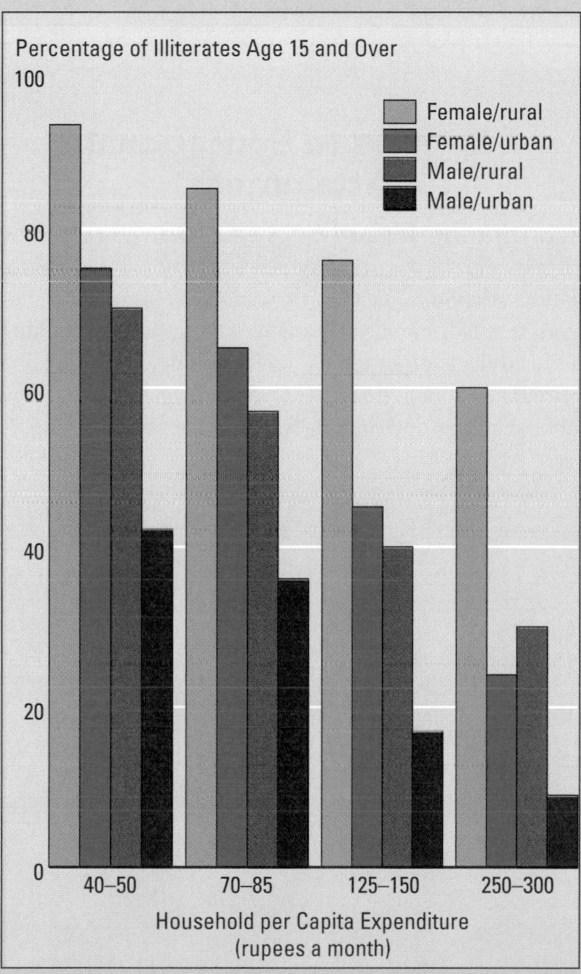

Source: Data from third quinquennial survey on employment and unemployment, Ministry of Planning, New Delhi, India.

Educational Attainment

In any discussion of education as a major social institution, the concept of **educational attainment** (number of years of school completed) holds a central place. Educational attainment is correlated with income, occupation, prestige, attitudes and opinions, and much else. It is essential, therefore, for social scientists to understand the impact of recent trends in school enrollment on the educational attainment of the population as a whole and of various subgroups of the population.

Table 11.2 shows the median number of years of school completed by the population as a whole in the decades since 1940, and Table 11.3 presents data on educational attainment by race, ethnicity, and sex for approximately the same time frame. Both tables apply to the population aged 25 and older. It is immediately clear that the average American today has much more education than the average American of the early 1940s. It is also clear that whites are more likely to complete high school than blacks or Hispanics and considerably more likely to attend college. The data indicate the increasingly high value placed on education, especially college education. People over age 75, who were born in the early decades of the twentieth century, have, on the average, considerably lower levels of educational attainment than their children and grandchildren (*Statistical Abstract,* 1999).

TABLE 11.2

Schooling Completed, by Decade, for U.S. Population as a Whole

Year	Median Years of School Completed
2000	12.9
1990	12.7
1980	12.5
1970	12.2
1960	10.6
1950	9.3
1940	8.6

Source: Data from Census Bureau.

Barriers to Educational Attainment

Continuing Racial Segregation. In the view of most jurists and sociologists who study U.S. race relations, the Supreme Court decision to end racial segregation in U.S. public schools was the most far-reaching judicial decision of the twentieth century. The decision stimulated the civil rights movement of the 1960s and abolished the doctrine of "separate but equal" schools

TABLE 11.3

Educational Attainment, by Race, Ethnicity, and Sex: 1947–1999

	Percent High School Graduate						Percent College Graduate					
	White		Black		Hispanic		White		Black		Hispanic	
Year	M	F	M	F	M	F	M	F	M	F	M	F
1999	84%	84%	77%	77%	56%	56%	29%	24%	14%	16%	11%	11%
1991	80	80	67	67	51	51	25	19	11	12	10	9
1987	77	77	63	64	52	50	25	17	11	10	10	8
1980	71	70	51	51	46	44	22	14	8	8	10	6
1970	57	58	32	35	NA	NA	15	9	5	4	NA	NA
1962	47	50	23	26	NA	NA	12	7	4	4	NA	NA
1947	33	37	13	15	NA	NA	7	5	2	3	NA	NA

*NA = Data not available.

Sources: Census Bureau, 1988, 1992; *Statistical Abstract,* 2000.

THE CENTRAL QUESTION: A CRITICAL VIEW

Can education in contemporary societies provide equality of opportunity for new generations of students?

Schools are among the few institutions that can significantly alter an individual's life chances. At least in theory, achievement in school can be based on merit, hard work, and motivation. Schools offer a chance to "level the playing field" and reduce the effects of economic inequality. In practice, however, schools vary in quality and in the amount of attention they can give to individual students. More affluent communities can provide better schools with higher standards for achievement, lower teacher–student ratios, more activities, and more diverse curriculums. Given these inequalities, it is no wonder there are so many different views about what schools should be doing and how they must change.

Public schools exist to provide education for all the members of a society, not just for those who can afford to pay for it. But consider the issues facing public schools in the United States: privatization and school choice, national standards for students and teachers, school prayer, school reform, violence in schools, desegregation, the role of athletics, appropriate curriculum, bilingual education, and many more. These topics often stimulate intense debate among people with different ideas about what education should accomplish and how schools should be organized and funded. The number of issues facing the schools, and the heat they generate in some communities, is staggering. To sort these issues out, it will help to stop and think about why public schools are the subject of such controversy.

One of the answers to this question is that ours is a democratic society in which public education is a major agency of socialization outside the family. Parents naturally have strong feelings about what experiences their children will be exposed to outside the home. If their own experiences in school were not entirely positive, as is often the case, or if they have strong views on moral issues, they are likely to have very strong feelings about what should and should not occur in the classroom.

How can any institution respond adequately to so many pressures and demands, especially in a time of decreasing budgets? As a sociologically informed person, you might decide, for example, that higher educational standards are the key to reducing social inequality. You will still have to decide whether individual schools should be allowed to set their own standards or whether there should be national standards. If you decide that there should be national standards, you will need to support your decision with sociological reasoning. Perhaps such standards will make it easier for poor districts to obtain the resources they need to meet national norms. Perhaps such standards will push teachers to demand more from their students. There are many other major educational policy issues to which you can apply your sociological skills, but you will have to be on the lookout for the trends and facts that will support your hunches. Above all, you will have to decide, by applying sociological insight and your own values, what aspects of public schools need to be changed and what existing educational practices and norms should be maintained. ■

for black and white Americans. But despite years of efforts to integrate public schools in the United States, racial segregation continues to present significant obstacles to equality of educational opportunity.

During the 1990s, the Supreme Court issued a number of rulings that have made it easier for school districts to avoid desegregation efforts and return to former patterns of segregation. Busing to achieve school desegregation is largely discredited as an effective strategy (Orfield, 1999). Minority black and Latino students tend to be concentrated in inner cities and older suburban communities, where the residential tax base is too low to permit the same quality of teaching, extracurricular activities, science laboratories, and other educational opportunities that are available in more affluent and less racially segregated communities.

In sum, it has proven to be far more difficult than anticipated to end school segregation. Most African American children still attend largely segregated schools, and segregation of Hispanic students is increasing as well. While there is a debate among African Americans and other minority groups about the wisdom of pursuing desegregation at all, most parents seek the best possible schooling for their children, and this desire steers them toward the better-endowed schools attended primarily by suburban white students. So the struggle to achieve desegregated schools in the name of equality of opportunity

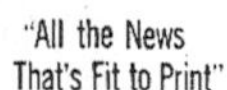
"All the News That's Fit to Print"

The New York Times.

LATE CITY EDITION
Fair and cool today. Mostly sunny, continued cool tomorrow.

Copyright, 1954, by The New York Times Company.

VOL. CIII . . . No. 35,178. NEW YORK, TUESDAY, MAY 18, 1954. FIVE CENTS

HIGH COURT BANS SCHOOL SEGREGATION; 9-TO-0 DECISION GRANTS TIME TO COMPLY

It took courageous action on the part of many African Americans—such as Elizabeth Eckford, the young girl shown in the first photo—to even begin to break down the barriers to racial desegregation in the South and, later, in the North.

and racial tolerance continues. Sociological research on this issue shows that as long as American communities remain segregated by race it will be extremely difficult to achieve desegregated schools (Massey & Denton, 1993; Myers & Wolch, 1995).

Tracking and Inequality. The rise of mass education gave the middle and lower classes greater opportunities for upward social mobility through educational attainment. But as early as the 1920s, many schools began to use "tracking" systems in which higher-achieving students were placed in accelerated classes while other students were shunted into vocational and other types of less challenging classes. Today tracking remains a major problem in public schools. Parents of "gifted" children seek educational challenges for them and do not want them to be held back by slower learners. However, tracking systems can make average students feel less valued, and there is a danger that gifted but alienated students will be labeled as nonachievers.

A major national study of more than 14,000 eighth-grade students in public schools showed quite clearly that white and Asian students are far more likely than black and Hispanic students to be tracked into high-ability groups, and that students from high socioeconomic backgrounds are also far more likely to be tracked into such groups. (Ability grouping is synonymous with tracking.) Some social scientists argue that these differences are a result of ability differences among racial and class groups; others assert that they are a consequence of race and class bias (Murray, 1993; Oakes, 1985). Since there are far more Hispanic and black students in the lower socioeconomic classes, there is little question that social factors outweigh biological ones in explaining these results. In any case, tracking separates children and is increasingly viewed as leading to educational inequalities. A current trend in educational practice, therefore, is to institute "detracking" programs and to provide highly gifted students with additional challenges through cooperative education (in which they have opportunities to teach others) and through after-school programs (Mansnerus, 1992).

Dropping Out. The educational attainment of various subgroups of the population is virtually identical up to the age of 13 (Jencks et al., 1972). The picture begins to change in the early secondary-school years, however. Until quite recently, the dropout rate for black and Hispanic students in high school was almost twice the rate for white students. In 1973, 14.2% of white students dropped out without finishing high school, while among black students the proportion was 26.5%. By 1998, the situation had improved considerably: In that year, 9% of white students and 17.2% of black students dropped out (Census Bureau, 2000). Table 11.4 shows that high school graduation rates (the opposite of dropout rates) vary among the states, with the highest in Alaska at 92.8% and the lowest in West Virginia at 75.1%.

The main reason students drop out of school is poor academic performance, but there are other reasons as well. Students often drop out because of the demands of work and family roles; many are married, or unmarried and pregnant, and/or working at regular jobs. Whatever the reason, the effects of dropping out can be serious. Dropouts have less chance of joining the labor force than high school graduates; whatever jobs they find tend to be low-paying ones. From 1974 to 1992, the income advantage from completing college increased for all individuals, regardless of gender or minority status. But black-white differentials are found at each level of attainment. White college graduates, for example, earned 23% more in 1992 than black college graduates of the same age (National Center for Educational Statistics, 1995).

Degree Inflation. The trend toward increasingly higher levels of educational attainment has had an unexpected effect known as "degree inflation." Employers have always paid attention to the educational credentials of potential employees, but today they require much more education than in the past. For example, in the early decades of the twentieth century a person could get a teaching job with a high school diploma; now a bachelor's and often a master's degree is usually required. The same is true of social work. And secretaries, who formerly could get by without a high school diploma, now are often required to have at least some college education and, in some cases, a college degree.

Degree inflation is discouraging to some students and prevents them from continuing their education. It also adds to the expense of education (both directly and indirectly, in terms of lost income) and hence prevents less advantaged students from undertaking advanced studies. Degree inflation also increases the amount of time that must be devoted to formal education. It therefore raises questions about the meaning of educational achievement—that is, the value of the time spent attaining educational credentials.

TABLE 11.4
High School Graduation Rates, by State (persons age 25 and over)

State	Percent	State	Percent
Alabama	81.1	Montana	88.8
Alaska	92.8	Nebraska	89.3
Arizona	83.1	Nevada	86.4
Arkansas	78.9	New Hampshire	86.5
California	80.4	New Jersey	87.4
Colorado	90.4	New Mexico	80.9
Connecticut	83.7	New York	81.9
Delaware	84.5	North Carolina	79.8
District of Columbia	82.8	North Dakota	84.9
Florida	82.7	Ohio	86.1
Georgia	80.7	Oklahoma	83.5
Hawaii	88.0	Oregon	86.2
Idaho	84.8	Pennsylvania	86.1
Illinois	85.4	Rhode Island	80.9
Indiana	82.9	South Carolina	78.6
Iowa	89.7	South Dakota	88.7
Kansas	87.6	Tennessee	79.1
Kentucky	78.2	Texas	78.2
Louisiana	78.3	Utah	91.0
Maine	88.9	Vermont	89.3
Maryland	84.7	Virginia	87.3
Massachusetts	85.1	Washington	91.2
Michigan	85.5	West Virginia	75.1
Minnesota	91.1	Wisconsin	86.8
Mississippi	78.0	Wyoming	90.7
Missouri	85.0		

Source: Adapted from *Statistical Abstract,* 2000.

Educational Achievement

Differences in levels of educational attainment are viewed as a sign that public education is not meeting the expectations of society in terms of the quantity of education provided to citizens. There is also controversy over the quality of education, or educational achievement as reflected in scores on standardized tests like the Scholastic Aptitude Test (SAT). **Educational achievement** refers to how much the student actually learns, measured by mastery of reading, writing, and mathematical skills. It is widely believed that the average level of educational achievement has declined drastically in the past two decades. In a 1983 report entitled "A Nation at Risk," the National Commission on Excellence in Education pointed to the decline in the average test scores of high school students since the mid-1960s and stated that the schools have failed to maintain high educational standards. In recent years, standardized scores in math have risen somewhat, but verbal scores (reading and vocabulary) have not. Many observers attribute the decline in test scores to a variety of social conditions such as too much television viewing and changing values related to family life.

Critics of excessive reliance on standardized tests might agree with prominent educational reformer Ted Sizer, who points out that "none of the major tests used in American elementary and secondary education correlates well with long-term success or failure. SAT scores, for example, suggest likely grades in the freshman year at college; they do not predict much thereafter" (1995, p. 58). For better or worse, however, the SAT and other standard tests continue to be used as primary measures of individual and school performance.

Another problem related to educational achievement is the high rate of "functional incompetency" among Americans. A surprisingly large number of adults are unable to read, write, keep a family budget, and the like. Although more and more people are obtaining a college education, many others are being left behind, particularly members of the lower social classes, people for whom English is a second language, and people with learning disabilities. As increasing amounts of education are required for better jobs, this cleavage between educational haves and have-nots becomes an ever more dangerous trend.

Recent research on school achievement offers even more disturbing evidence of deficiencies in the educational achievement of U.S. students, beginning at an early age. Tests of achievement by thousands of schoolchildren in comparable U.S. and Asian cities—Minneapolis, Chicago, Sendai (Japan), Beijing (China), and Taipei (Taiwan)—reveal that the mathematics scores of American first graders were lower than those of Asian first graders. In some American schools, first graders' scores were similar to those of Asian students, but by the fifth grade all the American students had fallen far behind. In computation, for example, only 2.2% of children in Beijing scored as low as the mean score for U.S. fifth graders. On a test consisting of word problems, only 10% of fifth graders in Beijing scored as low as the average U.S. student. Gaps also appeared in reading skills, although vast differences in written languages make it somewhat difficult to compare achievement in those skills (H. W. Stevenson, 1998; Tobin, Wu, & Davidson, 1989).

Americans are likely to suggest that these differences occur because citizens of Asian nations value education more than Americans do and that in Asian schools children are placed under great stress, learn by rote and drill, and spend long hours in school. The researchers found that only the first of these explanations is correct. It is true that Asian parents value education more than American parents do, but it is not true that Asian schoolchildren experience more stress, do more rote learning, and spend more time at their desks. In fact, the opposite is true: Asian children experience less stress and anxiety than American children, enjoy school more, and have far more opportunities for recreation and breaks from classroom work despite their longer school day. The frequency of recreation breaks appears to help them pay more attention to their classroom work than is typical of children in U.S. schools.

Close observation of teaching in the different schools revealed that Asian teachers tend to use work that the students do at their desks (seat work) as an opportunity to pay attention to individual students. Teachers in U.S. schools are less likely to circulate through the class and comment on individual students' efforts.

Finally, as shown in Figure 11.2, there are important differences in the way parents perceive schooling in Asia and the United States. Mothers in the United States tend to be far more positive than Asian mothers about their children's schoolwork at all grade levels. They are also more likely to believe that the school is doing an excellent job in educating their children. American parents tend to believe far more in the effects of innate ability and less in the value of effort in school than Asian parents do. This difference is especially important because it seems to lead U.S. parents to more readily accept their children's performance in school and to demand less of their children. Parents' belief in innate ability also appears to make them far less critical of the schools than they might otherwise be (H. W. Stevenson, 1992).

This research has many practical implications for social change in the schools. It is apparent, for example, that simply lengthening the school day in the United States would not have the desired results. Indeed, it might well have negative results without far greater provision for time when children can play and socialize with their peers as they do in Asian schools.

Education and Social Mobility

Studies by educational sociologists have consistently found a high correlation between social class and educational attainment and achievement, so much so that major efforts to address inequalities in educational opportunities were undertaken during the 1960s and 1970s. Nevertheless, there is evidence that the promise of equal education for all remains far from being fulfilled. Educational institutions have been subjected to considerable criticism by observers who believe that they hinder, rather than enhance, social mobility (Fullan, 1993). Christopher Jencks, for example,

FIGURE 11.2
Parents' Satisfaction With Academic Performance of Students and Schools in Selected U.S. and Asian Cities

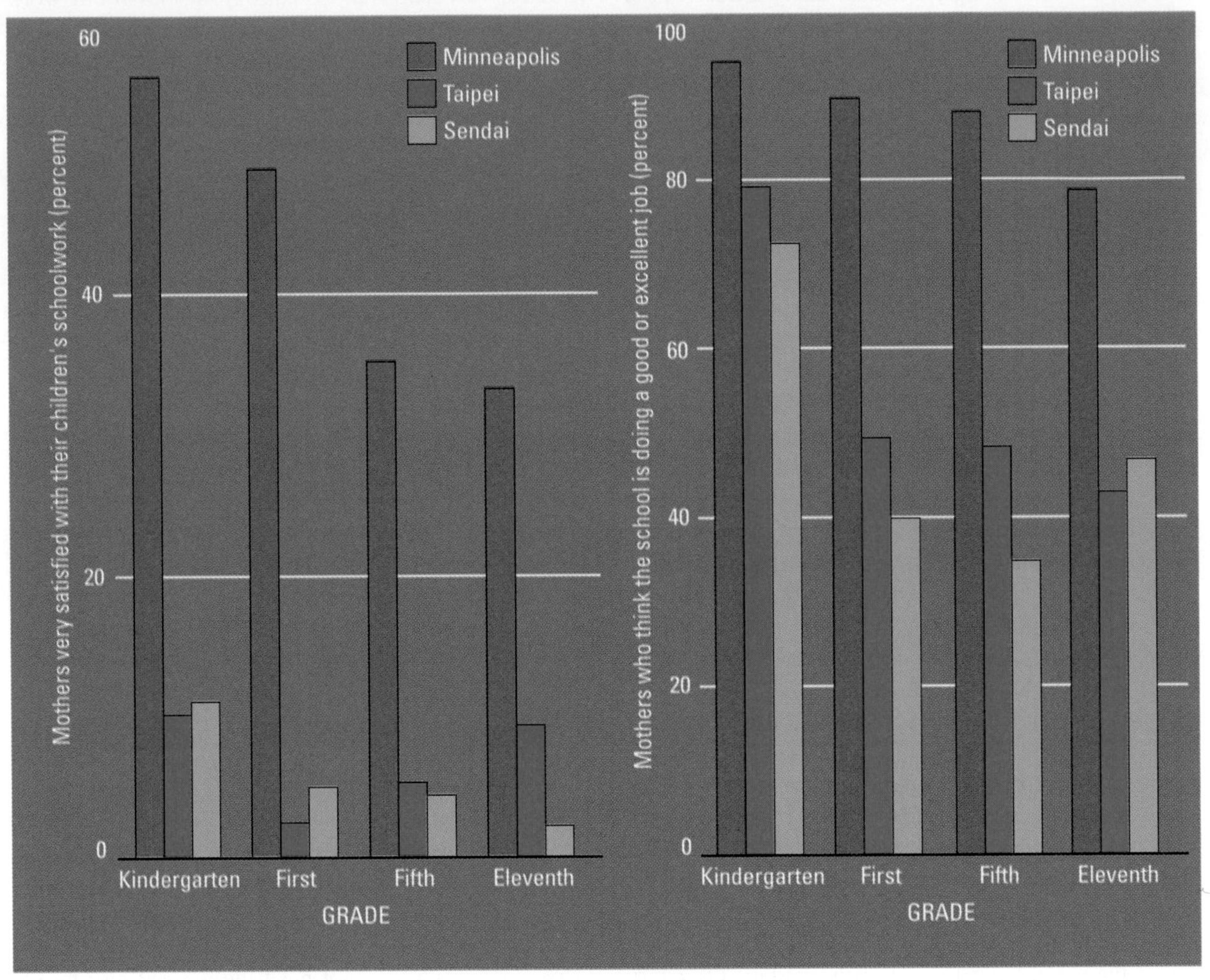

Source: From "Learning From Asian Schools," by H. W. Stevenson. Copyright © 1992 by Scientific American, Inc. All rights reserved.

argues that schools serve primarily to reproduce the existing stratification system. He asserts that "schools serve primarily as selection and certification agencies, whose job is to measure and label people, and only secondarily as socialization agencies, whose job is to change people" (Jencks et al., 1972, p. 135).

Jencks notes that the main criterion for granting a diploma or degree (in other words, certification) is usually the amount of time spent in school, not the skills learned. "Imagine," he writes, "what would happen to high school enrollment if states allowed anyone, regardless of age, to take a high school equivalency examination. Most capable students would probably leave high school by the time they were 16" (Jencks et al., 1972, pp. 135–136). In their study of the relationship between educational certification and inequality, Jencks and his researchers assumed that "the value of any given credential depends solely on how long it takes to acquire" (p. 136). And the chief factor that keeps some students in school and college longer than others, according to Jencks, is family background—in other words, social class. Members of the upper classes often regard college as an important transition to high-status occupations and therefore encourage their children to finish high school and attend college.

Inequality in the Classroom. In *The Urban School: A Factory for Failure* (1973), Ray C. Rist went even further, bluntly stating that "the system of public education in the United States is specifically designed to aid in the perpetuation of the social and economic inequalities found within the society" (p. 2)—despite the widespread belief that education increases social mobility. He went on to state:

> Schooling has basically served to instill the values of an expanding industrial society and to fit the aspirations and motivations of individuals to the labor market at approximately the same level as that of their parents. Thus it is that some children find themselves slotted toward becoming workers and others toward becoming the managers of those workers. (1973, p. 2)

This highly influential study of educational inequality inspired many additional studies that have attempted to discover the actual processes through which education reproduces patterns of inequality rather than correcting them. In one example of such research, Jay MacLeod (1987, 1995) studied teenagers in a poor community who came from the same housing projects and attended the same high school. The members of one group, the Hallway Hangers, were predominantly white. They often cut classes, misbehaved in the classes they did attend, and used drugs. The members of the other group, the Brothers, were largely black but came from the same neighborhood and the same social class as the Hallway Hangers. In contrast to the Hallway Hangers, the Brothers tended to study hard, attend classes, and behave well in class. They accepted the idea that achievement in school would lead to success in the outside world, an idea the Hallway Hangers rejected.

These findings show that the economic aspects of social class are insufficient to explain success and failure in school. One must also understand the social influences that shape the actual experiences of individual students. The Brothers, it turns out, came primarily from black families who believed in equality of educational opportunity. Most of the Hallway Hangers came from families that did not value education or believe their children had much of a chance to succeed through education. The Brothers' parents were strict with their sons and punished school failure. The parents of the Hallway Hangers gave their sons much more leeway and paid far less attention to what they were doing (or not doing) in school. Of course, these differences are specific to these particular groups. One could find situations in which the outcomes are quite different. The point here is that working-class and poor children do not automatically inherit their lower-class position through the failure of education to reach them or through their own failure to accept the value of education. The influence of specific family and school environments is of great importance (MacLeod, 1987, 1995; Mehan, 1992).

Other firsthand observations of educational settings shed light on the importance of culture in explaining how inequalities persist despite educational opportunity. For example, studies in schools where there are significant numbers of students who do not hear English spoken at home show that problems with learning are not caused by language differences alone. They may also result from different styles of expression and interaction in the non-English culture. Studies have shown, for example, that white middle-class teachers often use questions rather than the direct demands that students may be used to hearing at home—"Don't you want to try to work on your math?" instead of "Now sit down and work on your math" (Cazden, 1988; Delgado-Gaiton, 1987). And some groups, such as Native Americans, may not respond well to the competitive environment of the typical classroom but do far better in a class where cooperative work and sociability are valued (Philips, 1982).

Important as these cultural issues may be in explaining school achievement, economic class may be even more significant. It remains true that students from poor families are likely to attend schools with far fewer resources, less favorable student–teacher ratios, and more staff turnover than the schools attended by middle- and upper-class students (Karabel & Halsey, 1988). These class differences are also evident at the level of higher education.

Inequality in Higher Education

Inequality in higher education is primarily a matter of access—that is, ability to pay. Ability to pay is unequally distributed among various groups in society, and as a result, students from poor, working-class, and lower-middle-class families, as well as members of racial minority groups, are most likely to rely on public colleges and universities. Social-scientific evidence indicates that such inequalities are alleviated by federal aid to students from families with low or moderate income. It is likely that without Pell grants, work-study funds, and student loan programs, the proportion of students from such families would be far

FIGURE 11.3

Comparisons of Maximum Pell Grant and Tuition, Room, and Board at Public Four-Year Universities, 1973–1993

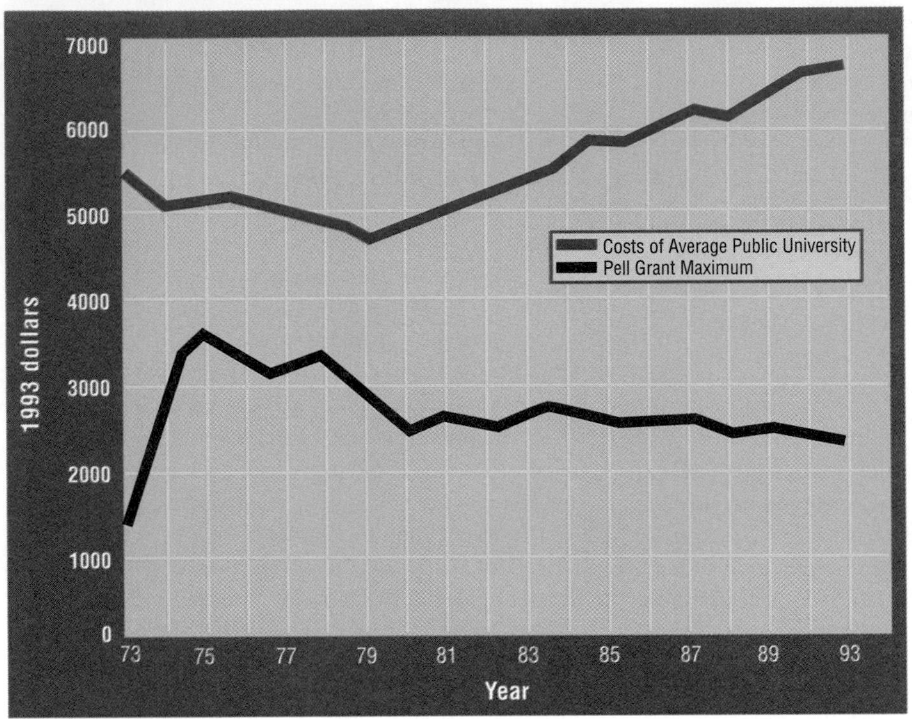

Source: The College Board, *Trends in Student Aid: 1984 to 1994* (Washington: Washington Office of the College Board, 1994).

lower than it currently is. Even with such aid, the proportion of low-income students in colleges and universities is lower than that of students from high-income families. Figure 11.3 shows that while the costs of tuition and student life at public universities has been increasing since the late 1970s, over the same period the maximum value of federal Pell grants decreased by about 22%.

Sandra Baum (1987) examined the educational careers of 2,000 students who were high school seniors in 1980. She found that college attendance rates were higher for students from high-income families than for students from low-income families (60% versus 46%). However, achievement in school seemed to account for an even larger difference. Students who scored high on achievement tests but came from low-income families had far higher rates of college attendance than low-scoring students from affluent homes. Thus, achievement in the primary school grades seems to be at least as important as family income in explaining a student's success in higher education—but remember that in the lower grades students from more affluent homes tend to achieve more than students from less-advantaged homes.

Baum and other researchers have also found that low-income students tend to be concentrated in two-year colleges and that students in such colleges are much more likely to drop out before completing a degree than students who enter four-year colleges. On the basis of findings like these, most educational researchers agree that without college assistance for needy students, differences in educational attainment and achievement would be much greater. And since public higher-education institutions remain the primary route to college degrees for low-income students, increased support for such institutions also tends to diminish inequality in access to higher education (Baum, 1987; Hansen & Stampen, 1987; U.S. Department of Education, 1994).

TABLE 11.5
Major Findings at Age 19 in the Perry Preschool Study

Category	Number Responding*	Preschool Group	No-Preschool Group
Employed	121	59%	32%
High school graduation (or its equivalent)	121	67%	49%
College or vocational training	121	38%	21%
Ever detained or arrested	121	31%	51%
Females only: teen pregnancies, per 100**	49	64	117
Functional competence (APL Survey: possible score 40)	109	24.6	21.8
% of years in special education	112	16%	28%

*Total *N* = 123.
**Includes all pregnancies.

Source: Data from Berrueta-Clement et al., 1984.

Education for Equality

The question of whether and in what ways education leads to social mobility remains open. Some social scientists view education as an investment like any other: The amount invested is reflected in the future payoff. This is known as human capital theory. In this view, differences in payoffs (jobs and social position) are justified by differences in investment (hard work in school and investment in a college education). However, critics of the educational system point out that the resources required to make such "investments" are not equally available to all members of society.

The question of whether the society as a whole, or only its more affluent members, will invest in its "human capital" through such means as preschool programs and student loans is a major public-policy issue. This is especially true as technological advances coupled with degree inflation increase the demand for educated people (Bell, 1973; Reich, 1992). Empirical evidence of the effectiveness of such investments may be seen in the results of an innovative study conducted across two decades by the High/Scope Educational Research Foundation (Berrueta-Clement, Schweinhart, Barnett, Epstein, & Weikart, 1984).

The subject of the study, begun in the 1960s, was a preschool program at the Perry Elementary School, located in a low-income black neighborhood of Ypsilanti, Michigan, a small industrial city on the outskirts of Detroit. The researchers randomly selected 123 three-year-old boys and girls and assigned them either to the preschool group or to another group that would not go to the preschool program. The latter children, like the majority of children at that time, would not be enrolled in school for another two years.

About twice a year for the next 20 years, the researchers traced the experiences of the experimental (preschool) and control (no-preschool) groups. The data in Table 11.5 show that there were important differences between the two groups as a consequence of the preschool experience. Members of the preschool group achieved more in school, found better jobs, had fewer arrests, and had fewer illegitimate children; in short, they experienced more success and fewer problems than members of the control group. These findings confirm what other, less well-known studies had shown. For example, studies by Benjamin Bloom (1976) and by Piaget and Inhelder (1969) had indicated that as much as 50% of variance in intellectual development takes place before the age of four.

The most significant contribution of studies like the High/Scope research is their documentation of the actual improvements that a good preschool program can produce in disadvantaged children. The High/Scope study was also able to translate its findings into dollar figures. A year in the Perry program cost about

$1,350 per child. The researchers showed that the "total social benefit" from the program (measured by higher tax revenues, lower welfare payments, and lower crime costs) was equivalent to $6,866 per person.

This research became a classic in the social-scientific study of education because it demonstrated the undeniable benefits of preschool education to society. It also appeared during a period when funding for public preschool programs was in danger of being drastically reduced. The findings of the High/Scope study gave congressional advocates of preschool education the evidence they needed to justify continued federal funding. In many instances, however, social scientists know what needs to be done to improve educational achievement and increase equality of educational opportunity but are frustrated by the resistance to change that is inherent in large-scale bureaucratic systems such as the public school systems of many states and municipalities.

THE STRUCTURE OF EDUCATIONAL INSTITUTIONS

A significant barrier to educational reform is the bureaucratic nature of school systems. We noted earlier that sociologists view the school as a specialized structure with a special socializing function and that it is also a good example of a bureaucratic organization. As any student knows, there is a clearly defined status hierarchy in most schools. At the top of the hierarchy in primary and secondary schools is the principal, followed by the assistant principal and/or administrative assistants, the counselors, the teachers, and the students. Although the principal holds the highest position in the system, his or her influence on students usually is indirect. The teacher, on the other hand, is in daily command of the classroom and therefore has the greatest impact on the students. In this section, we discuss several aspects of the structure of educational institutions and attempts to change those institutions.

Schools as Bureaucracies

As the size and complexity of the American educational system have increased, so has the tendency of educational institutions to become bureaucratized (Parelius & Parelius, 1987). The one-room schoolhouse is a thing of the past; today's schools have large administrative staffs and numerous specialists such as guidance counselors and special education teachers. Teachers themselves specialize in particular subject areas or grade levels. Schools are also characterized by a hierarchy of authority. The number of levels in the hierarchy varies, depending on the nature of the school system. In large cities, for example, there may be as many as seven levels between the superintendent and school personnel, making it difficult for the superintendent to control the way policies are carried out. Similarly, in any given school it may be difficult for the principal to determine what actually happens in the classroom.

The reforms sought by educational leaders are being implemented in schools throughout the United States. Large cities like Chicago and New York are instituting more opportunities for parents to get involved in school issues and are seeking to reduce the size of schools so that teachers, students, and administrators feel that they are part of a "learning community" in which they all know one another and are concerned with one another's well-being—conditions that may not be possible in larger schools, where bureaucratic rules and regulations often seem to take precedence over personal relations (Lee & Smith, 1993; Meier, 1995).

In many school districts, however, the pace of reform is slowed by bitter disagreements about how to fund the schools in an era of reduced public spending. There are also controversies over issues such as sex education, tolerance for homosexual students, distribution of condoms to prevent AIDS and teenage pregnancy, and school prayer. These can divide communities and prevent structural changes like the move toward smaller schools and more local responsibility for educational decisions (Fernandez, 1992; Rose, 1995).

The Classroom. In most modern school systems, the primary school student is in the charge of one teacher, who instructs in almost all subjects. But as the student advances through the educational structure, the primary school model (which evolved from the one-room school with a single teacher) is replaced by a "departmental" structure in which the student encounters a number of specialized teachers. The latter structure is derived largely from that of the nineteenth-century English boarding school. But these two basic structures are frequently modified by alternative approaches like the "open" primary school classroom, in which students are grouped according to their level of achievement in certain basic skills and work in these skill groups at their own pace. The various groups in the open classroom are given small group or individual instruction by one or more teachers rather than expected to progress at the same pace in every subject.

Open classrooms have not been found to produce consistent improvements in student performance, but they have improved the school attendance rates of students from working-class and minority backgrounds. Students in open classrooms tend to express greater satisfaction with school and more commitment to class work. The less stratified authority structure of the open classroom and the greater amount of cooperation that occurs in such settings may help students enjoy school more and, in the long run, cause them to have a more positive attitude toward learning.

Does Teaching Matter? Outside of parents and peers, teachers are potentially the most influential people in a person's life. Think of the lasting influence of the preschool and kindergarten teachers who, along with our parents, are our earliest guides to the world around us, or the elementary school teachers from whom we learn to read and do fractions, or the high school teachers who coach and cajole and mentor, and the college professors whose teachings may offer the promise of an adult career. In this steady progression of teachers, we encounter some who are better and some who are worse (often a very subjective perception), and it is natural to wonder how much of a difference teachers actually make in people's lives.

Among the studies that have attempted to assess teachers' lasting impact, one stands out as a classic (Pedersen & Faucher, 1978). It examined the impact of a single first-grade teacher on her students' subsequent adult status (see the Sociological Methods box on pages 324–325). The surprising results of this study have important implications. It is evident that good teachers can make a big difference in children's lives, a fact that gives increased urgency to the need to improve the quality of primary school teaching. The reforms carried out by educational leaders like James Comer and Deborah Meier suggest that when good teaching is combined with high levels of parental involvement, the results can be even more dramatic. Comer developed successful innovations in the schools of New Haven and Prince George's County, Maryland; Meier is the creator of innovative public schools in New York City and Boston that have become models for providing more choices within the public school system (Comer, 1984, 1985, 1987; Meier, 1995).

Teacher–Student Relationships. Sociologists have pointed out that the relationship between teacher and student is asymmetrical or unbalanced, with the teacher being in a position of authority and the student having little choice but to passively absorb the information provided by the teacher. In other words, in conventional classrooms there is little opportunity for the student to become actively involved in the learning process. The most influential contemporary French sociologist, Pierre Bourdieu (1988), emphasizes that teachers tend to form the strongest relationships with students from more advantaged families who can use language well and have stronger cultural backgrounds through early reading and family trips. These relationships often put less-affluent students at a disadvantage and may reinforce social-class distinctions within schools, which in principle are devoted to encouraging equality of opportunity (Swartz, 1997).

On their side, students often develop strategies for undercutting the teacher's authority: mentally withdrawing, interrupting, and the like (Csikszentmihalyi, Larson, & Prescott, 1977; J. Holt, 1965, 1967; Meier, 1995; Rose, 1995). Much current research assumes that students and teachers influence each other instead of assuming that the influence is always in a single direction from teachers to students.

The realization that education is a matter of enduring and reciprocal influences between teachers and students is a strong argument for continuing to seek ways to achieve greater racial and ethnic diversity in the schools. As the noted social psychologist Kenneth Clark argued in the studies cited in the Supreme Court's decision in *Brown* v. *Board of Education,* it is just as important for white students to be taught by nonwhite teachers as it is for minority students to be taught by white teachers. Through greater diversity in both students and faculty comes a wider range of educational experiences, which, in turn, foster greater tolerance and understanding.

Teacher Status. How different societies compensate teachers reveals a great deal about the relative status of teaching and schooling in those societies. The United States falls about in the middle of the range of teacher compensation among the major developed nations. Teachers are under increasing pressure to keep up with rapidly changing curricula, new teaching technologies, and new testing systems, but in the United States they must continually struggle to maintain their salaries and ensure that student–teacher ratios do not creep upward. For this reason, the

SOCIOLOGICAL METHODS

Tracing the Influence of a Remarkable Teacher

In an unusual study of teacher–student interaction, Eigil Pedersen and Therese Annette Faucher (1978) demonstrated the persisting value of an outstanding primary school teacher. The study is unusual because the social scientists were actually looking for a general pattern of low teacher expectations. Instead, they found an exception that demonstrated the powerful effects an exceptional teacher can have on students. In scientific research, an unexpected result that leads to a new insight is known as *serendipity.*

Pedersen and Faucher began their study of IQ and achievement patterns among disadvantaged children at a school that was marked by high rates of failure. A large percentage of the school's graduates failed in their first year of high school and dropped out. The researchers attempted to explain these failure rates in terms of the concept of the "self-fulfilling prophecy," the idea that if teachers expect students to do poorly, the students are likely to perform accordingly. But if the negative prophecy seems to work in many instances, can we find evidence that a *positive* self-fulfilling prophecy—the belief that students can perform well—will also work? This is what makes Pedersen and Faucher's study so interesting.

As they examined the IQ scores of pupils in the school's first-grade classes, the researchers found a clear association between changes in a pupil's IQ and the pupil's family background, first-grade teacher, and self-concept. Further examination of the school's records revealed a startling fact: The IQs of pupils in one particular teacher's first-grade class were significantly more likely to increase in subsequent years than the IQs of pupils in first-grade classes taught by other teachers. And pupils who had been members of that class were more than twice as likely to achieve high status as adults as pupils who had been members of other first-grade classes (see table).

What was so special about Miss A, as the outstanding teacher was labeled? First, she was still remembered by her students when they were interviewed 25 years after they had been in her class. More than three-quarters of those students rated her as very good or excellent as a teacher. "It did not matter what background or abilities the beginning pupil had; there was no way that the pupil was not going to read by the end of grade one." Miss A left her pupils with a "profound impression of the importance of schooling, and how one should stick to it" and "gave extra hours to the children who were slow learners" (p. 19). In nonacademic matters, too, Miss A was unusual:

unionization of teachers at all levels of formal education has been advancing far more rapidly than union membership in other white-collar professions (Arnold, 2000).

TECHNOLOGY AND SOCIAL CHANGE

Most of the technologies that are changing the contemporary world have actually been with us for much of the past century. Air travel, radio, television, telephones, automobiles, atomic energy, medical and biotechnology, and many more major technological systems continue to have a major impact on societies throughout the world. But the newest technologies, those based on the marriage of computers and telecommunications technologies—which we can refer to for simplicity as the Internet—are having the greatest impact on education in the advanced industrial nations, particularly the United States. In 1984, only 27% of students in the United States used computers at school. By 1997, the last year for which data are available, that figure had risen to almost 70% (*Statistical Abstract,* 2000). And although there is a gap in access to school computers between affluent and

Adult Status, by First-Grade Teacher

	First-Grade Teacher			
Adult Status	**Miss A**	**Miss B**	**Miss C**	**Others**
High	64%	31%	10%	39%
Medium	36	38	45	22
Low	0	31	45	39
Total	100%	100%	100%	100%
(*N*)*	(14)	(16)	(11)	(18)
Mean adult status	7.0	4.8	4.3	4.6

Note: "Adult status" was determined from interviews that included questions on occupational status and work history, highest grade completed, rent paid and number of rooms, and related indicators of social position.

*(*N*) = Number of students who could be located and interviewed 25 years later.

Source: Eigil Pedersen & Therese Annette Faucher, "A New Perspective on the Effects of First Grade Teachers on Children's Subsequent Adult Status," *Harvard Educational Review,* 1978, 48:1, 1–31.

When children forgot their lunches, she would give them some of her own, and she invariably stayed after hours to help children. Not only did her pupils remember her, but she apparently could remember each former pupil by name even after an interval of 20 years. She adjusted to new math and reading methods, but her success was summarized by a former colleague this way: "How did she teach? With a lot of love!" (pp. 19–20)

In summing up their findings, Pedersen and Faucher stated that their data "suggest that an effective first-grade teacher can influence social mobility" (p. 29). Their findings differ from those of Coleman (1976) and Jencks and colleagues (1972), who believe that there is little correlation between school experiences and adult status. Pedersen and Faucher agree that further research on the relationship between teacher effects and adult status is needed. However, "In the meantime, teachers . . . should not accept too readily the frequent assertion that their efforts make no long-term difference to the future success of their pupils" (1978, p. 30). As debates continue over how much difference good teaching can make and as governments and school systems wonder whether they should do more to reward good teaching, the results of this study bear careful review.

disadvantaged children, that gap has narrowed rapidly in the past few years due to policies like the federal "e-rate subsidy," intended to fund computer and Internet access in disadvantaged communities.

Sociological observers of this trend are often quite critical of what they think of as excessive "hype" about the benefits of computers for learning (Harmon, 2000). From his observations and interviews with students and teachers in a wide range of elementary and high schools, Larry Cuban (1999) finds that the official statistics about computers in classrooms are highly misleading. While it is true that 70% of American classrooms have computers, this does not mean that teachers and students are actually using them in effective ways—or at all in some instances. In the classrooms Cuban visited, computers were used for an average of only one hour a week. In some classrooms, where especially creative and technically competent teachers worked hard to integrate use of computers and the Internet into their teaching, the amount of time spent at the computer was far higher, but these were not typical classrooms or teachers. Cuban finds that there is a wide gap between availability of hardware and teachers' ability to use it in their teaching. Too few teachers have been trained in computer use or feel confident in their ability to adapt the new technologies

SOCIOLOGY AND SOCIAL JUSTICE

President Bush's choice of former football coach and school administrator Roderick Paige as Secretary of Education is widely viewed as a move to increase the quality of education, especially through standardized tests, which Paige has used effectively in Houston. But anyone critical of the cozy relationship between education and major sports is probably in for trouble. In Texas, as in many other states, school sports have become a form of civil religion that masses of people, especially men, regard as giving meaning to their lives.

One of the best studies of this phenomenon is *Friday Night Lights,* Buzz Bissinger's (1990b) journalistic study of high school football in the West Texas city of Odessa and an example of the sociological imagination at work. Here is how Bissinger (1990a) describes his book and the town's reaction to it:

> *Friday Night Lights* chronicles the fortunes of the Odessa Permian High Panthers during the 1988 football season. Permian is the most successful high school team in modern Texas history, with five state championships to its credit since 1965. The pride the Panthers have brought to Odessa, a lonely, hardscrabble oil town of 100,000 in the middle of the vacant west Texas plains, is real and in many ways quite wonderful. I think those feelings of pride are captured in the book. I also felt obliged, however, to write about another, far less glorious, side of the Permian program.
>
> That other side included some horrifyingly racist attitudes: The year I was there a coach on the Permian staff described a black player to me as a "big ol' dumb nigger." The book detailed the use of painkillers to enable players (remember, these are high school kids) to perform with broken ankles and hip pointers. The book also criticized a school that budgeted more money during the 1988–89 school year for rush-order films of Permian's games than for teaching materials for the English department. (p. 132)

Around the same time the book was published, Odessa's team was placed on probation for violating league practice rules. Banned from postseason playoffs and the excitement of statewide championship competition, Odessa's fans were distraught. Bissinger was thought to have betrayed the team and the community. Again in his words:

> There are many decent, right-thinking people in Odessa, but there are many others who have built their lives around Permian football and who have lost

to the classroom. In many cases, they do not believe that computers are useful tools for classroom learning and are content to let students adapt computers to their learning at home, which puts students without good home computers at a disadvantage (Cuban, 1994, 2000). Research in the United Kingdom reveals the same gap: Eighty percent of British classrooms have computers and access to the Internet, but only 45% of teachers say that they know how to use the equipment for teaching (Johnston, n.d.).

While teachers and school administrators are striving to keep up with accelerating demand for computer training, the technology continues to advance. Satellites permit wireless connections to individual

Bissinger's book *Friday Night Lights* explored the world of Texas high school football.

Is education taking a backseat to athletics in high schools such as those described in *Friday Night Lights?*

Cheerleaders and other spectators at a Permian High School football game.

> all perspective on what a game should be. I knew the ban on postseason play would shatter them. Although there was not a shred of evidence to suggest the book had anything to do with the league ruling, it was obvious I was to be the scapegoat.

The author's life was threatened, and he was advised to stay away from town. What bothered Bissinger most was that with all the problems of education in Odessa and thousands of places like it in America, education "takes a backseat to the great god of high school football." ■

laptops. New software promises students great gains in math and English. Community colleges push forward with online courses even though their effectiveness has not been evaluated. Major computer manufacturers tout the advantages of laptops and wireless connections to the Internet (Bigham & Bigham, 2000). As promising as all these technologies are, however, current research on student and teacher experiences with them has yet to produce convincing proof of their educational benefits. The most serious problem remains the lag in teachers' ability to incorporate the technologies in their classrooms (Seel & Casey, 2001).

SUMMARY

Education is the process by which a society transmits knowledge, values, norms, and ideologies and in so doing prepares young people for adult roles and prepares adults for new roles. Schools are bureaucratic organizations characterized by a clearly defined authority system and set of rules. Interactionists see the school as a distinctive set of interactions and patterns of socialization, whereas conflict theorists believe that the purpose of schools is to reproduce the society's existing stratification system.

The post–World War II baby boom caused a bulge in elementary school enrollments beginning in 1952 and an expansion of the college-age population in the 1960s. Since the 1970s, increasing numbers of adults have sought additional education.

Education in the United States is viewed as a means of transmitting democratic values, creating equality of opportunity, and preparing new generations of citizens to function in society. Education is also viewed as a tool for solving social problems. The United States has by far the broadest and most egalitarian approach to education in the developed nations. In the less developed nations, illiteracy and the "brain drain" are persistent problems.

Educational attainment, or the number of years of school a person has completed, is correlated with income, occupation, prestige, and attitudes and opinions. One effect of higher levels of educational attainment is "degree inflation," in which employers require more education of potential employees. Barriers to educational attainment include continuing racial segregation and dropping out.

Educational achievement refers to how much the student actually learns. Cross-cultural research has shown that deficiencies in the educational achievement of U.S. students are apparent from an early age.

Some observers believe that educational institutions hinder, rather than enhance, social mobility. Students who are able to obtain higher educational credentials usually come from the middle and upper classes. Students' school careers are affected by teachers' expectations, which in turn are affected by the teacher's knowledge of the student's family background. Inequality in higher education is primarily a matter of ability to pay.

As the size and complexity of the American educational system have increased, so has the tendency of educational institutions to become bureaucratized. Much current research assumes that students and teachers influence each other instead of assuming that the influence is always in a single direction, from teachers to students.

GLOSSARY

Education: The process by which a society transmits knowledge, values, norms, and ideologies and in so doing prepares young people for adult roles and adults for new roles. **(305)**

Educational attainment: The number of years of school an individual has completed. **(312)**

Educational achievement: How much the student actually learns, measured by mastery of reading, writing, and mathematical skills. **(316)**

QUESTIONS FOR THOUGHT AND DISCUSSION

1. What important changes on the global economic scene have contributed to the advantages enjoyed by people with higher education and to the disadvantages associated with limited education?

2. What sociological explanations can you offer for why public education is such a frequent target of criticism and demands for reform?

DIGGING DEEPER

Books

Schools and Societies (Steven Brint; Pine Forge Press, 1998). Extremely useful for comparative material on education as well as fine chapters on classroom dynamics.

The Power of Their Ideas: Lessons for America From a Small School in Harlem (Deborah Meier; Beacon, 1995). The portrait of a model school that has played a seminal role in the current school reform movement, by one of the most innovative educators in the world today.

Savage Inequalities: Children in America's Schools (Jonathan Kozol; Crown, 1991). A chilling view of the persistent inequalities of race and social class that bar children and teachers in the nation's urban public schools from achieving educational success.

The Sociology of Education, 4th ed. (Jeanne H. Ballantine; Prentice Hall, 1997). A comprehensive text in the sociology of education, with many additional bibliographic sources.

Ain't No Makin' It: Leveled Aspirations in a Low-Income Neighborhood, rev. ed. (Jay MacLeod; Westview Press, 1995). A brilliant firsthand description of how education often reproduces inequality from one generation to the next.

Journals

Social Problems. A sociological journal devoted to research and policy analysis related to major social problems. Publishes good articles on family research.

Sociology of Education. A quarterly journal published by the American Sociological Association. Presents recent research on education and human social development.

EXPLORING SOCIOLOGY ON THE INTERNET

The Carnegie Foundation
www.carnegie.org/

A major sponsor of strategic research on education and educational policy in the United States. A good place to start searching for research studies.

Education Resources Information Center (ERIC)
www.aspensys.com/eric/index.html

The world's largest database for research and information about schools and education.

U.S. Department of Education
www.ed.gov/

Provides information about federal education programs and initiatives, with links to other major education sites.

United Nations Educational, Scientific, and Cultural Organization (UNESCO)
www.unesco.org/

The UN's major education agency. Sponsors conferences and research programs on issues of global education, literacy, community development, professional development, and much more.

National Center on Educational Statistics
www.ed.gov/NCES/pubs/ce/index.html

Offers a wealth of data on school performance, especially in the United States.

Chapter 12

ANNUIT COEPTIS
MDCCLXXVI
NOVUS ORDO SECLORUM
THE GREAT SEAL
VOTE

Economics and Politics

Can economic and political institutions bring about positive social change in a globalizing world?

War hero, senator, runner-up candidate for the 2000 Republican presidential nomination, staunch advocate of campaign reform—few people in the United States are as familiar with the links between economic and political interests as Arizona Senator John McCain. For years, he and his Democratic colleague Russ Feingold have been pushing for major reform in the way candidates fund their political campaigns. They advocate a ban on "soft money"—that is, contributions to political parties from private individuals, corporations, and labor unions that exceed a specific minimum amount. Existing campaign finance laws limit direct contributions to individual campaigns. To circumvent these limits, large donations are made to the political parties, which then spend the money as they see fit for specific campaigns. The 2000 presidential election was reported to have cost a total of over $2.2 billion, much of which came from major soft-money contributions. The Republicans raised $239 million in soft money and the Democrats $217 million.

For many years, sociologists and political scientists have been showing that campaign donations unfairly favor wealthy donors by giving them undue influence over the executive and legislative branches of government. This influence also creates cynicism among voters, who are becoming less likely to believe that major reform is possible. Most major Republican leaders, including President George W. Bush, have opposed banning soft money because they usually receive more of it. That is why Senator McCain is often viewed as a maverick bent on destroying a system that works well for some of the nation's most powerful political leaders, regardless of party.

McCain and his allies want to limit the enormous influence that very wealthy individuals, corporations, political action committees, and labor unions have on American electoral politics. The majority of the voters seem to agree with them. Most polls show that 70% or more of U.S. voters want to see limits on the influence of big money in politics. The majority of U.S. senators also favor the McCain–Feingold limits on soft-money contributions. The McCain forces promised to introduce campaign reform proposals immediately in the 2001 legislative session. But the proposal has powerful opponents who favor the status quo, so at this writing its fate is quite uncertain. What is certain, however, is that Senator McCain is a tough opponent of the vast influence of soft money in U.S. politics and is not likely to give up the cause very quickly (Ingraham, 2001). ■

Arizona Senator John McCain, one of the nation's staunchest supporters of campaign finance reform, has a difficult time convincing American voters that any new laws will diminish the influence of huge contributors and powerful interests in a political system increasingly dependent on television advertising.

THE ENDURING IMPORTANCE OF ECONOMICS AND POLITICS

Few aspects of human life are more intertwined than economics and politics. Entire sectors of an advanced industrial economy may rise or fall rapidly; thousands of jobs may be lost or gained according to the changing political fortunes of a nation. The threat of war, for example, may lead to rapid expansion in the aerospace and shipbuilding industries. Greater prospects for world peace may force the same industries to lay off workers in their efforts to adapt to a peacetime economy. So many aspects of contemporary life—immigration policy, the passage of laws regulating or deregulating commerce and manufacturing, the politics of road construction, environmental protection, worker safety, affirmative action in the workplace, and countless other major social issues—emerge at the intersection of economics and politics. Yet surprisingly few people really understand or accept this fact, especially when it affects their own freedom to act as they wish.

Our economic lives and our political lives are closely related through laws, regulations, and public policies of all kinds. There are always winners and losers when laws are enforced, and some people feel that their rights or needs have been trampled on. In consequence, there will always be people who long for a simpler social world, one in which they are free to do what they wish without having to worry about rules and regulations. But the idea of a return to a simpler, unregulated past is just wishful thinking. While societies may make their political and economic systems more fair and efficient, there is no way of avoiding the trend toward ever greater economic and political interdependence.

In this chapter, we first consider the sociology of markets and economic activity and then turn to an examination of changing political institutions. After discussing markets and world economic systems, we examine some of the major economic changes occurring in advanced economies like that of the United States (partly as a result of global economic and social change) and how these changes are affecting working people and the organization of the workplace. Later sections explore issues of politics, particularly the central concepts of power, legitimacy, and authority as these are applied in our own and other political institutions. We end by considering some major perspectives on politics and political institutions.

MARKETS AND THE DIVISION OF LABOR

The well-known economist Paul Samuelson (1980) has defined *economics* as "the study of how people and society end up choosing, with or without the use of money, to employ scarce productive resources that could have alternative uses—to produce various commodities and distribute them for consumption, now or in the future, among various persons and groups in society" (p. 2). Sociologists are also concerned with how individuals and societies make choices involving scarce resources like time, talent, and money. But sociologists do not assume that scarcity or supply and demand are the only reasons for making such choices. They are deeply concerned with showing how the norms of different cultures affect economic choices.

Sociologists who study economic institutions are also interested in how markets for goods and services change as people learn to use new technologies and to divide up their labor in increasingly specialized ways. They also look at how competition for profits may lead to illegal activities and cause governments to try to regulate economic institutions. In addition, sociologists attempt to show how the labor market is organized and how professions develop.

The Nature of Markets

Markets are economic institutions that regulate exchange behavior. In a market, different values or prices are established for particular goods and services, values that vary according to changing levels of supply and demand and are usually expressed in terms of a common measure of exchange, or currency. A market is not the same thing as a marketplace. As an economic institution, a market governs exchanges of particular goods and services throughout a society. This is what we mean when we speak of the "housing market," for example. A marketplace, on the other hand, is an actual location where buyers and sellers make exchanges. Buyers and sellers of jewelry, for instance, like to be able to gather in a single place to examine the goods to be exchanged. The same is true for many other goods, such as clothing and automobiles.

Market transactions are governed by agreements or contracts in which a seller agrees to supply a particular item and a buyer agrees to pay for it. Exchanges based on contracts are a significant factor in the development of modern societies. As the social theorist

Niklas Luhmann (1982) observes, the use of contracts "makes 'impersonal' relations possible: It neutralizes the relevance of the other roles of the participants" (p. 199), such as kinship and other personal relationships, that govern exchanges in nonmarket situations. In contractual relations, for example, the fact that people are friends or kin does not, in principle, change the terms of their agreement and the need to repay debts.

Among hunting-and-gathering peoples and in relatively isolated agrarian societies before the twentieth century, markets in the modern sense of the term did not exist. The idea of buying foodstuffs or other objects with currency was foreign to these people. In social-scientific terms, a society cannot be said to have a fully developed market economy if many of the commodities it produces are not exchanged for a common currency at prices determined by supply and demand. The spread of markets into nonmarket societies began during the period of European conquest and colonialism and was accelerated by the desire among tribal and peasant peoples to obtain the goods produced by industrial societies.

Markets and the World Economic System

In the late fifteenth and early sixteenth centuries, according to sociologist Immanuel Wallerstein (1974), "there came into existence what we may call a European world-economy. . . . It was not an empire, yet it was as spacious as a grand empire and shared some features with it. But it was different, and new. It was a kind of social system the world had not really known before." The new system was based on economic relationships, not on political empires; in fact, it encompassed "empires, city-states, and the emerging 'nation-states'" (p. 15).

Great empires had been a feature of the world scene for at least 5,000 years before the dawn of the modern era. But the empires of China, India, Africa, the Mediterranean, and the Middle East were primarily political rather than economic systems. Wallerstein argues that because the great empires dominated vast areas inhabited by peoples that lacked military and political power, they were able to establish a flow of economic resources from the outlying regions to the imperial centers. The means used were taxation, tribute (payments for protection by the imperial army), and trade policies in which the outlying societies were forced to produce certain goods for the imperial merchants.

But this system—exemplified most clearly in the case of the Roman empire—required a huge military and civil bureaucracy, which absorbed much of the imperial profit. Local rebellions and wars continually increased the expense of maintaining imperial rule. Political empires thus can be viewed as a primitive means of economic domination. "It is the social achievement of the modern world," Wallerstein comments, "to have invented the technology that makes it possible to increase the flow of the surplus from the lower strata to the upper strata, from the periphery to the center, from the majority to the minority" (p. 16) without the need for military conquest.

What technologies made the new world system possible? They were not limited to tools of trade, such as the compass or the oceangoing sailing vessel, or to tools of domination, such as the Gatling machine gun. These technologies also included organizational techniques for bringing land, labor, and local currencies into the larger market economy: ways of enclosing and dividing up land in order to charge rent for its use; financial and accounting systems that led to the creation of new economic institutions like banks; and many others. We discuss the role of technology in social change more thoroughly in Chapter 14, but it is important to note here that the term **technology** refers not only to tools but also to the procedures and forms of social organization that increase human productive capacity (Polanyi, 1944).

Although the European colonial powers (and the United States) often used political and military force to bring isolated societies into their markets, in the twentieth century they could allow their former colonies to gain independence yet still maintain economic control over them. This occurred because the economies of the colonial societies had become dependent on the technologies and markets controlled by the Western powers. Today former colonies are developing independent economic systems, but their ability to compete effectively in world markets is limited by the increasing power of **multinational corporations,** or *multinationals.* These are economic enterprises that have headquarters in one country and conduct business activities in one or more other countries (Barnet, 1980).

Multinationals are not a new phenomenon. Trading firms like the Hudson's Bay Company and the Dutch East India Company were chartered by major colonial powers and granted monopolies over the right to trade with native populations for furs, spices, metals, gems, and other valued commodities. Thus, exploitation of the resources of colonial territories has

STUDY CHART 12.1
Political-Economic Ideologies

Ideology	Description	Example
Mercantilism	An economic philosophy based on the belief that the wealth of a nation can be measured by its holdings of gold or other precious metals and that the state should control trade.	European nations under feudalism
Laissez-Faire Capitalism	An economic philosophy based on the belief that the wealth of a nation can be measured by its capacity to produce goods and services and that these can be maximized by free trade.	England at the time of the industrial revolution.
Socialism	An economic philosophy based on the concept of public ownership of property and sharing of profits, together with the belief that economic decisions should be controlled by the workers.	The Soviet Union before 1989; China and Cuba today
Democratic Socialism	An economic philosophy based on the belief that private property may exist at the same time that large corporations are owned by the state and run for the benefit of all citizens.	Holland and the Scandinavian nations
Welfare Capitalism	An economic philosophy in which markets determine what goods will be produced and how, but the government regulates economic competition.	The United States

been directed by multinational corporations for over two centuries. Modern multinationals do not generally have monopolies granted by the state, yet these powerful firms, based primarily in the United States, Europe, and Japan, are transforming the world economy by "exporting" manufacturing jobs from nations in which workers earn high wages to nations in which they earn far less. This process, which is particularly evident in the shoe, garment, electronics, textile, and automobile industries, has accelerated the growth of industrial working classes in the former colonies while greatly reducing the number of industrial jobs in the developed nations.

Changes in worldwide production patterns challenge Wallerstein's thesis that there is a world economic order dominated by the former colonial powers. In particular, industrial nations like the United States and Japan are increasingly losing manufacturing jobs while industrializing nations like Brazil and Korea are gaining them. But the economic survival of millions of rural workers in these emerging nations continues to be heavily influenced by changing markets in the highly developed nations.

POLITICAL-ECONOMIC IDEOLOGIES

The earliest social scientists were deeply interested in understanding the full importance of modern economic institutions. In fact, since the eighteenth century almost all attempts to understand large-scale social change have dealt with the question of how economic institutions operate. But the age-old effort to understand economic institutions has not been merely an academic exercise. The fate of societies throughout the world has been and continues to be strongly influenced by theories concerning how economic institutions operate, or fail to do so, in nation-states with different types of political institutions. The major economic ideologies thus are also political ideologies, and economics is often called political economics. In this section, we review three economic ideologies—mercantilism, capitalism, and socialism—and some variations of them. (The political-economic ideologies discussed in this section are summarized in the study chart above.)

Mercantilism

The economic philosophy known as **mercantilism,** which was prevalent in the sixteenth and seventeenth centuries, held that a nation's wealth could be measured by the amount of gold or other precious metals held by the royal court. The best economic system, therefore, was one that increased the nation's exports and thereby increased the court's holdings of gold.

The mercantilist theory had important consequences for economic institutions. For example, the guilds, or associations of tradespeople, that had arisen in medieval times were protected by the monarch, to whom they paid tribute. The guilds controlled their members and determined what they produced. In this way, they were able to produce goods cheaply, and those goods were better able to compete in world markets, thereby increasing exports and bringing in more wealth for the court. But the workers were not free to seek the best jobs and wages available. Rather, the guilds required them to work at assigned tasks for assigned wages, and the guildmasters fixed the price of work (wages), entry into jobs, and all working conditions. Land in mercantilist systems was not subject to market norms either. As in feudalism, land was thought of not as a commodity that can be bought and sold—that is, a commodity subject to the market forces of supply and demand—but as a hereditary right derived from feudal grants.

Laissez-Faire Capitalism

The ideology of **laissez-faire capitalism** attacked the mercantilist view that the wealth of nations could be measured in gold and that the state should dominate trade and production in order to amass more wealth. The laissez-faire economists believed that a society's real wealth could be measured only by its capacity to produce goods and services—that is, by its resources of land, labor, and machinery. And those resources, including land itself, could best be regulated by free trade in world markets (Halevy, 1955; A. Smith, 1910/1776).

The ideology of laissez-faire capitalism also sought to free workers from the restrictions that had been imposed by the feudal system and maintained under mercantilism. Thus, it is no coincidence that the first statement of modern economic principles, Adam Smith's *The Wealth of Nations,* was published in 1776. Revolution was in the air—not only political revolution but the industrial revolution as well. And some of the most revolutionary ideas came from the pens of people like Adam Smith, Jeremy Bentham, and John Stuart Mill, who understood the potential for social change contained in the new capitalist institutions of private property and free markets.

Private property (as opposed to communal ownership) is not merely the possession of objects but a set of rights and obligations that specify what their owner can and cannot do with them. The laissez-faire economists believed that the owners of property should be free to do almost anything they liked with their property in order to gain profit. Indeed, the quest for profit would provide the best incentive to produce new and cheaper products. This, in turn, required free markets in which producers would compete to provide better products at lower prices.

These economic institutions are familiar to us today, but the founders of laissez-faire capitalism had to struggle to win acceptance for them. No wonder they thought of themselves as radicals. In fact, their economic and political beliefs were so opposed to the rule of monarchs and to feudal institutions like guilds that they could readily be seen as revolutionary. They believed that the state should leave economic institutions alone (which is what *laissez-faire* means). In their view, there is a natural economic order, a system of private property and competitive enterprise that functions best when individuals are free to pursue their own interests through free trade and unregulated production. As Smith put it, "Every individual . . . intends only his own gain, and he is in this, as in many other cases, led by an invisible hand to promote an end which was no part of his intention" (quoted in Halevy, 1955, p. 90). One of the most famous images in the history of the social sciences, Smith's "invisible hand" refers to the pervasive influence of the forces of supply and demand operating in markets for goods and services. Laissez-faire economists believe that in the long run these forces will improve the lives of people everywhere.

Socialism

As an economic and political philosophy, **socialism** began as an attack on the concepts of private property and personal profit. These aspects of capitalism, socialists believed, should be replaced by public ownership of property and sharing of profits. As we have noted in earlier chapters, this attack on capitalism was motivated largely by horror at the atrocious living conditions resulting from the industrial revolution.

The early socialists of the late eighteenth and early nineteenth centuries (Robert Owen in England and Henri de Saint-Simon in France) thought of economics as the "dismal science" because it seemed to excuse a system in which a few people were made rich at the expense of the masses of workers. They detested the laissez-faire economist's defense of low wages and wondered how workers could benefit from the industrial revolution instead of becoming "wage slaves." They proposed the creation of smaller-scale, more self-sufficient communities that would produce modern goods but would do so within a cooperative framework.

Later in the nineteenth century, Karl Marx viewed these ideas as utopian dreams. He taught that the socialist state must be controlled by the working class, led by their own trade unions and political parties, which would do away with markets, wage labor, land rent, and private ownership of the means of production. These aspects of capitalism would be replaced by socialist economic institutions in which the workers themselves would determine what should be produced and how it should be distributed.

Marx never completed his blueprint of how an actual socialist society might function. That chore was left to the political and intellectual leaders of the communist revolutions—Lenin, Leon Trotsky, Rosa Luxemburg—and, finally, to authoritarian leaders like Joseph Stalin, Mao Zedong, and Fidel Castro. These leaders believed that all markets and all private industry must be eliminated and replaced by state-controlled economic planning, collective farms, and worker control over industrial decision making. Unless capitalist economic institutions were completely rooted out, they believed, small-scale production and market dealings would give rise to a new bourgeois class.

In the socialist system as it evolved under the communist regimes of the Soviet Union and China, centralized planning agencies and the single legal party, the Communists, had the authority to set goals and organize the activities of the worker collectives, or soviets. Party members and state planners also devised wage plans that would theoretically balance the need to reward skilled workers against the need to prevent the huge income inequalities found in capitalist societies. In Soviet-style societies, markets were not permitted to regulate demand and supply; this vital economic function was supposedly performed by government agencies. Societies that are managed in this fashion are said to have *command economies:* The state commands economic institutions to supply a specific amount (a quota) of each product and to sell it at a particular price. (Note that not all command economies are dominated by communist parties. Germany under the Nazis and Italy under the fascists were also command economies.)

With the collapse of the Soviet political empire, the economies of Russia and the other nations of the former Soviet Union (as well as its satellites in Eastern Europe) are undergoing a transition to capitalist economic institutions. It has become evident that command economies provided these nations with overdeveloped industrial infrastructures and limited capability to compete in world markets for goods and services. Indeed, it has been said that the command economic system of the former Soviet Union gave it "the most impressive nineteenth century industrial infrastructure in the world, 75 years too late" (Ryan, 1992, p. 22).

The Soviet system was notorious for the inefficiency of its economic planning and industrial production. Under the command system, factory managers must continually hoard supplies, or raid supplies destined for similar factories in other regions, in order to meet their production quotas. Or they must trade favors with other factory managers to obtain supplies, a situation that creates a hidden level of exchange based on bribes and favoritism. And because there are no free markets, goods that are desired by the public often are not available simply because the planners have not ordered them. In fact, in the Soviet-style economies there were always clandestine markets that operated outside the control of the authorities and supplied goods and services to those with the means to pay for them. These underground markets thereby increased inequality in those societies and generated greater public disillusionment with command economies and one-party communist rule.

The massive social upheavals of 1989 and 1990 brought an end to the communist command economic system, but they also initiated a period of great economic and social uncertainty in the former Soviet-dominated nations. At present, efforts to privatize their economies and use market forces to regulate the supply of goods and services, including labor, have begun to produce economic growth in those nations, but these rapid changes have also produced new forms of insecurity. In Russia and Ukraine, for example, the rise of political leaders in league with business entrepreneurs affiliated with organized crime groups produces new forms of corruption. This has the effect of discouraging outside investment in the Russian economy, and without such investment the nation cannot develop essential infrastructure projects like roads, power plants, and hospitals. The election of Vladimir Putin, a stern leader with a law-enforcement background, may put the brakes on the rampant kleptocracy that characterized the previous administration (Cohen, 2000; Freeland, 2000; Nesterenko, 2000).

Democratic Socialism

A far less radical version of socialism than that attempted in the Soviet empire is known as **democratic socialism.** This economic philosophy is practiced in the Scandinavian nations, especially Sweden, Denmark, and Norway, as well as in Holland and to a lesser extent in Germany, France, and Italy. It holds that the institution of private property must continue to exist because people want it to and that competitive markets are needed because they are efficient ways of regulating production and distribution. But large corporations should be owned by the nation or, if they are in private hands, required to be run for the benefit of all citizens, not just for the benefit of their stockholders. In addition, economic decisions should be made democratically.

Democratic socialists look to societies like Sweden and Holland for examples of their economic philosophy in practice (Harrington, 1973). In Sweden, workers can invest their pension benefits in their firm and thereby gain a controlling interest in it. This process is intended to result in socialist ownership of major economic organizations.

In the United States, there is a long history of conflicted relations between workers and owners of capital. Cooperative systems in which authority and even ownership are shared are developing, but more slowly than in the social democracies of Europe. Social scientists who study the U.S. economy note that in recent years there has been a reaction among owners of capital against the principles of democratic socialism, especially those that stress cooperation between labor and management and the right of workers to organize unions and engage in collective bargaining. In a recent survey of workers and managers' attitudes toward cooperation in decision making at work (a social-democratic ideal), 75% of the managers interviewed said that they would much rather deal with workers as individuals than as an organized group. They admitted that it would hurt their careers a great deal if their workers decided to form a union (another social-democratic ideal; Freeman & Rogers, 1999).

Welfare Capitalism

Emerging to some extent as a response to the challenge posed by the Russian Revolution (which called attention to many of the excesses of uncontrolled or laissez-faire capitalism), welfare capitalism represents a new way of looking at relationships between governmental and economic institutions. **Welfare capitalism** affirms the role of markets in determining what goods and services will be produced and how, but it also affirms the role of government in regulating economic competition (for example, by attempting to prevent the control of markets by one or a few firms).

Welfare capitalism also stresses the role that governments have always played in building the roads, bridges, canals, ports, and other facilities that make trade and industry possible. Expanding on this role, the theory of welfare capitalism asserts that the state should also invest in the society's human resources—that is, in the education of new generations and the provision of a minimum level of health care. Welfare capitalism also guarantees the right of workers to form unions in order to reach collective agreements with the owners and managers of firms regarding wages and working conditions. It creates social-welfare institutions like Social Security and unemployment insurance. And in order to stimulate production and build confidence in times of economic depression, welfare capitalism asserts that governments must borrow funds to finance large-scale public works projects like the construction of the American interstate highway system during the 1950s.

The theory of welfare capitalism is associated with the writings of John Maynard Keynes, Joan Robinson, John Kenneth Galbraith, and James Tobin, all of whom contributed to the revision of laissez-faire economic theory. Welfare capitalism dominated American economic policy from World War II until the 1970s, when a succession of economic crises—inflation, energy shortages, and unemployment—turned the thoughts of many Americans once again toward laissez-faire capitalism. Today, however, the increasingly evident gap between the haves and the have-nots in American society has given rise to renewed interest in enhancing the institutions of the welfare state—unemployment insurance, health insurance, low-income housing, public education, and others. At the same time, fear of governmental control over people's lives, and its costs in the form of higher taxes, is generating a revolt against the "welfare state" in many segments of the population in the United States

FIGURE 12.1

Household Net Worth Among the Middle-Aged and Elderly, 1984 and 1999 (thousands of 1999 dollars)

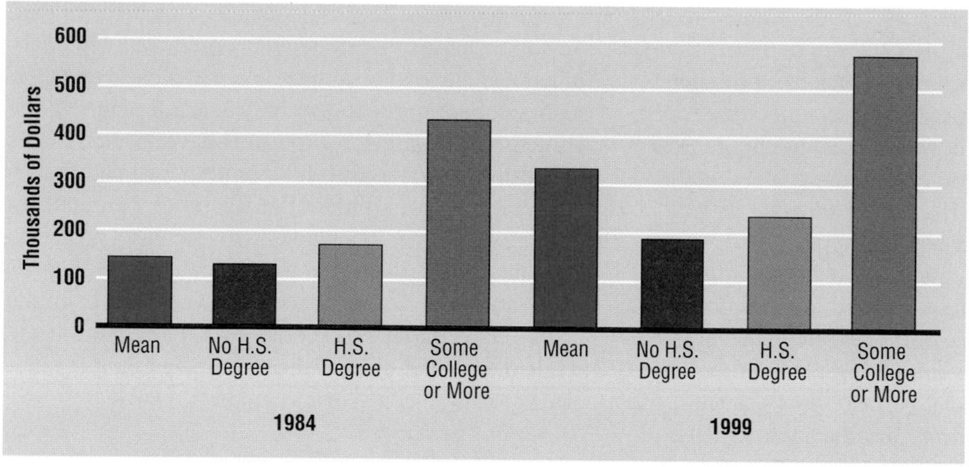

Source: *Newsletter,* Panel Study of Income Dynamics (Ann Arbor: Institute for Social Research, University of Michigan, April 2000).

and other industrial nations. This trend has resulted in conservative electoral victories and unprecedented reductions in social spending. The 2000 presidential election demonstrated that the U.S. electorate is quite evenly divided about the need for security from economic stress versus the desire for lower taxes and reductions in the scope of government bureaucracies.

The United States: A Postindustrial Society?

Although many people in the United States see the failure of socialism in the Soviet bloc as a victory for capitalism, sociologists and poll takers have called attention to some underlying problems in the U.S. economy. As demographer Valerie Kincade Oppenheimer observes, "Since about 1970, the average real earnings position of young men has deteriorated considerably. This has been true not only for high school dropouts but also for high school and college graduates" (1994, p. 331). The declining incomes of U.S. workers over the past quarter-century have contributed to the growth in dual-earner households.

Figure 12.1 presents the mean net worth—the total value of all possessions, savings, and other assets—for households with heads age 45 or over, that is, people the same age as the parents or grandparents of contemporary college students. Note that net worth increases dramatically with the education of the head of the household. This was true in 1984 and is even more true today, especially for those who have attended college.

The disparity between the incomes of people with professional and technologically sophisticated jobs and those of people with less intellectually demanding jobs is likely to increase as economic growth centers on computers, telecommunications, and biological technologies. Jobs that require lower levels of education and training may be more onerous, but the number of people who can perform them is relatively large. The basic economic laws of supply and demand therefore drive up the wages of highly educated workers while those of less-educated workers are driven downward.

In an influential study, *The Coming of Post Industrial Society,* sociologist Daniel Bell (1973) reviewed changes in American economic institutions indicating that we are undergoing a transition from an industrial to a postindustrial society. A postindustrial society "emphasizes the centrality of theoretical knowledge as the axis around which new technology, economic growth, and the stratification of society will be organized" (p. 116). As societies undergo the transition to postindustrial economic institutions, an "intellectual technology" based on information arises alongside machine technology. Industrial production does not disappear, but it becomes less important. New industries devoted to providing knowledge and information become the primary sources of economic growth (Block, 1990; Montagna, 1977; Piore & Sabel, 1984).

Changes in the distribution of jobs in the goods-producing and service-producing sectors of the American economy clearly show the growing importance of services, especially those employing professional and technical workers. For Bell, this change is inevitable. Industries decline, but new ones appear. At the beginning of the twentieth century, for example, more than 70% of Americans were employed in agriculture and related occupations. Now this figure is below 3%. Similar transformations are taking place in manufacturing today, and they can be made less painful if the society is willing to invest in education and the retraining of workers in declining industries like mining, steel production, or manufacturing for military purposes (Heilbroner, 1990).

Other social scientists view these changes less optimistically. Some note that declines in industrial production have pushed thousands of skilled workers into lower-paying jobs in the service sector—for example, in the fast-food industry or as janitors or security guards (Aronowitz, 1984; Harrison, Tilly, & Bluestone, 1986). Other social scientists are concerned that the gains made by blacks and women in access to better-paying industrial jobs since World War II are being erased. William Julius Wilson (1978) has commented that "access to the means of production is increasingly based on educational criteria . . . and thus threatens to solidify the position of the black underclass" (p. 151).

European sociologists have criticized the theory of postindustrial society on even broader grounds, saying that it tends to promote belief in the inevitability of an increasingly impoverished working class with even more reason to demand a more equal distribution of wealth (Touraine, 1971). In the United States, the theory of postindustrial society has been criticized on the ground that even though manufacturing accounts for a declining share of the labor force, the society still depends on manufactured goods of all kinds. Critics argue that the United States cannot relinquish its manufacturing base simply because a theory says it is outmoded (Cohen & Zysman, 1987).

Can the loss of manufacturing jobs be slowed in economies that are said to be in their "postindustrial" phase? Those who are concerned about this question often press for higher tariffs on imported manufactured goods in order to protect the jobs of workers in the United States. It is also argued that higher tariffs might stimulate foreign companies to locate their manufacturing plants in nations where wages are higher. To some degree, this strategy has produced new manufacturing jobs; for example, major Japanese and German automobile producers have located plants in the United States, and U.S. producers have located plants in Europe and Asia. But the incentive for manufacturers to locate their production facilities in countries where wages are low enough to increase profits is even more powerful. On a global scale, these changes have immense consequences as workplaces in the advanced economies become ever more automated while the pace of industrialization and economic modernization accelerates elsewhere in the world.

SOCIOLOGICAL PERSPECTIVES ON THE WORKPLACE

Industrial sociology is concerned with the social organization of work and the types of interactions that occur in the workplace. Like sociologists in other fields, industrial sociologists use the functionalist, conflict, and interactionist perspectives in their research. The results of that research have been used to support various approaches to labor–management relations.

Human Relations in Industry: A Functionalist Perspective

The *human relations* perspective on management is associated with the research of Elton Mayo (1945) and his colleagues at Western Electric's Hawthorne plant and in aircraft and metal production plants. Their goal was to use experimental methods and observation of workers and managers on the job in an attempt to understand how the factory's formal organization and goals are affected by patterns of informal organization within the workplace. This was in sharp contrast with earlier approaches to labor–management relations, especially the *scientific management* approach developed early in the twentieth century by Frederick W. Taylor.

Scientific management, or Taylorism as it is often called, was one of the earliest attempts to apply objective standards to management practices. After rising through the ranks from laborer to chief engineer in a large steel company, Taylor turned his attention to the study of worker productivity. He noticed that workers tend to adhere to informal norms that require them to limit their output. For example, he found that the rate of production in a machine shop was only about one-third of what might normally be expected. Taylor decided to use the authority of management to speed up the workers. He fired stubborn men, hired "green hands" who did not know the norms of the experienced workers, and experimented with ways of breaking down the labor of each worker into its components. Every job, he claimed, could be scientifically studied to determine how it could be performed most efficiently. Such "time and motion studies," combined with piecework payment systems that induced workers to produce more because they were paid for each unit produced above a set number, became the hallmarks of scientific management (Miller & Form, 1964). Taylor's principles were quickly incorporated into the managerial practices of American businesses, but because they demanded more effort from workers, they were often resisted by workers and their unions (Braverman, 1974).

Mayo's experiments at the Hawthorne plant and elsewhere were intended to determine what conditions would foster the highest rates of worker productivity. His observations convinced him that increased productivity could be obtained by emphasizing teamwork among workers and managers, rather than through pay incentives or changes in such variables as lighting, temperature, and rest periods (Bendix, 1974). Mayo's research showed that the attention given to the workers was what mattered, not the various experimental conditions. But Mayo drew another, more important conclusion from the Hawthorne experiments:

> The major experimental change was introduced when those in charge sought to hold the situation humanly steady . . . by getting the cooperation of the workers. What actually happened was that six individuals became a team and the team gave itself wholeheartedly and spontaneously to cooperating in the experiment. (Mayo, 1945, pp. 72–73)

Although Mayo's research focused on the interactions between workers and managers, the human relations approach that grew out of that research can be said to represent the functionalist perspective on the workplace because it stresses the function of managerial efforts in increasing worker productivity. The functionalist perspective is also illustrated by William F. Whyte's (1949) classic study of the restaurant industry. In a small restaurant, the organization's structure is simple: "There is little division of labor. The owner and employees serve together as cooks, countermen, and dishwashers." But when the restaurant expands, a number of supervisory and production occupations are added to its role structure. According to Whyte, this magnifies old problems and gives rise to new ones:

> In a large and busy restaurant a waitress may take orders from 50 to 100 customers a day (and perhaps several times at each meal) in addition to the orders (much less frequent) she receives from her supervisor. When we add to this the problem of adjusting to service pantry workers, bartenders, and perhaps checkers, we can readily see the possibilities of emotional tension—and, in our study, we did see a number of girls break down and cry under the strain. (1949, p. 304)

Whyte discovered that tension and stress could be reduced, and customers served more happily, if the restaurant was organized in such a way that lower-status employees were not required to give orders directly to higher-status ones. For example, waiters and waitresses should not give orders directly to cooks but should place their orders with a pantry worker or use a system of written orders. By changing the organization of statuses and roles in the restaurant (and, thus, the way they functioned as a social system), it would be possible to increase work satisfaction and output.

Conflict at Work

The human relations approach seeks to improve cooperation between workers and managers in order to achieve the organization's goals. Industrial sociologists who take a conflict perspective feel that this approach automatically condones the goals of managers and fails to consider more basic causes of worker–management problems such as class conflict. This results in continual "fine-tuning" of organizational structures rather than in more thorough reforms.

In a study of a midwestern metal products factory, Michael Burawoy (1980) found that even after changes were made on the basis of the human relations approach, the workers continued to limit their output. The workers called this "making out" and saw it as a way of maximizing two conflicting values: their pay and their enjoyment of social relations at work. But instead of wondering why the workers refused to produce more, Burawoy asked why they worked as hard as they did. He concluded that the workers were actually playing into the hands of the managers and owners because neither they nor the unions ever questioned management's authority to control basic production decisions. The norms of the shop-floor culture, of which "making out" is an example, caused the workers to feel that they were resisting the managers' control. As a result, the workers did not feel a need to engage in more direct challenges to the capitalist system itself.

Conflict-oriented industrial sociologists also study how class and status relations outside the workplace influence shop-floor cultures. For example, in an analysis of work relations in a large midwestern steel mill, William Kornblum (1974) described how ethnic and racial conflicts among whites, blacks, and Chicanos are carried into the plant itself. These conflicts diminish as the steelworkers work together in teams over periods of many years, but the basic class differences that divide workers from managers remain and are institutionalized in the norms of union–management bargaining.

A fundamental problem for conflict sociologists is that class conflict must be shown to exist; it cannot be assumed to exist simply because Marxian theory defines workers and business owners as opposing classes. In fact, there is considerable evidence that, despite their misfortunes, workers do not view their interests as automatically opposed to those of owners and managers. As workers and supervisory personnel lost their jobs in the severe recession of the early 1990s, they often expressed discontent with the nation's leadership and with company managers. Generally, however, the anguish of the unemployed tends to be turned inward, taking the form of depression rather than militant class consciousness. Class consciousness may be popularized in the lyrics of a Bruce Springsteen song—"They're closing down the textile mill across the railroad track./Foreman says these jobs are going, boys/and they ain't coming back to your hometown"—but it has not produced much overt class conflict or new social movements (Bensman & Lynch, 1987).

Continuing changes in the American economy through the 1990s and into the new millennium are resulting in demands for new social policies. This trend is especially evident in calls for new approaches to health insurance and higher wages for workers who are fully employed but cannot escape from poverty due to extremely low wages (Barker & Christensen, 1998). As can be seen in Figure 12.2, the American economy is highly successful in creating new jobs, but many of those jobs pay lower wages and provide fewer benefits than the jobs that were lost (Harrison, 1990; Uchitelle, 1993).

Changing Patterns of Job Security. The shift toward nonstandard work arrangements is one of the most important trends in work in the United States and other industrialized nations. By *nonstandard work arrangements,* sociologists refer to work that is not full time and often does not imply a direct contract between workers and their employers. *Part-time work* is employment for about 30% less time than the standard definition of full time (for example, 40 hours per week in the United States, 35 hours in France, 36 in Germany). Nearly one in five workers in the United States currently works part-time—by far the most widely used form of nonstandard work in this country. About 16% of European workers are part-timers. Thirty years ago, the majority of Americans who worked part-time did so voluntarily, but today the majority would prefer to be working full-time (Kalleberg, Reskin, & Hudson, 2000).

Temporary work may be full- or part-time but is generally limited to a specific time period. Since the 1970s, the number of temporary workers has been increasing by about 11% a year, from 0.3% of the U.S. labor force in 1972 to nearly 2.5% in 1998, a rate of growth considerably faster than the overall rate of job creation (Siegal & Sullivan, 1997). Temporary work is also growing rapidly in other nations, but it is often arranged by nonprofit agencies to accelerate employment of young people or hard-to-place workers, rather than by private temp companies as in the United States.

Another important trend is toward *contract work,* in which the employer supervises the worker's performance but at a client's workplace. A person may be working at General Motors or the Bank of America but actually be employed by a construction firm, an advertising agency, or a computer firm. Companies that purchase contract labor have limited responsibilities for the workers and can use this arrangement to avoid increasing their own staff, with the attendant costs of benefits, the possibility of unionization, and other personnel issues. Many employers (about 33% in the United States) also hire *independent contractors,* who are typically technical or creative professionals. Independent contractors do not have a long-term employment relationship with the company but are paid to fulfill a specific contract on a project basis.

FIGURE 12.2
Changes in the U.S. Job Market
Share of Nonfarm Employment in Manufacturing and Services Industries (Annual Averages), 1945–1996.

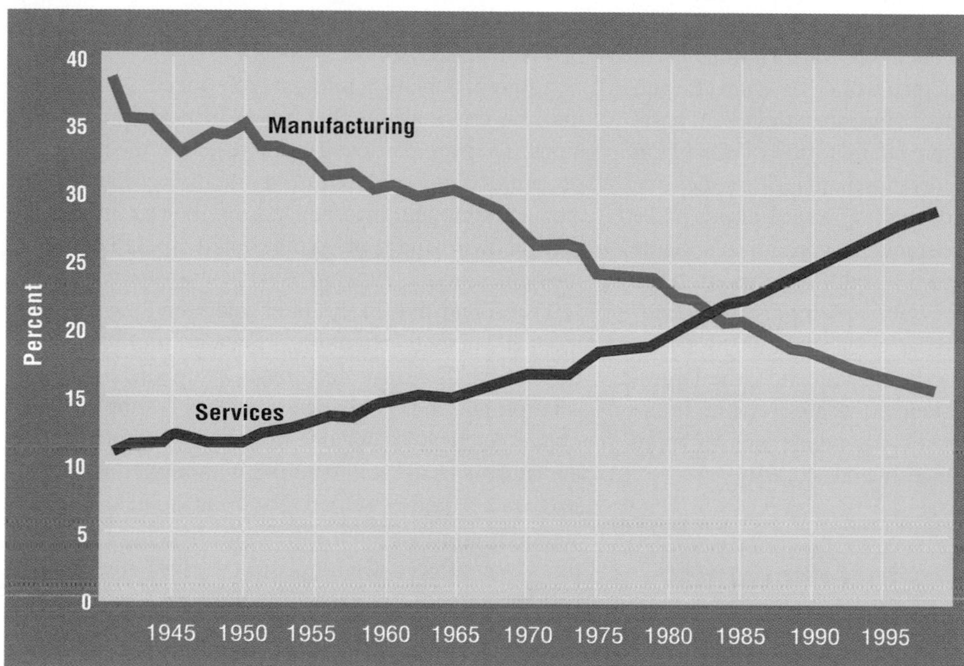

Percent of Nonfarm Employment in Each Sector*

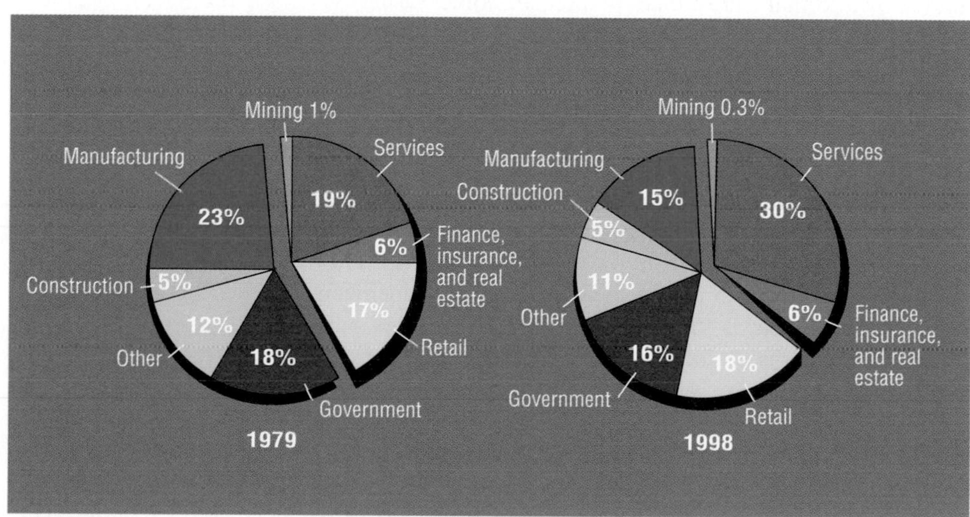

*Because of rounding, percentages do not add to 100.
Source: *Statistical Abstract,* 1999.

All these and other forms of nonstandard work relationships are creating far more diversity in work relationships than existed in the early decades of the twentieth century. Part-time and other arrangements always existed, but they are becoming more common each year. As service work has outpaced manufacturing, union strength has declined, since unions were highly concentrated in factory work. This has allowed employers more leeway to experiment and to cut costs by resorting to these nonstandard forms of work. On the workers' side, the arrangements often deprive workers of secure health benefits; in fact, almost 40% of those employed in nonstandard work arrangements in service businesses have no health benefits. In retail sales, the largest of the service employment categories, the majority work at wages that are not adequate to support families. In consequence, the largest number of people living below the poverty line in the United States are in households where one or more members works full-time (Ellwood, 2000).

Professions: An Interactionist Perspective

Factory workers, temps, and other lower-status employees are not the only subjects of research by sociologists who study work and employment. Because they are a growing segment of modern economic institutions, professionals have also been studied extensively (Abbott, 1988). A **profession** is an occupation with "a body of knowledge and a developed intellectual technique . . . transmitted by a formal educational process and testing procedures. A code of professional ethics governs each profession and regulates relations with colleagues, clients, and the public [and] most professions are licensed by the state" (Montagna, 1977, p. 197). This definition could be applied to a variety of occupations. For example, doctors and attorneys have long been considered professionals, but what about nurses and stockbrokers? Sociologists have addressed this question in their research on the "professionalization" of occupations—that is, on the way different occupations attempt to achieve the status of professions.

According to Everett C. Hughes (1959), a profession is a set of role relationships between "experts" and "clients," in which the professional is an expert who offers knowledge and judgments to clients. Often the professional must assume the burden of the client's "guilty knowledge": The lawyer must keep secret the client's transgressions, and the physician must hide any knowledge of the sexual behavior or drug use of famous and not-so-famous patients. Professions thus "rest on some bargain about receiving, guarding, and giving out communications" (Hughes, 1959, p. 449; also see Abbott, 1988, 1993).

Interactionist sociologists pay special attention to the role relationships that develop in various professions. They are particularly interested in the processes of *professional socialization*—learning the profession's formal and informal norms. For example, in a well-known study of professional socialization in medical schools, Howard Becker and his colleagues (1961) found that in the later years of their training, medical students' attention is devoted more and more to learning the informal norms of medical practice. The formal science they learn in class is supplemented by the practical knowledge of the working doctor, and the student's skill at interacting with higher-status colleagues become extremely important to achieving the status of a professional physician. In a later study on the same subject, Charles Bosk (1979) showed that resident surgeons in a teaching hospital are forgiven if they make mistakes that the older surgeons believe to be "normal" aspects of the learning process, but not when their mistakes are repeated or are thought to be due to carelessness. The point of these and similar studies is that role performance is learned through interaction. Role expectations vary from one profession to another, as do the ways in which these expectations are experienced by members of the profession (Abbott, 1988).

The way professionals conduct their affairs is often subject to regulation and fierce politics, as we can see in the case of doctors and nurses. The rise of managed care, in which large health maintenance companies try to manage the patient's care, potentially detracts from the health professional's autonomy in decision making and is causing great unrest in the medical professions. How these professions adjust, by mobilizing their members and their potential supporters in communities outside the medical profession, will no doubt determine much of the future of managed care in the United States (Sullivan, 1999).

In almost all situations of conflict involving professions, questions arise about the extent to which professionals (as well as other workers and entrepreneurs) must be responsible to the public. This is a subject of political debate and decision making, at

FIGURE 12.3
Political Institutions of the United States

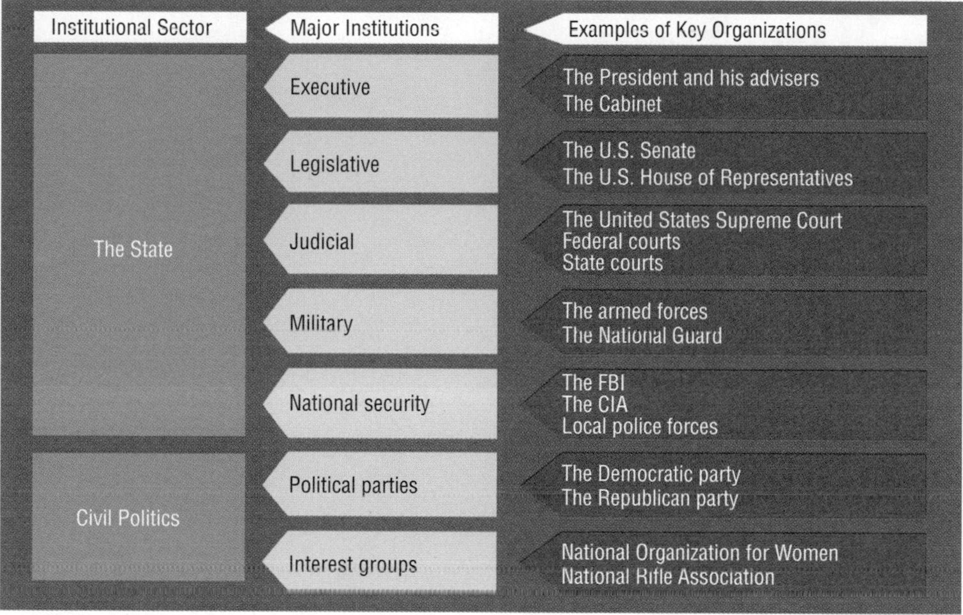

least in democratic nations. As noted at the beginning of the chapter, in a world of increasing interdependence, political issues are never very far removed from economics. (See the Global Social Change box on page 346.) Whether the issue is how to lift people out of poverty, or how to get more people covered by health insurance plans, or how to reform campaign finance, both economic and political institutions are involved. We now turn, therefore, to political institutions and how they, too, are changing.

POLITICS, POWER, AND AUTHORITY

In any society, Harold Lasswell (1936) argued, politics determines "who gets what, when, and how." Different societies develop their own political institutions, but everywhere the basis of politics is competition for power. **Power** is the ability to control the behavior of others, even against their will. To be powerful is to be able to have your way even when others resist (Mills, 1959). The criminal's power may come through a gun and the threat of injury—but many people wield power over others without any threat of violence, merely through the agreement of the governed. We call such power *authority.* **Authority** is institutionalized power—power whose exercise is governed by the norms and statuses of organizations. Those norms and statuses specify who can have authority, how much authority is attached to different statuses, and the conditions under which that authority can be exercised.

Political institutions are sets of norms and statuses that specialize in the exercise of power and authority. The complex set of political institutions—judicial, executive, and legislative—that operate throughout a society form the **state** (see Figure 12.3). This chapter will concentrate primarily on these explicitly political institutions, but keep in mind that in modern societies many institutions that are active in politics are not part of the state. Labor unions, for example, are economic institutions because they represent their members in bargaining with business owners and managers, but they play an active role in politics when they support particular political candidates or when they lobby for government-funded benefits for workers and their families.

When we look at the way in which conflicts occur over scarce resources like wealth, power, and prestige, we are looking at politics. From this perspective, there

GLOBAL SOCIAL CHANGE

The Political and Economic Impact of Immigrants

These photos represent some of the most important economic and political consequences of the recent wave of immigration to the United States. But these consequences are not limited to the immigrants themselves, and they affect other countries besides the United States. Remittances of funds to parents and other relatives in immigrants' countries of origin create a flow of capital that is of immense importance to those countries. On a global basis, economists and demographers estimate that the flow of immigrant remittances rivals in importance the cash flows generated by petroleum exports. There are many regions in Mexico, the Caribbean, Latin America, Africa, and Asia where cash remittances from immigrants are accelerating the development of local economies and contributing to movements for democratization (Dobjas & Freeman, 1992). One of the more inspiring aspects of this global flow of funds is that much of it is sent by workers earning extremely low wages, often in the least desirable jobs available in their new country.

The photo of immigrant men clustered around a car illustrates immigrants' dependence of the market for casual labor. These men, many of whom are undocumented aliens, are hoping that the man who has stopped to talk to them will give them a chance to work that day. Such scenes are common in the early morning in areas of recent immigrant settlement. But such casual labor is often highly exploitive, because in return for giving the workers (who may be undocumented or have problems of substance abuse) a chance to work, the employer may pay substandard wages and offer no insurance in case of injury, even though the kind of work involved may pose serious health risks.

These adverse working conditions, and the hostility immigrants often encounter from citizens who see them as economic competitors, often give rise to social movements among the new immigrants. The Asian American workers shown here, for example, are demonstrating against the clauses in the 1996 welfare bill that threaten to deprive immigrants of health and welfare benefits. Anti-immigrant sentiment in many parts of the United States and Europe has generated major political movements among the immigrants. In the United States, for example, recent immigrants are becoming citizens at record rates, not only to avoid anti-immigrant hostility but also in the hope that as citizens they will be able to make their needs and wishes felt through democratic processes.

These Asian-American workers are demonstrating against the clauses in the 1996 welfare reform bill that threaten to deprive immigrants of health and welfare benefits.

Latino men wait on a street corner early in the morning in the hope of being chosen for a day's work, off the books and usually for long hours, with almost no chance of advancement or insurance benefits.

STUDY CHART
Weber's Three Types of Authority

Type	Definition	Example
Traditional Authority	Authority that is hereditary and is legitimated by traditional values, particularly people's ideas of the sacred.	Tribal chiefs; absolute monarchs
Charismatic Authority	Authority that comes to an individual through a personal calling, often claimed to be inspired by supernatural powers, and is legitimated by people's beliefs that the leader does indeed have God-given powers.	Joan of Arc; Mahatma Gandhi
Legal Authority	Authority that is legitimated by people's beliefs in the supremacy of the law; obedience is owed not to a person but to a set of impersonal principles.	Presidents; prime ministers

can be a politics of the family, in which its members vie for attention, respect, use of the family car, money, and other resources. There can be a politics of the school, in which teachers compete for benefits such as smaller class size or better students. And, of course, the politics of government at all levels of society determines how the society's resources are allocated among various groups or classes. This view of politics, which looks at competition for power and at conflict over the use of power in a variety of settings, was reflected in the works of Karl Marx and Max Weber, two pioneering sociologists who devoted much of their work to the analysis of politics and change in political institutions. According to Marx, the way power is distributed in a society's institutions is a feature of its system of stratification. People and groups that have power in economic institutions, for example, are most likely to have power in political institutions as well (Washburn, 1982). However, much research in political sociology is devoted to the analysis of competition for power in political parties and electoral campaigns (Janowitz, 1968). A central problem of political sociology is whether the electoral process can sustain legitimate authority in a society in which power and wealth are stratified.

Legitimacy and Authority

A basic dilemma that every political institution must solve is how to exercise legitimate authority—that is, how to govern with the consent and goodwill of the governed. "You can't sit on bayonets," goes an old political expression. This is a way of saying that although a state can exercise its power through the use of police force or coercion, eventually this will not be sufficient to govern a society. When the Soviet Union abandoned the use of force to control its satellite states in Eastern Europe, the governments of those nations (Romania, Czechoslovakia, East Germany) were soon toppled by popular revolts. The communist parties in those nations had governed for too long without the consent of the governed—that is, without legitimacy.

But why do people consent to be governed without the use of force? For political sociologists since Weber's time, the answer to this question begins with the concept of legitimacy. As defined by political sociologist Seymour Martin Lipset (1981), **legitimacy** is the capacity of a society "to engender and maintain the belief that the existing political institutions are the most appropriate for the society" (p. 64). In other words, legitimacy results from citizens' belief in the norms that specify how power is to be exercised in their society. Even if they disagree with some aspects of their political institutions or dislike their current leaders, they still hold to an underlying belief in their political system.

Traditional, Charismatic, and Legal Authority. Max Weber (1947) recognized that shared beliefs about legitimate authority may differ widely in different groups or societies. These differences result in three basic types of authority—traditional, charismatic, and legal—that have important effects on the way societies cope with social change. (See the study chart above.)

THE CENTRAL QUESTION: A CRITICAL VIEW

Can economic and political institutions bring about positive social change in a globalizing world?

Those who study the impacts of globalization in the poorest nations have cause to be highly pessimistic. Massive protests in Seattle in 1999 and Prague, Czechoslovakia, in 2000 raised many questions about the benefits and the consequences of economic globalization. In Seattle especially, the protesters who disrupted the meetings of the World Trade Organization were extremely diverse. Some were protesting against free-trade rules that would move jobs from the United States to nations with lower wage rates. Others were protesting against the spread of American fast-food and other franchise businesses to other nations. Still others were protesting against the mounting debt of Third World nations, which makes it difficult for these nations to invest in health care and education for their populations. Many protesters were incensed by the fact that global patterns of business activity seem to be spawning new forms of human exploitation and misery. Defenders of free markets and the rapid globalization of trade point out that economic growth will bring about improvements in the well-being of people all over the world, but the protesters were not convinced.

Critical social scientists share these concerns. Kevin Bales, for example, is the world's leading authority on new forms of human slavery. His sociological case studies of various forms of human bondage in rich and poor nations demonstrate that there has been a dramatic rise in new forms of slavery. Much of this slavery is stimulated by rapid population growth in poor regions of the world, by rampant corruption and political instability fueled by arms buildups, and by the impact of global economic change. These changes often deprive peasants of their land and forces them to sell their children into various forms of slavery, including sex slavery, agricultural slavery, and bondage to sweatshop manufacturers. The United Nations and human rights groups that monitor global patterns of slavery agree on a very rough estimate of about 27 million people in slavery throughout the world. The labor of these enslaved people is thought to generate some $13 billion in profits for their owners.

In the Dominican Republic, enslaved Haitians harvest the sugar that flows into the food we eat.

Bales's research shows that global manufacturing and trading corporations often do not know, or do not acknowledge, that the competitive cost of their products often reflects savings based on slave labor. For example, charcoal produced in Brazilian forests by slave laborers is sold to Brazilian steelmakers, which helps lower the price of the steel

Much of India's agricultural production is done by bonded laborers, such as these women gathering fodder.

In many parts of the developing world, manufacturing is still human-powered, often by children in debt bondage, as in this can-producing workshop in Bangladesh.

exported by global trading corporations. "Workers making computer parts or televisions in India can be paid low wages in part because food produced by slave labor is so cheap. This lowers the cost of the goods they make, and factories unable to compete with their prices close in North America and Europe" (Bales, 1999, p. 24). In the United States, there have been numerous discoveries of bonded or slave laborers working in manufacturing sweatshops or as domestic servants. "Slave labor anywhere," Bales observes, "threatens real jobs everywhere" (1999, p. 24).

Sex slavery is one of the most rapidly growing and pernicious forms of the new slavery. Indonesia, Malaysia, the Philippines, and Thailand are world centers of this phenomenon, which is directly related to globalization. The ability of people from more affluent regions or cities to travel by jet to such cities as Bangkok, for example, has stimulated a rapid increase in the sale of children into sex slavery by destitute parents. Sociologist Lin Lean Lim (1998) has assembled a major study of the economic and social bases of prostitution and sex tourism in Southeast Asia. Although it is difficult to determine the numbers of people involved, Lim estimates that in Thailand alone, between 200,000 to 300,000 women and young children work as prostitutes, with another 100,000 Thai women serving as prostitutes outside Thailand. How many of these are actually slaves is debatable, but the buying and selling of young boys and girls into the sex industry is common.

While not all workers in the sex industry are enslaved, there is no doubt that sex slavery is a growing phenomenon. Given the threat of AIDS and violence in the world of sex workers, it is not surprising that the findings of sociological research are stimulating nongovermental agencies and official bodies like the United Nations to make greater efforts to publicize the problem and crack down on those who profit from slavery, sexual or otherwise. ■

In tribal and feudal societies, Weber observed, **traditional authority** prevails. The rulers usually attain authority through heredity, by succeeding to power within a ruling family or clan. Traditional authority is legitimated by the people's ideas of the sacred, and traditional leaders are thought to derive their authority from God. The absolute monarchs of Europe, for example, were believed to rule by "divine right."

Traditional leaders usually embrace all the traditional values of the people they rule. This gives immense scope to their authority but makes it difficult for them and their systems of government to adapt to social changes that challenge those values. Thus, the absolute monarchs of Europe had difficulty adapting their methods of governing to the growing power of new social classes like the proletariat and the bourgeoisie. The traditional chiefs of African societies today find it difficult to retain their authority as people in those societies gain more education and wealth and abandon their traditional village and agrarian ways of life.

Charismatic authority, Weber's second type, finds its legitimation in people's beliefs that their leader has God-given powers to lead them in new directions. A charismatic leader, such as Jesus or Joan of Arc, comes to power not through hereditary succession but through a personal calling, often claimed to be inspired by supernatural powers. The charismatic leader usually develops a new social movement that challenges older traditions. Charismatic leaders may appear in all areas of life, "as prophets in religion, demagogues in politics, and heroes in battle. Charismatic authority generally functions as a revolutionary force, rejecting the traditional values and rebelling against the established order" (Blau & Scott, 1962, pp. 30–31).

Charismatic authority tends to bring about important social changes but is usually unstable because it depends on the influence of a single leader. Moreover, charisma tends to become institutionalized as the charismatic leader's followers attempt to build the values and norms of their movement into an organization with a structure of statuses and roles and its own traditions. This process occurred in the early history of Christianity as the disciples built organizations that embodied the spirit and teachings of Christ.

Weber's third type of authority, **legal authority,** is legitimated by people's beliefs in the supremacy of the law. This type of authority "assumes the existence of a formally established body of social norms designed to organize conduct for the rational pursuit of specified goals. In such a system obedience is owed not to a person—whether a traditional chief or a charismatic leader—but to a set of impersonal principles" (Blau & Scott, 1962, p. 31). The constitutions of governments establish legal forms of authority. All formal organizations, not only modern nation-states but also factories, schools, military regiments, and so on, have legal forms of authority.

Ideally, legal forms of authority adhere to the principle of "government of laws, not of individuals." We may be loyal to a leader, but that leader, in turn, must adhere to the laws or regulations that establish the rights and obligations of the leader's office. In practice, individual leaders often abuse their office by using their authority for private gain, as we have seen in cases of government corruption ranging from Watergate to the savings-and-loan scandal. But legal systems and the rule of law cannot long maintain their legitimacy if official wrongdoing is not punished.

Of Weber's three basic types, legal systems of authority are the most adaptable to social change. In legal systems, when the governed and their leaders face a crisis, they can modify the laws to adapt to new conditions. This does not always occur easily or without political conflict, however, especially when momentous changes are under way.

Political Culture

Because the stability of political institutions depends so directly on the beliefs of citizens, we can readily see that political institutions are supported by cultural norms, values, and symbols, such as the Statue of Liberty. These are commonly referred to as the society's *political culture.* When we look at other societies, it is clear that their political cultures differ quite markedly from ours. Americans justify a political system based on competitive elections by invoking the values of citizen participation in politics and equality of political opportunity. Chinese Communist party leaders justify their single-party political system by downplaying the value of citizen participation in politics and asserting the need for firm leaders and a centralized state that will hold the society together and reduce inequalities. Since there is widespread skepticism about the ability of most people to participate in the political process, political sociologists believe that it will be difficult to promote democracy and the rule of law in many parts of the world (Mestrovíc, 1991; Moore, 1968; H. Smith, 1976).

Nationalism

Perhaps the strongest and most dangerous political force in the world today is nationalism. **Nationalism** is the belief of a people that they have the right and the duty to constitute themselves as a nation-state. Religion, language, and a history of immigration or oppression are among the shared experiences that can cause a people to feel that they ought to have their own state. In a recent study of nationalism throughout the world, sociologist Michael Hechter (2000) finds that while people with distinct cultures may have shared a sense of nationhood across the centuries, only when their self-governance is threatened are political leaders able to rally the people behind the banner of national sovereignty. The leaders of centralized nation-states seek to consolidate territorial rule by promoting nationalism. Threatened nations, like the republics that came into being after the disintegration of the Soviet empire in 1989, try to protect themselves from such encroachments. For Hechter, this explains why nationalism as a force for global social change is a relatively recent phenomenon. It became a major social force only with industrialization and the rise of rapid communications systems and, most important, the development of centralized states in the nineteenth century.

The situation of the French-speaking population of Quebec is a good example. Their language sets the Quebecois apart from English-speaking Canadians. So does their history of discrimination in English-dominated workplaces. In 1995, the people of Quebec narrowly defeated a separatist referendum, but it is still possible (though ever more unlikely) that nationalist sentiments in Quebec may yet result in the formation of a new nation on the North American continent (Wallace, 1999).

In the case of Quebec, where political institutions and the rule of law are well established, such a change would not necessarily be traumatic. But elsewhere in the world the rise of ethnic hostility and nationalism threaten to produce chaos. As Warren Christopher, at the time serving as secretary of state in the first Clinton administration, noted in a statement before the House Foreign Relations Committee, "If we don't find some way that the different ethnic groups can live together in a country, how many countries will we have? We'll have 5,000 countries rather than the 100-plus we have now" (quoted in Blinder & Crossette, 1993, p. A1). While the numbers may be exaggerated, few political sociologists doubt that new nations will be created as a consequence of nationalist conflicts in coming years.

The Nation-State in the New Millennium

For almost two centuries, nation-states have dominated the world political scene. England, Germany, Russia, the United States, and many others are so familiar to us that we take their sovereignty and their global influence for granted. And so we often think that all the world's populations are included in viable nation-states. But two opposing trends challenge this assumption. First, in large parts of the world in which nation-states were created by English, German, French, U.S., or other colonial rulers, nations are disintegrating. Ethnic hatreds, religious fundamentalism, corruption and kleptocracy, AIDS and other diseases, and border wars seriously threaten the future of the nations of Africa and Southeast Asia. Even in the United States, the experience of the 2000 presidential election was a shocking signal of how deeply divided the nation is on ideological grounds, although its stability is not yet in danger.

The other, and seemingly far more positive, trend away from national sovereignty is the creation of supranational entities like the European Union (EU) and the North Atlantic Free Trade Agreement (NAFTA). The EU is creating vast changes in how Europeans understand the meaning of citizenship and nation. While the nations of Europe are far from abandoning national sovereignty, the creation of a common currency, laws that permit free trade and free movement of people across their borders, and a European Parliament are signs that a new regional power is truly emerging among the once competitive and bickering nations of twentieth-century Europe (Serfaty, 2000).

Unfortunately, the trend toward crises of legitimacy and the increasing vulnerability of populations to war and economic disaster is having an even greater effect on the world political-economic scene.

Crises of Legitimacy

Societies occasionally undergo political upheavals like those that have recently occurred in Nigeria, Sierra Leone, Congo (formerly Zaire), the former Yugoslavia, and the former Soviet Union. When these periods of unrest and instability involve enough of the citizens to such a degree that they challenge the legitimacy of the nation's political institutions, sociologists call the situation a *crisis of legitimacy.* For example, the Declaration of Independence was written during a crisis in which the American colonists challenged the legitimacy of British rule. In modern South Africa, where the black majority now rules with the cooperation and participation of the white minority, it still remains to be seen whether political legitimacy can be maintained in the face of extremes of wealth and poverty.

In sum, as we scan the globe today, it is evident that the world's political ecology includes many states that are extremely stable and many others in which political instability threatens to develop into a crisis of legitimacy.

POLITICAL INSTITUTIONS IN MODERN SOCIETIES

Citizenship and Political Participation

If you travel abroad, you must obtain a passport. The passport serves as official proof of citizenship; without it you would not be allowed to return to the nation of which you are a citizen. The international system of passports, visas, and border checks is how nation-states control the movement of people across their borders. Migration from less affluent to more affluent nations is held in check by these techniques. Within many nations, also, citizens must carry passports or identity cards as proof of citizenship. But citizenship is far more than a technique for controlling the movement of populations; it is a central feature of political life in modern societies (Janowitz, 1978; Shils, 1962).

Rights of Citizenship. The term **citizenship** refers to the status of membership in a nation-state. Like all statuses, citizenship is associated with a specific set of rights and obligations (Bendix, 1969; Weber, 1968). As political sociologist Reinhard Bendix put it, "A core element of nation-building is the codification of the rights and duties of all adults who are classified as citizens" (1969, p. 89). In other words, the roles of citizens in the society's political life must be made clear to all.

In feudal societies, most of the members of the society did not participate in political life. The needs of various groups in those populations were represented, if at all, by the edicts of powerful landholders, generals, and clergymen. In modern nation-states, as a result of major social movements, this form of representation has been replaced by a form based on citizen participation, in which representatives elected by the citizens are entitled to vote on important public issues. But as societies adopted the principle of citizen participation, conflicts arose over the question of who would be included among a society's citizens. This is illustrated by the violent struggle over citizenship and political participation for blacks in South Africa, led by Nelson Mandela and the African National Congress, that finally resulted in a new constitution and majority rule.

The rights of citizenship include much more than the right to vote. T. H. Marshall (1964) defined these rights as follows:

- *Civil rights* such as "liberty of person, freedom of speech, thought and faith, the right to own property and to conclude valid contracts, and the right to justice"
- *Political rights* such as the right to vote and the right of access to public office
- *Social rights* ranging from "the right to a modicum of economic welfare and security to the right to share to the full in the social heritage and to live the life of a civilized being according to the standards prevailing in the society" (pp. 71–72).

This list does not necessarily imply that these rights actually exist in a given society or that they are shared equally by all of the society's citizens. In the United States, for example, before 1920 women were considered citizens but were denied the right to vote.

Participation in Local Politics. The question of who is entitled to full participation in politics is a key issue at the local level as well as at the national level. At the national level, participation is vital to a group's position in society, as illustrated by the struggles of women and blacks to win the right to vote. But in the cities and towns and communities in which daily life is lived, the same guarantee of full participation plays a role in social mobility. W. E. B. Du Bois' study, *The Philadelphia Negro,* includes this poignant petition by black community leaders to the city's mayor:

> We are here to state to your excellency that the colored citizens of Philadelphia are penetrated with feelings of inexpressible grief at the manner in which they have thus far been overlooked and ignored by the Republican party in this city. . . .

> We are therefore here, sir, to earnestly beseech of you as a faithful Republican and our worthy chief executive, to use your potent influence as well as the good offices of your municipal government, if not inconsistent with the public weal, to procure for the colored people of this city a share at least, of the public work and the recognition which they now ask for and feel to be justly due to them. (1967/1899, p. 374)

At the dawning of the twentieth century, Philadelphia, like most American cities, was dominated by the leaders of the local Republican party organization, who were among the town's richest and most powerful citizens. Immigrant workers and the poor were effectively excluded from political participation. The party that represented their interests, the Democratic party, was unable to gain power in the city. But the Great Depression drastically altered this pattern. Fearing that the misery created by the Depression would lead to unrest among the lower classes that could result in social disorder and even revolution, upper- and middle-class Americans began to look more favorably on the Democrats' call for the creation of new social-welfare institutions. Such institutions would ease the plight of the poor and the unemployed without fundamentally changing capitalist economic institutions. At the same time, the poor and the working class began to exercise the right to vote in increasing numbers. The result was the political turnabout that brought the Democratic candidate, Franklin D. Roosevelt, to the presidency in 1932.

In 1994, the Republican party gained a sweeping victory that gave it control of the House of Representatives, largely because of its new strength in suburban and small-town communities. Both major parties today, however, are faced with declines in political participation. Clearly, the problem of making political participation meaningful to more citizens will continue to be a subject of sociological research in coming years (Katznelson, 1981; Putnam, 1995).

Political Structures

The central problem of contemporary politics, according to Seymour Martin Lipset (1979), is this: "How can a society incorporate continuous conflict among its members and social groups and yet maintain social cohesion and legitimacy of state authority?" (p. 108). This question has taken on particular urgency since World War II, when the colonial empires of Europe crumbled in the face of nationalistic movements and dozens of new nations were created. Writing in 1963, Edward Shils wondered how modern political institutions could emerge in societies in which most of the citizens lived in traditional communities. "All the founders of the new states," he commented, "face the problems of establishing an effective government and staffing it" (p. 3). Moreover, they faced the problems of creating "a rationally conducted administration, a cadre of leaders grouped in the public form of a party system . . . and a machinery of public order" (p. 3). Because these are some of the many political institutions one expects to find in any modern state, it is helpful to examine in greater detail their structure and the problems associated with their development.

A Rational Administration. Modern nations are governed by elected and appointed officials whose authority is defined by laws. To the extent possible, they are expected to use their authority for the good of all citizens rather than for their own benefit or for that of particular groups. They are forbidden to use their authority in illegitimate ways—that is, in ways that violate the rights of the citizens.

Modern political institutions usually specify some form of separation of powers. The authors of the U. S. Constitution, for example, were careful to create a system of checks and balances among the nation's legislative, executive, and judicial institutions so that abuses of authority by one could be remedied by the others. Thus, when agents of the Nixon campaign organization were caught breaking into the Democratic party headquarters during the 1972 presidential election campaign and the president and some of his advisers were found to have covered up their role in this and other illegal activities, the impeachment proceedings carried out by the legislative and judicial branches of the government led to the president's resignation.

In many other nations, abuses of state authority are far more common. The power of rulers is unchecked by other political institutions, and although citizens may question the legitimacy of their rule, they are powerless to prevent the rulers' use of coercion in violation of their rights. Such states are often ruled by **demagogues,** or leaders who use personal charisma and political symbols to manipulate public opinion. Demagogues appeal to the fears of citizens and essentially trick them into giving up their rights of political participation. Hitler's ability to sway the suffering German masses made him the outstanding example of demagoguery in the twentieth century.

A Party System. Organizations of people who join together in order to gain legitimate control of state authority—that is, of the government—are **political parties.** Parties may be based on ideologies, or they may simply represent competing groups with the same basic values. Many American political sociologists assert that nations must make certain that other political parties are able to compete with the ruling party (Janowitz, 1968; Lipset, 1981; Oberschall, 1973). Failure to protect the existence of an opposition party or parties leads to **oligarchy**—rule by a few people who seek to stay in office indefinitely rather than for limited terms of office.

Parties that seek legitimate power and accept the rule of other legitimate parties form a "loyal opposition" that monitors the actions of the ruling party, prevents official corruption, and sustains the hopes of people whose needs are not adequately met by the ruling party. Revolutionary political parties, it should be noted, do not view the state as legitimate. They therefore do not agree to seek authority through legitimate procedures like elections. For this reason, they tend to be banned by most governments. On the other hand, many modern nations, including the United States, have banned or repressed nonrevolutionary communist parties, largely because of their opposition to private property and their sympathy for the communist-dominated regime of the former Soviet Union.

Institutions for Maintaining Order. We have seen that states control the use of force within their borders. They also seek to protect their territories against attacks by other states. For these purposes, most states maintain police forces and armies. But sometimes the state's leaders have difficulty controlling these institutions, and quite often military factions seize power in what is called a *coup d'etat* (or simply *coup*).

A coup usually results in the establishment of an oligarchy in which the state is ruled by a small elite that includes powerful members of the military. The Latin American *junta* is a special type of oligarchy in which military generals rule, usually with the consent of the most powerful members of the nation's nonmilitary elite. Their rule is commonly opposed by members of the intellectual professions, especially journalists, professors, writers, and artists. The dissent expressed by these individuals often leads to censorship and further repression of nonmilitary political institutions.

Regimes that accept no limits to their power and seek to exert their rule at all levels of society, including the neighborhood and the family, are known as **totalitarian regimes;** Nazi Germany and the Soviet Union under Stalin are examples. Such regimes cannot exercise total power without the cooperation of the military. It would be incorrect, however, to attribute the existence of oligarchies and totalitarian regimes to the power of the military alone. On many occasions, military leaders have led coups that deposed the state's existing rulers and then turned over state power to nonmilitary institutions.

Nevertheless, in modern societies military institutions have become extremely powerful. Even before the end of World War II, Harold Lasswell (1941) warned that nations might be moving toward a system of "garrison states—a world in which the specialists in violence are the most powerful group in society" (p. 457). This theme was echoed by Dwight D. Eisenhower, one of the nation's celebrated soldier-presidents. As he was leaving office, Eisenhower warned that the worldwide arms build-up, together with the increasing sophistication of modern weapons, was producing a "military-industrial complex." By this, he meant that the military and suppliers of military equipment were gaining undue influence over other institutions of the state.

Democratic Political Systems

In contrast to oligarchies and totalitarian regimes, democratic societies offer all of their citizens the right to participate in public decision making. Broadly defined, **democracy** means rule by the nation's citizens (the Greek *demos,* from which *democracy* is derived, means "people"). In practice, democratic political institutions can take many different forms as long as the following conditions are met:

1. The political culture legitimizes the democratic system and its institutions.
2. One set of political leaders holds office.
3. One or more sets of leaders who do not hold office act as a legitimate opposition (Lipset, 1981).

The two most familiar forms of democratic political rule are the British and the American systems. In the British *parliamentary system,* elections are held in which the party that wins a majority of the seats in the legislature "forms a government," meaning that the leader of the party becomes the head of the government and appoints other party members to major offices. Once formed, the government generally serves for a specified length of time. If no party gains a majority of the legislative seats, a coalition may be formed in which smaller parties with only a few seats can bargain for positions in the government. Such a system encourages the formation of smaller parties.

In the American *representative system,* political parties attempt to win elections at the local, state, and national levels of government. In this system, the president is elected directly, and the party whose candidate is elected president need not have a majority of the seats in the legislature. Success therefore depends on the election of candidates to national office. In the United States, the two major political parties—the Republicans and the Democrats—have developed the resources and support necessary to achieve this. It is difficult for smaller parties to gain power at the national level because voters do not believe such parties have much chance of electing national leaders—and even if these candidates do win a few legislative seats, they will not be asked to form coalitions, as they would in a parliamentary system.

PERSPECTIVES ON POLITICAL INSTITUTIONS

Our description of democratic political institutions leaves open the question we posed earlier: "How can a society incorporate continuous conflict among its members and social groups and yet maintain social cohesion and legitimacy of state authority?" (Lipset, 1979, p. 108). One answer to this question is that a society can resolve conflicts through democratic processes. But this leads us to ask what conditions allow democratic institutions to form and, once formed, what ensures that they actually function to reduce the inequalities that engender conflict. The broadest test of democratic institutions is not whether they are embodied in formal organizations like legislatures and courts but whether they are able to address the problems of inequality and injustice in a society.

There are at least three schools of sociological thought regarding these questions. The first, derived from the functionalist perspective, asserts that democratic political institutions can develop and operate only when certain "structural prerequisites," such as a large middle class, exist in a society. The second school of thought, often referred to as the **power elite model,** is based on the conflict perspective. It is highly critical of the functionalist view, supporting its criticism with evidence of the ways in which so-called democratic political institutions actually operate to favor the affluent. A third position, known as the **pluralist model,** asserts that the existence of a ruling elite does not mean that a society is undemocratic, as long as there are divisions within the elite and new groups are able to seek power and bargain for policies that favor their interests.

Structural Prerequisites

Seymour Martin Lipset (1981, 1994) argues that democratic political institutions are relatively rare, because if they are to exist and function well, the society must have attained a high level of economic and cultural development. To prove his theory, Lipset surveyed data on elections, civil rights, freedom of the press, and party systems in 48 nation-states, and subsequently continued his research in many other countries (Lipset, 1994). He found that the presence of these institutions was correlated with a nation's level of economic development, its degree of urbanization, the literacy of its citizens, and the degree to which its culture values equality and tolerates dissent. Table 12.1 and the Sociological Methods box on pages 358–359 present these findings in more detail.

In this cross-cultural research and in his research on democracy in the United States, Lipset attempted to show that the growth of a large middle class is essential to democracy. The middle-class population tends to be highly literate and, hence, able to make decisions about complex political and social issues. Moreover, middle-class citizens feel that they have a stake in their society and its political institutions. Accordingly, they often support policies that would reduce the class and status cleavages—the distinctions between the haves and the have-nots—that produce social conflict.

Sociologists who study voting behavior tend to support Lipset's thesis that the stability of democratic institutions rests on structural features that diminish conflict in a society. But their research has revealed something less than complete stability. Morris Janowitz (1978) made this observation:

> Since 1952, there has been an increase in the magnitude of shifts in voting patterns from one national election to the next. . . . Increasingly important segments of the electorate are prepared to change their preference for president and also to engage in ticket splitting [voting for candidates of different parties]. (p. 102)

TABLE 12.1

A Comparison of European and Latin American Democracies and Dictatorships, by Wealth, Industrialization, Education, and Urbanization

A. Indices of Wealth

	Per Capita Income in $	Thousands of Persons per Doctor	Persons per Motor Vehicle	Telephones per 1,000 Persons	Radios per 1,000 Persons	Newspaper Copies per 1,000 persons
European and English-speaking stable democracies	685	0.86	17	205	350	341
European and English-speaking unstable democracies	305	1.4	143	58	160	167
Latin American democracies and unstable dictatorships	171	2.1	99	25	85	102
Latin American stable dictatorships	119	4.4	274	10	43	43

B. Indices of Industrialization

	Percentage of Males in Agriculture	Per Capita Energy Consumed
European stable democracies	21	3.6
European dictatorships	41	1.4
Latin American democracies	52	0.6
Latin American stable dictatorships	67	0.25

C. Indices of Education

	Percentage Literate	Primary Education Enrollment per 1,000 Persons	Post-Primary Enrollment per 1,000 Persons	Higher Education Enrollment per 1,000 Persons
European stable democracies	96	134	44	4.2
European dictatorships	85	121	22	3.5
Latin American democracies	74	101	13	2.0
Latin American stable dictatorships	46	72	8	1.3

D. Indices of Urbanization

	Percent in Cities Over 20,000	Percent in Cities Over 100,000	Percent in Metropolitan Areas
European stable democracies	43	28	38
European dictatorships	24	16	23
Latin American democracies	28	22	26
Latin American stable dictatorships	17	12	15

Source: *Political Man: The Social Bases of Politics,* by Seymour Martin Lipset (Baltimore: Johns Hopkins University Press, 1981, expanded and updated edition). Reprinted by permission of the author.

These shifts in voting behavior may represent new alignments of voters that could lead to major changes in the nation's public policies. The important point, however, is that these realignments, while they influence which parties and political leaders gain or lose power, do not affect the process of democratic competition itself.

The Power Elite Model

However important elections are to the functioning of democratic institutions, there are strong arguments against the idea that they significantly affect the way a society is governed. Some social scientists find, for example, that political decisions are controlled by an elite of rich and powerful individuals. This "power elite" tolerates the formal organizations and procedures of democracy (elections, legislatures, courts, etc.) because it essentially owns them and can make sure that they act in its interests no matter what the outcome of elections may be. C. Wright Mills (1956), the chief proponent of this point of view, has described the power elite as follows:

> The power elite is composed of men whose positions enable them to transcend the ordinary environments of ordinary men and women. They are in positions to make decisions having major consequences. . . . They are in command of the major hierarchies and organizations of modern society. They rule the big corporations. They run the machinery of the state. . . . They direct the military establishment. They occupy the strategic command posts of the social structure. . . .
>
> The power elite are not solitary rulers. . . . Immediately below the elite are the professional politicians of the middle levels of power, in the Congress and in the pressure groups, as well as among the new and old upper classes of town and city and regions. Mingling with them in curious ways . . . are those professional celebrities who live by being continually displayed. (p. 4)

When it was first published, Mills's *The Power Elite* created a stir among sociologists. Mills challenged the assumption that societies that have democratic political institutions are in fact democratic. He asserted instead that party politics and elections are little more than rituals, with real power exercised by a ruling elite of immensely powerful military, business, and political leaders that can put its members into positions of authority whenever it wishes to do so. Mills's claim was significant for another reason as well: Although Mills appreciated Marx's views on the role of class conflict in social change, he did not believe that the working class could win power without joining forces with the middle class. He therefore attempted to demonstrate the existence of a ruling elite to an educated public, which would then, he hoped, be able to see through the rituals of political life and make changes through legitimate means.

Other sociologists have questioned the power elite thesis on methodological grounds. For example, Talcott Parsons (1960) noted that the power elite was supposed to act behind the scenes rather than publicly. Its actions could not be observed, and therefore the power elite thesis could not be either proved or disproved and hence was not scientifically sound.

Numerous adherents of the power elite thesis have attempted to show that a ruling elite does indeed exist and that its activities can be observed. One of the best known of these researchers is Floyd Hunter (1953), whose classic studies of Atlanta's "community power structure" attempted to show that no more than 40 powerful men were considered to have the ability to make decisions on important issues facing the city and its people. Most of these men were conservative, cost-conscious business leaders. The Hunter study stimulated many attempts to find similar power structures in other cities and in the nation as a whole (Domhoff, 1978, 1983). But although the term *power structure* has found its way into the language of politics, these studies have been strongly criticized for basing their conclusions on what people say about who has power rather than on observations of what people with power actually do (Walton, 1970).

The Pluralist Model

The power elite thesis has not gone unchallenged. In another famous study of politics this one in New Haven, Connecticut—Robert Dahl (1961) found that different individuals played key roles in different types of decisions. No single group was responsible for all of the decisions that might affect the city's future. No power elite ruled the city, Dahl argued. Instead, there were a number of the members of the elite that interacted in various ways on decisions that affected them. In situations in which the interests of

SOCIOLOGICAL METHODS

Diagramming Social Change

The accompanying chart is an example of how a diagram can effectively communicate ideas about the directions and influences of social change. This diagram presents the social structures from which democratic political institutions arise (listed on the left) and some possible further consequences of the emergence of those political institutions (on the right).

The arrows in the diagram indicate either conditions that contribute to democracy or possible consequences of the emergence of democratic institutions. In this example, as in most social-scientific analysis, the author, political scientist Seymour Martin Lipset, is careful not to make direct assertions about causality. He does not state, for example, that an open class system (measured by upward and downward social mobility) inevitably leads to the emergence of democratic regimes. Instead, social mobility and the other aspects of social structure listed on the left are "necessary but not sufficient" preconditions for the emergence of democracy. That is, they must exist to some degree in a society, but they are not sufficient to "cause" democracy. For that to happen, other conditions must also be present, such as the leadership of people who are willing to sacrifice their desire for personal power in the interest of creating democratic political institutions. George Washington, for example, is revered as a pioneer of democracy, because when he had the chance to become a monarch, he chose to relinquish personal power and lead the newly independent colonies toward the creation of a constitutional government.

Note that the arrows on the right side of the diagram suggest that democratic rule can be associated with the emergence of social phenomena that run counter to democracy. Bureaucracy, for example, is a common consequence of democratic regimes. Laws passed by democratic processes tend to spawn administrations designed to deliver services, provide protection, enforce regulations, and much more. These bureaucracies, in turn, may stifle democracy by replacing the will of the majority with rules, regulations, and procedures. Note, however, that these excesses can be corrected by the actions of democratic institutions (for example, further legislation). The arrows merely indicate that democratic rule has certain counter-democratic consequences.

In an expansion of the comparative research on which this diagram is based, Lipset makes the following comment:

> Not long ago, the overwhelming majority of the members of the United Nations had authoritarian systems. As of the end of 1993, over half, 107 out of 186 countries, have competitive elections and various guarantees of political and individual rights—that is, more than twice the number two decades earlier in 1970. (1994, p. 1)

But Lipset, a veteran of many decades of close observation of democratic rule in nations throughout the world, is quite cautious about predicting a trend toward the emergence of democratic regimes. Too many nations, like Nigeria and Russia, have turned toward more authoritarian rule after promising attempts to develop democratic political institutions. Above all, Lipset warns:

> [If nations] can take the high road to economic development, they can keep their political houses in order. The opposite is true as well. Governments that defy the elementary laws of supply and demand will fail to develop and will not institutionalize genuinely democratic systems. (1994, p. 1)

Conditions and Consequences of Democracy

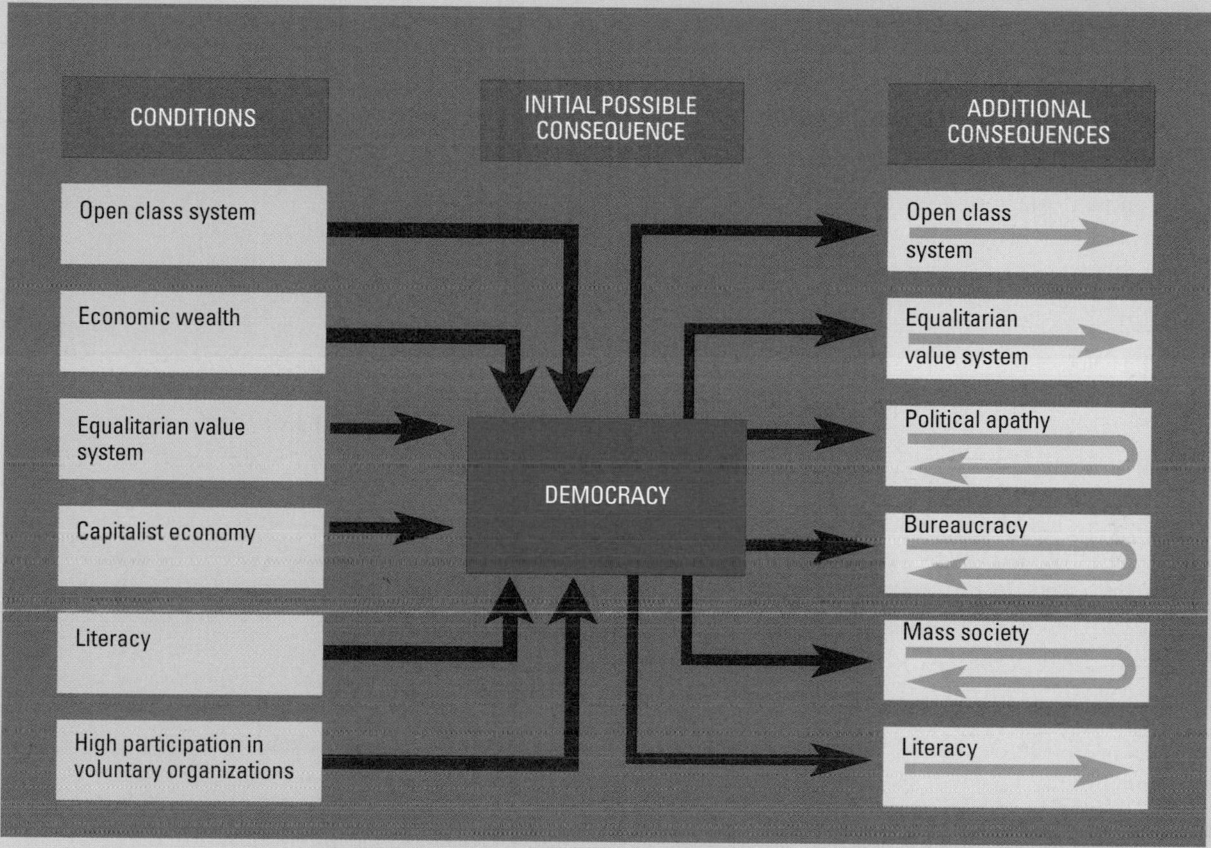

Source: From *Political Man: The Social Bases of Politics,* by Seymour Martin Lipset (Baltimore: Johns Hopkins University Press, 1981, expanded and updated edition). Reprinted by permission of the author.

numerous groups were involved, a plurality of decision makers engaged in a process of coalition building and bargaining.

The pluralist model calls attention to the activities of interest groups at all levels of society. **Interest groups** are not political parties; they are specialized organizations that attempt to influence elected and appointed officials on specific issues. These attempts range from **lobbying**—the process whereby interest groups seek to persuade legislators to vote in their favor on particular bills—to making contributions to parties and candidates who will support their goals. Trade unions seeking legislation that would limit imports, organizations for the handicapped seeking regulations that would give them access to public buildings, and ethnic groups seeking to influence U.S. policy toward their country of origin are among the many kinds of interest groups that are active in the United States today.

In recent decades, as the number of organized interest groups has grown, so has the complexity of the bargaining that takes place between them and the officials they want to influence. Social scientists who study politics from the pluralist perspective frequently wonder whether the activities of these groups threaten the ability of elected officials to govern effectively. Thus, a study of a federal economic development project in Oakland, California, found that the personnel in the agency responsible for such projects had to learn to deal with supporting and opposing interest group leaders and elected officials throughout the life of the project. It took six years of constant negotiation on many unanticipated issues—affirmative action in employment, environmental concerns, design specifications—before a firm could win a contract to build an airplane hangar (Pressman & Wildavsky, 1984).

No area of life in a democracy is without conflict, either actual or potential, nor is it without efforts to resolve conflict through legitimate political processes like voting and court actions. This observation is illustrated by the mention of affirmative action, environmental regulations, and labor–management relations in the example just given—all of these are clearly controversial areas. In 1996, for example, California voters approved a resolution to ban certain types of affirmative action. Courts and legislatures in many other states are deciding whether to follow California's example. And this is only one of many examples of political efforts to resolve conflicts that arise out of what are essentially economic as well as political issues.

TECHNOLOGY AND SOCIAL CHANGE

Polling technologies and practices took a huge hit on election night 2000. After years of reliance on pre-election and election-day exit polls, the major television networks and newspapers, which often commission the polls, are no doubt going to be under pressure to change their practices. There were many problems, but the most egregious errors occurred during the election itself. *Exit polls,* interviews conducted in strategically chosen voting districts as voters exit the polling places, are usually highly accurate. Eventual election outcomes can be predicted well before all the votes have been counted if pollsters know how a small sample of voters actually voted and can compare their voting behavior to past elections in the same districts. But the Bush–Gore election was so close, and the actual technologies of voting in Florida were so problematic, that the exit polls turned out to be incapable of actually predicting the vote. Unfortunately, that did not stop the polling professionals from trying. Early in the evening of election day, Florida exit polls indicated that Gore would win that state. This news was widely broadcast and might have discouraged voters who had not yet come to the polls from actually voting. In 2002, therefore, Congress will probably consider legislation to ban the practice of releasing exit polls in states where the polls are still open. The time difference between the East and West Coasts will complicate the debate.

A second major problem with the polling techniques involved the tendency for people, especially women, to say they would vote one way and then actually vote another way. This is not a new problem for sociologists. There is never a certainty that opinions and behavior will match, only a likelihood that they will. In this election, as Table 12.2 shows, men were much more likely to vote for Bush than for Gore, whereas the reverse was true for women. The gender gap was a major factor in Gore's eventual victory in the popular vote. But it was impossible to actually predict this outcome from the pre-election polls. They showed that women were more favorable to Gore than to Bush, but by a small margin, only 1%, that could easily have been due to sampling error rather than to a real difference in how women and men felt about the candidates. But clearly, when it came time to actually vote, women favored Gore by almost 11 percentage points. Women's rights issues, including abortion, must have figured prominently in their final decisions, but they were not admitting their real thoughts to pollsters before the election. That is a major reason why the polls were so close before election day and why the science of political polling is so inexact in a very close race like this one. In future elections, when polling experts point out the influence of sampling errors and other possible distortions in their results, their clients in the political parties and the media will no doubt take them far more seriously then they have in the past.

TABLE 12.2

Results of Pre-election and Exit Polls in the 2000 Presidential Election

	Last Pre-election Poll			Exit Poll		
	Bush	**Gore**	**Nader**	**Bush**	**Gore**	**Nader**
Men	50%	37%	6%	53%	42%	3%
Women	43%	44%	4%	43%	54%	2%

Sources: Benke & Elder, 2000; Connelly, 2000.

SOCIOLOGY AND SOCIAL JUSTICE

Senators McCain and Feingold have plenty of supporters in their fight for campaign finance reform. Probably none are as committed to the cause as Granny D. Her story is an inspiration to all in the movement for limiting the influence of money and elite power in the U.S. campaign process.

Just before the 2000 presidential election, 90-year-old Doris "Granny D" Haddock pounded out the last few miles of her 3,200-mile walk across the United States to mobilize people behind the effort for campaign finance reform. A woman of great sociological imagination, Granny D had followed the 1999 Rose Bowl parade in Pasadena, California, and just kept walking. She averaged 10 miles each day she walked. She wore through four pairs of sneakers and four straw hats, and celebrated her eighty-ninth and ninetieth birthdays on the road. She spent four days hydrating in a hospital after crossing the Mojave Desert. At times, she was urged to quit, or at least to take a ride across the desert. When she finally got to Washington, she spoke at the Lincoln Memorial of Lincoln's desire for a government "of the people, by the people, for the people," a dream of sovereignty, she says, that requires public financing of elections (Lynch, 2000). ■

"Granny D," from her Web site, www.grannyd.com.

Former President Clinton making a pitch for campaign funds. Today U.S. political candidates spend the majority of their time seeking donations to fund their contests.

SUMMARY

Economics is the study of how individuals and societies choose to employ scarce resources to produce various commodities and distribute them among various groups in the society. A *market* is an economic institution that regulates exchange behavior. The spread of markets throughout the world began during the late fifteenth century as a result of the development of new *technologies* that facilitated trade. Today world markets are dominated by *multinational corporations,* economic enterprises that have headquarters in one country and conduct business activities in one or more other countries.

Mercantilism is an economic philosophy based on the belief that the wealth of a nation can be measured by its holdings of gold, so the best economic system is one that increases the nation's exports and thereby increases its holdings of gold. *Laissez-faire capitalism* argues that a society's wealth should be measured only by its capacity to produce goods and services—that is, its resources of land, labor, and machinery. *Socialism* arose out of the belief that private property and personal profit should be replaced by public ownership of property and sharing of profits. *Democratic socialism* holds that private property must continue to exist but that large corporations should be owned by the nation or, if they are in private hands, must be run for the benefit of all citizens. In *welfare capitalism,* markets determine what goods and services will be produced and how, but the government regulates economic competition.

In the United States, the transition from an economy based on the production of goods to one based on the provision of services has resulted in the displacement of thousands of skilled workers into lower-paying jobs in the service sector.

The scientific management approach to labor–management relations attempts to increase productivity by determining how each job can be performed most efficiently and by using piecework payment systems to induce workers to produce more. The human relations approach is based on the recognition that cooperation between workers and managers is an important ingredient in worker satisfaction and output. Conflict theorists study how social class and status, both at work and outside the workplace, influence relations between workers and managers. An important trend in industrialized nations is toward nonstandard work arrangements such as temporary and contract work.

Interactionist theorists have devoted considerable study to professionalization, or the way in which occupations attempt to gain the status of *professions,* and to the processes of professional socialization (in other words, learning the formal and informal norms of the profession).

Politics determines "who gets what, when, and how." The basis of politics is competition for *power,* or the ability to control the behavior of others, even against their will. *Authority* is institutionalized power, or power whose exercise is governed by the norms and statuses of organizations. Sets of norms and statuses that specialize in the exercise of power and authority are *political institutions,* and the set of political institutions that operate in a particular society forms the *state.*

Legitimacy is a society's ability to engender and maintain the belief that the existing political institutions are the most appropriate for that society. It is the basis of a society's political culture—the cultural norms, values, and symbols that support and justify its political institutions. According to Max Weber, the political cultures of different societies give rise to three different types of authority: *traditional, charismatic,* and *legal.* Perhaps the strongest and most dangerous political force in the world today is *nationalism,* the belief of a people that they have the right and the duty to constitute themselves as a nation-state. Today many nations are disintegrating as a result of conditions such as ethnic hatreds and border wars; at the same time, supranational entities like the European Union are encouraging greater cooperation among member nations.

Citizenship is the status of membership in a nation-state. The question of who should be included among a society's citizens (and thus who is entitled to full participation in politics) is a key issue at the local level as well as at the national level.

Modern nations are governed by elected and appointed officials whose authority is defined by laws. In order to prevent abuses of authority, modern political institutions usually specify some form of separation of powers in which abuses by one institution can be remedied by others. *Political parties* are organizations of people who join together in order to gain legitimate control of state authority. Parties that accept the rule of other legitimate parties form a "loyal opposition" that monitors the actions of the ruling party and prevents the emergence of *oligarchy,* or rule by a few people who stay in office indefinitely. Regimes that accept no limits to their power and seek to exert their rule at all levels of society are known as *totalitarian regimes.*

Democracy means rule by the nation's citizens: Citizens have the right to participate in public decision making, and those who govern do so with the explicit consent of the governed. In the British parliamentary system, elections are held in which the leader of the party that wins a majority of the seats in the legislature becomes the head of government and appoints other party members to major offices. In the American representative system, the party whose candidate is elected president need not have a majority of the seats in the legislature.

Functionalist theorists assert that certain "structural prerequisites" must exist in a society for democratic political institutions to develop and operate. Among these are high levels of economic development, urbanization, and literacy, as well as a culture that tolerates dissent. The *power elite model* holds that the presence of democratic institutions does not necessarily mean that a society is democratic; political decisions are actually controlled by an elite of rich and powerful individuals. This view is challenged by the *pluralist model,* which holds that political decisions are influenced by a variety of *interest groups* through a process of coalition building and bargaining.

GLOSSARY

Market: An economic institution that regulates exchange behavior through the establishment of different values for particular goods and services. **(333)**

Technology: Tools, procedures, and forms of social organization that increase human productive capacity. **(334)**

Multinational corporation: An economic enterprise that has headquarters in one country and conducts business activities in one or more other countries. **(334)**

Mercantilism: An economic philosophy based on the belief that the wealth of a nation can be measured by its holdings of gold or other precious metals and that the state should control trade. **(336)**

Laissez-faire capitalism: An economic philosophy based on the belief that the wealth of a nation can be measured by its capacity to produce goods and services (in other words, its resources of land, labor, and machinery) and that these can be maximized by free trade. **(336)**

Socialism: An economic philosophy based on the concept of public ownership of property and sharing of profits, together with the belief that economic decisions should be controlled by the workers. **(336)**

Democratic socialism: An economic philosophy based on the belief that private property may exist at the same time that large corporations are owned by the state and run for the benefit of all citizens. **(338)**

Welfare capitalism: An economic philosophy in which markets determine what goods will be produced and how, but in which the government regulates economic competition. **(338)**

Profession: An occupation with a body of knowledge and a developed intellectual technique that are transmitted by a formal educational process and testing procedures. **(344)**

Power: The ability to control the behavior of others, even against their will. **(345)**

Authority: Power whose exercise is governed by the norms and statuses of organizations. **(345)**

Political institution: A set of norms and statuses pertaining to the exercise of power and authority. **(345)**

State: The set of political institutions operating in a particular society. **(345)**

Legitimacy: The ability of a society "to engender and maintain the belief that the existing political institutions are the most appropriate for that society." **(347)**

Traditional authority: Authority that is hereditary and is legitimated by traditional values, particularly people's ideas of the sacred. **(350)**

Charismatic authority: Authority that comes to an individual through a personal calling, often claimed to be inspired by supernatural powers, and is legitimated by people's beliefs that the leader does indeed have God-given powers. **(350)**

Legal authority: Authority that is legitimated by people's beliefs in the supremacy of the law; obedience is owed not to a person but to a set of impersonal principles. **(350)**

Nationalism: The belief of a people that they have the right and the duty to constitute themselves as a nation-state. **(351)**

Citizenship: The status of membership in a nation-state. **(352)**

Demagogue: A leader who uses personal charisma and political symbols to manipulate public opinion. **(353)**

Political party: An organization of people who join together to gain legitimate control of state authority. **(354)**

Oligarchy: Rule by a few people who stay in office indefinitely rather than for limited terms. **(354)**

Totalitarian regime: A regime that accepts no limits to its power and seeks to exert its rule at all levels of society. **(354)**

Democracy: A political system in which all citizens have the right to participate in public decision making. **(354)**

Power elite model: A theory stating that political decisions are controlled by an elite of rich and powerful individuals even in societies with democratic political institutions. **(355)**

Pluralist model: A theory stating that no single group controls political decisions; instead, a plurality of interest groups influence those decisions through a process of coalition building and bargaining. **(355)**

Interest group: An organization that attempts to influence elected and appointed officials regarding a specific issue or set of issues. **(360)**

Lobbying: The process whereby interest groups seek to persuade legislators to vote in their favor on particular bills. **(360)**

QUESTIONS FOR THOUGHT AND DISCUSSION

1. What are some of the outstanding features of industrial and postindustrial societies? Does industry disappear in postindustrial societies? What kinds of changes are occurring in postindustrial societies that affect the careers of young workers?

2. Contrast the views of a laissez-faire capitalist and a social democrat with reference to issues of health and advertising regulations, the role of government in the economy, and environmental protection. How has the end of the Cold War affected debates over these issues?

DIGGING DEEPER

Books

Disposable People: New Slavery in the Global Economy (Kevin Bales; University of California Press, 1999). An important study of the darkest side of economic globalization.

Lean and Mean: The Changing Landscape of Corporate Power in the Age of Flexibility (Bennett Harrison; Basic Books, 1994). An analysis of the origins of corporate "downsizing" in a global economy and its consequences for stratification.

The Winner-Take-All Society (Robert H. Frank & Philip J. Cook; Free Press, 1995). An influential analysis of the U.S. economy that shows that as more Americans compete for fewer and bigger prizes, income inequality and economic waste are the consequences.

Out to Work (Alice Kessler Harris; Oxford University Press, 1982). A history of wage-earning women in the United States with much insight into the impact of changing economic institutions on women and their careers.

Post Industrial Possibilities: A Critique of Economic Discourse (Fred Block; University of California Press, 1990). A sociologist skilled in economic and sociological analysis illuminates the quandaries and possibilities of postindustrial society.

"Politics as a Vocation," in ***From Max Weber*** (Hans Gerth & C. W. Mills, eds.; Oxford University Press, 1946). A fascinating sociological statement about what it takes to live the life of a professional politician and what kinds of people and professions are best suited to that life.

Politics Against Markets (Gosta Esping-Anderson; Princeton University Press, 1985). Describes the politics of welfare states and the emergence of social-democratic institutions that place limits on the capitalist economies of the Scandinavian nations.

Continental Divide: The Values and Institutions of the United States and Canada (Seymour Martin Lipset; Canadian-American Committee, 1989). Lipset is the dean of American political sociologists. In this empirical study comparing the United States and Canada, he shows why Canada has had more success in adapting its political institutions to address social problems.

Other Sources

Current Population Reports: Economic Characteristics of Households in the United States. A quarterly report published by the Bureau of the Census. Presents monthly averages of household income and participation in cash and noncash transfer programs.

Handbook of Economic Statistics. Published annually by the Central Intelligence Agency. Compares economic statistics for communist, Organization for Economic Cooperation and Development (OECD), and selected other countries.

World Almanac and Book of Facts. Published annually by Funk & Wagnalls, Mahwah, NJ.

America Votes. Published biennially by the Elections Research Center, Washington, DC. Presents data on both federal and state elections.

EXPLORING SOCIOLOGY ON THE INTERNET

International Monetary Fund
www.cgs.edu/acit/help/imf.html

Provides access to a great deal of comparative data on how the world's nations are developing—or not developing—along a variety of important economic and social dimensions.

The World Bank
www.worldbank.org/

Organization for Economic Cooperation and Development (OECD)

www.oecd.org/

Two sources of extensive data on social indicators such as health and education in different nations and regions of the world.

WWW Virtual Library on Economics
http://netec.wustl.edu/WebEc.html

Offers innumerable resources on the many applications of economic and political-economic research throughout the world.

The New York Times
www.nytimes.com/

An unparalleled source for political events and trends in the world. To receive the full range of available materials, you will have to register at this address.

Stanford University, Hoover Institution on War, Revolution, and Peace
www-hoover.stanford.edu/

Analyzes social, political, and economic change in the world.

African Studies Department at the University of Pennsylvania
www.sas.upenn.edu/African_Studies/AS.html

A valuable site for those interested in the economics and politics of a rapidly changing continent.

The White House
www.whitehouse.gov

A comprehensive site that features an Interactive Citizens' Handbook providing information about the U.S. federal government. The site also offers a list of commonly requested federal services and ways to communicate with the president and vice president via electronic mail.

United States Federal Judiciary
www.uscourts.gov

Provides a clearinghouse for information from and about the judicial branch of the U.S. government. The site offers guidelines for understanding the federal court system and publishes press releases about recent decisions.

Chapter 13

Population, Urbanization, and Public Health

Can we control population growth and achieve higher levels of health worldwide?

Tidi was 10 years old when photographer, teacher, and social scientist Wendy Ewald came to her village in Gujarat, in the desert of northwestern India near the Pakistani border. Tidi used a camera lent her by Ewald to photograph her village. Naturally, she took photos of the people closest to her, like these women and children of her social caste, the Harijan or Untouchable caste. Here is a small passage from Tidi's reflections on her photographs:

> I have a mommy, father, two brothers, and three sisters. One brother is married and the other is going to marry. My father laughs a lot, but I don't like him when he beats us. My small brother and cousin are at home. One younger brother and sister got smallpox right after they were born. When they died, we buried them on the roadside and marked the grave with a red cloth. I can't find it now. The sweepers must have taken it away.
>
> I get enough food to eat—even during the drought. We went to the city then and lived in our uncle's house for six months. All the Harijans went. My father did construction work. My mother helped him and washed dishes and clothes for a rich family. I don't mind washing dishes, but I'd like to go to school.
>
> I had to leave my studies because of the housework. Anyhow, soon we have to marry. I'll be happy then. My husband will get me a scooter. I'll get on it and wander. I want sons, as many as God wants to give me—two, three, or four. I'll play with my children and scold them when they're mischievous. When I'm older I'll look after my husband's house. (quoted in Ewald, 1996)

Ewald's photos and Tidi's comments help put a human face on world population issues. With every tick of the second hand, another four or five people are born somewhere in the world. During each of those seconds, about two people die—including some who where just born. Tidi's baby brother and sister were among these. Of the three who live, two are added to the population of the world's poorest nations. The study of population issues often seems extremely bleak because of facts like these. But Tidi's energy and hope brighten the picture by giving us a sense that both rich and poor countries are populated by people with the same hopes and dreams for their futures. And the study of how populations grow and change can reveal a great deal about how to develop effective social policies to address problems of world poverty and crises in public health like the world AIDS epidemic.

Many issues related to population and public health are explored in this chapter. We begin by introducing some basic concepts that are essential to the study of population growth. Then we consider the consequences of the dramatic shift from rural to urban settlement and present key concepts in the study of urbanization. We conclude with an examination of some current topics of medical sociology as they relate to the effects of population growth and urbanization. ■

Tidi and her Harijan (Untouchable caste) neighbors. "I don't mind washing dishes [for the rich family for whom she worked in the city], but I'd like to go to school."

PEOPLE, CITIES, AND URBAN GROWTH

The human population continues to spread throughout the world. Although the vast majority of the world's people live in rural villages, in modern societies people typically live in cities or metropolitan areas rather than on farms or in small towns. In fact, 90% of Americans live within 25 miles of a city center (Beale, 2000; Berry, 1978). The study of population growth and the growth of cities as centers of demographic and social change has long been a major field in sociology. In this section, therefore, we will examine how increasing population size and other important changes in populations are connected with the rise of cities and metropolitan areas throughout the world.

The Population Explosion

In the contemporary world, the growth of populations and the rapid increase in the number of cities and metropolitan regions are directly linked. Eighty percent of the world's population lives in the less-developed regions of the world, where rural poverty drives millions of people each year to cities that are growing at astounding rates. In the metropolitan regions of the developed nations, the issues are different, but they also involve changes in the population composition of cities and suburbs (Brockerhoff, 2000).

Before we can study the phenomenon of urbanization, we must understand the impact of population growth on the formation and expansion of cities. The rate of growth of the world's population increased dramatically in the twentieth century, giving rise to the often-used term *population explosion.* The world's population is currently estimated at 6 billion, and according to recent United Nations forecasts, it will increase by another 1.5 to 2 billion in the next 20 years. The United States began the twentieth century with 76.2 million people. It ended the century with 275 million people, or 3.6 times as many as in 1900 (Beale, 2000).

At present, about 90 million new people are added to the world population each year. Most forecasts predict that this rapid growth will begin to decline by midcentury, but by then there could well be more than 10 billion people on a very crowded planet. Demographers point out, however, that crop production, famine and war, contraception, economic development, and many other variables affect population. The potential effects of these variables make it extremely difficult to predict trends in population growth. It is certain, though, that rapid population growth, especially in the poorer regions of the world, will continue in coming decades and will pose severe challenges to human survival and well-being (Bongaarts & Bulatao, 1999). The pressures of population growth on resources of food, space, and water have produced changes in economic and social arrangements throughout history, but never before have those changes occurred as rapidly or on as great a scale as they are occurring today.

Malthusian Population Theory. Is there a danger that the earth will become overpopulated? The debate over population growth is not new. For almost two centuries, social scientists have been seeking to determine whether human populations will grow beyond the earth's capacity to support them. The earliest and most forceful theory of overpopulation appeared in Thomas Malthus's *An Essay on Population* (1927–1928/1798). Malthus attempted to show that population size normally increases far more rapidly than the food and energy resources needed to keep people alive. Couples will have as many children as they can afford to feed, and their children will do the same. This will cause populations to grow *geometrically* (2, 4, 8, 16, 32, etc.). Meanwhile, available food supplies will increase *arithmetically* (2, 3, 4, 5, 6, etc.) as farms are expanded and crop yields increased. As a result, population growth will always threaten to outstrip food supplies. The resulting poverty, famine, disease, war, and mass migrations will act as natural checks on rapid population growth.

History has proved Malthus wrong on at least two counts. To begin with, we are not biologically driven to multiply beyond the capacity of the environment to support our offspring; Malthus himself recognized that people could limit their reproduction through delay of marriage or celibacy. (Today, many forms of contraception are readily available in developed nations but far less so in areas of rapid population growth.) The second fault of Malthusian theory is its failure to recognize that technological and institutional changes could expand available resources rapidly enough to keep up with population growth. This has occurred in the more affluent regions of the world, where improvements in the quality of life have tended to outstrip population growth. Improvements in agricultural technology have also increased the yield of crops in some of the less-developed parts of the world, such as India. But rates of population growth and exhaustion of environmental resources (firewood, water, grazing land) are highest in the poorest nations. Will reductions in population growth rates and increases in available resources also occur there? Or will the Malthusian theory prove correct in the long run? The theory known as the *demographic transition* provides a framework for studying this question. (To understand how sociologists measure population change, see the Sociological Methods box on pages 374–375.)

The Demographic Transition. The **demographic transition** is a set of major changes in birth and death rates that has occurred most completely in urban industrial nations in the past 200 years. Beginning in the second half of the eighteenth century and continuing until the first half of the twentieth century, there was a marked decline in death rates in the countries of northern and western Europe. Improvements in public health practices were one factor in that decline. Even more important were higher agricultural yields owing to technological changes in farming methods, as well as improvements in the distribution of food as a result of better transportation, which made cheaper food available to more people (Matras, 1973; Vining, 1985). At the same time, however, birthrates in those countries remained high. The resulting gap between birthrates and death rates produced huge increases in population. It appeared that the gloomy predictions of Malthus and others would be borne out.

In the second half of the nineteenth century, birthrates began to decline as couples delayed marriage and childbearing. As a result of lower birthrates, the gap between birthrates and death rates narrowed and population growth slowed (see Figure 13.1, which illustrates the demographic transition in Sweden between 1691 and 1963). This decrease in population growth rates occurred at different times in different countries, but the general pattern was the same in each case: a stage of high birthrates and death rates (the *high growth potential* stage) followed by a stage of declining death rates (the *transitional growth* stage) and, eventually, by a stage of declining birthrates (called the stage of *incipient decline* because it is possible for population growth rates to decrease at this point).

In the second, or transitional growth, stage of the demographic transition, the population not only grows rapidly but undergoes changes in its age composition. Because people now live longer, there is a slight increase in the proportion of elderly people in the population. There is also a marked increase in the proportion of people under age 20 as a result of significant

FIGURE 13.1
The Demographic Transition: Sweden, 1691–1963
The peaks in birthrates and death rates in the early 1800s are due to social unrest and war. The drop in deaths and simultaneous rise in births in the early 1700s were a result of peace, good crops, and the absence of plagues.

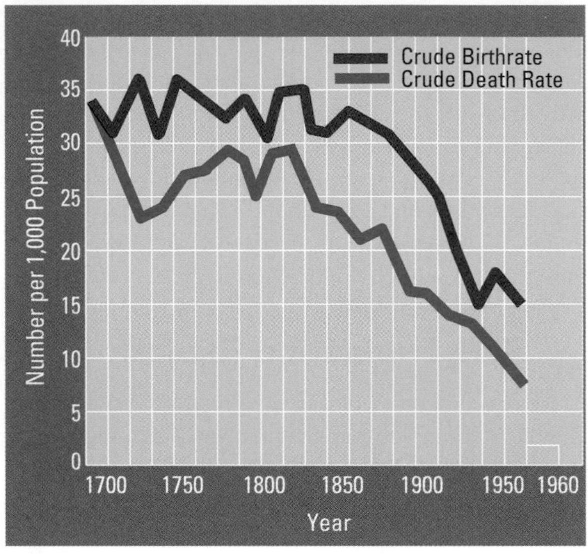

Source: Judah Matras, *Population and Societies,* Englewood Cliffs, N.J., © Prentice Hall, 1973. Permission of Armand Colin Éditeur.

decreases in infant and child mortality. This is the stage in which many less-developed countries find themselves today: Death rates dropped in the twentieth century because of improved medical care and public health measures, as well as increased agricultural production. Yet in these societies birthrates have remained high, causing phenomenal increases in population, especially in the younger, more dependent age groups.

No population has entered the third stage of the demographic transition without limiting its birthrate. This can be achieved by encouraging couples to marry later and postpone childbearing or by preventing pregnancies or births through various birth control techniques. In advanced industrial societies, many couples use both approaches, making their own decisions about whether and when to have children. In other societies, such as China and India, the state has attempted to limit population growth by promoting birth control through educational programs or, in the case of China, by imposing penalties on couples who have more than a prescribed number of children. Such measures have had some success, but often at a high political cost to national leaders, causing demographers to question whether the less-developed countries will be able to complete the demographic transition.

Note that economic and social development is essential if the demographic transition is to occur. Death rates cannot decrease, or food supplies increase, without progress in social institutions like public health, medicine, and transportation. People in more-developed societies tend to limit their family size because they seek economic advancement and wish to delay marriage and childbearing until they can support a family.

In many highly industrialized nations, on the other hand, population growth rates have fallen below the rate necessary to maintain existing levels. If a fertility rate of two children or less per couple became the norm for an entire population, the growth rate would slow to zero or even a negative rate. The United States, New Zealand, Japan, Australia, and Canada all have total fertility rates of 2.0 or less (Livernash & Rodenberg, 1998). Similar low rates are appearing in many European nations. This is a highly significant development; if it continues over a generation or more, it could result in negative rates of natural increase and, possibly, other consequences such as slower economic development and labor shortages. Continued low fertility could also result in increased immigration from countries with high birthrates to countries with lower birthrates.

A full exposition of the processes and politics of population control is not possible here. Suffice it to say that most sociologists would agree that low population growth rates are due primarily to a combination of delay of marriage, celibacy, and use of modern birth control techniques. This trend is beginning to occur in many less-developed nations. Figure 13.2 illustrates the demographic transition in Singapore, a nation in Southeast Asia that has undergone this important population change more recently. Note that the decline in deaths, which took almost a century and a half in Western industrial nations, took less than half a century in Singapore. The decline in births began in the late 1950s, spurred by rapid economic development, the new desire for smaller families, and widespread availability of contraceptives. Whereas in Western nations like Sweden this phase of declining births unfolded over more than a century, in Singapore this phase occurred in only about 20 years (between about 1955 and 1975). Although world population growth remains a serious problem, the demographic transition in formerly poor nations like this one shows that population control efforts along with economic development can have dramatic effects (K. Davis, 1991).

FIGURE 13.2
Crude Birthrates and Death Rates: Singapore, 1930–1988

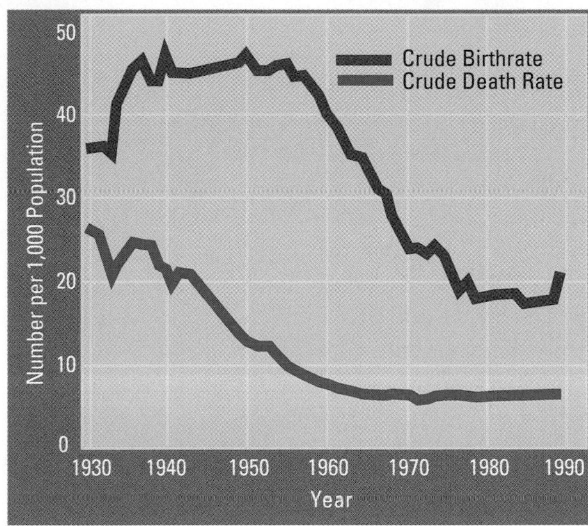

Source: K. Davis, 1991.

Population Growth and Urbanization.

Studies of the demographic transition in Europe (Friedlander, 1969; Laslett, 1972, 1983) have concluded that its specific course in any given society depended on a complex combination of factors: higher age at marriage and fewer couples marrying, use of birth control techniques, increased education, migration to other countries, and rural–urban migration. In each case, however, the bulk of the population growth was absorbed by the cities. Rapid urbanization thus to some extent is an outgrowth of the demographic transition.

Urbanization refers to the proportion of the total population that is concentrated in urban settlements. Although urbanization and the growth of cities have occurred together, it is important to distinguish between the two. Cities can continue to grow even after the majority of the population is urbanized and the society's dominant institutions (e.g., government agencies, major markets, newspapers, television networks) are located in its urban centers.

SOCIOLOGICAL METHODS

Measuring Population Growth

Populations change as a consequence of births, deaths, out-migration, and in-migration. The relationships of these variables to a society's total population are expressed in the basic demographic equation:

$$P_t = P_o + (B - D) + (M_i - M_o)$$

where

P_o = the census count for an earlier period
P_t = the census count for the later period
B = total births between P_o and P_t
D = total deaths between P_o and P_t
M_o = out-migration between P_o and P_t
M_i = in-migration between P_o and P_t

Once they know the absolute values of the terms in this equation, demographers usually convert them into percentages in order to compare populations of different sizes.

The most basic measures of population change are *crude rates,* or the number of events of a given type (e.g., births or deaths) that occur in a year divided by the midyear population (Bogue, 1969). Thus, the **crude birthrate (CBR)** is the number of births occurring during a year in a given population divided by the midyear population, and the **crude death rate (CDR)** is the number of deaths occurring during a year divided by the midyear population. These fractions are usually expressed as a rate per thousand persons. They are "crude" because they compare the total number of births or deaths with the total midyear population when in fact not all members of the population are equally likely to give birth or to die.

The **rate of reproductive change** is the difference between the CBR and the CDR for a given population. It is a measure of the natural increase of the population; that is, it measures increases due to the excess of births over deaths and disregards in- and out-migration. At present there are several nations in which the rate of reproductive change is zero or less, meaning that there is no natural population growth. Germany, for example, had a CBR of 10 and a CDR of 11 in 1995, resulting in an annual rate of increase of –0.1%. In Austria, the CBR was 12 and the CDR 10, for an annual rate of increase of 0.1%. In the United States, the rate of population growth is about 0.7%, representing an increase of about 2 million people per year. These rates are in dramatic contrast with the annual growth rates of countries like Syria and Guatemala, which are above 3%. Rates of population growth, including *doubling time*—the number of years in which the population will double at the current rate of increase—for selected countries are listed in Table A.

We can easily see from Table B that an annual rate of population growth of only 1% will lead to an increase of almost 270% in a century. Since World War II, the world's population has been increasing at a rate of more than 1.5%—which means that in 2000 it passed the projected figure of 6 billion.

As a result of the trend toward urbanization, the vast majority of people in the developed nations live in the human-built environment of cities and suburbs; relatively few live in the "country" environment of rural areas. Among the many consequences of this shift is the tendency of large numbers of people to settle in coastal or riverfront areas where the risk of environmental disasters, such as earthquakes and flooding, is very high. (See the Mapping Social Change box on page 377.)

In the developing nations, a large majority of the people live in rural villages, but that picture too is changing. Figures 13.3 and 13.4 show that the world's urban population is projected to overtake the rural population by about 2020, given current trends. In the developing nations, rural growth rates are decreasing rapidly, from 1.11% between 1975 and 2000 to a projected 0.02% between 2000 and 2025. But the urban growth rate of 2.4% projected for the next 25 years, while much lower than before, will apply to a

TABLE A
Population Growth, Selected Countries

Country	Population (millions)	Annual Growth Rate* (percent)	Doubling Time** (years)
India	1,002.1	1.8	39
China	1,264.5	0.9	79
Brazil	170.1	1.5	45
Bangladesh	128.1	1.8	38
Nigeria	123.3	2.8	24
Pakistan	150.6	2.8	25
Indonesia	212.4	1.6	44
Russia	145.2	–0.6	—
Mexico	99.6	2.0	36
United States	275.6	0.6	120

*Annual rate of natural increase.

**Number of years in which the population will double at current rate of increase.

Source: Population Reference Bureau, 2000.

TABLE B
Relationship of Population Growth per Year and per Century

Population Growth per Year (percent)	Population Growth per Century (percent)
1	270
2	724
3	1,922

Source: Data from Worldwatch Institute.

population base of about 1.94 billion urban dwellers, for a projected increase of 1.6 billion people. This population increase, which will include large numbers of people who have just made the transition from rural to urban living, will present the greatest public health and urban management challenges in the third world well into the twenty-first century (Brockerhoff, 2000).

Urbanization contributes to the lower birthrates that are characteristic of the third stage of the demographic transition. As Louis Wirth pointed out, "The decline in the birthrate generally may be regarded as one of the most significant signs of the urbanization of the Western world" (1968/1938, p. 59). In the city, a variety of factors lead to the postponement of marriage and childbearing. For one thing, living space is limited. For another, newcomers to the city must find jobs before they can even think about marrying, and often they lack the ties to family and kin groups that might encourage them to marry and have children.

FIGURE 13.3
Urban and Rural Population, Less-Developed Countries, 1950–2025

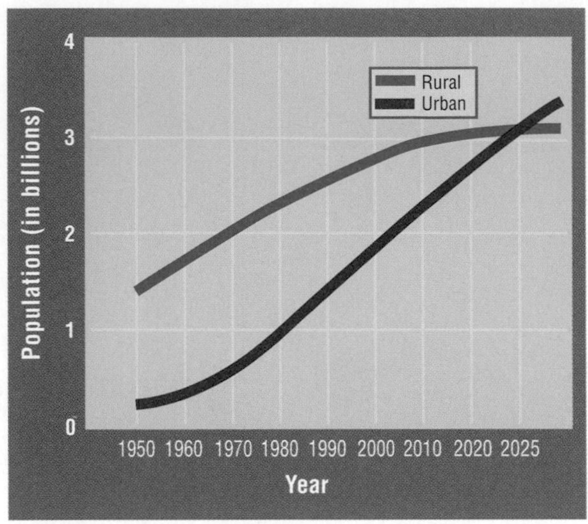

Source: Brockerhoff, 2000.

Studies of declining birthrates in non-Western countries find that—contrary to the experience of cities in the West, where urbanization has brought declining birthrates—the populations of some cities in Asia and Africa are increasing faster than those of rural areas. The reason seems to be that rapid change in rural areas, especially the mechanization of agriculture, pushes rural people to the cities, where they often live in shantytowns and village-like settlements. In those settlements, the rural tradition of large families is not immediately altered. In addition, because they have greater access to health care during pregnancy and childbirth, the infant death rate may be lower among the migrants than in rural villages. However, when these populations become part of the urban economy, they too begin to limit their family size (Lowry, 1990; W. C. Robinson, 1963).

In sum, urbanization is closely linked with rapid increases in population, yet the nature of life in cities tends to limit the size of urban families. Cities grow primarily as a result of migration, but new migrants do not find it easy to form families. Thus, in his research on the changing populations of Western industrial cities, Hauser (1957) showed that birthrates were

FIGURE 13.4
Population Growth Rates in Urban and Rural Areas, Less- and More-Developed Countries, 1975–2000 and 2000–2025

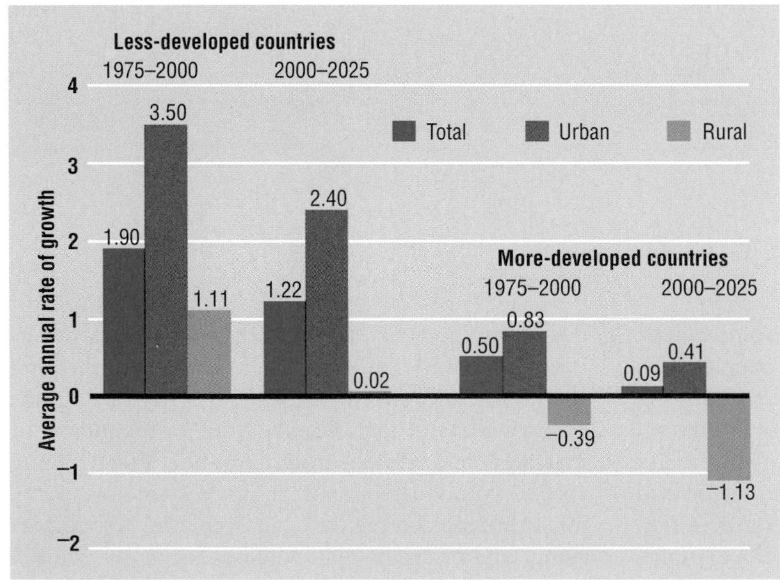

Source: Brockerhoff, 2000.

MAPPING SOCIAL CHANGE

Urban Locations and Environmental Risk

Millions of people live in highly unstable physical environments where there is a high risk of natural disasters like the recent earthquakes in Seattle, Washington, and Kobe, Japan. This map shows the regions of the earth where earthquake activity is greatest. Along the West Coast of the United States and Latin America, in the Valley of Mexico, in much of Japan, and on the Pacific Rim, where rates of urban growth are among the highest in the world, the risk of devastating earthquakes is an undeniable fact of life. Or is it?

Our knowledge of where and how earthquakes occur has improved considerably in the last 30 years, but the cities in question began growing rapidly long before the risks were well understood. People continue to migrate to these risky environments, often denying that the risks actually apply to them. But denial alone is not an adequate explanation. Social policies that encourage the growth of such cities—through favorable mortgage and insurance terms, for example—also play a significant role. Social policies that require better preparation for earthquakes, better design and construction of buildings and freeways, and advance planning for disaster relief are among the consequences of recent disasters. But the desirable coastal locations of vulnerable cities will result in continued population growth and an increasing need for effective planning to avert disasters.

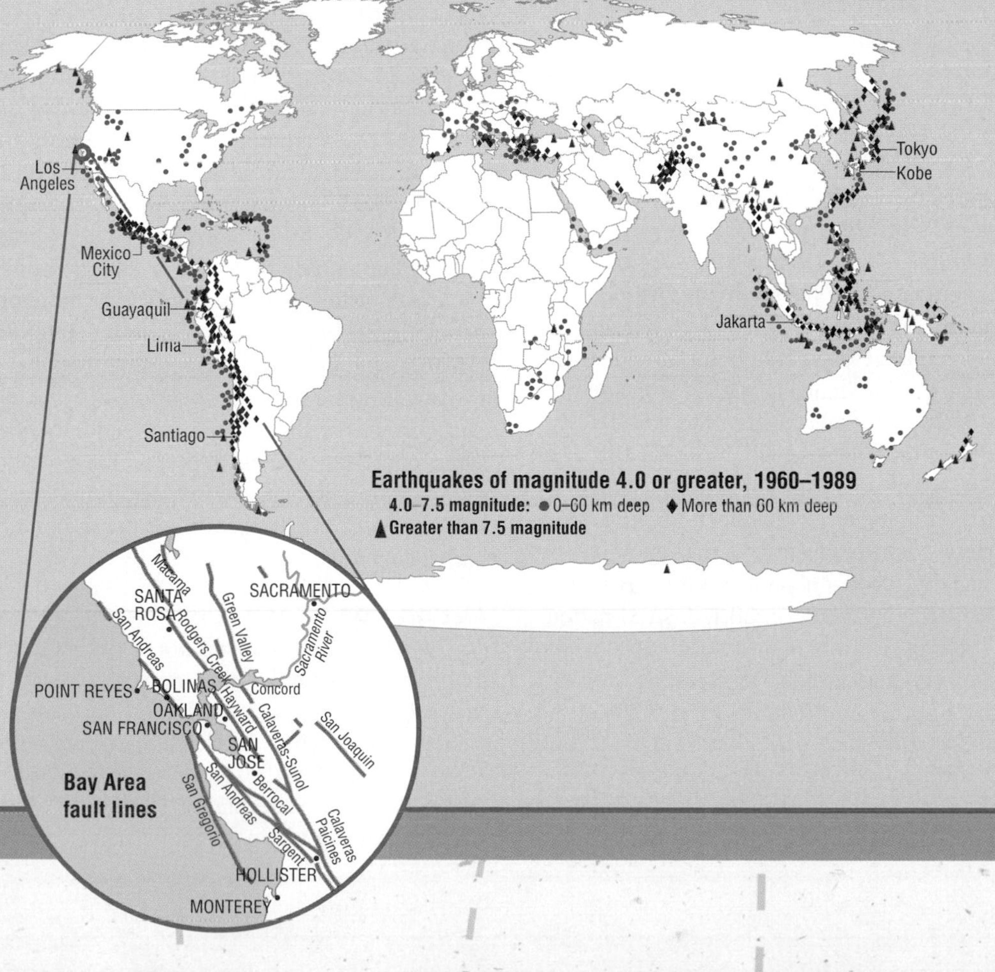

Los Angeles
Mexico City
Guayaquil
Lima
Santiago
Tokyo
Kobe
Jakarta

Earthquakes of magnitude 4.0 or greater, 1960–1989
4.0–7.5 magnitude: ● 0–60 km deep ◆ More than 60 km deep
▲ Greater than 7.5 magnitude

Bay Area fault lines
SANTA ROSA
SACRAMENTO
Maacama
Rodgers Creek
Green Valley
Sacramento River
San Andreas
POINT REYES
BOLINAS
Concord
OAKLAND
Hayward
Calaveras-Sunol
SAN FRANCISCO
SAN JOSE
San Joaquin
San Gregorio
Berrocal
Sargent
Calaveras Paicines
HOLLISTER
MONTEREY

lowest in the areas of the city that had the highest proportions of new migrants. There is also mounting evidence from archaeology that child sacrifice in ancient cities may have helped control urban populations (Browne, 1987). The eventual result of large-scale migration to cities may be a slowdown in population growth: As an increasing proportion of a society's population lives in cities, the rate of growth of the population as a whole tends to drop.

The Growth of Cities

In Chapter 4 we noted that cities became possible when agricultural populations began to produce enough extra food to support people who were not directly engaged in agriculture, such as priests, warriors, and artisans. Changes in the technology of food production made it possible for increasingly larger populations to be supported by the same number of agricultural workers. This has been a central factor in the evolution of cities, but as we will see shortly, the most dramatic increases in urban populations have occurred only in the past 150 years.

The Urban Revolution. The increasing tendency of people throughout the world to live in cities has been referred to as the *urban revolution*. The extent of this "revolution" can be grasped by comparing a few figures. In 1800, only 3% of the world's people lived in cities with populations above 5,000, and of this proportion, a mere 2.4% lived in cities with populations above 20,000. Between 1800 and 1970, a period during which the world's population increased fourfold, the percentage of people living in cities with 5,000 or more inhabitants increased elevenfold, whereas the percentage of people living in cities with 100,000 or more inhabitants increased almost fourteenfold. By 1970, fully one-third of the world's population lived in cities (Matras, 1973; Vining, 1985).

The speed with which urbanization has changed the size and layout of cities is remarkable. "Before 1850 no society could be described as predominantly urbanized, and by 1900 only one—Great Britain—could be so regarded. Today . . . all industrial nations are highly urbanized, and in the world as a whole the process of urbanization is accelerating rapidly" (K. Davis, 1968, p. 33). In fact, by 1985, 42% of the world's population lived in urban areas, and by 2000 nearly half of the world's population lived in urban areas—40% of the population in developing countries and 78% in developed countries (Brockerhoff, 2000). Africa currently has the fastest rate of urbanization; and at the turn of the twenty-first century about 70% of the population of Latin America lived in urban areas—a rate of urbanization comparable to that of North America and Europe.

These data indicate not only that increasing percentages of the world's population are living in cities but also that the cities themselves are larger than ever before. The growth of cities in the twentieth century gave rise to the concept of the **metropolitan area,** in which a central city is surrounded by a number of smaller cities and suburbs that are closely related to it both socially and economically. Most people in the United States live in large metropolitan areas.

At the same time, it should be noted that many metropolitan regions in older advanced nations like the United States (especially Greater New York), Great Britain (Greater London), and Germany (Rhein–Ruhr) reached their peak of growth decades ago and are now growing slowly, if at all. In contrast, as Table 13.1 shows, with only a few exceptions the largest urbanized areas in the world today are located in Latin America and Asia. In these areas, many metropolitan regions are experiencing explosive growth and attracting vast populations of urban newcomers, many of whom are living as squatters on vacant land.

Rapid urbanization occurring throughout the world brings together diverse groups of people in cities that often are not prepared to absorb them. The problems caused by such urbanization are immense. They include housing, educating, and caring for the health of newcomers; preventing gang violence and intergroup hatred; and many other difficult tasks. Moreover, as the world becomes ever more urbanized, populations become increasingly interdependent. Urban populations are supported, for example, by worldwide agricultural production, not just by the produce grown in the surrounding countryside. In the same way, the problems of one major city or one large urbanizing region can no longer be thought of as isolated from the problems of the older, more affluent urbanized regions.

Urban Societies. Urbanization produces *urban societies*. By this we mean not only that cities are the cultural and institutional centers of a society but also that urban life has a pervasive influence on the entire society (Durkheim, 1964/1893; Weber, 1962/1921; Wirth, 1968/1938). Today the United States is spanned by interstate highways that link the nation's rapidly growing urban and suburban places and carry

TABLE 13.1

The World's Largest Megacities, 1970–2015

The United Nations (UN) coined the term *megacities* in the 1970s to designate all urban agglomerations with a population of 8 million or more. In the 1990s, the UN raised the population threshold to 10 million, following the practice of institutions such as the Asian Development Bank. The UN estimates that there are 19 megacities in the world at the beginning of the twenty-first century.

Population (in millions)			
1970		**2015**	
1. Tokyo, Japan	16.5	1. Mumbai (Bombay), India	28.2
2. New York, United States	16.2	2. Tokyo, Japan	26.4
3. Shanghai, China	11.2	3. Lagos, Nigeria	23.2
4. Osaka, Japan	9.4	4. Dhaka, Bangladesh	23.0
5. Mexico City, Mexico	9.1	5. São Paulo, Brazil	20.4
6. London, England	8.6	6. Karachi, Pakistan	19.8
7. Paris, France	8.5	7. Mexico City, Mexico	19.2
8. Buenos Aires, Argentina	8.4	8. Delhi, India	17.8
9. Los Angeles, United States	8.4	9. New York, United States	17.4
10. Beijing, China	8.1	10. Jakarta, Indonesia	17.3

Source: Brockerhoff, 2000.

traffic through rural areas at high speeds. Waterways, forests, hills, and valleys are channeled and cut and bulldozed to make way for expanding settlements. Once considered far from the urban scene, national parks and forests now receive millions of visitors from the metropolitan centers. And in an urban society more and more people, even those living in isolated rural communities, share in the mass culture of that society—television and radio programs, movies, books and magazines, all of which stress themes that appeal to people who are familiar with metropolitan living. In an urban society, not everyone lives in the cities, but no one can escape the pervasive influence of urban centers.

THE URBAN LANDSCAPE

Throughout the twentieth century, sociologists devoted considerable study to urbanization and the changes that accompany it. The ways in which the growth of cities and metropolitan regions alters the surface of the planet are part of the sociological study of urbanization. In this section, we examine some of those processes.

Urban Expansion

The effects of urbanization began to be felt in American society after the Civil War (1861–1865). As the population began to push into the West, waves of immigrants from Ireland, Germany, Italy, and many other parts of central and southern Europe streamed into the cities of the East. In the 1840s and 1850s, for example, approximately 1.35 million Irish immigrants arrived in the United States, and in just 12 years—from 1880 to 1892—more than 1.7 million Germans arrived (Bogue, 1985). In earlier chapters, we discussed the impact of the great migrations from China and Korea and Latin America, the importation of slaves from Africa, and the large numbers of people of all races and ethnic groups who continue to arrive in U.S. cities. The point here is that for well over a century North American cities have been a preferred destination for people from all over the world; as a result, they have received numerous waves of newcomers since their period of explosive growth in the nineteenth century.

The science of sociology found early supporters in the United States and Canada partly because the cities in those nations were growing so rapidly.

It often appeared that North American cities would be unable to absorb all the newcomers who were arriving in such large numbers. Presociological thinkers like Frederick Law Olmsted, the founder of the movement to build parks and recreation areas in cities, and Jacob Riis, an advocate of slum reform, urged the nation's leaders to invest in improving the urban environment, building parks and beaches, and making better housing available to all (Cranz, 1982). As we saw in Chapter 1, these reform efforts were greatly aided by sociologists who conducted empirical research on the social conditions in cities. In the early twentieth century, many sociologists lived in cities like Chicago that were characterized by rapid population growth and serious social problems. It seemed logical to use empirical research to construct theories about how cities grow and change in response to major social forces as well as more controlled urban planning.

The founders of the Chicago school of sociology, Robert Park and Ernest Burgess, attempted to develop a dynamic model of the city, one that would account not only for the expansion of cities in terms of population and territory but also for the patterns of settlement and land use within cities. They identified several factors that influence the physical form of cities. Among them are "transportation and communication, tramways and telephones, newspapers and advertising, steel construction and elevators—all things, in fact, which tend to bring about at once a greater mobility and a greater concentration of the urban populations" (Park, 1967/1925a, p. 2). The important role of transportation is described in one of Park's essays:

> The extent to which . . . an increase of population in one part of the city is reflected in every other depends very largely upon the character of the local transportation system. Every extension and multiplication of the means of transportation connecting the periphery of the city with the center tends to bring more people to the central business district, and to bring them there oftener. This increases the congestion at the center; it increases, eventually, the height of office buildings and the values of the land on which those buildings stand. The influence of land values at the business center radiates from that point to every part of the city. (1967/1926, pp. 57–58)

Park and Burgess saw urban expansion as occurring through a series of "invasions" of successive zones or areas surrounding the center of the city. For example, migrants from rural areas and other societies "invaded" areas where housing was cheap. Those areas tended to be close to the places where the migrants worked. In turn, people who could afford better housing and the cost of commuting "invaded" areas farther from the business district, and these became the Brooklines, Gold Coasts, and Greenwich Villages of their respective cities. This model of urban expansion has come to be known as the *concentric-zone model.*

Studies by Park, Burgess, and other Chicago school sociologists showed how new groups of immigrants tended to become concentrated in segregated areas within inner-city zones, where they encountered suspicion, discrimination, and hostility from ethnic groups that had arrived earlier. Over time, however, each group was able to adjust to life in the city and to find a place for itself in the urban economy. Eventually, many of the immigrants were assimilated into the institutions of American society and moved to desegregated areas in outer zones; the ghettos they left behind were promptly occupied by new waves of immigrants (Kasarda, 1989). Many people believe that the present wave of immigration to the United States will follow the same pattern. The children and grandchildren of the new immigrants will gradually add to the diversity of cities and towns throughout the nation. But recent research indicates that this trend is developing very slowly, if at all. Asian and Latin American immigrants and their children remain heavily concentrated in "gateway" cities like New York, Los Angeles, Miami, Chicago, and San Francisco (Feagin, 2000; Frey, 1998).

The Park and Burgess model of growth in zones and natural areas of the city can still be used to describe patterns of growth in cities that were built around a central business district and that continue to attract large numbers of immigrants. But this model is biased toward the commercial and industrial cities of North America, which have tended to form around business centers rather than around palaces or cathedrals, as is the case in so many other parts of the world. Moreover, it fails to account for other patterns of urbanization, such as the rise of satellite cities and the rapid urbanization that occurs along commercial transportation corridors.

Satellite and Edge Cities. Outside the city of Detroit lies the town of River Rouge, long famous as a center of steel and automobile production. Outside the city of Toronto lies Hamilton, also a smoky manufacturing town where much of Canada's steel is produced. And outside Chicago is Gary, Indiana, another major center of heavy industrial production. Outside New

FIGURE 13.5
Strip Development

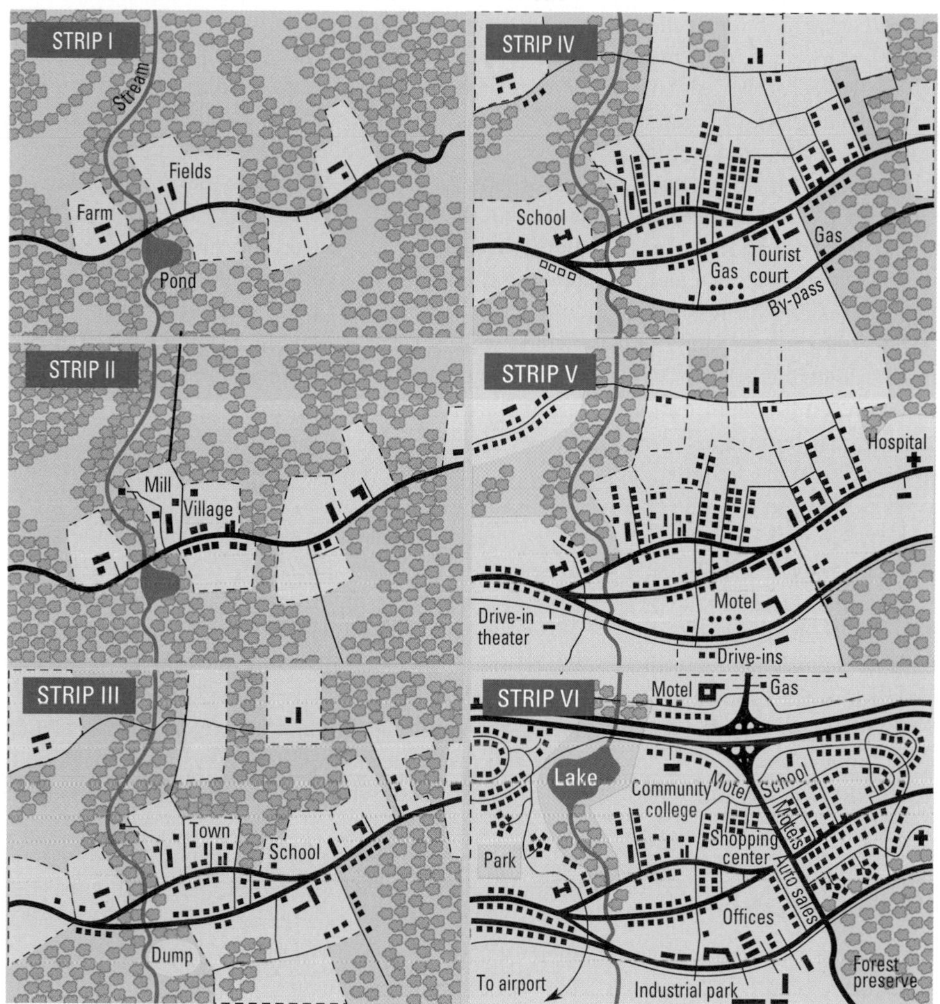

Source: Clay, 1980.

York City there are many other *satellite cities,* some devoted to heavy industries whose needs for space, rail and water service, and energy make it impossible to locate them in the central business districts.

Other satellite cities are devoted to less environmentally stressful industries and are sometimes called *edge cities* to distinguish them from the older industrial satellites (Garreau, 1991). Fort Worth, Texas, for example, is a major center for white-collar industries that are leaving older central-city locations in search of cheaper space and a well-educated labor force. Still others, like Tysons Corner, Virginia, near Washington, D.C., have grown around large shopping centers and mall complexes. In the second half of the twentieth century, the growth of these and other satellite cities was accelerated by public investment in the interstate and metropolitan highway systems, a point to which we will return shortly in discussing the rise of metropolitan urban systems (Baldassare, 1986; Clay, 1994; Fishman, 1987).

Strip Development. The growth of satellite cities specializing in the production of a particular commodity or product is typical of the period of rapid industrial growth that occurred before World War II. A more current model of urbanization is known as *strip development;* it is shown in schematic form in Figure 13.5. Strip I represents a typical nineteenth- or early twentieth-century farming area bisected by a road and a stream. In time, the bridge over the stream and the path along the stream became a road that intersected the original road (Strip II), creating an intersection that stimulated the growth of a village with a mill; families living in the village depended on wages from the mill. As the village grew into a town, population growth and increasing automobile traffic created the need for wider roads and bypass roads, as shown in Strips III and IV. Specialization as an "automobile convenient" strip, with the development of more motels, a drive-in theater, and drive-in businesses, can be seen in Strip V. At this time, the residential functions along the strip began to disappear and the area assumed an increasingly commercial character.

Strip VI shows the strip's present stage of development. A spur of the limited-access highway system creates new growth around a cloverleaf (including a new community college situated between the artificial lake and the shopping center). Note that along the original main road are office buildings and a modern industrial park rather than the smaller retail businesses that once defined the town center. Indeed, as the strip has developed and surrounding roads have been enlarged to form part of the metropolitan highway system, the town center itself has largely disappeared (Clay, 1980). The strip development model thus describes the incorporation of smaller communities and towns into a larger metropolitan area.

Metropolitan Areas

Megalopolis. After 1920, new metropolitan areas developed largely as a result of the increasing use of automobiles and the construction of a network of highways covering the entire nation (Flink, 1988). The shift to automobile travel brought former satellite cities within commuting distance of the major industrial centers, thereby adding to the size of those metropolitan areas. In the South and Southwest, new metropolitan areas developed. In fact, in recent years these have become the fastest growing urban areas in the nation.

Since World War II, sociologists have been studying an increasingly important urban phenomenon: the emergence of large multinuclear urban systems. The term **megalopolis** is used to describe these vast complexes, whose total population exceeds 25 million. Jean Gottmann (1978) pointed out that a megalopolis is not simply an overgrown metropolitan area; rather, it is a system of cities distributed along a major axis of traffic and communication. According to Gottmann, there are six megalopolises in the world today: the American Northeastern megalopolis, the Great Lakes megalopolis, the Tokaido megalopolis in Japan, the megalopolis in England (the London area), the megalopolis of northwestern Europe (extending from Amsterdam to the Ruhr), and the Urban Constellation in China (centered on Shanghai). Four others are developing rapidly: the Rio de Janeiro–São Paulo complex in Brazil, the Milan–Turin–Genoa triangle in northern Italy, the Valley of Mexico, and the urban swath extending from San Diego to the San Francisco Bay area.

A megalopolis is characterized by an "intertwined web of relationships between a variety of distinct urban centers . . . expressed partly in a physical infrastructure consisting of highways, railways, waterways, telephone lines, pipelines, water supply and sewage systems crisscrossing the whole area, and partly in more fluid networks, such as the flows of traffic, the movement of people and goods, the flows of telephone calls [and] of mail" (Gottmann, 1978, p. 57). Despite their interdependence, however, "the sizes and specializations of the various . . . components [of a megalopolis] are extremely varied, as demonstrated by the diverse characteristics of the cities, towns, villages, suburban and rural areas that form the vast system" (p. 57). Therefore, a megalopolis can best be described as a huge social and economic mosaic.

The recent history of Los Angeles provides a good example of the development of a megalopolis. Between 1960 and 1970, the population of Los Angeles increased by more than 2 million, double the growth of Chicago and more than that of New York and San Francisco put together (R. A. Smith, 1968). Today the Los Angeles metropolitan area continues to grow, although at a somewhat slower rate. As a result of its extraordinary growth, the region must struggle to control the effects of air pollution from more than 3 million automobiles. It must also struggle to furnish adequate supplies of water for its residents. In a review of conflicts between Angeleno leaders and the residents of smaller towns near water sources, sociologist John Walton (1992) has shown that control over water and other natural resources, especially by powerful groups that dominate the regional real estate markets, is a key to understanding the history not only of the Los Angeles metropolitan region but of the entire desert West.

Decentralization. One effect of the growth of megalopolitan areas is *decentralization,* in which outlying areas become more important at the expense of the central city. This trend is not new. In the 1960s and 1970s, large numbers of middle-income city dwellers moved to suburban areas while the poor remained in the central cities. Business and industry also moved to the suburbs, giving rise to widespread speculation that vibrant central cities would become a thing of the past. Recently, however, the old central cities of some metropolitan areas—such as New York, Philadelphia, Chicago, Boston, and Baltimore—have shown renewed vitality. Far from decaying, they have become major financial and cultural centers serving the needs of huge populations. On the other hand, medium-sized cities like Gary, Indiana, and Paterson, New Jersey, have suffered, because their central business districts have little to offer suburban dwellers in the way of financial services like banking and insurance or cultural attractions like theaters and symphony orchestras (Kasarda, 1988).

An important feature of megalopolitan areas is their diversity. These huge urban regions include many different kinds of communities: ethnic communities in both central cities and suburbs, middle-class "bedroom" suburbs, industrial towns, areas devoted to truck gardening or dairy farming, "second home" communities (e.g., beachfront areas), and so on. Each meets the economic and cultural needs of a specific urban population. Sociologists have devoted considerable study to these urban communities, and we discuss their findings in the next section.

URBAN COMMUNITIES

"The city," wrote Robert Park, is more than a set of "social conveniences—streets, buildings, electric lights, tramways, and telephones, etc.; something more, also, than a mere constellation of . . . courts, hospitals, schools, police, and civil functionaries of various sorts. The city is, rather, a state of mind, a body of customs and traditions. . . . It is a product of nature, and particularly of human nature" (1967/1925a, p. 1).

Social scientists who study cities have devoted a great deal of attention to the tension between "community" and "individualism" as it relates to life in cities. Country dwellers have been thought of as "happily ensconced in warm, humanly rich and supportive social relationships: the family, neighborhood, town," whereas city dwellers are "strangers to all, including themselves. They are lonely, not emotionally touching or being touched by others, and consequently set psychically adrift" (Fischer, 1976, p. 19). On the other hand, country dwellers are sometimes viewed as "stifled by conventionality, repressed by the intrusion and social control of narrow-minded kin, neighbors, and townsmen," whereas city dwellers are "free to develop individual abilities, express personal styles, and satisfy private needs" (p. 20). These views of the city are obviously contradictory, and much research has been devoted to the question of how urban life affects individuals and communities. In this section, we look at some of the findings of that research and the theories of urban life that have been proposed on the basis of those findings.

The Decline-of-Community Thesis

Early studies of the nature and effects of urban life were dominated by efforts to evaluate the differences between rural and urban societies. They tended to reach rather gloomy conclusions. We have already noted (Chapter 4) that Ferdinand Tönnies described the process of urbanization as a shift from gemeinschaft (a community based on kinship ties) to gesellschaft (a society based on common interests). Émile Durkheim reached a similar conclusion: Small rural communities are held together by ties based on shared ideas and common experiences, whereas urban societies are held together by ties based on the interdependence of people who perform specialized tasks. Both Tönnies and Durkheim believed that urban life weakens kinship ties and produces impersonal social relationships.

In a 1905 essay, "The Metropolis and Mental Life," Georg Simmel focused on the effects of urban life on the minds and personalities of individuals. According to Simmel, cities bombard their residents with sensory stimuli: "Horns blare, signs flash, solicitors tug at coattails, poll takers telephone, newspaper headlines try to catch the eye, strange-looking and strange-behaving persons distract attention" (quoted in Fischer, 1976, p. 30). The urban dweller is forced to adapt to this profusion of stimuli, which Stanley Milgram (1970) termed *psychic overload.* The usual way of adapting is to become calculating and emotionally distant. Hence the image of the city dweller as aloof, brusque, and impersonal in his or her dealings with others.

This view of the effects of urban life found further expression in the work of Louis Wirth, especially in his essay "Urbanism as a Way of Life"

(1968/1938). Wirth began by defining the city as "a relatively large, dense, and permanent settlement of socially heterogeneous individuals" (p. 28). He then attempted to show how these characteristics of cities produce psychological stress and social disorganization. The primary psychological effect of urban life, according to Wirth, is a weakening of the individual's bonds to other people. Without such bonds, the individual must deal with the crises of life alone; often the result is mental illness. In other cases, the city dweller, again because of the absence of close ties to friends or kin, lacks the restraints that might prevent him or her from engaging in antisocial behaviors.

Wirth linked social disorganization to the diversity that is characteristic of cities. Unlike rural residents, city dwellers work in one place, live in another, and relax in yet another. They divide their social lives among coworkers, neighbors, friends, and kin. Their jobs, lifestyles, and interests are extremely varied. As a result, no single group, be it the family, the friendship group, or the neighborhood, controls their lives. In Wirth's view, this absence of social control produces anomie or normlessness (see Chapter 5). Urban dwellers frequently do not agree on the norms that should govern their lives, and therefore they are likely to either challenge existing norms or ignore them. Consequently, instead of being controlled by the norms of primary groups, the lives of city dwellers are controlled by impersonal agencies like banks and police forces.

One consequence of the impersonality of urban life, some argue, is greater callousness among city dwellers. An often-cited example is the case of Kitty Genovese, a young woman who was murdered while 38 residents of nearby apartments, though they heard her cries, did nothing. After that episode, many commentators called attention to the callous character of city dwellers. Subsequent research on bystander apathy revealed that the presence of many other people tends to diffuse the sense of responsibility. We are less likely to take action if we have reason to believe that someone else will do so (Hunt, 1985; Latané & Darley, 1970). In this sense, the bigger the city and the more one is surrounded by strangers, the more likely it is that such behavior will occur—but this does not mean that city dwellers are alienated from one another when they are among people they know, or that when alone they would not help someone in trouble. The fact that one is likely to feel less responsibility for others when one is surrounded by strangers may be a condition of city life, but it is hardly evidence for the decline-of-community thesis.

The Persistence of Urban Communities

The idea that urbanization leads to the decline of community has been criticized on a number of grounds. Rural life is nowhere near as pleasant as some urban sociologists have assumed it to be; evidence of this is the almost magical attraction that cities often have for rural people. At the same time, urban social disorganization is not as extensive as the early urban sociologists believed. Many city dwellers maintain stable, intimate relationships with kin, neighbors, and coworkers. Moreover, urban life is not necessarily stressful or anomic.

Subcultural Theory. A more recent view of urban life sees the city as a "mosaic of social worlds" or intimate social groups. Numerous studies have shown that the typical urban dweller does not resemble the isolated, anomic individual portrayed by Simmel and Wirth. In fact, communities of all kinds can be found in cities. Many urban dwellers, for example, are members of ethnic communities who have not become fully assimilated into American society and are unlikely to do so. They may be the children or grandchildren of immigrants who formed ethnic enclaves within large cities in the late nineteenth and early twentieth centuries, or they may be recent immigrants themselves, trying to build a new life in a strange culture.

But group ties among urban dwellers are not based solely on ethnicity. They may be based on kinship, occupation, lifestyle, or similar personal attributes (Fischer, 1976, 1987; Hummin, 1990; Suttles, 1972). Thus, many cities contain communities of college students, elderly people, homosexuals, artists and musicians, wealthy socialites, and so on. Although the members of any given group do not always live in the same neighborhood, they are in close touch with one another much of the time. Their sense of community is based not so much on place of residence as on the ability to come together by telephone or e-mail, in special meeting places like churches or synagogues, or even in restaurants and bars (Duneier, 1992; Fischer, 1976, 1987; Oldenburg, 1989).

An example of this point of view, known as *subcultural theory,* is Illsoo Kim's (1981) detailed study of the Korean community in New York City. This growing ethnic community has developed since the passage of the Immigration Act of 1965, which eliminated the nationality quotas established earlier in the twentieth century. Largely because of population pressure in South Korea (which has led to overcrowded cities and high rates of unemployment), more than 650,000 Koreans have immigrated to the United States since 1970, with more than 100,000 finding new homes in the city of New York.

Like all immigrants, the Koreans have had to create a new way of life. They have had to find new ways of making a living, and they have had to adapt to a new culture. The first problem has been solved mainly by opening small businesses, the second by establishing neighborhoods where the immigrants can maintain their own culture while they and their children learn the values and norms of Western culture. In summing up the situation of Korean immigrants in New York, Kim points out that "until they completely master the American language, education, and culture, Koreans will be forced to rely on one another" (p. 319). For this reason, they are likely to maintain their own community within the city for at least two generations. However, in time their commitment to education and a better life will cause them to become more fully assimilated into American society, and their ties to the Korean community will be weakened.

It seems reasonable to conclude that the effects of urban life on communities and individuals are more complex than Durkheim, Simmel, and Wirth suggested. Certainly, social disorganization occurs in cities, but so does social reorganization. Communities that are uprooted by urban renewal eventually may be re-formed in other parts of the city or its suburbs. The new community may be less homogeneous than the old one, but it is a community nonetheless.

The Suburbs. Like city dwellers, suburban dwellers have been said to lack the close attachments that are thought to characterize rural communities. In fact, before the late 1960s most social scientists had a rather dismal view of suburban life. Suburbs had grown rapidly after World War II as large numbers of middle-class Americans left the central cities in search of a less crowded, more pleasant lifestyle. Many suburban dwellers were corporate employees, and in the 1950s and early 1960s they became the subject of a widely held stereotype. The suburbs, it was said, were "breeding a new set of Americans, as mass-produced as the houses they lived in, driven into a never-ending round of group activity ruled by the strictest conformity. Suburbanites were incapable of real friendships; they were bored and lonely, alienated, atomized, and depersonalized" (Gans, 1967, pp. xxvii–xxviii).

In *The Levittowners,* a classic study of a new suburban community in New Jersey, Herbert Gans (1967, 1976) challenged this stereotype. He found that the residents of Levittown lost no time in forming attachments to one another. At first, they associated only with their neighbors, but before long they formed more extensive associations based on shared interests and concerns. Moreover, far from being bored and isolated, a majority of the Levittowners were satisfied with their lives and felt that Levittown was a good place to live. Gans concluded that "new towns are merely old social structures on new land" (1967, p. vii). In other words, the contrasts between central-city and suburban life are generally exaggerated. Other researchers have reached similar conclusions. They note that as suburban communities age and newcomers buy homes from people whose children are grown, these neighborhoods go through ethnic, racial, and generational changes that are not so different from the patterns occurring inside the cities (K. Jackson, 1985).

For the past few decades, the phenomenon of "white flight" from older inner-city communities to newer and presumably safer suburban communities has been a subject of concern among urban sociologists and policy makers (W. J. Wilson, 1987). Recent data show, however, that African Americans and other minority groups are also fleeing the central cities at an accelerating rate. Demographer William Frey notes that "minority suburbanization took off in the 1980s both as the black middle class came into its own and as more assimilated Latinos and Asians translated their moves up the socioeconomic ladder into a suburban lifestyle" (quoted in De Witt, 1994, p. A1). This outward migration threatens to leave central-city communities with even greater concentrations of poor people. In cities with large numbers of immigrants, this discouraging trend is countered by increasing racial, ethnic, and class diversity (Huang, 1994), but in other cities the trend toward greater concentration of poor people inside the city and more affluent people in the suburbs is a growing problem. Frey's research in Washington, D.C., shows, for example, that educated middle-class African Americans are flocking to the suburbs, leaving more concentrated poverty in the communities they leave behind. (See Table 13.2.)

TABLE 13.2
Educational Attainment of Residents, Washington, D.C.

Schooling	In the Suburbs	In the City
Only to the eighth grade	18,896 5.2%	31,209 11.4%
Some high school	48,672 12.7%	66,653 24.3%
High school graduate only	103,799 28.3%	76,392 27.8%
Some college	112,477 30.7%	57,341 20.9%
College graduate	84,815 23.1%	43,189 15.8%

Source: Data from University of Michigan.

CITIES AND SOCIAL CHANGE

In the preceding section, we stressed the presence of a wide variety of communities and subcultures within cities. We noted that contemporary social scientists see the city as a place where many different communities coexist and thrive, rather than as a place where people are isolated and do not have a feeling of belonging to a particular social group or community. At the same time, however, there is no escaping the fact that various urban communities are often in conflict. For example, ethnic and racial communities may clash in violent confrontations over such issues as the busing of children to achieve school integration. In this section, we examine the origins and implications of such conflicts.

Social change in an urban society is likely to be felt most deeply by city dwellers, for it is in the cities that people are most densely congregated. In recent decades, this has been the case in North America as manufacturing jobs have been "exported" to Asia and Latin America. As a result of this shift, American cities are undergoing massive social change. "America's major cities are different places today from what they were in the 1960s," concludes John Kasarda (1989, p. 28), one of the nation's foremost urban sociologists. New modes of transportation, new communication technologies such as satellites and computers, and new industrial technologies (e.g., automation and recycling) are transforming our cities from production and distribution centers to administrative, financial, and information centers. "In the process," writes Kasarda, "many blue-collar jobs that once constituted the economic backbone of cities and provided employment opportunities for their poorly educated residents have either vanished or moved." Many of those jobs have been replaced by "knowledge-intensive white-collar jobs with educational requirements that exclude many with substandard education" (1989, p. 28).

Inequality and Conflict

As noted earlier, American cities and metropolitan regions have always attracted streams of migrants and immigrants. Migrants arrive from other parts of the nation—blacks from the rural South, Chicanos from the Southwest. Immigrants come from foreign lands—Haiti, Poland, China, Korea, Italy, and many others. Until the 1970s, except during economic depressions or recessions, the new arrivals had no difficulty finding work in the mills and factories that produced textiles, clothing, steel, rubber, glass, cars, and trucks. And much of their education was gained while working. Formal schooling was not as crucial to job success as it is today. The settlement house movement, pioneered by women activists like Jane Addams, was extremely influential in extending the promise of social mobility to these populations during the first half of the twentieth century.

Today, although there are still many manufacturing jobs, it is much more difficult to make a good living at such a job. Within the cities, where the decline in manufacturing jobs has been greatest, the number of low-status jobs, especially in restaurants and other low-paying services, has increased. So has the number of highly paid jobs in management and the professions, positions that require high levels of education. At the same time, more than 20 million legal immigrants have entered the country since 1970, usually settling in the cities. The result, often, has been fierce competition between immigrants and more established residents.

The net effect of these trends is a growing gap between the haves and the have-nots, especially in large cities (Feagin, 2000; Silk, 1989). In many poor urban neighborhoods, for example, wealthier city dwellers are buying the buildings in which the poor live, forcing them to move elsewhere. In neighborhoods where this occurs, the shops and stores that once catered to lower-income residents cannot afford the higher rents paid by stores that serve the wealthy newcomers and are forced out of business. This process, whereby poor and dilapidated neighborhoods are renovated by higher-income newcomers while poor residents and merchants are pushed out, is known as *gentrification.*

Status Conflict

Urban renewal projects have sometimes been labeled "urban removal" because they remove poor people from decaying neighborhoods and force them to find homes elsewhere. Recently urban renewal has fallen into disfavor and *redevelopment* has become popular. Redevelopment does not involve the destruction of entire neighborhoods, but it may have similar effects. For example, the redevelopment of New York City's Times Square by the Disney Corporation and other powerful groups, both public and private, was intended to restore a sleazy area to its former glory as an entertainment center. The goal was to make the area attractive to "decent" people and to eliminate the prostitutes, drug and pornography dealers, and petty criminals that had made Times Square a symbol of vice throughout the world (Kornblum & Williams, 1998; Mollenkopf, 1985).

Not everyone agreed that Times Square should be redeveloped. Owners of legitimate businesses in the area argued that they were being attacked unfairly, and they fought the redevelopment plan in the courts. This is a typical instance of *status conflict,* in which different groups vie for territory, occupational advantages, and other benefits that will enhance their prestige in the neighborhood or community.

Some urban sociologists see the city as divided into "defended neighborhoods" or territories (Suttles, 1972). This concept was originally developed by Park and Burgess, who viewed such neighborhoods as a type of "natural area" within the city. The defended neighborhood is a territory that a certain group of people consider to be their "turf" or base, which they are willing to defend against "invasion" by outsiders. Neighborhood defense is a common element of urban conflict, although the means used to defend neighborhoods may differ. For example, in wealthy suburban neighborhoods it may take the form of zoning regulations that establish minimum lot sizes of an acre or more. Such regulations effectively defend the neighborhood against invasion by people who cannot afford the large lots or by developers who would like to erect apartment buildings (Perin, 1977). In less-affluent neighborhoods, defense is often conducted by neighborhood improvement groups, which sometimes engage in vigilante action when they fear racial "invasion" (Hamilton, 1969). In poor neighborhoods, defense is very often the province of street-corner gangs.

Sociologists who study urban areas in the industrialized nations of western Europe and North America often point to a widening gap between the conditions of life in affluent communities and in neighborhoods characterized by chronic poverty and recent job loss. But another major concern for residents of poor neighborhoods centers on health. The increase in the number of families that lack medical insurance, the spread of diseases that were thought to be under control (e.g., tuberculosis and other respiratory illnesses), the AIDS crisis, and the lack of agreement over how to address the problems of extending health care to poor communities—all are important social issues that reveal the need for change in health care institutions. These issues also highlight the contribution of research by sociologists who study health and medicine in urban societies, a subject to which we now turn.

PUBLIC HEALTH IN URBAN SOCIETIES

Throughout most of human history, limitations on food production, together with lack of medical knowledge, have placed limits on the size of populations. Dreadful diseases like the bubonic plague have actually reduced populations. In England, the plague, known as the Black Death, was responsible for a drastic drop in the population in 1348 and for the lack of population growth in the seventeenth century (Wrigley, 1969).

Until relatively recently, physicians were powerless either to check the progress of disease or to prolong life. In fact, they often did more damage than good—their remedies were more harmful than the illnesses they were intended to cure. As Lewis Thomas (1979) stated:

> Bleeding, purging, cupping, the administration of infusions of every known plant, solutions of every known metal, every conceivable diet including total fasting, most of these based on the weirdest imaginings about the cause of the disease, concocted out of nothing but thin air—this was the heritage of medicine up until a little over a century ago. (p. 133)

THE CENTRAL QUESTION: A CRITICAL VIEW

Can we control population growth and achieve higher levels of health worldwide?

There is no simple answer to this question, and in fact none of us may live long enough to learn its answer. This chapter shows that one of the answers depends on what happens in urban regions of the world, because it is there that populations are growing fastest. This was the case in the United States for most of the twentieth century, so the U.S. experience in making cities healthier and more livable places is critical in this regard. But feminist sociologists point out that the role of women in achieving livable cities has been neglected.

In her recent research on women and urban communities, Daphne Spain (2000) shows that early in the twentieth century, while men insisted that "women's work" must remain within the boundaries of the home, in fact women were transforming America's urban landscape. Spain's research focuses on the activities of women in urban spaces following the Civil War. Immigrant populations were dramatically increasing in the northern states. Southern blacks migrated north in vast numbers. Europeans, Irish, and Russians sought freedom and opportunity in the industrial cities of the United States. While the most powerful men wheeled and dealed in the world of real estate, construction, and urban banking, building skyscrapers and industrial empires, it was the women who made the cities livable. Women did the "municipal housekeeping." Their efforts made cities safe and welcoming for their inhabitants.

Without such women as Jane Addams—the founding mother of the professional field of social work—Spain argues that the chaotic growth of cities might easily have led to more incidents such as the 1919 Chicago riot, in which blacks and immigrant whites fought pitched battles during almost a week of bloodshed. Addams created the first volunteer-staffed "settlement house" in one of Chicago's most impoverished immigrant neighborhoods. She and the dedicated women and men of Hull House helped give immigrants a fighting chance.

Spain's research analyzes the rise of volunteer and settlement movements in Chicago, Boston, Philadelphia, and New York as well as in smaller midwestern cities. Hull House was the inspiration for many of these movements. The residential staff at Hull House in the 1920s consisted of about 65 people, some of whom lived and worked there for at least 20 years. They were people from many walks of life and included famous philosophers, construction workers, doctors, lawyers, musicians, poets, and politicians. Spain and others who have conducted research on Hull House find that many of the progressive ideals of urban reformers who struggle against inequality, urban gangs, addiction, and other problems of urban society can be traced back to Addams and other founders of the social work movement of the early twentieth century (Johnson, 1989).

Jane Addams reading to children in the Mary Crane Nursery in the early 1930s.

These Italian immigrant boys, proudly showing off their new clubhouse on Halsted Street in Chicago in the 1920s, were organized into a Hull House club called the Silver Rivers, an alternative to street gangs.

Thomas's point is that before the nineteenth century, when scientists finally began to understand the nature of disease, physicians based their treatments on folklore and superstition. In fact, with few exceptions, the practice of healing, like many other aspects of science, was closely linked to religion. In ancient Greece, people who suffered from chronic illnesses and physical impairments would journey to the temple of Asklēpios, the god of healing, in search of a cure. In medieval times, pilgrims flocked to the cathedral at Lourdes in France (as many still do today) in the belief that they would thereby be cured of blindness, paralysis, or leprosy. Not until Louis Pasteur, Robert Koch, and other researchers developed the germ theory of disease did medicine become fully differentiated from religion. Their discoveries, together with progress in internal medicine, pathology, the use of anesthesia, and surgical techniques, led to the twentieth-century concept of medicine as a scientific discipline (Cockerham, 1998).

Social Epidemiology

During the nineteenth century, scientific research resulted in the discovery of the causes of many diseases, but at first this progress led physicians to do less for their patients rather than more: They began to allow the body's natural healing processes to work and ceased to engage in damaging procedures like bloodletting. At the same time, they made major strides toward improving public health practices. They learned about hygiene, sterilization, and other basic principles of public health, especially the need to separate drinking water from waste water.

The growing body of knowledge about how diseases are spread gave rise to the field of study known as *epidemiology*—that is, the scientific study of epidemics and diseases. Research on the relationships between disease and social conditions came to be known as *social epidemiology.*

New knowledge gained from research in social epidemiology, and the relatively simple but revolutionary innovations that followed from them, occurred before the development of more sophisticated drugs and medical technologies and contributed to a demographic revolution that is still under way in some parts of the world. Suddenly rates of infant mortality decreased dramatically, births began to outnumber deaths, and life expectancy increased. As mentioned earlier, this change resulted not from the highly sophisticated techniques of modern medicine but largely from improvements in public health practices, such as separating supplies of drinking water from sewage systems (McKinlay & McKinlay, 1977; Rockett, 1999). In fact, these simple technologies have had such a marked effect on infant survival that the rate of infant mortality in a society is often used as a quick measure of its social and economic development (see Figure 13.6). Nations with extremely high rates of infant mortality, such as Afghanistan and Liberia, are examples of places where modern public health systems and the institutions of modern medicine either have not developed or have broken down as a result of civil war and social chaos.

In sum, as medical science progressed toward greater understanding of the nature of disease and its prevention, new public health and maternal care practices contributed to rapid population growth. In the second half of the nineteenth century, such discoveries as antiseptics and anesthesia made possible other life-prolonging medical treatments. In analyzing the effects of these technologies, sociologists ask how people in different social classes gain access to them and how they can be more equitably distributed among the members of a society.

Global Inequality and Mortality Rates

Death is not a believer in equality. Some categories of people avoid it longer and more effectively than others. Death rates vary according to many social variables, especially age, sex, socioeconomic status, and above all, region of the world (McFalls,1998). From a global perspective, however, perhaps the most significant change in human history has been the extension of human life. In the United States, life expectancy improved from an average of 47 years in 1900 to an estimated 77 years in 2000 (*Statistical Abstract,* 1999). Elsewhere in the world, life expectancies are much lower. In much of Asia and Africa below the Sahara, for example, life expectancies are still in the 40s, and in highly unstable nations like Sierra Leone and Liberia they are actually decreasing.

When children live through infancy, life expectancy also takes a great leap forward. In 1900, when the average American could expect to live to about 47, the rate of death in childhood was about 72 per 100,000, just about what infant mortality rates are in India or the poorest nations of Latin America and

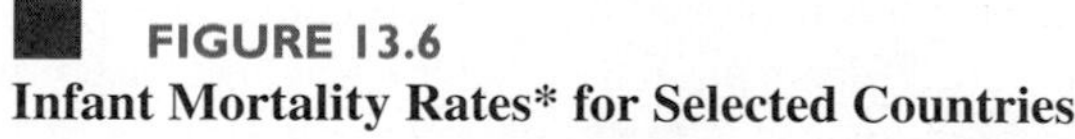

FIGURE 13.6
Infant Mortality Rates* for Selected Countries

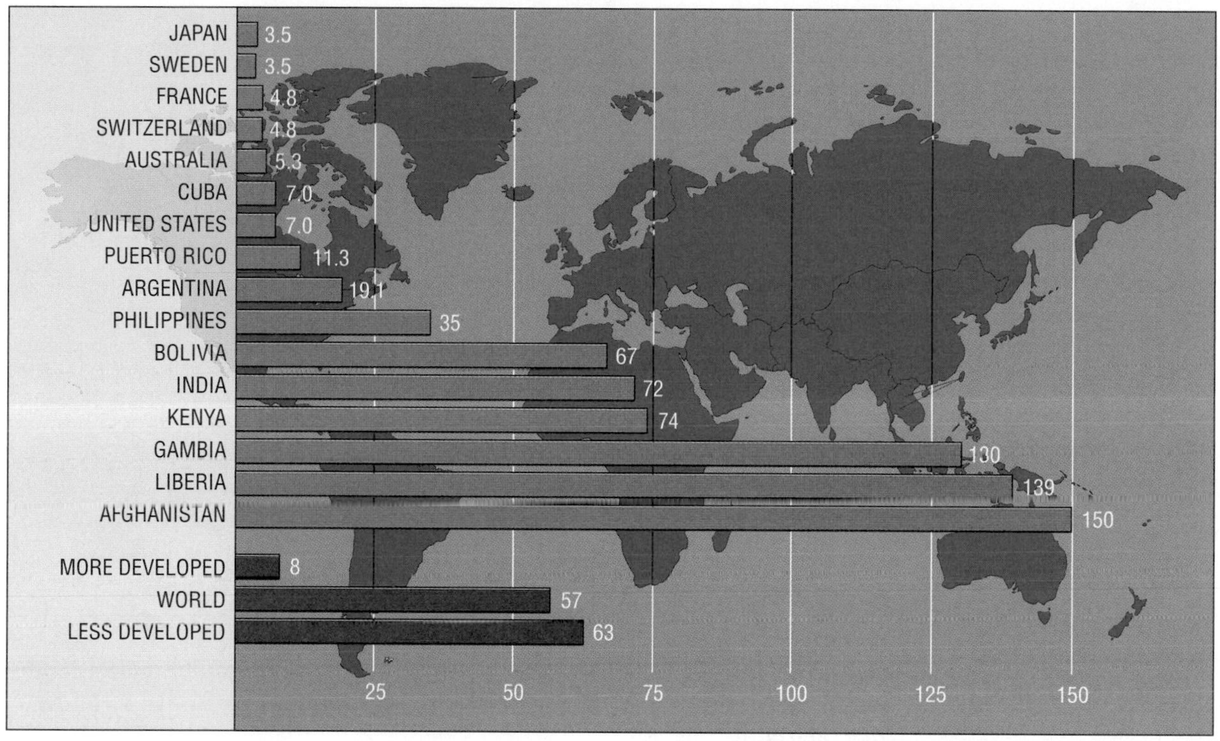

*The infant mortality rate is the annual number of deaths of infants under 1 year of age per 1,000 live births in a given year.
Source: Data from Population Reference Bureau, 2000.

Africa (see Figure 13.6). Today the U.S. infant mortality is still higher than those for many other affluent nations because so many poor American women do not receive adequate health care before or during pregnancy. But another major factor in reducing infant mortality is education of women. This relationship is vividly illustrated in Figure 13.7 (Gelbard, Haub, & Kent, 1999).

Facts like these have spurred feminist sociologists to help build a global social movement to address gender inequality. Addressing gender inequality in families and communities has become a primary economic and social development strategy. The stratification systems that deprive women of opportunities to vote, become literate, own property, and control their reproductive lives also lead to high rates of childbirth and infant mortality.

Still another factor in variations in mortality rates is disease. Infectious diseases are among the most frightening causes of death because they often defy modern medical technologies or, in the case of Third World nations, kill masses of people. Throughout the world, infectious and parasitic diseases are not disappearing. The AIDS crisis alerted the world to this fact, and to the emergence of new diseases as well. Since 1973, 28 new disease-causing microbes have been identified. A new strain of cholera killed thousands of people in Asia and Africa. Outbreaks of plague, diphtheria, cholera, and Ebola make headlines, and older diseases like tuberculosis and malaria are on the increase. (See Figure 13.8.) Death from infectious diseases account for more than one-fourth of all deaths worldwide. In Africa, HIV/AIDS has dramatically reduced the life expectancies of entire nations and is likely to cause more than 30 million deaths in the coming decades (see the Global Social Change box on pages 394–395).These new medical emergencies are occurring for many reasons, but the main reason may

FIGURE 13.7
Mother's Education and Infant Mortality in Selected Countries

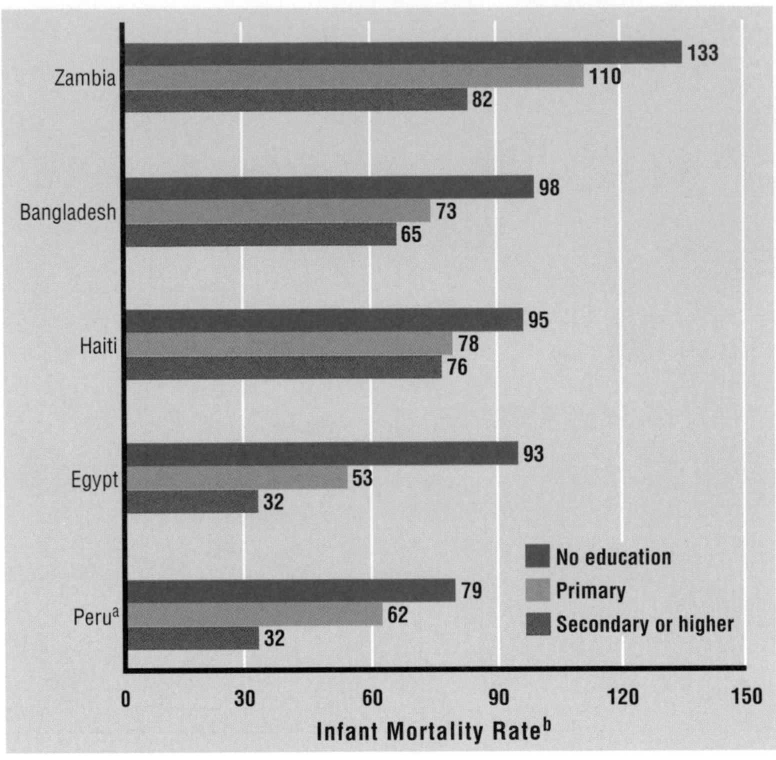

[a]Excludes women with higher than secondary educations.
[b]The infant mortality rate is the number of deaths to children under age 1 per 1,000 live births in a given year.
Source: Gelbard, Haub, & Kent, 1999.

be the reduced effectiveness of human immune systems among impoverished populations and increasing stress on the world's supplies of water (*Population Today,* August 1997).

Medical Sociology. The relatively new field of *medical sociology* has emerged in response to the development of medicine as a major institution of modern societies. Many sociologists are employed by health care institutions, and some medical schools have established faculty positions for sociologists. These trends are further evidence of the increasing role of sociology in assessing the effects of technological change on other aspects of society.

In recent years, medical sociologists have faced a new and serious challenge: helping society cope with the ethical issues that arise as it becomes increasingly possible to prolong human life by artificial means. Procedures such as heart transplants are extremely expensive and cannot possibly be made available to all patients who need them. Are they to be limited to those who can pay for them? If not, how should the patients who will benefit from such procedures be chosen? Medical sociologists are frequently asked to conduct research that will affect decisions of this nature.

Medical sociology is a rapidly growing subfield of sociology because so many of the basic health problems of societies are social as well as biological. In urbanized nations like the United States, there are immense problems in the institutions of health care, insurance plans, hospital organization, the application of new technologies, and much more. As people in the more affluent nations live longer, their medical needs and problems change, with more people experiencing

FIGURE 13.8

Selected Outbreaks of Infectious and Parasitic Diseases Other Than HIV/AIDS, July 1996–June 1997

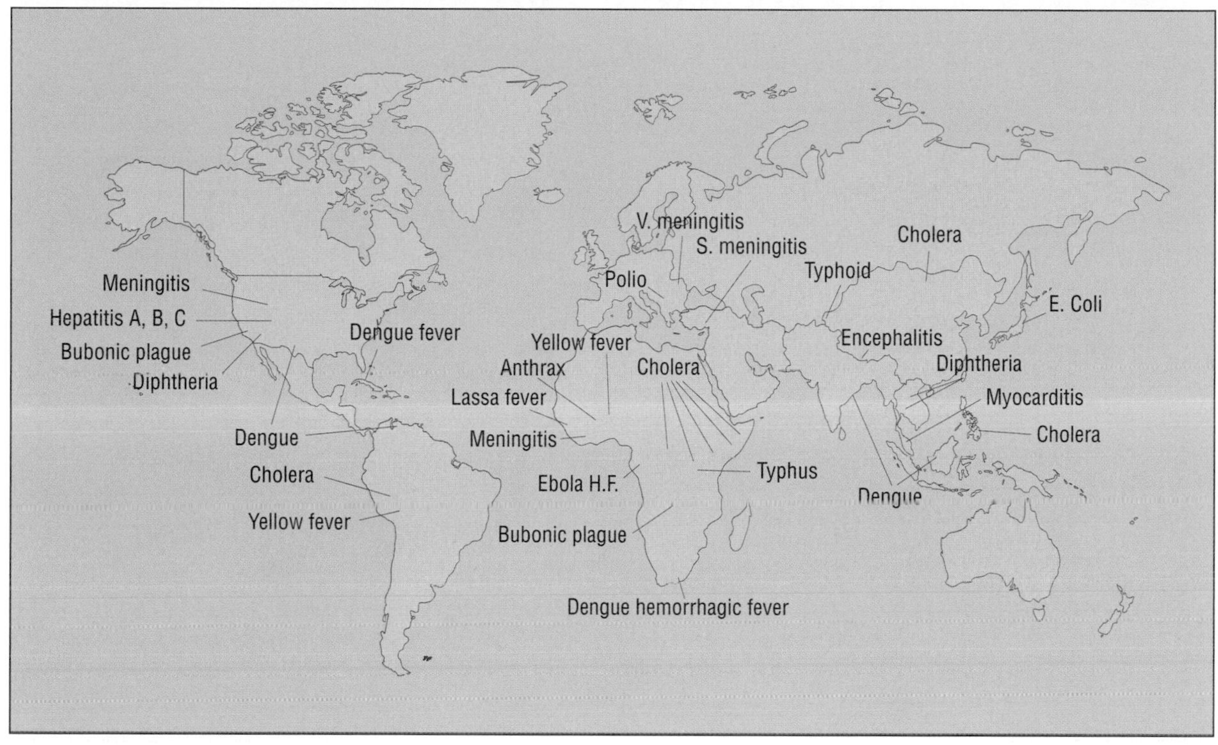

Source: Woodall, 1997.

prolonged illnesses that require intense medical care. Hospices for the dying are an example of a social solution to such problems. But in the developing nations, especially those that are too poor to afford modern medical institutions, the social problems related to health are even more urgent, as observed in the worldwide AIDS crisis. Clearly, there is a need for more trained medical personnel in these regions of the world, but there is also a great need for sociologically trained professionals with an understanding of the social issues involved in improving the health of societies undergoing rapid population growth and extremely high rates of rural–urban migration.

TECHNOLOGY AND SOCIAL CHANGE

Today the technologies available for the diagnosis and treatment of serious illnesses are often described as "miracles of modern medicine." The MRI scanner, for example, allows hospital technicians to observe a patient's internal organs without the use of X-rays; renal dialysis is used to prevent patients from dying of kidney failure; open-heart surgery is practically a routine operation, with the patient kept alive during the process by an external heart–lung machine; other surgical procedures involve the use of laser beams and fiber optics to perform delicate operations. All of these technologies require the use of extremely expensive equipment and highly trained personnel.

The development of increasingly sophisticated and costly medical technologies, together with the practice of requiring patients (or their insurance companies) to pay for hospital services, has led to a crisis in American medical care. The high and continually rising cost of medical care has become a major public issue, as has the fact that some groups in the population are unable to obtain adequate care. Some critics claim that the American health care system is suffering from *hypertrophy,* by which they mean that it has expanded to a size and complexity at which it has

GLOBAL SOCIAL CHANGE

The Worldwide AIDS Epidemic

AIDS is a worldwide epidemic that highlights issues of global social change. The disease also reveals the gap between women's and men's economic and political standing throughout the world.

The World Health Organization (WHO) estimates that almost 22 million people are infected with HIV, the human immunodeficiency virus that leads to AIDS. Women now account for 40% of all new AIDS cases, compared to 10% a decade ago. An estimated 7 to 8 million women of childbearing age are thought to be HIV positive, and between 5 and 10 million children have lost their mothers to AIDS; a high proportion of those children are HIV positive themselves.

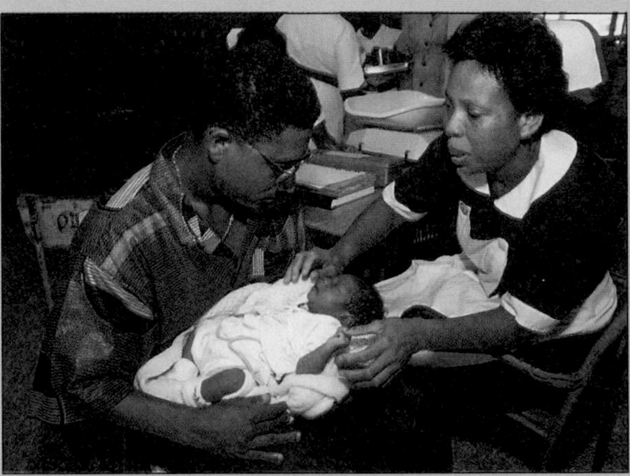

A South African nurse comforts a father and his infant, who has tested positive for HIV.

A study by WHO shows that the number of children born with the HIV virus has decreased throughout the world in the past year or two. Unfortunately, this improvement has not been uniform. There are nations in Africa and Asia where the number of children contracting the disease from their mothers continues to increase. About 10.5 million children under age 5 died from all diseases in 1999. Of those 10.5 million deaths, 3.8 million occurred in Africa, with a further 2.5 million in India and 750,000 in China. The authors of the study warn that the full impact of the HIV/AIDS epidemic on child death rates has not yet been felt. They say that the disease has the potential to dramatically slow, if not reverse, gains in childhood survival in some African and Asian countries (Ahmad, Lopez, & Inoue, 2000).

become dysfunctional. In their view, excessive emphasis on technological progress has created a situation in which the needs of the patient are subordinated to those of the providers of health care.

According to Paul Starr (1982), the problems of the American health care system stem from the way in which medical institutions evolved. As medical knowledge increased and technological advances were made, physicians developed narrow specialties and hospitals invested in specialized equipment. The physicians referred their patients to the hospitals for sophisticated medical testing and treatment. At the same time, the institution of health insurance emerged in response to demands for a more equitable distribution of health care. Insurance companies or the government began to pay for the services

In the United States, AIDS initially spread most rapidly among homosexual men and among men and women who share needles during intravenous drug use. But now the disease is spreading most rapidly among women involved in prostitution. And as is true in many Third World nations with high rates of AIDS infection, in the United States, the incidence of AIDS is highest among the poor, many of whom are members of minority groups.

In the more affluent nations, the advent of more effective treatments involving combinations of new drugs has slowed the rate of HIV infection and in some cases may prevent the onset of AIDS-related illnesses. But the new drug treatments are extremely costly. Many poor people with AIDS are not receiving these treatments (Ayala, 1996). And the situation is even worse in impoverished regions of Africa, where the incidence of the disease is highest and availability of the new drugs, or the ability to pay for them, is almost entirely lacking.

Education, especially of women at risk of AIDS infection, is the key to reducing the spread of the deadly disease. In consequence, WHO strongly recommends the following measures:

- Prevent HIV infection in women by protecting the human rights of women and girls, increasing educational opportunities for girls and young women, increasing women's access to economic activities, and other measures.
- Reduce the impact of HIV/AIDS on women by including women living with HIV/AIDS in the development of HIV/AIDS policy, prevention, and care programs; encouraging voluntary, confidential testing; and supporting programs that work with families of women living with the disease.
- Care for women with HIV/AIDS by providing appropriate health and welfare services, ensuring that women have access to contraceptive measures, increasing access to child-care and other support services, and similar policies (WHO, 1994).

As these recommendations suggest, however, in many regions of the world, AIDS education is closely related to a set of social issues: how to protect women's rights; how to enable them to participate fully in economic and political institutions; and how to provide adequate childcare, maternal health care, and social services. Women throughout the world are organizing grassroots social movements calling for increased attention to these issues. ■

provided by hospitals. Physicians and hospitals became highly interdependent, so much so that they began to "assert their long-run collective interests over their short-run individual interests" (Starr, 1982, p. 230). Their collective interests involve continued investment in complex technologies, with the result that medical care is becoming more and more expensive. The rising cost of medical care makes it more difficult for the poor, the elderly, and other groups to afford high-quality care and heroic life-preserving measures. In the closing years of the twentieth century, efforts to make health care less costly while preserving the insurance systems that provide medical coverage for the majority of Americans proved to be one of the most controversial and rancorous issues in U.S. political life.

SOCIOLOGY AND SOCIAL JUSTICE

Documentary photographer Wendy Ewald (1996) believes that social justice depends on our ability to put a human face on social issues. She knows that when we encounter the people of the Third World as individual human beings who strive to better their lives against all the odds of poverty and ignorance, we forge a human connection to their situation that may yield important results. The photos of Tidi and the female members of her caste, and the photos of children using the cameras Ewald lends them to document their lives, are included in her book, *I Dreamed I Had a Girl in My Pocket.* Ewald is dedicated to positive social change for the world's children. She has taught in Appalachia, in Colombia, on Canadian Native American reservations, and in Mexico and South Africa. Although not a professional sociologist, she has a highly developed sociological imagination. Here is how she describes her research method:

> I gave the children assignments to expand their ideas of picture-making while staying close to what they knew. I asked them to photograph their festivals, their families, themselves, their village, and their animals. I asked them to restage and photograph their dreams and fantasies. When the subject of dreams came up, the difference between the children who were in school and those who weren't became evident again. Most of the children who spent their days working in the houses or the fields claimed not to remember their dreams or fantasies. (1996, p. 14). ■

"Grandfather is smoking on the ledge with the dog."

"Hasmukh, Chandrakant, Harshad, and Dasrath learning to use the camera."

SUMMARY

The rate of population growth increased dramatically in the twentieth century, resulting in what is known as the population explosion. Rapid population growth, especially in the poorer regions of the world, will continue in coming decades and will pose severe challenges to human survival and well-being.

The *demographic transition* takes place in three stages: (a) high birthrates and death rates, (b) declining death rates, and (c) declining birthrates. These stages are accompanied by changes in the age composition of the population.

Urbanization is closely linked with rapid increases in population. Cities grow primarily as a result of migration, which is often caused by population increases in rural areas. The increasing tendency of people throughout the world to live in cities has been referred to as the urban revolution. The growth of cities should be distinguished from urbanization, which refers to the proportion of the total population concentrated in urban settlements. The end result of urbanization is an urban society, in which urban life has a pervasive influence on the entire society.

An early model of urban expansion was the concentric-zone model, in which a central business district is surrounded by successive zones devoted to light manufacturing, workers' homes, higher-class apartment buildings and single-family homes, and a commuters' zone. This model fails to account for other patterns of urbanization, such as the rise of satellite cities and the rapid urbanization that occurs along commercial transportation corridors. A more current model, known as strip development, describes the incorporation of smaller communities and towns into a larger *metropolitan area.*

Metropolitan areas have expanded greatly since the mid-twentieth century. In some areas, this growth has created large multinuclear urban systems, or *megalopolises.* One effect of the development of such areas is decentralization, in which outlying areas become more important at the expense of the central city.

Early studies of urban life tended to conclude that it weakens kinship ties and produces impersonal social life, as well as "psychic overload" and anomie. More recently these conclusions have been criticized by researchers who have found that many city dwellers maintain stable, intimate relationships with kin, neighbors, and coworkers and that urban life is not necessarily stressful or anomic. Subcultural theory sees the city as a mosaic of social worlds or intimate social groups. Suburban dwellers also have been found to be far less bored and isolated than was previously supposed.

Conflicts between communities within cities may arise out of different class interests or the conflicting goals of different status groups. Some urban sociologists see the city as divided up into "defended neighborhoods."

Until relatively recently, physicians were powerless either to check the progress of disease or to prolong life. Scientific research led to the discovery of the causes of many diseases and to the development of the field known as social epidemiology. Mortality rates and life expectancies vary by region of the world due to social inequalities, especially inequalities between men and women. In recent years, medical sociologists have been faced with the challenge of helping society cope with the ethical issues that arise as it becomes increasingly possible to prolong human life by artificial means.

GLOSSARY

Demographic transition: A set of major changes in birth and death rates that has occurred most completely in urban industrial nations in the past 200 years. **(372)**

Urbanization: A process in which an increasing proportion of a total population becomes concentrated in urban settlements. **(373)**

Crude birthrate: The number of births occurring during a year in a given population divided by the midyear population. **(374)**

Crude death rate: The number of deaths occurring during a year in a given population divided by the midyear population. **(374)**

Rate of reproductive change: The difference between the crude birthrate and the crude death rate for a given population. **(374)**

Metropolitan area: A central city surrounded by a number of smaller cities and suburbs that are closely related to it both socially and economically. **(378)**

Megalopolis: A complex of cities distributed along a major axis of traffic and communication. **(382)**

QUESTIONS FOR THOUGHT AND DISCUSSION

1. What sociological forces and social changes explain how societies move from the second to the third phase of the demographic transition? (Hint: Think about urbanization, industrialization, delay of marriage, birth control, improved public health, etc.)

2. What social changes made it possible for nations like the United States and Canada to make the rapid transition from agrarian to urban societies? What forces account for the decline of some inner-city communities and the spread of urban growth into the suburbs?

DIGGING DEEPER

Books

Western Times and Water Wars: State, Culture, and Rebellion in California (John Walton; University of California Press, 1992). A seminal work on the relationship of protest movements and collective behavior to metropolitan growth; describes the struggle to control vital natural resources in rapidly growing but arid regions.

Recent Social Trends in the United States, 1960–1990 (Theodore Caplow, Howard M. Bahr, John Modell, & Bruce A. Chadwick; McGill-Queens University Press, 1991). A compilation of statistical tables that document trends in population, urbanization, and other significant aspects of social change, based on data from the U.S. Census Bureau and other government agencies, and from major polling organizations, including the General Social Survey and the Gallup Poll.

The Social Construction of Communities (Gerald Suttles; University of Chicago Press, 1972). An important theoretical statement; shows how sociology explains phenomena like territoriality and the formation of neighborhoods and communities.

Close-Up: How to Read the American City, 2nd ed. (Grady Clay; University of Chicago Press, 1980). A brilliant, humorous, and extremely well-illustrated guide to the physical ecology of American urbanization in the second half of the twentieth century.

The Urbanization of the Third World (Joseph Gugler, ed.; Oxford University Press, 1988). A collection of original essays by leading urban researchers, tracing most of the outstanding trends and problems of urbanization in the developing areas of the world.

Streetwise: Race, Class, and Change in an Urban Community (Elijah Anderson; University of Chicago Press, 1990). A brilliant and original study of life in two adjacent neighborhoods, one populated primarily by lower-income minority households, the other mixed in terms of class and race but undergoing gentrification. Through detailed ethnographic research, the author documents changes in public behavior, norms of sexual conduct, drug use and control, and family life.

The Social Transformation of American Medicine (Paul Starr; Basic Books, 1982). A sociological history that shows why American medical institutions have become increasingly cumbersome and costly.

The Healing Experience: Readings on the Social Context of Health Care (William Kornblum & Carolyn D. Smith, eds.; Prentice Hall, 1994). An anthology that explores medical and health-related problems from the perspectives of well-known health care providers and medical sociology researchers.

Journals

Population Reference Bulletin. Offers extremely valuable syntheses and overviews of global population, urban, and epidemiological issues.

Population and Development Review. Presents international research on population change and the effects of population variables on other aspects of social and economic development. A valuable source of comparative data and studies of population problems in the third world.

Urban Affairs Quarterly. A leading journal of original research on urban social change.

Journal of Health and Social Behavior. A quarterly journal that presents sociological research on problems of human health and illness. Also features articles on change in social institutions and organizations as a consequence of new technologies.

EXPLORING SOCIOLOGY ON THE INTERNET

The Population Council
www.popcouncil.org/

One of the best places to begin looking at the wealth of resources on the Internet that deal with population growth, population control, and studies of population processes.

The Urban Institute
www.urban.org/

One of the nation's premier think tanks concerned with urban problems. Its Web site contains many of its publications.

Women's Health Interactive
www.womens-health.com/

Offers interactive sessions on a wide variety of critical health issues affecting women throughout the world.

The Latinos Web
www.catalog.com/favision/resource.htm

Offers a wide selection of research data on migrations and population changes among Latinos, with world trends and links as well.

UNAIDS: The Joint United Nations Programme on HIV/AIDS
www.unaids.org/

A good place to begin searching for health resources on a global basis. The AIDS page provides many links to other health resources.

National Association of Community Action Agencies
www.nacaa.org/

A home page that includes a section called "National Dialogue on Poverty" where you can find analyses and the recommendations of different communities and think tanks from a public forum held in 1996.

National Institutes of Health
www.nih.gov

Health information, press releases about scientific discoveries, and a link to scientific resources are all found on this site. The site also has links to major support services including CancerNet, AIDS information, and clinical alerts.

Chapter 14

Collective Behavior, Social Movements, and Social Change

How do social movements that enlist individual participation bring about social change?

As a Peace Corps volunteer serving as an instructor in physics and chemistry during the early 1960s, I lived in a tribal village on the outskirts of Abidjan, the capital of Ivory Coast. During the day I taught modern science to advanced students in a nearby college. In the evenings, however, the members of the Ebrié tribe with whom I lived taught me about living in a totally different social world, one that was experiencing even more rapid social change than my own. My best friend in the village worked in a modern insurance corporation in the center of the rapidly growing city. In the evenings he returned home, put on his African clothes, and had dinner in the family compound with one of his wives and her children. His three wives and their 14 children all lived together in relative harmony. I became fascinated with trying to understand this complex extended family and the village in which they lived, as well as the ways in which social forces were changing the Ebrié tribe and its people. Gradually I realized that I needed to shift my studies from the physical to the social sciences.

Social change affects people's lives all over the world, sometimes for the better, sometimes for the worse. As I sat under the palm trees with my Ebrié friends, we endlessly debated issues of social change. In many ways, our debates were about the difference between change and progress. "Before the white man came," my friends told me, "we had time but no watches. Now we have watches but no time." The village primary school was known in Ebrié as the "paper house" where children struggled to master the three R's in order to be able to earn money, or "white man's paper." But there was also general agreement that many of the technologies introduced by the Europeans were of immense value. Health care innovations that could prevent debilitating diseases like malaria and schistosomiasis were considered a great blessing. Yet the most prestigious Western invention from the villagers' viewpoint was the camera—for with it, they exclaimed, "our children can see their ancestors."

This chapter begins with some definitions of key terms in the study of social change and then moves to a consideration of the most powerful social forces that change societies: war, modernization, and social movements. We will see that modernization rarely if ever follows a straight upward path but is fraught with difficulties, including inequalities of wealth and power that often produce social unrest and revolution. Mass behavior and social movements therefore may be the most important forces of social change. The chapter closes with a discussion of how social change and social movements are experienced in everyday life in the United States and elsewhere. ■

THE NATURE OF SOCIAL CHANGE

Throughout this book, we deal with social changes of all kinds. The term **social change** refers to variations over time in the ecological ordering of populations and communities, in patterns of roles and social interactions, in the structure and functioning of institutions, and in the cultures of societies. We have seen that such changes can result from social forces building within societies **(endogenous forces)** as well as from forces for change exerted from the outside **(exogenous forces).**

The Ebrié do not have a written language. The villagers speak their history to each other and to their children as they gather in the evenings around their home fires. When I was living with them, they often told stories about how, in the distant past before the Europeans came, a tribal village sometimes grew too large. When that happened, severe conflict resulted. Factions formed and fought over land or other resources. Soon there would be a major split: The stronger faction would stay and keep the existing village, while the less powerful faction would establish a new village on unused land along the lagoon. This is an example of people adjusting to endogenous change caused by population increase, a very common cause of social change.

The Ebrié also told stories about recent changes in their lives, changes brought on mainly by colonial conquest and, more recently, by their own efforts to create a new nation and maintain its social and economic growth. Colonial rule and rebellion against it are examples of social change produced by exogenous forces, particularly the influence of a powerful conquering society attempting to impose its culture on conquered peoples.

The people of Ivory Coast and other societies that are undergoing rapid social change do not see the future as a matter of becoming Westernized. Rather, they strive to blend their own culture with certain aspects of Western culture. They are aware of a decline in their sense of community and mourn the weakening of their culture, but they actively embrace the aspects of modernity that will permit more of them to lead longer and perhaps more enjoyable lives. For them, there is no question of returning to an earlier state. Change is inevitable. They want to help make the change, however, and to do this, they understand that they must compete in modern social institutions like schools and businesses while at the same time they attempt to preserve their village life as best they can.

In the years since I left the Ebrié and became a sociologist, a great deal has happened to them and to the nations of Africa. The Ebrié of Ivory Coast have continued to prosper and become educated citizens of a developing nation. But their kin in coastal Liberia and Sierra Leone have experienced years of civil war, political repression, conquest, and terror. The economies of these African nations have worsened; the schools have been closed for months at a time; the rule of law has often deteriorated into a matter of who has the power to enforce his will. Similar calamities have dealt major setbacks to modernization in Nigeria, Rwanda, and Congo (formerly Zaire). At this writing, Ivory Coast, too, is threatened by the twin plagues of political instability and disease. The AIDS epidemic is taking a great toll on the nation's population, while conflict between Christians in the south and Muslims in the north threatens the nation's political institutions. To better understand the great forces of social change and how they are experienced, for better or worse, we turn now to a discussion of war, modernization, and related social forces.

FORCES OF SOCIAL CHANGE

War, modernization, and social movements are among the most powerful and pervasive forces that produce social change at every level of social life. Of course, as we have noted, social change is caused by many other social forces, especially technological innovation and population growth or mobility. But we will see in this section that war, modernization, and social movements also stimulate these other forces of change. For instance, many of the most significant technological innovations we experienced in the twentieth century were developed in response to the crisis of war. And many of the greatest movements of people over the planet were first set in motion by the disruptions caused by warfare or modernization (Chirot, 1986, 1994a, 1994b; Janowitz, 1978). But wars do not start by themselves. Often there are major social movements, such as communism or fascism or movements for national liberation or religious change, that mobilize huge numbers of people who, under some conditions, may eventually engage in war.

War and Conquest

War is among the greatest and is certainly the most violent of the forces that produce social change. Ironically, the deadliest forms of warfare are associated with the rise of modern civilizations. As societies have become more advanced in their command of technology and in their social organization, the devastation caused by war has increased. The wars fought by so-called primitive societies were frequently ritual affairs; the combatants often withdrew from the field after a single skirmish. Although not all preindustrial warfare was so ritualistic, relatively few combatants were killed because the technologies for killing were so limited compared with those available today.

In the Middle Ages, often viewed as a warlike time owing to the influence of knights and crusaders, the rate of fatalities among warriors in battle was about 2%. In contrast, in World War I the proportion was 40%. Modern warfare is increasingly dangerous not only for the combatants but for civilians as well. In World War I, about half of those who died were civilians, but in the Vietnam War the toll rose to more than 75% (Galtung, 1985; McNeill, 1982).

Any evaluation of war as a mechanism of social change must consider three broad questions: First, what are the ecological effects of war on human populations? Second, how do wars help shape the consciousness and culture of a people? Third, how does war change the institutions of societies?

The Ecological Impact of War. Casualties and conquest are the major ecological effects of war. People all over the world were shocked by televised scenes of oil fires and polluted coastlines during the 1991 Persian Gulf War, but the most significant ecological impacts of the Gulf War were felt by human populations. Like an epidemic of cholera or bubonic plague, war accounts for extraordinary and rapid declines in population. Pitirim Sorokin (1937) estimated that between the years 1100 and 1925 about 35.5 million people died in European wars alone. World War I claimed the lives of about 8.4 million soldiers and about 1.4 million civilians, and in World War II about 17 million military personnel and about 34 million civilians died. It is estimated that the Soviet Union lost about 15 million people during World War II and that in China about 22 million perished. Germany lost 3.7 million, Japan about 2.2 million, and the United States slightly less than 300,000 (Beer, 1981).

When millions of soldiers are killed, entire populations are unbalanced for more than a generation. Many women are widowed or remain single and either do not have children or raise their children alone. One effect of this imbalance is reduced population pressure on food and other resources, but at the same time

When the Plains Indians and the U.S. Cavalry fought bloody battles during the period of westward expansion in the nineteenth century, warfare was brutal and deadly, especially for the less well-equipped Indian warriors. As combat has become less a matter of brute force and more dependent on brains and technology, the possibility that well-trained women can compete as warriors has become a reality.

there are labor shortages and economic disarray due to the loss of so many skilled workers.

War also results in large-scale shifts in population and rapid acceleration of economic change. For example, the western parts of the United States and Canada experienced their most rapid growth as a result of mobilization for war during the first half of the twentieth century. New dams, new electric power plants, new factories to produce all kinds of goods were built. San Diego, Los Angeles, San Francisco, Portland, Seattle, and Vancouver all experienced massive population growth, as did many inland centers of industry and agriculture. The end of World War II saw continued growth in the western states as young families who had come west during the war decided to settle there permanently.

For the losers in war, the ecological consequences of defeat are usually far more dramatic. Population loss, economic subjugation, the imposition of a foreign language and culture, and forced movement to new towns and industrial areas are common. During more than a century of genocidal wars waged by whites against Native Americans in the nineteenth and early twentieth centuries, the consequences for the losers were death, expulsion, and banishment to reservations. For tribal peoples of Africa, invasion, war, and conquest led to colonial rule and rapid social change, often imposed through taxation, labor gangs, military draft, and similar means. In areas of the world that are torn by war today, such as the former states of Yugoslavia (especially Bosnia) and African countries like Zaire, Somalia, and Liberia, war has brought genocidal ethnic conflicts, mass expulsions ("ethnic cleansing"), and years of abject poverty in squalid refugee camps (Glenny, 2000).

The Cultural Impact of War. War changes a society's culture by stamping the memories of chaos and cruelty, heroism and camaraderie on entire generations. Years after a major war, its effects on values and norms continue to be felt (Schuman & Scott, 1989). One need only think of the impact of the American Civil War on the former Confederate states—an impact that remains strong today and can be seen in Civil War memorials, rebel yells, the conduct of interracial relations, and North–South animosities. Recent sociological research shows that even for generations that did not experience war firsthand, the memory or threat of warfare is perceived as extremely important. Table 14.1 shows that people will cite specific wars first among "national events or changes that seem especially important" to them—even if they themselves were not alive during the wars they mention.

Wars change entire cultures by increasing contacts between different societies. In the case of modern nations like Japan, Italy, Germany, and Korea, all of which have experienced military occupation by foreign powers, the cultural consequences include the acculturation of new norms and behaviors. For example, baseball was introduced to Japan by U.S. occupation troops after World War II and is now that nation's most popular spectator sport. Indeed, the influence of North American culture, conveyed through movies, sports, consumer goods, and language, spread rapidly throughout Southeast Asia as a result of World War II and its lingering political consequences.

War's Impact on Social Institutions. The structure of a society, especially its major social institutions, may be drastically changed by war and preparation for war. The mobilization of large numbers of people and the marshaling of new technologies for military purposes have a centralizing effect on social institutions. In the United States, for example, the growth of large research universities in the 1960s was accelerated by huge investments in applied science and technology after the Soviet Union became the first nation to launch a space satellite. Universities that were capable of developing new science programs grew rapidly, and their administrations gained greater power. The power and influence of the national government have also grown, often at the expense of local governmental institutions, as a consequence of the two world wars and the arms race. Providing for national defense is extremely expensive and requires that the central government be granted increased taxing powers.

The French sociologist Raymond Aron (1955) has called the twentieth century "the century of total war" because of the capacity of warfare to shape the destiny of entire regions and because of the unprecedented power of nuclear and other weapons. Aron's phrase also captures the transforming power of modern war: its ability to alter societies. In the United States, for example, there is no doubt that the mobilization necessary to fight two world wars, major regional wars in Korea and Vietnam, and innumerable smaller skirmishes in Africa, the Middle East, Latin America, and Asia contributed to the controversial growth of the federal government during the twentieth century. The impetus for creating many of the institutions of the welfare state, such as child support, was the need to care for war widows and their children. And the massive growth of the federal budget deficit in the second half of the twentieth century could be

TABLE 14.1

U.S. Respondents' Rank Ordering of Important Events as They Perceive Them

	Age					
Event/Change	**18–29**	**30–39**	**40–49**	**50–59**	**60–69**	**70 Plus**
World War II (265)	14	16	24	29	30	23
Vietnam (144)	18	18	13	2	4	1
Space exploration (93)	8	6	8	10	6	8
Kennedy assassination (62)	3	8	10	3	1	1
Civil rights (77)	7	7	5	7	6	3
Nuclear war, threat of (55)	6	5	6	4	2	3
Communication/transportation (46)	1	4	4	5	3	9
Depression (43)	3	3	2	5	7	13
Computers (23)	2	1	2	3	3	0
Terrorism (43)	4	2	0	1	1	0
Moral decline (28)	2	2	2	2	4	1
Women's rights (20)	1	2	3	0	2	1
Other event/change (357)	30	26	22	29	33	37
	100	100	100	100	100	100
	(289)	(312)	(200)	(167)	(165)	(110)

Note: Numbers in parentheses refer to number of respondents.

Source: Schuman & Scott, 1989.

attributed in large measure to the expenditure of billions of dollars on military institutions (Galbraith, 1995).

Elsewhere in the world, the dominance of military institutions and the fragility of legal and governmental institutions produce "garrison states" in which economic growth and the rule of law are subordinated to the needs of the military (Lasswell, 1941). The end of the Cold War has led to political instability in central Europe and in many areas once dominated by the former Soviet Union, but it has also caused unprecedented economic and social changes in Western Europe. Political instability in many regions of the world is used to justify continued military spending; in the United States, military spending still exceeds $250 billion annually even while other public spending is being drastically reduced. Environmentalists and other opponents of high levels of military spending argue that individuals cannot comprehend the magnitude of military spending unless it is presented in terms of trade-offs indicating what alternative social choices would cost. Table 14.2 presents some examples of such trade-offs.

Modernization

A second major source of social change is the set of trends that are collectively known as **modernization.** This term encompasses all the changes that societies and individuals experience as a result of industrialization, urbanization, and the development of nation-states. These processes occurred during a period of two or more centuries in the Western nations and Japan, but they are taking place at a far more rapid rate in the former colonial societies that are today's new nations.

The concept of modernization does not assume that change is irreversible. For example, the rise of nation-states throughout the world during the past few centuries does not imply the end of loyalties based on a sense of "peoplehood" that conflict with the sense of shared citizenship in a nation (see Chapter 3). The bloodshed in the former Yugoslavia and in many places on the African continent, or in the former Soviet Union, bears witness to the strength of feelings about peoplehood as opposed to loyalty to the modern nation-state with its laws and governments.

TABLE 14.2
Trade-Offs Between Military and Social or Environmental Priorities

Military Priority	Cost	Social/Environmental Priority
Trident II submarine and F-18 jet fighter programs	$100,000,000,000	Estimated cost of cleaning up the 10,000 worst hazardous-waste dumps in the United States
Stealth bomber program	$68,000,000,000	Two-thirds of estimated costs to meet U.S. clean-water goals.
Two weeks of world military expenditure	$30,000,000,000	Annual cost of the proposed UN Water and Sanitation Decade
Three days of global military spending	$6,500,000,000	Cost to fund Tropical Forest Action Plan over five years
Development cost for Midgetman ICBM	$6,000,000,000	Annual cost to cut sulfur dioxide emissions by 8–12 million tons/year in the United States to combat acid rain
Two days of global military spending	$4,800,000,000	Annual cost of proposed UN Action Plan to halt third world desertification over 20 years
Ten days of European Economic Community military spending	$2,000,000,000	Annual cost to clean up hazardous-waste sites in 10 European Union countries
One Trident submarine	$1,400,000,000	Global five-year child immunization program against six deadly diseases, preventing 1 million deaths a year
Two months of Ethiopian military spending	$50,000,000	Annual cost of proposed UN Anti-Desertification Plan for Ethiopia
One nuclear-weapon test	$12,000,000	Cost for installation of 80,000 hand pumps to give third world villages access to safe water
One-hour operating cost, B-1B bomber	$21,000	Cost of community-based maternal health care in 10 African villages for 10 years

Source: Adapted from Renner, 1989.

Nevertheless, the term *modernization* summarizes most of the major changes, for better or worse, that societies throughout the world are experiencing, albeit at differing rates and with varying amounts of social disruption. Neil Smelser (1966) associates modernization with the following set of changes:

1. *In the realm of technology,* a developing society is changing from simple and traditionalized techniques toward the application of scientific knowledge.
2. *In agriculture,* the developing society evolves from subsistence farming toward the commercial production of agricultural goods. This means specialization in cash crops, purchase of nonagricultural products in the market, and often agricultural wage labor.
3. *In industry,* the developing society undergoes a transition from the use of human and animal power toward industrialization proper, or men working for wages at power-driven machines, which produce commodities marketed outside the community of production.
4. *In ecological arrangements,* the developing society moves from the farm and village toward urban concentrations. (pp. 110–111; emphasis added)

These processes can take place simultaneously, but this is not always the case. Many societies mechanize their agriculture and begin to produce cash crops for foreign markets before their cities and urban forms of employment have begun to grow rapidly. This was the case, for example, in Sri Lanka, Indonesia, and many of the newer African nations.

Smelser and others who study modernization have shown that "technical, economic, and ecological changes ramify through the whole social and cultural fabric" (1966, p. 111). In the political sphere of life, we see the authority systems of the village giving way to domination by the institutions of nation-states. In the area of education, as societies attempt to produce workers who can meet the needs of new industries, new educational institutions are established. In the area of religion, there is a decrease in the strength of organized religions. Families change as traditional extended families adapt to new economic institutions that demand greater mobility.

Patterns of inequality in societies also change. Older patterns of gender inequality are modified (and often replaced by new forms of inequality) as women are in greater demand to fill positions in new economic institutions. And the emergence of a new class, the wage workers, increases the power of the common people, usually adding to their determination to become better educated and to participate more fully in political life. None of these changes is inevitable or irreversible; workers, for example, may see their unions "busted" in times of recession or economic change. But in the long run all of these trends are likely to appear in a modernizing society.

The term *modernization* should be used cautiously, in its sociological sense rather than as a value judgment about different societies. It does not mean that we can judge life in modern societies as better or more satisfactory or more humane than life was for people in societies like the one the Ebrié once knew. As noted earlier, modern societies have developed the capacity to cause more destruction and human suffering than any simpler society could possibly have caused. And we have seen in other chapters that many of the advantages enjoyed by the most modern nations have come at a high cost to simpler, less-modern societies.

Modernization in the Developing Nations. Social scientists often use the term *third world* to refer to nations that have won independence from colonial dominance in the decades since World War II. If the "first world" is that of the capitalist nation-states and the "second world" that of the former communist nations, the third world nations are those that are not aligned with either of these "worlds" but are united in their need to survive in an environment dominated by more politically or economically powerful nations.

But the term *third world* can be misleading, since in the past 30 years many of these nations have made strides toward modernity. Moreover, trends in global social and economic change have produced impoverished areas—sometimes called the "fourth world"—in affluent nations where illegal immigrants or stigmatized minorities work under conditions similar to those found in the third world (Sassen, 1991). For this reason, we prefer to use the term *developing nations* or *modernizing nations*. A **developing nation** is one that is undergoing a set of transformations whose effect is to increase the productivity of its people, their health, their literacy, and their ability to participate in political decision making. (See the Sociological Methods box on pages 410–411.)

AIDS and the Failure of Modernization. Since it was identified 20 years ago, AIDS has come to rank among history's worst epidemics. About 21 million people have died of the disease, and an estimated 36 million are now infected (Altman, 2001). Especially in many regions of Africa, HIV/AIDS represents a spectacular failure of modernization. Figure 14.1 shows that South Africa and many of its neighboring states suffer from rates of HIV infection between 15% and 36%, with other areas along the western coast not much better off. Due to the ravages of the disease, entire nations are experiencing dramatic declines in many indicators of health and welfare. In these regions, the high cost of effective medical treatment and failure to mobilize the population to change risky sexual behaviors are major reasons for the unchecked spread of the disease. But Western technology has also failed to live up to its promises. Twenty years ago, scientists and political leaders in the United States and Europe were confident that a vaccine against the HIV virus would soon be developed. Now scientists must admit that while far more is known about the disease and its causes, there is still no vaccine or cure.

Reliance on technological innovation has often led to delays in public education and major AIDS awareness campaigns, especially in the third world. In fact, the exceptions to this glum assessment are of special sociological importance because they highlight the importance of social approaches to fighting the disease. In Brazil, to cite the best example, enlightened political leaders and their allies in community organizations have created outreach programs to reach gay populations, drug users, and all other seg-

FIGURE 14.1

A Virus's Reach

HIV rates vary greatly across Africa, which is the most infected continent.

Source: Centers for Disease Control and Prevention.

ments of the population at risk for contracting HIV. In addition, they have begun to produce lower-cost anti-AIDs drugs despite lack of cooperation from powerful U.S. and European drug manufacturers. Their rationale is that fighting the disease must have top priority; questions of international licensing and patents can be settled after the spread of the disease has been brought under control. As a consequence, Brazil has done more to lower its rate of AIDS than any other major developing nation and has created a model that African nations would do well to follow (Csillag, 1999).

Antimodernist Movements. The theory of modernization, as we have described it, implies that modernization will occur in a similar fashion in every society. But the differing experiences of the developing nations call this view into question. Not only do we often see the industrialization of agriculture (i.e., the growth of huge mechanized farms) without the rise of industrial cities, or the growth of cities without a decline in the strength of organized religions or the emergence of modern educational institutions, but we also see the rise of antimodernist social movements in some of these nations (Germani, 1973). Events in the Islamic world are a case in point. In Algeria, Pakistan, Iran, Saudi Arabia, Libya, Sudan, and other Islamic nations, a fundamentalist religious movement has gained strength in the past few decades. This resurgence of traditional Islamic beliefs and practices denies that modernization must be accompanied by the rejection of religious faith, by the separation of religion and government, or by more democratic political participation; these and other aspects of the Western version of modernity are being strongly challenged by Islamic fundamentalism.

Studies of these antimodernist movements are being conducted throughout the world. For example, in her studies of Algerian women confronting the antimodernist movement sweeping through the Arab nations, the Algerian American sociologist Marnia Lazreg (1994) has found that these movements are misnamed. In reality, she concludes, they are actually radical attempts to seize political power in which the movement's leaders use the religious fervor of the masses to achieve their goal.

The rise of antimodernist movements is not limited to the Islamic nations. Similar movements can be seen in the United States. Conservative groups plead for a return to more traditional values, and some radical groups advocate a return to self-sufficient communities that would engage in farming on a small scale. The effects of such movements on a society's institutions show that modernization does not necessarily follow a single direction or imply a single set of changes (e.g., the decline of religious faith, the rise of science, or the growth of industrial cities).

Another challenge to this view of modernization is posed by the fact that the world's resources of raw materials, water, and energy are far less plentiful than they once were. Today there are serious doubts about whether those resources are adequate to permit the poor nations to become developed to anywhere near the extent that the Western nations have, or whether the rich nations can continue to grow as they have in the past.

Modernization and Dependency. Some sociologists argue that the development of the more advanced modern nations actually impedes development

SOCIOLOGICAL METHODS

Measuring Macrosocial Change: Comparative Indicators

When social scientists compare the pace of social change and development in different nations, they often use a standard set of measures known as *indicators of social change*. These indicators include the variables listed in the accompanying table. As you can see, some of these variables measure major aspects of a population's growth potential and well-being (e.g., life expectancy, literacy, food intake in calories per day per person), as well as the extent to which the population is urbanized. Other measures focus on macro-level variables such as energy consumption and military expenditures.

What is not evident in the table are the large investments in data collection that must be made in order to generate these comparative indicators of development. Very often such research is supported by funds from the United Nations or the World Bank. If a nation does not produce the data needed to calculate these key indicators, it may be disqualified from consideration for development loans or other types of investments. To demonstrate their need for such aid, nations must have social scientists who are capable of producing and understanding the necessary quantitative data.

Among the many lessons that could be drawn from the table, consider the energy consumption variable. This indicator shows some remarkable differences between the United States, the world's biggest consumer of energy, and the developing nations included in the table. Although many of these nations—like India, Pakistan, and Egypt—are in tropical or arid zones where heating fuel is not a necessity of life, their extremely low energy consumption rate per capita is also an indication of the lack of motorized transport, the low level of energy use for cooking, and the enormous progress that still needs to be made in bringing electricity to villages and rural areas. More rapidly developing nations like Cuba and Mexico have correspondingly higher rates of energy use—but the figure for the United States is far beyond that for any other nation. With less than a tenth of the world's population, the United States consumes almost a third of world energy production each year. Such figures are frequently cited in debates over energy consumption and global warming, in which representatives of developing nations argue that the United States must curtail its enormous rate of energy use before admonishing other nations about their pollution problems. ■

in the newer nations, or at least channels it in directions that are not always beneficial. In a general statement of this theory, André Gunder Frank (1966) questions the idea that the less-developed societies are merely at an earlier stage of modernization than the advanced nations. He cites the development of one-crop economies in many parts of Latin America as evidence of how social forces in the developed nations actually transform the tropical countryside. According to Frank, when peasants give up subsistence agriculture and trading in local markets because their land has been absorbed into huge banana or coffee plantations, the result is a form of *underdevelopment* that did not exist before, one in which the peasantry is transformed into a class of landless rural laborers (Gimenez, 1998).

Immanuel Wallerstein (1974) has proposed a more general theory that he calls *world system theory*.

Indicators of Social Change in Selected Nations

	Bolivia	Chile	Cuba	Egypt	El Salvador	India	Mexico	Pakistan	Turkey	United States
Percentage of population living in urban areas	61.2	85.8	76.0	43.0	54.8	26.8	71.3	32.4	68.8	76.4
Percentage of population literate	79.5	95.2	95.7	51.4	74.1	52.0	89.6	35.0	82.3	95.5
Daily per capita caloric intake	2,192	2,769	2,291	3,327	2,577	2,388	3,136	2,475	3,593	3,642
Daily newspaper circulation per 1,000 population	69	101	122	43	53	21	115	22	44	238
Energy consumption per capita (mil. Btu)	—	61	38	29	—	13	59	12	46	351
Military expenditure as percentage of GNP	2.3	3.8	1.6	5.7	1.1	2.4	1.0	6.0	4.1	3.8
Percentage of population under age 15	40.6	3.8	1.6	5.7	1.1	2.4	1.0	6.0	4.1	3.8
Life expectancy at birth (years), male	59.6	71.8	72.8	65.4	65.5	59.1	66.5	67.7	69.5	73.9
Life expectancy at birth (years), female	62.9	77.8	77.7	69.5	72.4	60.3	73.1	71.8	74.4	79.4

Sources: Data from *Britannica Book of the Year,* 1998; *Statistical Abstract,* 2000.

In this theory, he divides the world into core states, semiperipheral areas, and peripheral areas. The **core states** include the United States, England, France, Russia, Germany, and Japan—which are the most technologically advanced nations and which dominate the banking and financial functions of the world economy. The **semiperipheral areas** are places like Spain and Portugal, the oil-producing nations of the Middle East, and Brazil and Mexico. In these areas, industry and financial institutions are developed to some extent, but they remain dependent on capital and technology provided by the core states. The **peripheral areas** include much of Africa, Asia, and Latin America. They supply basic resources and labor power to the core states and the semiperipheral areas. This world system, Wallerstein asserts, is based on various forms of economic domination and does not require political repression as well.

Wallerstein's theory has the drawback of suggesting that the so-called core states do not include any areas of production that may resemble the peripheral, dependent regions more than they do the fully modernized nations. There is evidence that even the most modern nations contain such regions. For example, in a study of sharecropping in California, Miriam Wells (1984, 1996) showed that large California berry growers have been dividing their land into small plots and renting them to low-income farm laborers. Wells contends that the growers have adopted this strategy as a means of avoiding the higher costs of unionized farm labor. She reports: "The sharecropper is responsible for maintaining the plots, for harvesting and packing the fruit, and for hiring and paying whatever labor is necessary to accomplish these tasks" (1984, p. 17). In return, the sharecropper receives from 50% to 55% of the proceeds minus the costs of handling, loading, hauling, and marketing the crop. Wells points out that modernization theory views sharecropping as an obsolete form of production, yet it can reappear even in the most advanced societies under certain conditions, such as lack of machinery for harvesting (Wells, 1996). Once again, therefore, we see that modernization is not a unilinear process with inevitable outcomes for every society or nation. People in all societies, including our own, experience social change in an infinite number of ways. (See the Global Social Change box on page 413.)

Collective Behavior and Social Movements

A third major force of social change occurs when people act collectively to initiate social movements. For instance, knowing that in solidarity there is strength, farm workers who fall further into poverty even as they work long hours in the fields have attempted to organize unions to promote their goals. The elderly, too, have organized groups to lobby and demand greater attention to their needs. Social movements may also arise out of more spontaneous events like riots or unorganized protests such as wildcat strikes. Through these various collective actions, people seek to change society in ways they see as positive.

The term **collective behavior** refers to a continuum of unusual or nonroutine behaviors that are engaged in by large numbers of people. At one extreme of this continuum is the spontaneous behavior of people who react to situations they perceive as uncertain, threatening, or extremely attractive. The violent fighting between blacks and whites during the Chicago race riot of 1919 is one example of spontaneous collective behavior. Another is the spontaneous outpouring of grief among millions of people in Britain and elsewhere in the world after the death of Princess Diana in 1997. Such behaviors are not governed by the routine norms that control behavior (Smelser, 1962).

At the other extreme of the continuum of collective behaviors are rallies, demonstrations, marches, protest meetings, festivals, and similar structured events. These activities involve large numbers of people in nonroutine behaviors, but they are organized by leaders and have specific goals. When workers in a union plan a strike, for example, their picketing and rallies are forms of organized collective behavior whose purpose is to demonstrate their solidarity and their determination to obtain their demands. When blacks marched in commemoration of those who died in the 1919 race riot, the event was organized to build solidarity and publicize a new determination to resist racism. In such cases, the organization that plans the event and uses collective behavior to make its feelings or demands known is a *social movement.*

Social movements are intentional efforts by groups in a society to create new institutions or reform existing ones. Such movements often grow out of more spontaneous episodes of collective behavior; once they are organized, they continue to plan collective events to promote their cause (Blumer, 1978; Genevie, 1978). Some movements, like the antiabortion movement, resist change; others, like the labor movement, have brought about far-reaching changes in social institutions. Still others, like the gay and lesbian movement, seek to gain rights for gays and lesbians that other groups have already won. In this sense, they are movements for full inclusion in society by people who feel that they are discriminated against. Many social movements in a multiethnic and culturally diverse society challenge discrimination and deviant labels (Tarrow, 1994).

The American labor movement provides many examples of the ways in which spontaneous episodes of collective behavior and the social movements associated with them can change the course of a society's development. During the nation's stormy transition from an agrarian to an industrial society, workers fought against the traditional right of employers to establish individual wage rates and to hire and fire workers as they pleased. They demanded instead the right to organize unions that could negotiate a collective wage rate (collective bargaining) for each category of workers.

GLOBAL SOCIAL CHANGE

Singing Out Against Anomie

Bruce Springsteen sings about the feelings of loss that people experience when their factories are torn down. The murdered singer Selena captured the hearts of millions by singing popular Tejano songs that described her people's trials in coping with new social and economic conditions along the rapidly industrializing Texas–Mexico border. And now Susan Aglukark, a popular Inuit singer from the frozen shores of Hudson Bay, has become the first of her people ever to sign a major recording contract. EMI Music Canada has signed her to produce more of her haunting songs about the struggle of aboriginal people in the Inuit-controlled region of Nunavut, which extends well above the Arctic Circle in the Northwest Territories.

Aglukark's Inuit or Eskimo name is Uuliniq, which means "scarred from burns." Although she is not literally scarred—the name is purely symbolic—the same symbolic name could easily be applied to all of Canada's 35,000 Inuit people. Like people living at the edges of industrialization and urbanization throughout the world, they have experienced severe disruptions in their traditional way of life. They can no longer maintain their existence as a hunting-and-gathering people, not only because of depleted supplies of game but also owing to the far-reaching effects of changing lifestyles and technologies. For example, their need for gasoline to power generators and snowmobiles and their desire for televisions and prepared foods like coffee and sugar make it impossible for them to maintain their traditional culture.

As a result of these social changes, many of Aglukark's people are experiencing anomie or normlessness. Caught between the modern and the traditional, not knowing how to find work or where to go for advice, they often turn to alcohol and drugs, become angry and abusive, or even commit suicide. Aglukark's music has elements of rock and country gospel, but the words may deal with subjects such as a girlhood friend who has taken her own life. She hopes that her songs will help young people everywhere deal with the feelings that lead them to despair and sometimes to suicide. Since this is a particular problem in aboriginal communities, she is concerned with showing her people that "life is the most beautiful thing."

Susan Aglukark.

Susan Aglukark is proud to be identified with the newly formed Inuit region of Nunavut. She notes that the need to cope with change has challenged her people for many generations: "The trouble was in the past, it was forced on us. At least now we may have some control over our future" (quoted in Farnsworth, 1994, p. A4).

The Amazonian Indian tribes of Brazil fear that despite their protests the destruction of the forests on which their societies depend will continue unabated. Losing their traditional hunting and fishing grounds will mean the loss of their way of life and their culture.

They also demanded better working conditions and benefits. Mass picketing, sit-down strikes, and pitched battles between workers and company agents or the police were everyday events in the 1890s.

Often the workers joined in walkouts, rioting, or other spontaneous collective actions when faced with mine disasters or intolerable working conditions. From these episodes of bitter conflict emerged the modern labor movement. In the ensuing decades, culminating in the period of New Deal legislation in the 1930s, the workers' demand for collective bargaining was *institutionalized.* By this we mean that the right to join unions and to bargain collectively was incorporated into the nation's laws. Labor unions thereby became recognized organizations and collective bargaining the recognized means of settling labor disputes. The unions, in turn, had to discipline their workers by specifying when collective behavior such as strikes and other actions could be used.

By the end of World War II, labor unions and collective bargaining had become legitimate social institutions. In recent decades, however, drastic declines in industrial employment in North America and Europe, combined with renewed attacks on labor unions by employers, have led to a decrease in the ranks of organized labor. But major social movements that affect the lives of millions of people often go through cycles of growth and decline followed by a rebirth of activity. It remains to be seen whether this will be true of the labor movement (Geoghegan, 1991).

Spontaneous collective behavior, from which social movements often arise, can take many forms. Such behaviors range from the demonstrations and riots that mark major revolutions to the fads, fashions, crazes, and rumors that sweep through modern societies with such rapidity that what is new and shocking one day will be a subject of nostalgia the next. Some of these types of collective behavior become associated with social movements, but many do not. For example, in the early 1970s the craze known as streaking (in which one or more people ran nude in front of a large gathering) called attention to other forms of public nudity. Suddenly the number of beaches, parks, and recreation areas where people were going nude increased dramatically (Douglas & Rasmussen, 1977; Fortin & Ecker, 1978), and people who believed in the goodness of nude living joined forces with the older naturist movement to sponsor "nude-ins" and other collective events that publicized their beliefs. In the early 1990s, groups of women in the Northeast protested bans on topless sunbathing at beaches and won some concessions from local authorities. Despite these sporadic protests, the naturist movement has not given rise to organizations dedicated to expressing nudist values; there is no nudist movement comparable to the civil rights movement or the women's movement.

Crowds and Masses. Those who study spontaneous forms of collective behavior often begin by examining the structure of such behavior—that is, by discovering whether the people who engage in a particular kind of behavior are in close proximity to one another or whether they are connected in a more indirect way. In this regard, it is useful to distinguish between crowds and masses. A **crowd** is a large number of people who are gathered together in close proximity to one another (e.g., at a demonstration or a football game). A **mass** is more diffuse; it does not occur in a physical setting. A mass is a large number of people who are all oriented toward a set of shared symbols or social objects (Lofland, 1981); an example is the audience for a particular television program. Collective behavior can occur in crowds, in masses, or in both at once.

Crowd behavior may turn into mass behavior; spontaneous collective behavior may generate a social movement. For example, when the stock market crashed in 1929 at the beginning of the Great Depression, crowds of panic-stricken investors spilled onto the streets of the financial districts of New York, Chicago, and San Francisco. There, within sight of one another, people found their fears magnified. And those fears quickly spread through the mass of Americans. The panic started among people who had made investments on credit, but it spread to people who merely had savings in local banks. Crowds of terrified savers descended on the banks, which were unable to handle the sudden demand for withdrawals because they themselves had been investing in stocks that were suddenly worthless. As a result, millions of Americans lost all their savings.

In the aftermath of the stock market crash and the resulting mass panic, social movements arose to seek reforms that would protect investors against similar episodes in the future. During the depression that followed the crash, citizens' fears about their lack of retirement or pension savings led to the development of new institutions like Social Security and bank deposit insurance. Today the American Association of Retired Persons, an organization that represents the interests of older people who once protested as an unorganized mass, is the largest single lobbying group in American politics.

Types of Social Movements. There are as many types of social movements as there are varieties of spontaneous collective behavior. Just think of how many "causes" can motivate people to take organized action. We have a social-welfare movement, an antitax movement, a civil rights movement, a women's movement, an environmental protection movement, a consumer movement, an animal rights movement, and on and on. Within each of these movements, numerous organizations attempt to speak for everyone who supports the movement's goals. Actually, within any large social movement there are likely to be a number of different social movement organizations (sometimes referred to as SMOs) that may have different, and at times competing, ideas about how to achieve the movement's goals (E. Goode, 1992). As we will see, the way such organizations operate and the goals they seek depend to a large extent on the characteristics of the social movement itself. (See the Study Chart on page 416.)

Social movements can be classified into four general categories based on the goals they seek to achieve:

1. *Revolutionary movements* seek to overthrow existing stratification systems and institutions and replace them with new ones (Hopper, 1950). The Russian Bolsheviks who founded the Communist party were a revolutionary social movement that sought to eliminate the class structure of Russian society. They also believed that the institutions of capitalism—the market and private property—must be replaced with democratic worker groups or *soviets* whose efforts would be directed by a central committee. Meanwhile, certain other existing institutions, especially religion and the family, had to be either eliminated or drastically altered.
2. *Reformist movements* seek partial changes in some institutions and values, usually on behalf of some segments of society rather than all. The labor movement is basically reformist. It seeks to alter the institutions of private property by requiring the owners of businesses to bargain collectively with workers concerning wages and working conditions and to reach an agreement that will apply to all the workers in the firm. Collective bargaining replaces the old system of individual contracts between workers and employers, but it does not totally destroy the institution of private property.
3. *Conservative movements* seek to uphold the values and institutions of society and generally resist attempts to change them, unless their goal is to undo undesired changes that have already occurred. The conservative movement in the United States seeks to reinforce the values and functions of capitalist institutions and to support the traditional values of such institutions as the family and the church.
4. *Reactionary movements* seek to return to the institutions and values of the past and, therefore, to do away with some or all existing social institutions and cultural values (Cameron, 1966). The Ku Klux Klan is part of a reactionary social movement that seeks a return to the racial caste system that was supported by American legal institutions (laws, courts, and the police) until the 1954 Supreme Court decision that declared the "separate but equal" doctrine unconstitutional.

STUDY CHART
Types of Social Movements

Type	Description	Example
Revolutionary	Seeks to overthrow and replace existing stratification systems and institutions.	Bolsheviks
Reformist	Seeks partial changes in some institutions and values, usually on behalf of some segments of society rather than all.	Labor movement
Conservative	Seeks to uphold the values and institutions of society and resists attempts to change them.	Conservative movements
Reactionary	Seeks to return to the institutions and values of the past and do away with existing institutions and values.	Ku Klux Klan
Expressive	Devoted to the expression of personal beliefs and feelings.	Punk movement
Millenarian	Combines elements of revolutionary and expressive movements.	People's Temple

It should be pointed out, however, that although they cover some of the movements that have had the greatest impact on modern societies, the preceding four categories do not include all the possible types of movements. Herbert Blumer (1969a), for example, identified a fifth category: *expressive social movements,* or movements devoted to the expression of personal beliefs and feelings. Those beliefs and feelings may be religious or ethical or may involve an entire lifestyle, as in the case of the punk movement of the 1980s. An important aspect of expressive social movements is that their members typically reject the idea that their efforts ought to be directed at changing society or the behavior of people who do not belong to their movement. Their quest is for personal expression, and if others choose to "see the light," that is welcome.

Some of the various New Age groups and organizations that are now popular in the United States fit quite well into the expressive social movement category. Non-Native Americans who appropriate some aspects of Native American spirituality (including the use of sweat lodges or the practice of vision quests) but do not proselytize are involved in an expressive social movement. So are men who follow the teachings of poet Robert Bly, practicing collective drumming to bond together and express suppressed emotions. But, like the followers of any social movement, those in expressive social movements may sometimes have extreme beliefs that can lead to tragedy, as in the 1996 case of Jessica Dubroff, the 7-year-old pilot who wanted to become the youngest person to fly across the continent—and whose parents encouraged her to follow her vision at all costs. Such advice resulted in her death, as well as the deaths of her father and her flight instructor.

Other social scientists who study social movements identify still more categories or develop subcategories that combine elements of those just presented. For example, in *The Pursuit of the Millennium* (1961), Norman Cohen analyzed what are known as *messianic* or *millenarian* movements, which are both revolutionary and expressive. A millenarian movement promises its followers total social change by miraculous means and envisions a perfect society of the future, "a new Paradise on earth, a world purged of suffering and sin, a Kingdom of the Saints" (p. xiii). Most millenarian movements are secular, but the term is derived from the Christian concept of the millennium, the return of Christ as the messiah who would save the world after a thousand years. Often millenarian movements begin as intensely dedicated groups with a charismatic leader (someone who seems to possess special powers) and a few devoted followers.

Examples of millenarian movements include Heaven's Gate, whose members committed suicide in the spring of 1997 in order to "enter a higher level of existence," and the People's Temple, whose 900 members committed mass suicide in Jonestown, Guyana, in 1978. In his insightful study of the events leading up to the People's Temple tragedy, sociologist John Hall (1987) demonstrates that popular interpretations of this shocking story are inadequate. The conclusion reached in most previous accounts is that the movement's leader, Jim Jones, was a deranged personality who acted as the Antichrist and led his misguided followers to their deaths. Such accounts, according to Hall, may be comforting expressions of moral outrage, but by exaggerating the influence of an individual personality they lead to wrong ideas about how such tragedies can occur.

In his analysis of the Jonestown tragedy, Hall shows that the members of the People's Temple were part of a cohesive social movement. Like the Heaven's Gate group, they felt alienated from the larger society and saw the Temple as their community and often as their family. It is only by understanding the strength of their attachments to one another and to their belief in Jones' vision of a better world to come that one can make sense of what appears to be a senseless, insane episode of collective behavior. Faced with what they perceived as imminent attacks on the religious community they were building in the South American jungle, some Temple members worked themselves into a frenzy in which they became convinced that by drinking their homemade poison of cyanide-laced Kool-Aid they were "stepping over" into a better world. Others may or may not have felt these emotions; we cannot ever know. Probably some people thought the occasion was a practice suicide. In any case, Hall's study shows that Jones planned the act of "collective martyrdom" but that the vast majority of the Temple members followed to the end because of their unshaken belief in the goals of their social movement.

Both the People's Temple and the Heaven's Gate mass suicides point out the immense influence a social movement organization can have on its members. This attraction may be felt by members of many kinds of movements, but of course not all social movement organizations demand the sacrifices required by millenarian movements like Heaven's Gate. In the next section, we look more closely at some explanations of the similarities and differences among various types of social movements and the collective behavior associated with them.

SOCIAL MOVEMENT THEORY

Early theories of collective behavior were derived from the notion that hysteria or contagious feelings like hatred or fear could spread through masses of people. One of the first social thinkers to develop a theory of collective behavior was Gustave LeBon. An aristocratic critic of the emerging industrial democracies, LeBon believed that those societies were producing "an era of crowds" in which agitators and despots are heroes, "the populace is sovereign, and the tide of barbarism mounts" (1947/1896, pp. 14, 207). LeBon thus developed the first modern theory of crowd behavior.

LeBon attributed the strikes and riots that are common in rapidly urbanizing societies to the fact that people are cut off from traditional village social institutions and jammed into cities, creating a mass of strangers. This view of early industrial societies was clearly exaggerated; research has shown that urban newcomers seek to form new social attachments, often drawing upon networks of people who came to the city from the same rural area or small town. They may join in mass demonstrations when faced with sudden unemployment, for example, but often it is difficult to mobilize them into action. In studying the behavior of masses of people and the social movements that arise out of their collective behavior, therefore, it is helpful to think of these behaviors as occurring along a continuum.

A Continuum of Collective Behavior

Collective behaviors and social movements can be placed along a continuum according to how much they change the societies in which they occur. At one end of the continuum are revolutions; at the other are fads or crazes. In the twentieth century, there were three revolutions that had a major impact on world history: the communist revolutions in Russia and China and the fascist revolution that brought Hitler to power in 1933. These can be placed at the high end of the continuum. At the low end are the numerous fads and crazes that may excite us for a while but usually do not bring about lasting change in the major structures of society (Turner, 1974). Some people become so completely involved in a fad or craze (as occurred, for example, among collectors of Beanie Babies) that they devote all their energies to the new activity.

THE CENTRAL QUESTION: A CRITICAL VIEW

How do social movements that enlist individual participation bring about social change?

"I don't go to demonstrations or meetings. What difference does it all make?"

"I don't care about voting. What difference can my vote make?"

Sociologists hear comments like these all the time. But those who study social movements and their effects are extremely critical of such attitudes. They know that when significant numbers of individuals join a social movement that demands change, the results can change the course of history.

In Eastern Europe in 1989, mass meetings and weeks of large-scale demonstrations in Czechoslovakia, East Germany, Hungary, Poland, and Romania brought about the fall of the communist regimes in those countries. Similar demonstrations and the mobilization of hundreds of thousands of protestors spurred the collapse of the Soviet Union in 1991 (Giugni, 1998). More recently, mass protests in Belgrade brought down the authoritarian regime of Slobodan Milosevic and opened up new opportunities for democratic institutions in Serbia.

People who participated in the U.S. civil rights movement were often accused of "going too fast," of "trying too hard" and upsetting people's peace and quiet. But the civil rights movement of the 1960s and 1970s is now credited with ending the most blatant forms of American apartheid. More recently, social movements on the right of the American political spectrum, such as those against abortion and gay rights and in favor of prayer in schools and federal support for private schools, were extremely influential in the 2000 presidential election.

Clearly, individual participation in social movements can bring about social change. But what features of the movements people join are most closely associated with successful outcomes? Sociologist William Gamson's research, which compares many different types of social movements active in the United States between 1800 and 1945, offers some answers:

- Groups espousing a single issue and demands related to that issue were more successful than groups with multiple-issue demands.

Can one man stop a line of armored tanks? This televised image of a man who stepped in front of a tank during the demonstrations for more democratic rule in Tiananmen Square in Beijing, China, in 1989 quickly became an image of individual courage standing against the armed power of the state.

- The use of violence and disruptive tactics was associated with success but also brought violence against the movements.
- Successful groups tended to be more bureaucratized, centralized, and unfactionalized (Gamson, 1990; Gamson & Wolfsfield, 1993).

This last finding conflicts with the common belief that social movements that are less bureaucratically organized are more effective. In fact, the data show that as social movements grow, they need formal organization and leaders chosen for their skill in building organizations. It might seem that because such organizations (e.g., labor unions, environmental or political organizations) are well organized, an individual member cannot make a difference. But this is exactly the attitude that social movements devoted to changing the status quo must combat. ■

Between the two extremes of revolutions and fads are social movements that capture the attention of masses of people and have varying effects on the societies in which they occur. The women's movement, the civil rights movement, and the environmental protection movement are examples of movements with broad membership in many nations and the continuing power to bring about social change. The gay liberation movement, the antiabortion movement, and the evangelical movement are examples of more narrowly focused movements that are concerned with a single issue or set of issues but also exert powerful pressure for change in the United States and other societies. Then there are movements that may involve large numbers of people for a brief time but have only limited effects on society as a whole. An example is the movement to aid famine victims in Africa. In the United States, concern about hunger in Ethiopia originated among a small number of people, but when rock music promoters enlisted the help of major artists and the music industry to produce the benefit record "We Are the World" and the Live Aid Concert, millions of Americans took notice of the situation, and by buying records and attending concerts they contributed to famine relief. These more transient social movements serve to raise consciousness about a particular social condition, but they do not usually bring about lasting change.

Theories of collective behavior and social movements often seek to explain the origins and effects of revolutionary and reform movements (Kimmel, 1990; McAdam, McCarthy, & Zald, 1988). In the remainder of this section, we discuss several such theories. We will examine the nature of revolutions and revolutionary movements and will attempt to explain why major revolutions and revolutionary social movements have arisen and what factors have enabled them to succeed. Note that these are macro-level theories: They explain revolutions and the associated social movements as symptoms of even larger-scale social change.

Theories of Revolution

Sociologists often distinguish between "long revolutions," or large-scale changes in the ecological relationships of humans to the earth and to one another (e.g., the rise of capitalism and the industrial revolution), and revolutions that are primarily social or political in nature (Braudel, 1984; Wolf, 1984a, 1984b). The course of world history is shaped by long revolutions, whereas individual societies are transformed by social revolutions like those that occurred in the United States, France, and Russia.

Political and Social Revolutions. Sociologist Theda Skocpol (1979) makes a further distinction between political and social revolutions. **Political revolutions** are transformations in the political structures and leadership of a society that are not accompanied by a full-scale rearrangement of the society's productive capacities, culture, and stratification system. **Social revolutions,** on the other hand, sweep away the old order. They not only change the institutions of government but also bring about basic changes in social stratification. Both political and social revolutions are brought about by revolutionary social movements as well as by external forces like colonialism.

The American Revolution, according to Skocpol's theory, was a political revolution. The minutemen who fought against British troops in the Battle of Concord were a political revolutionary group that did not at first seek to change American society in radical ways. On the other hand, the revolutions that destroyed the existing social order in Russia and China in the twentieth century were social revolutions. Their goal was to transform the class structure and institutions of their societies.

Skocpol's analysis further shows that the actual outcomes of social revolutions depend on the kinds of coalitions formed among revolutionary parties and larger social classes. The communist revolutionaries in Russia, for example, did not join forces with the peasants as the Chinese communist revolutionaries did. One consequence was that in the former Soviet Union the peasants were persecuted during the revolution; their land was taken from them and many were forced to live on collective farms. These disruptions of the society's agrarian base severely weakened the economy. In China, however, while there were many disruptions of rural life after the communists took power, food production continued to be carried out in villages and small landholdings. Peasant communist leaders replaced feudal landowners, but agricultural production did not suffer as it did in the former Soviet Union.

Why do revolutionary social movements arise, and what makes some of them content mainly with seizing power while others call for a complete reorganization of society? For most of the twentieth century, the answer was some version of Marxian conflict theory. This theory uses detailed knowledge of existing societies to predict the shape of future ones. According to Marx, the world would become capitalist; then capitalist markets would come under the control of monopolies; impoverished workers and colonial peoples would rebel in a generation of mass social movements and revolutions; and a new, classless society would be created in which the workers would own the means of production. (Marx's theory of class conflict leading to revolution is discussed in Chapters 5 and 6.)

Relative Deprivation. The idea that the increasing misery of the working class would lead workers to join revolutionary social movements was not Marx's only explanation of the causes of revolution. He also believed that under some conditions, "although the enjoyments of the workers have risen," their level of dissatisfaction could rise even faster owing to the much greater increase in the "enjoyments of the capitalists, which are inaccessible to the worker" (Marx & Engels, 1955, vol. 1, p. 94). This is a version of the theory known as **relative deprivation** (Stouffer, Suchman, DeVinney, Star, & Williams, 1949). According to this theory, the mere presence of deprivation (i.e., poverty or misery) does not by itself explain why people join revolutionary social movements. Instead, it is people's feelings of deprivation relative to others. We tend to measure our own well-being against that of others, and even if we are doing fairly well, if they are doing better we are likely to feel a sense of injustice and, sometimes, extreme anger. It is this feeling of deprivation relative to others that gives rise to revolutionary social movements.

Alexis de Tocqueville came to the same conclusion in his study of the causes and results of the French Revolution. He was struck by the fact that the revolution did not occur in the seventeenth century, when the economic conditions of the French people were in severe decline. Instead, it occurred in the eighteenth century, a period of rapid economic growth. Thus, Tocqueville concluded that "revolutions are not always brought about by a gradual decline from bad to worse. Nations that have endured patiently and almost unconsciously the most overwhelming oppression often burst into rebellion against the yoke the moment it begins to grow lighter" (1955/1856, p. 214). The Zapatista rebellion in the Mexican state of Chiapas is an example of a revolutionary movement that began among very poor Indian peasants but became a broader movement during a time of relative economic prosperity (Vilas, 1993). Now that the Mexican government itself has undergone a major change toward greater democracy, the Zapatistas face important decisions about whether to remain a revolutionary movement or make their peace with the government and possibly take a place in the leadership of Chiapas.

Charisma and Leadership

In addition to theories citing political, social, and economic influences, leadership also plays an important part in the success or failure of revolutionary social movements. The leaders of such movements are often said to have almost supernatural powers to inspire and motivate masses of followers. Max Weber (1968) called this ability **charisma.** A charismatic leader appears to possess extraordinary "gifts of the body and spirit" that mark him or her as specially chosen to lead. The voices that Joan of Arc heard, which convinced her to take up arms to save France, gave her the inspired commitment to her cause that is a central feature of charisma. Mahatma Gandhi, probably the twentieth century's greatest example of a charismatic leader, swayed the Indian masses with his almost mystical spirituality and courage. Similarly, the exceptional gifts of oratory, faith, and energy displayed by Martin Luther King, Jr., allowed him to emerge as the most influential leader of the American civil rights movement. Often a movement's followers themselves attribute special powers to their leader. This process may produce additional myths about the leader's powers and add to whatever personal magnetism the leader originally possessed. For example, Hitler was no doubt a fiery orator, but his followers' frenzied reception of many of his ideas contributed to the perception that he possessed exceptional charisma.

The charisma of the leader may eventually pose problems for the movement. Every social movement must incorporate the goals and gifts of its leaders into the structure of the movement and eventually into the institutions of society. But how can this be done without losing track of the movement's original purpose and values? Scholars who follow Weber's lead in this area of research have termed this problem the *institutionalization of charisma* (Shils, 1970). Weber recognized that the more successful a movement becomes in taking power and assuming authority, the more difficult it is to retain the zeal and motivation of its charismatic founders. Once the leaders have power and can obtain special privileges for themselves, they tend to resist continued efforts to do away with inequalities of wealth and power. Perhaps the most instructive example of this problem is the institutionalization of the ideals of Marx and Lenin in the organization of the Bolshevik party in Russia after the revolution of 1917.

"Workers of the world, unite," Marx urged in *The Communist Manifesto* (Marx & Engels, 1969/1848);

"you have nothing to lose but your chains." The vision of a society in which poverty, social injustice, repression, war, and all the evils of capitalism would be eliminated formed the core of Marxian socialism. The founders of the Bolshevik party realized, however, that revolutions are not made with slogans and visions alone (Schumpeter, 1950). The unification of the workers required an organization that not only would carry forward the revolution but also would take the lead in mobilizing the Russian masses to work for the ideals of socialism. The charismatic leader who built this organization was Vladimir Ilyich Lenin.

Under the leadership of Lenin (and far more ruthlessly under his successor Josef Stalin), any Bolshevik who challenged party discipline was exiled or killed. Because the goals of the Bolsheviks were so radical and the need for reform so pressing, Lenin and Stalin argued that the party had to exert total control over its members in order to prevent spontaneous protests that might threaten the party's goals. By the same logic, other parties had to be eliminated and dissenters imprisoned. Such efforts by an elite to exert control over all forms of organizational life in a society are known as *totalitarianism.* All social movements whose leaders believe that the goals they seek justify whatever means are used to achieve them risk becoming totalitarian because they sacrifice the rule of law and the ideal of democratic process to gain power (Harrington, 1987; Howe, 1983).

Unfortunately, threats to the rule of law and democratic processes do not stem only from totalitarian movements or power-mad leaders with great charisma like Adolf Hitler. The rise of militias in the United States, violent anti-immigrant movements in Europe, and the excesses of the Nixon administration, which led to the Watergate scandal, are all examples of how movements can arise that threaten to reverse any progress that has been made toward peaceful self-government and rule by laws rather than by powerful individuals. This is the kind of problem that leads people to claim that "history repeats itself," an issue that is often discussed by social scientists who devote themselves to the study of social change.

THEORETICAL PERSPECTIVES ON SOCIAL CHANGE

Sociologists have often attempted to develop models of social change that span many generations and predict the future of whole societies or civilizations. Of course, none of these theories can be tested using data from actual experience. As Robert Nisbet (1969) has observed, "None of us has ever seen a civilization die, and it is unimaginable, short of cosmic disaster or thermonuclear holocaust, that anyone ever will" (p. 3). Instead, he continues,

> We see migrations and wars, dynasties toppled, governments overthrown, economic systems made affluent or poor; revolutions in power, privilege, and wealth. We see human beings born, mating, child-rearing, working, worshipping, playing, educating, writing, philosophizing, governing. We see generation succeeding generation, each new one accepting, modifying, rejecting in different proportions the works of preceding generations. We see, depending upon our moral or esthetic disposition, good and evil, greatness and meanness, tragedy, comedy, and bathos, nobility and baseness, success and failure. (p. 3)

Nisbet's point is that we can trace trends in all of these areas, but it is extremely difficult to develop theories that can explain them all and, more important, predict the decline of existing societies and civilizations or the rise of new ones. Even when we believe we are witnessing the birth of a new society out of the chaos of revolution or war, it takes many generations to distinguish what is truly new, in terms of culture and social structure, from what has been carried over from the past.

Popular theories of social change include ideas of a "golden age" when human life was kinder and gentler and people lived together in greater harmony. Along with these ideas of a beautiful past or golden age often comes the notion that history is cyclical and that what is past or changed will someday return. Among social scientists, however, the two most powerful theoretical perspectives are the more familiar ones of conflict theory and functionalism.

Conflict Perspectives

Cyclical theories of social change, with their cycles of rise and fall and their brief golden ages, may seem to confirm the common notion that history repeats itself. This notion is erroneous, however. Societies may experience similar events, such as wars or revolutions, at different times in their history, but the actual populations and issues involved are never exactly the same. More important, cyclical theories fail to deal with changes in social institutions or class structures. This is where conflict models of social change are most useful. They argue that conflict among groups with different amounts of power produces social change, which leads to a new system of social stratification, which in turn leads to further conflict and further change.

"The history of all hitherto existing societies," wrote Karl Marx and Friedrich Engels in *The Communist Manifesto* (1969/1848), "is the history of class struggles." As we saw in Chapter 6, social classes are defined by how people make their living or gain their wealth. Marx argued that struggles between classes (e.g., workers versus managers and company owners) are the source of social change in every period of history. In any society, the main conflicting classes will be the exploiters and the exploited: those who control the means of production and those whose labor power is necessary for making those means of production actually produce. The exploited workers could become a revolutionary class—that is, one that could bring about an entirely new social order. But this can occur only when changes in the means of production—new technologies like the factory system, for example—make older classes obsolete.

We know that revolutionary class conflict has not occurred in many capitalist societies, nor did the former communist societies of Eastern Europe or the Soviet Union succeed in eliminating worker exploitation. Yet as we look around the world at the struggles between the haves and the have-nots, the rulers and the ruled, the rich nations and the poorer nations, we cannot help but apply Marxian categories and test Marx's theory of social change over and over again.

Some modern conflict theorists depart from the Marxian view of social change, finding conflict among many different kinds of groups and in every social institution. For Ralf Dahrendorf (1959), this conflict produces social change at all times, but the change is not always revolutionary. We cannot change our laws, our bureaucracies, or even our families, for example, without first experiencing conflicts among various group and individual interests. (An example of such a conflict is the half-humorous, half-serious debate that occurred in 1992 between Vice President Dan Quayle and television character Murphy Brown over what is meant by "family values.") In most cases, however, it is only when the deprivation experienced by whole classes or status groups is extreme that conflict is likely to be violent and to produce the unrest that could end in revolutionary social change.

Functionalist Perspectives

From a functionalist perspective, social change occurs as a result of population growth, changes in technology, social inequality, and efforts by different groups to meet their needs in a world of scarce resources. There is no prediction of cyclical rise and fall or of unilinear changes like those we find in classical Marxian theory. Instead, the functionalist model sees change as occurring on so many fronts that it seems incredible that society can exist at all.

One of the dominant figures in functionalist social theory, Talcott Parsons (1960), developed a *homeostatic* model of society. As change occurs, said Parsons, a society's institutions attempt to restore it to something approaching equilibrium. Conflict is minimized through the emergence of legitimate governing institutions; decisions are made about who governs and with what form and degree of authority. Adjustments are also made in economic institutions: New occupational roles develop; old roles decline; wages and status rankings such as occupational prestige explain who gets what rewards. Cultural institutions, schools, the arts, the media, and religious institutions maintain the shared values that support our feeling that our government is legitimate, that a certain amount of inequality is required to maintain individual initiative, and that opportunities are distributed as well as can be expected.

Parsons and other functionalist theorists do not contend that efforts to adapt to change, to create an integrated, well-functioning social system, always work. The integrated functioning of social institutions can be disrupted, sometimes quite severely, when some institutions experience rapid change while others are slow to adapt. Technological innovations in health care, for instance, make possible longer lives for some individuals at great expense, but other economic and political institutions have been slow to develop norms for distributing these costly benefits among the members of society and finding ways of paying for them.

Applying the Perspectives

As we have seen throughout this book, the functionalist and conflict perspectives are among the basic conceptual tools of sociology. They are most useful at the global or macro level of analysis. But they can also be used, along with the interactionist perspective, to explain social change at the micro and middle levels of social analysis. Sociologist Daniel Chirot (1994) successfully applies all three of these perspectives in his analysis of social change.

How is it, Chirot asks, that Europe, which was a backward region in comparison to the Middle East or North Africa in the first millennium of the Current

Era, became the dominant region of the world for a long period of the second millennium? Chirot applies the functionalist perspective in showing that technological change, especially the invention of the plow and other agricultural innovations, worked to increase productivity but decrease political centralization. No one state was strong enough to entirely dominate or suppress local applications of technology and the profit-seeking activity of particular nations or trading companies. In China, by contrast, the state was strong enough to dominate, and this application of the conflict perspective illustrates how different the European situation was. The interactionist perspective is applied at a more micro level: In a context of growing business activity and competition, interactions among people based on rationality, calculation, and reliance on empirical evidence became strong features of the cultures of most European nations.

Chirot's analysis suggests that technological change, especially major events like the development of steam power or, later, nuclear power, leads to new forms of social organization and places new stresses on existing cultures to adapt to change. Chirot does not see change as occurring in one direction necessarily. Thus, he does not foresee steady progress toward democracy everywhere in the world. Poor regions of the world, which cannot afford new technologies to grow their economies or keep their populations healthy, are often centers of autocratic rule, as we have seen in the case of Congo in central Africa. Nations that depend on one source of wealth, especially oil, often fail to develop large middle classes that make demands on leaders for more democratic rule, and as a consequence oil nations often remain highly autocratic as well.

As we apply all the sociological perspectives to understanding social change, we see that in Western societies, including the United States, social change often involves people in deeply troubling struggles to understand and cope with new situations. Consider only two areas of life in the United States—gender relations and race relations—that are subjects of profound controversy because of the pace of change over the past century.

Gender Roles and the Family. Clearly, one of the most significant social changes in Western societies in the past quarter century has been the changing definition of women's roles. Although the entry of women into the labor force is only one indicator of this development, it is an important aspect. It should be noted, of course, that there have always been subgroups in the population, especially African Americans and immigrants, for whom women's wages were necessary to the family's survival. And even among women who "kept house" toward the turn of the twentieth century, some 20% took in lodgers and earned cash in this fashion (Aldous, 1982; Modell & Hareven, 1973). Today, however, over 70% of all married women with children are at work in offices, factories, and other workplaces, compared with only about 4% in 1890 (*Statistical Abstract,* 2000).

For both women and men, these changes often produce feelings of guilt and stress. One writer on the subject (Fallows, 1985) has urged young mothers to reject career goals so that they can raise their children themselves. On the basis of observations of day-care centers, she has concluded that day care is a poorly developed institution that cannot substitute for maternal care. Arguments of this nature are common. But other studies show that high-quality day care does not impede a child's development. Moreover, advocates of gender equality question the assumption that child rearing should remain the primary role of women. This assumption has the effect of depriving women of opportunities to contribute their skills and talents to society. If day care is inadequate, a more practical response would be to improve it through public funding, education of child-care workers, and the like, rather than urging women to return to their traditional role as homemakers (Bergmann & Hartmann, 1995; Browne Miller, 1990; England, 1986).

Race Relations in a Postindustrial Society. Another significant area of social change is race relations. When we look at the bitter racial strife occurring in Rwanda or Bosnia, we often congratulate ourselves on the progress our society has made toward racial and ethnic equality for African Americans, Hispanics, and other minority groups. Racial discrimination in the United States, which until fairly recently was supported by laws in many parts of the nation and by informal norms elsewhere, has decreased a great deal as a result of the civil rights movement of the 1960s. Since that stormy decade, African Americans have made gains in all of our society's major institutions. Voting rights, greater access to education and jobs, achievement in sports, and full civil rights for African Americans are often taken for granted. Yet, as the 1992 Los Angeles riots clearly demonstrated, much needs to be done to address racial inequality. In the words of philosopher and sociologist Cornel West (1992):

> The verdict that sparked the incidents in Los Angeles was perceived to be wrong by the vast majority of Americans. But whites have often failed to

> acknowledge the widespread mistreatment of black people, especially black men, by law-enforcement agencies, which helped ignite the spark. The Rodney King verdict was merely the occasion for deep-seated rage to come to the surface. This rage is fed by the "silent" depression ravaging the country—in which real weekly wages of all American workers since 1973 declined nearly 20%, while at the same time wealth has been upwardly distributed.
>
> The exodus of stable industrial jobs from urban centers to cheaper labor markets here and abroad, housing policies that have created "chocolate cities and vanilla suburbs," white fear of black crime and the urban influx of poor Spanish-speaking and Asian immigrants—all have helped erode the tax base of American cities just as the federal government has cut its supports and programs. The result is unemployment, hunger, homelessness, and sickness for millions. (p. 24)

It is worth noting here, however, that West's critical observations signal some of the ways in which inequalities of wealth and opportunity overlap with inequalities of race and tend to exacerbate the problem of race relations in American society.

Many other areas of daily life in which rapid change is occurring could be analyzed for their impact on aspects of social life. Religion, for example, is giving rise to conflicts in many societies as populations are torn between science and rationality on the one hand and the desire for orthodoxy and moral certainty on the other. Communications and the mass media comprise another cultural institution that is producing vast changes in the way people think and the kinds of information they are exposed to, changes that raise the issue of freedom of expression versus protection of innocent people, especially children, from exposure to values or experiences that some consider undesirable. These examples—and many others that we could mention—provide further evidence of the importance of using the concepts and methods of sociology to study social change.

TECHNOLOGY AND SOCIAL CHANGE

In coming decades, the populations of the United States and many other urban industrial nations will continue to age. By 2025, about a third of the U.S. population will be over age 50, and the proportion over age 80 will be growing rapidly. Technological innovations like gene therapy, organ transplants, and new drugs will be used to treat the diseases of the elderly more effectively, making their lives far more active and productive. But the same technologies that will be keeping people alive longer will also be adding more frail elderly people to the population. And this will increase the burden on younger people who are still active in the workforce.

The spread of automobiles, electrical appliances, and television (which increases people's desire for things they do not yet possess) will bring major changes in people's ways of life. They risk also increasing the quantity of so-called greenhouse gases in the earth's atmosphere, causing global warming and massive climate change.

Advancing knowledge of nuclear energy technologies around the world makes it ever more likely that a rogue state or a terrorist group will gain access to weapons of mass destruction. As a consequence, the promise of greater stability due to the end of the Cold War and the fall of the Soviet empire has not been fulfilled.

These examples suggest some important points to keep in mind about technology and social change. First, all technologies can be used both for good and for evil. This is just as true for nuclear energy as it is for the internal combustion engine, which can drive a tank as well as a family car. Second, technologies often produce lags between the use of a new device or process and society's ability to control its use for constructive purposes. Where societies must cooperate to control the effects of technological change, as in the case of global warming, the challenge created by gaps between uses and their effects is even greater and perhaps more threatening to human life. A third sociological insight is that in our own era the pace of technological change is so rapid that it is almost impossible to predict how social organization can adapt to its consequences. This often leads people in the societies that are the most technologically innovative, especially the United States, to believe that technology will solve all problems somehow. Critical sociologists call this the fallacy of the "techno fix." They warn against the dangerous ideas that any problem, from global warming to world hunger, will ultimately be fixed by technological means rather than by new forms of social organization or behavior (Muller, 2000; Orr, 1998).

SOCIOLOGY AND SOCIAL JUSTICE

For decades after achieving independence in the early 1960s, Ivory Coast was considered one of the more stable and economically attractive West African nations. In the past decade or so, its neighbors, Sierra Leone and Liberia, have suffered endless civil war between competing and highly corrupt and violent military dictators. Refugees escaping the violence have added to the difficulty of maintaining political and economic stability in Ivory Coast. Worse yet, like many nations created during the colonial era by foreign conquering powers, Ivory Coast includes people of quite different cultures and religions who do not necessarily trust one another. Conflict between Muslims in the north and Christian ethnic groups in the south could result in severe political instability and even civil war.

Throughout the postcolonial world, there are grave doubts about the viability of entire nation-states based on the weakening of central governments, a decline in the rule of law, and a decrease in the possibility of legitimate democratic leadership. At the level of cities and villages, this instability is extremely dangerous. Adjame, a community in Abidjan, Ivory Coast's capital, was once an Ebrié village like the one I lived in decades ago. Adjame has become an urban slum where people are overcrowded, housing is extremely rickety, and disease, including AIDS, has become a major problem. Thus, the decline in the stability of nation-states, in Africa and elsewhere, endangers the future of villages and cities within those states as well. ■

Ebrié villagers: children playing in a dugout canoe; women carrying out domestic chores in the lagoon behind the village. At right, a Harrist church in an Ebrié village. The Harrist religion combines major themes of Christianity with African dance and art.

SUMMARY

Social change refers to variations over time in the ecological ordering of populations and communities, in patterns of roles and social interactions, in the structure and functioning of institutions, and in the cultures of societies. Such changes can result from forces building within societies *(endogenous forces)* as well as from forces exerted from the outside *(exogenous forces).*

A major force producing social change is war. The primary ecological effects of war are casualties and conquest. War also results in large-scale shifts in population and rapid acceleration of economic change. It can affect a society's culture and may drastically change a society's institutions.

A second major source of social change is *modernization,* whose consequences include the application of scientific knowledge, commercial production of agricultural goods, the use of power-driven machines, and urbanization. The spread of AIDS in developing nations represents a spectacular failure of modernization.

Developing nations are undergoing a set of transformations whose effect is to increase the productivity of their people, their health, their literacy, and their ability to participate in political decision making. Wallerstein's world system theory divides the world into *core states, semiperipheral areas,* and *peripheral areas.*

The term *collective behavior* refers to a continuum of unusual or nonroutine behaviors that are engaged in by large numbers of people. At one end of the continuum is the spontaneous behavior of people reacting to situations they perceive as uncertain, threatening, or extremely attractive. At the other end are structured events like rallies, marches, or festivals, which also involve large numbers of people in nonroutine behaviors but are organized by leaders and have specific goals. The set of organizations that plan such events is a social movement.

Social movements have been classified into four general types based on the goals they seek to achieve. Revolutionary movements aim to overthrow existing stratification systems and social institutions; reformist movements seek partial changes in some institutions and values; conservative movements attempt to uphold the existing values and institutions of society; and reactionary movements seek to return to the institutions and values of the past. In addition, there are expressive social movements, or movements devoted to the expression of personal beliefs and feelings, and millenarian movements, which are both revolutionary and expressive.

Sociologists often distinguish between "long revolutions," or large-scale changes in the ecological relationships of humans to the earth and to one another, and revolutions that are primarily social or political. *Political revolutions* are transformations in the political structures and leadership of a society that are not accompanied by a full-scale rearrangement of the society's productive capacities, culture, and stratification system. *Social revolutions* not only change the institutions of government but also bring about basic changes in social stratification.

According to Marx, revolutions would occur as a result of the spread of capitalism: Impoverished workers and colonial peoples would rebel against the capitalists and create a new classless society. Also, Marx and Tocqueville pointed to the role of *relative deprivation,* noting that the feeling of deprivation relative to others—not the presence of deprivation itself—may give rise to revolutionary social movements.

Successful leaders of social movements are often said to have almost supernatural powers to inspire and motivate their followers. Max Weber called this ability *charisma.* Over time, however, those leaders' goals must be incorporated into the structure of the movement, a process that is referred to as the institutionalization of charisma. However, the more successful the movement, the more difficult it is to maintain the zeal of its founders. In extreme cases, the process can end in totalitarianism, or efforts by an elite to control all forms of organizational life in the society.

Conflict theorists argue that conflict among groups with different amounts of power produces social change, which leads to a new system of social stratification, which in turn leads to further conflict and further change. From a functionalist perspective, social change occurs as a result of population growth, changes in technology, inequalities among classes and status groups, and efforts by different groups to meet their needs in a world of scarce resources.

The entry of large numbers of women into the labor force has upset the traditional norms of family life, and changes in the structure of the economy have severely affected the employment prospects of African Americans and Hispanics. It is clear that social changes of many kinds will continue to exert powerful forces on all societies.

GLOSSARY

Social change: Variations over time in the ecological ordering of populations and communities, in patterns of roles and social interactions, in the structure and functioning of institutions, and in the cultures of societies. **(403)**

Endogenous force: Pressure for social change that builds within a society. **(403)**

Exogenous force: Pressure for social change that is exerted from outside a society. **(403)**

Modernization: A term used to describe the changes societies and individuals experience as a result of industrialization, urbanization, and the development of nation-states. **(406)**

Developing nation: A nation that is undergoing a set of transformations in its institutions whose effect is to increase the productivity of its people, their health, their literacy, and their ability to participate in political decision making. **(408)**

Core state: A technologically advanced nation that has a dominant position in the world economy. **(411)**

Semiperipheral area: A state or region in which industry and financial institutions are developed to some extent but that remains dependent on capital and technology provided by other states. **(411)**

Peripheral area: A region that supplies basic resources and labor power to more advanced states. **(411)**

Collective behavior: Nonroutine behavior that is engaged in by large numbers of people responding to a common stimulus. **(412)**

Social movement: Organized collective behavior aimed at changing or reforming social institutions or the social order itself. **(412)**

Crowd: A large number of people who are gathered together in close proximity to one another. **(414)**

Mass: A large number of people who are all oriented toward a set of shared symbols or social objects. **(414)**

Political revolution: A set of changes in the political structures and leadership of a society. **(419)**

Social revolution: A complete transformation of the social order, including the institutions of government and the system of stratification. **(419)**

Relative deprivation: Deprivation as determined by comparison with others rather than by some objective measure. **(420)**

Charisma: A special quality or "gift" that motivates people to follow a particular leader. **(420)**

QUESTIONS FOR THOUGHT AND DISCUSSION

1. Collective behavior—for example, riots, wildcat strikes, or spontaneous demonstrations—often gives rise to social movements. Can you give some examples to support this observation? What are some key differences between a social movement and an episode of collective behavior?

2. In the poorer parts of the contemporary world, delays or difficulties in economic development often seem to be related to failures of political institutions. A good example of this observation would be contemporary Nigeria. Can you explain this observation further?

DIGGING DEEPER

Books

How Societies Change (Daniel Chirot; Pine Forge Press, 1994). A review of theories of change by one of the leading sociologists of social change; reflects on the many meanings of the failure of Soviet-style regimes.

Exploring Social Change, 3rd ed. (Charles L. Harper; Prentice Hall, 1998). A comprehensive text on social change, with critical chapters on modernization, postmodernism, development, and world system theory.

The Wretched of the Earth (Frantz Fanon; Grove Press, 1968). A classic analysis of the psychology of colonial peoples and their yearning for independence even in the face of violence and bloodshed.

Technology, the Economy, and Society: The American Experience (Joel Colton & Stuart Bruchey, eds.; Columbia University Press, 1987). A collection of articles assessing the role of technology in transforming everything from agriculture, industry, and the structure of corporations to politics, law, the military, education, and religion.

Frontiers in Social Movement Theory (Aldon Morris & Carol McClurg Mueller, eds.; Yale, 1992). A collection of recent studies of social movements that includes examples from most new areas of research in this rapidly changing field.

Social Movements in an Organizational Society (Meyer N. Zald & John McCarthy, eds.; Transaction, 1987). A thorough analysis of competing explanations of the evolution of social movements.

Women's Movements in America: Their Successes, Disappointments, and Aspirations (Rita J. Simon & Gloria Danziger, eds.; Praeger, 1991). Assesses the relative success of a major social movement in securing positive change for women in the United States; contains excellent bibliographical material and references to other works.

Journals

Politics and Society. A critical journal devoted to theories and research on social change.

Economic Development and Cultural Change. A journal of economic and sociological writing about the interrelationships between culture and economics and their influence on social change in developing nations.

World Development Report. Published annually by the World Bank. A valuable source of comparative data on social and economic development for all the world's nations and territories.

Other Sources

General Social Survey: Cumulative Codebook. Published by the National Opinion Research Center (NORC), University of Chicago. An annual survey that provides measures of the demographic characteristics and behaviors of a large sample of Americans. Available through the data archives of most colleges and universities, along with other research reports published by the NORC.

Statistical Yearbook. Published annually by the United Nations, Department of International Economic and Social Affairs. Presents general socioeconomic statistics.

EXPLORING SOCIOLOGY ON THE INTERNET

National Association of Community Action Agencies
www.nacaa.org/

A good place to look for what people at the grassroots level are doing to bring about positive social change.

American Sociological Association, Section on Social Movements
www.ssc.wisc.edu/

Provides links to Web sites that cover activism and offers an almost unlimited introduction to the many social movements that use the Web to advertise and organize.

Social Science Research Council
www.ssrc.org/

A professional social science organization that has sponsored research on social change for many decades.

Sage Publications
www.sagepub.com/

An online catalog of books and research monographs dealing with all aspects of social change, including works by sociologists in many nations.

Project on Social Research for Social Change, Cornell University
www.cornell.edu

Provides information about grassroots social movements, with links to many related sites.

The AFL-CIO Home Page
www.aflcio.org

An example of an older social movement organization that is adapting to new conditions and technologies.

U.S. Census Bureau: The Official Statistics
www.census.gov

The official site for the U.S. Census Bureau continually updates labor and social statistics. The current U.S. population count and current economic indicators are popular features of the site.

GLOSSARY

Accommodation The process by which a smaller, less-powerful society is able to preserve the major features of its culture even after prolonged contact with a larger, stronger culture.

Acculturation The process by which the members of a civilization incorporate norms and values from other cultures into their own.

Achieved status A position or rank that is earned through the efforts of the individual.

Age cohort A set of people of about the same age who are passing through the life course together.

Age grade A set of statuses and roles based on age.

Ageism An ideology that justifies prejudice and discrimination based on age.

Agencies of socialization The groups of people, along with the interactions that occur within those groups, that influence a person's social development.

Agents of socialization Individuals who socialize others.

Animism A form of religion in which all forms of life and all aspects of the earth are inhabited by gods or supernatural powers.

Anomie A state of normlessness.

Anticipatory socialization Socialization that prepares an individual for a role he or she is likely to assume later in life.

Ascribed status A position or rank that is assigned to an individual at birth and cannot be changed.

Assimilation The process by which culturally distinct groups in a larger civilization adopt the norms, values, and language of the host civilization and are able to gain equal statuses in its groups and institutions.

Authority Power that is considered legitimate both by those who exercise it and by those who are affected by it.

Behaviorism A theory that states that all behavior is learned and that this learning occurs through the process known as conditioning.

Bisexuality Sexual orientation toward either sex.

Bureaucracy A formal organization characterized by a clearly defined hierarchy with a commitment to rules, efficiency, and impersonality.

Capitalism A system for organizing the production of goods and services that is based on markets, private property, and the business firm or company.

Caste A social stratum into which people are born and in which they remain for life.

Charisma A special quality or "gift" that motivates people to follow a particular leader.

Charismatic authority Authority that comes to an individual through a personal calling, often claimed to be inspired by supernatural powers, and is legitimated by people's beliefs that the leader does indeed have God-given powers.

Church A religious organization that has strong ties to the larger society.

Citizenship The status of membership in a nation-state.

Civilization A cultural complex formed by the identical major cultural features of a number of societies.

Civil religion A collection of beliefs and rituals that exist outside religious institutions.

Class A social stratum that is defined primarily by economic criteria such as occupation, income, and wealth.

Class consciousness A group's shared subjective awareness of its objective situation as a class.

Closed society A society in which social mobility does not exist.

Closed stratification system A stratification system in which there are rigid boundaries between social strata.

Collective behavior Nonroutine behavior that is engaged in by large numbers of people responding to a common stimulus.

Community A set of primary and secondary groups in which the individual carries out important life functions.

Conditioning The shaping of behavior through reward and punishment.

Confidentiality The promise that the information provided to a researcher by a respondent will not appear in any way that can be traced to that respondent.

Conflict theory A sociological perspective that emphasizes the role of conflict and power in society.

Control group In an experiment, the subjects who do not experience a change in the independent variable.

Controlled experiment An experimental situation in which the researcher manipulates an independent variable in order to observe and measure changes in a dependent variable.

Core state A technologically advanced nation that has a dominant position in the world economy.

Correlation A specific relationship between two variables.

Counterculture A subculture that challenges the accepted norms and values of the larger society and establishes an alternative lifestyle.

Crime An act or omission of an act that is prohibited by law.

Crowd A large number of people who are gathered together in close proximity to one another.

Crude birthrate The number of births occurring during a year in a given population, divided by the midyear population.

Crude death rate The number of deaths occurring during a year in a given population, divided by the midyear population.

Cult A new religion.

Cultural evolution The process by which successful cultural adaptations are passed down from one generation to the next.

Cultural relativity The recognition that all cultures develop their own ways of dealing with the specific demands of their environments.

Culture All the modes of thought, behavior, and production that are handed down from one generation to the next by means of communicative interaction rather than by genetic transmission.

***De facto* segregation** Segregation created and maintained by unwritten norms.

Deference The respect and esteem shown to an individual.

***De jure* segregation** Segregation created by formal legal sanctions that prohibit certain groups from interacting with others or place limits on such interactions.

Demagogue A leader who uses personal charisma and political symbols to manipulate public opinion.

Demeanor The way in which individuals present themselves to others through body language, dress, speech, and manners.

Democracy A political system in which all citizens have the right to participate in public decision making.

Democratic socialism An economic philosophy based on the belief that private property may exist at the same time that large corporations are owned by the state and run for the benefit of all citizens.

Demographic transition A set of major changes in birth and death rates that has occurred most completely in urban industrial nations in the past 200 years.

Denomination A religious organization that is on good terms with the institution from which it developed but must compete with other denominations for members.

Dependent variable The variable that a hypothesis seeks to explain.

Developing nation A nation that is undergoing a set of transformations in its institutions whose effect is to increase the productivity of its people, their health, their literacy, and their ability to participate in political decision making.

Deviance Behavior that violates the norms of a particular society.

Differential association A theory that explains deviance as a learned behavior that is determined by the extent of a person's association with individuals who engage in such behavior.

Differentiation The processes whereby sets of social activities performed by one social institution are divided among different institutions.

Discrimination Behavior that treats people unfairly on the basis of their group membership.

Downward mobility Movement by an individual or group to a lower social stratum.

Dramaturgical approach An approach to research on interaction in groups that is based on the recognition that much social interaction depends on the desire to impress those who may be watching.

Dyad A group consisting of two people.

Education The process by which a society transmits knowledge, values, norms, and ideologies and in so doing prepares young people for adult roles and adults for new roles.

Educational achievement Mastery of basic reading, writing, and computational skills.

Educational attainment The number of years of school an individual has completed.

Ego According to Freud, the part of the human personality that is the individual's conception of himself or herself in relation to others.

Endogamy A norm specifying that a person brought up in a particular culture may marry within the cultural group.

Endogenous force Pressure for social change that builds within a society.

Equality of opportunity Equal opportunity to achieve desired levels of material well-being and prestige.

Equality of result Equality in the actual outcomes of people's attempts to improve their material well-being and prestige.

Ethnic group A population that has a sense of group identity based on shared ancestry and distinctive cultural patterns.

Ethnocentrism The tendency to judge other cultures as inferior to one's own.

Ethnomethodology The study of the underlying rules of behavior that guide group interaction.

Exogamy A norm specifying that a person brought up in a particular culture may marry outside the cultural group.

Exogenous force Pressure for social change that is exerted from outside a society.

Experimental group In an experiment, the subjects who are exposed to a change in the independent variable.

Expulsion The forcible removal of one population from a territory claimed by another population.

Extended family An individual's nuclear family plus the nuclear families of his or her blood relatives.

Family A group of people related by blood, marriage, or adoption.

Family of orientation The nuclear family in which a person is born and raised.

Family of procreation The nuclear family a person forms through marriage or cohabitation.

Feral child A child reared outside human society.

Field experiment An experimental situation in which the researcher observes and studies subjects in their natural setting.

Folkways Weakly sanctioned norms.

Formal organization A group that has an explicit, often written, set of norms, statuses, and roles that specify each member's relationships to the others and the conditions under which those relationships hold.

Frequency distribution A classification of data that describes how many observations fall within each category of a variable.

Functionalism A sociological perspective that focuses on the ways in which a complex pattern of social structures and arrangements contributes to social order.

Gemeinschaft A term used to refer to the close, personal relationships of small groups and communities.

Gender The culturally defined ways of acting as a male or a female that become part of an individual's personal sense of self.

Gender role A set of behaviors considered appropriate for an individual of a particular gender.

Gender socialization The specific aspects of socialization whereby people learn the behaviors and attitudes considered appropriate for males and females in their culture.

Generalized other A person's internalized conception of the expectations and attitudes held by society.

Genocide The intentional extermination of one population by a more dominant population.

Gesellschaft A term used to refer to the well-organized but impersonal relationships among the members of modern societies.

Group A collection of people who interact with one another on the basis of shared expectations regarding the others' behavior.

Hermaphrodite A person whose primary sexual organs have features of both male and female organs, making it difficult to categorize the individual as either male or female.

Heterosexuality Sexual orientation toward the opposite sex.

Homogamy The tendency to marry a person who is similar to oneself in social background.

Homosexuality Sexual orientation toward the same sex.

Horticultural society A society whose primary means of subsistence is raising crops, which it plants and cultivates, often developing an extensive system for watering the crops.

Human ecology A sociological perspective that emphasizes the relationships among social order, social disorganization, and the distribution of populations in time and space.

Hypersegregation Segregation of a racially or ethnically distinct population that results in segregation indexes of over 80% and profoundly affects the segregated population's life chances.

Hypothesis A statement that specifies a relationship between two or more variables that can be tested through empirical observation.

Id According to Freud, the part of the human personality from which all innate drives arise.

Ideas The ways of thinking that organize human consciousness.

Identification The social process whereby an individual chooses role models and attempts to imitate their behavior.

Ideologies Systems of values and norms that the members of a society are expected to believe in and act upon without question.

Impression management The strategies one uses to "set a stage" for one's own purposes.

Independent variable A variable that the researcher believes causes a change in another variable (i.e., the dependent variable).

Informal organization A group whose norms and statuses are generally agreed upon but are not set down in writing.

Informed consent The right of respondents to be informed of the purpose for which the information they supply will be used and to judge the degree of personal risk involved in answering questions, even when an assurance of confidentiality has been given.

In-group A social group to which an individual has a feeling of allegiance; usually, but not always, a primary group.

Institution A more or less stable structure of statuses and roles devoted to meeting the basic needs of people in a society.

Institutional discrimination The systematic exclusion of people from equal participation in a particular institution because of their group membership.

Interactionism A sociological perspective that views social order and social change as resulting from all the repeated interactions among individuals and groups.

Interest group An organization that attempts to influence elected and appointed officials regarding a specific issue or set of issues.

Intergenerational mobility A change in the social class of family members from one generation to the next.

Intragenerational mobility A change in the social class of an individual within his or her own lifetime.

Jim Crow The system of formal and informal segregation that existed in the United States from the late 1860s to the early 1970s.

Kinship The role relations among people who consider themselves to be related by blood, marriage, or adoption.

Labeling A theory that explains deviance as a societal reaction that brands or labels as deviant people who engage in certain behaviors.

Laissez-faire capitalism An economic philosophy based on the belief that the wealth of a nation can be measured by its capacity to produce goods and services (in other words, its resources of land, labor, and machinery) and that these can be maximized by free trade.

Laws Norms that are written by specialists, collected in codes or manuals of behavior, and interpreted and applied by other specialists.

Legal authority Authority that is legitimated by people's belief in the supremacy of the law; obedience is owed not to a person but to a set of impersonal principles.

Legitimacy The ability of a society to engender and maintain the belief that the existing political institutions are the most appropriate for that society.

Life chances The opportunities that an individual will have or be denied throughout life as a result of his or her social-class position.

Life course A pathway along an age-differentiated, socially created sequence of transitions.

Life expectancy The average number of years a member of a given population can expect to live beyond his or her present age.

Lifetime negative experiences Experiences that cause long-term stress, such as the death of a child or spouse.

Linguistic-relativity hypothesis The belief that language determines the possibilities for thought and action in any given culture.

Lobbying The process whereby interest groups seek to persuade legislators to vote in their favor on particular bills.

Macro-level sociology An approach to the study of society that focuses on the major structures and institutions of society.

Market An economic institution that regulates exchange behavior through the establishment of different values for particular goods and services.

Mass A large number of people who are all oriented toward a set of shared symbols or social objects.

Master status A status that takes precedence over all of an individual's other statuses.

Material culture Patterns of possessing and using the products of culture.

Megalopolis A complex of cities distributed along a major axis of traffic and communication.

Mercantilism An economic philosophy based on the belief that the wealth of a nation can be measured by its holdings of gold or other precious metals and that the state should control trade.

Metropolitan area A central city surrounded by a number of smaller cities and suburbs that are closely related to it both socially and economically.

Micro-level sociology An approach to the study of society that focuses on patterns of social interaction at the individual level.

Middle-level sociology An approach to the study of society that focuses on relationships between social structures and the individual.

Minority group A population that, because of its members' physical or cultural characteristics, is singled out from others in the society for differential and unequal treatment.

Modernization A term used to describe the changes societies and individuals experience as a result of industrialization, urbanization, and the development of nation-states.

Monotheistic A term used to describe a theistic belief system centered on belief in a single all-powerful God.

Mores Strongly sanctioned norms.

Multinational corporation An economic enterprise that has headquarters in one country and conducts business activities in one or more other countries.

Nationalism The belief of a people that they have the right and the duty to constitute themselves as a nation-state.

Nation-state The largest territory within which a society's political structures can operate without having to face challenges to their sovereignty.

Natural selection The relative success of organisms with specific genetic mutations in reproducing new generations with the new trait.

Nonterritorial community A network of relationships formed around shared goals.

Normative order The array of norms that permit a society to achieve relatively peaceful social control.

Norms Specific rules of behavior.

Nuclear family Two or more people related by blood, marriage, or adoption who share a household.

Objective class In Marxian theory, a social class that has a visible, specific relationship to the means of production.

Oligarchy Rule by a few people who stay in office indefinitely rather than for limited terms.

Open society A society in which social mobility is possible for everyone.

Open stratification system A stratification system in which the boundaries between social strata are easily crossed.

Out-group Any social group to which an individual does not have a feeling of allegiance; may be in competition or conflict with the in-group.

Participant observation A form of observation in which the researcher participates to some degree in the lives of the people being observed.

Pastoral society A society whose primary means of subsistence is herding animals and moving them over a wide expanse of grazing land.

Patriarchy The dominance of men over women.

Peer group An interacting group of people of about the same age that has a significant influence on the norms and values of its members.

Percent analysis A mathematical operation that transforms an absolute number into a proportion as a part of 100.

Peripheral area A region that supplies basic resources and labor power to more advanced states.

Pluralist model A theory stating that no single group controls political decisions; instead, a plurality of interest groups influence those decisions through a process of coalition building and bargaining.

Pluralistic society A society in which different ethnic and racial groups are able to maintain their own cultures and lifestyles while gaining equality in the institutions of the larger society.

Political institution A set of norms and statuses pertaining to the exercise of power and authority.

Political party An organization of people who join together to gain legitimate control of state authority.

Political revolution A set of changes in the political structures and leadership of a society.

Polytheistic A term used to describe a theistic belief system that includes numerous gods.

Power The ability of an individual or group to change the behavior of others.

Power elite model A theory stating that political decisions are controlled by an elite of rich and powerful individuals even in societies with democratic political institutions.

Prejudice An attitude that prejudges a person on the basis of a real or imagined characteristic of a group to which that person belongs.

Primary group A social group characterized by intimate, face-to-face associations.

Privacy The right of a respondent to define when and on what terms his or her actions may be revealed to the general public.

Profane A term used to describe phenomena that are not considered sacred.

Profession An occupation with a body of knowledge and a developed intellectual technique that are transmitted by a formal educational process and testing procedures.

Projection The psychological process whereby we attribute to other people behaviors and attitudes that we are unwilling to accept in ourselves.

Psychosocial risk behaviors Behaviors that are detrimental to health, such as smoking and heavy drinking.

Race An inbreeding population that develops distinctive physical characteristics that are hereditary.

Racism An ideology based on the belief that an observable, supposedly inherited trait is a mark of inferiority that justifies discriminatory treatment of people with that trait.

Rate of reproductive change The difference between the crude birthrate and the crude death rate for a given population.

Recidivism The tendency for a person convicted of a crime to commit another crime after a term of imprisonment.

Relative deprivation Deprivation as determined by comparison with others rather than by some objective measure.

Religion Any set of coherent answers to the dilemmas of human existence that makes the world meaningful; a system of beliefs and rituals that serves to bind people together into a social group.

Religiosity The depth of a person's religious feelings.

Resocialization Socialization whose goal is to correct patterns of social learning that are considered detrimental.

Rite of passage A ceremony marking the transition to a new stage of a culturally defined life course.

Ritual A formal pattern of activity that symbolically expresses a set of shared meanings.

Role The way a society defines how an individual is to behave in a particular status.

Role conflict Conflict that occurs when in order to perform one role well a person must violate the expectations associated with another role.

Role expectations A society's expectations about how a role should be performed, together with the individual's perceptions of what is required in performing that role.

Role strain Conflict that occurs when the expectations associated with a single role are contradictory.

Role taking Trying to look at social situations from the standpoint of another person from whom one seeks a response.

Sacred A term used to describe phenomena that are regarded as extraordinary, transcendent, and outside the everyday course of events.

Sample A set of respondents selected from a specific population.

Sample survey A survey administered to a selection of respondents drawn from a specific population.

Sanctions Rewards and punishments for abiding by or violating norms.

Scapegoat A convenient target for hostility.

Scientific method The process by which theories and explanations are constructed through repeated observation and careful description.

Secondary group A social group whose members have a shared goal or purpose but are not bound together by strong emotional ties.

Sect A religious organization that rejects the beliefs and practices of existing churches; usually formed when a group leaves the church to form a rival organization.

Secularization A process in which the dominance of religion over other institutions is reduced.

Segregation The ecological and institutional separation of races or ethnic groups.

Semiperipheral area A state or region in which industry and financial institutions are developed to some extent but that remains dependent on capital and technology provided by other states.

Sex The biological differences between males and females, including the primary sex characteristics that are present at birth (i.e., the presence of specific male or female genitalia) and the secondary sex characteristics that develop later (facial and body hair, voice quality, etc.).

Sexism An ideology that justifies prejudice and discrimination based on sex.

Sexuality The manner in which a person engages in the intimate behaviors connected with genital stimulation, orgasm, and procreation.

Significant other Any person who is important to an individual.

Simple supernaturalism A form of religion in which people may believe in a great force or spirit but do not have a well-defined concept of God or a set of rituals involving God.

Slavery The ownership of one racial, ethnic, or politically determined group by another group that has complete control over the enslaved group.

Social change Variations over time in the ecological ordering of populations and communities, in patterns of roles and social interactions, in the structure and functioning of institutions, and in the cultures of societies.

Social control The set of rules and understandings that control the behavior of individuals and groups in a particular culture.

Social Darwinism The notion that people who are more successful at adapting to the environment in which they find themselves are more likely to survive and reproduce.

Social group A set of two or more individuals who share a sense of common identity and belonging and who interact on a regular basis.

Socialism An economic philosophy based on the concept of public ownership of property and sharing of profits, together with the belief that economic decisions should be controlled by the workers.

Socialization The processes whereby we learn to behave according to the norms of our culture.

Social mobility Movement by an individual or group from one social stratum to another.

Social movement Organized collective behavior aimed at changing or reforming social institutions or the social order itself.

Social revolution A complete transformation of the social order, including the institutions of government and the system of stratification.

Social strata Invisible layers, or strata, in a society that may be more or less closed to entry by people from outside any given layer.

Social stratification The process whereby the members of a society are sorted into different statuses.

Social structure The recurring patterns of behavior that create relationships among individuals and groups within a society.

Society A population that is organized in a cooperative manner to carry out the major functions of life.

Sociobiology The hypothesis that all human behavior is determined by genetic factors.

Socioeconomic status (SES) A broad social-class ranking based on occupational status, family prestige, educational attainment, and earned income.

Sociological imagination According to C. Wright Mills, the ability to see how social conditions affect our lives.

Sociology The scientific study of human societies and human behavior in the groups that make up a society.

Spatial mobility Movement of an individual or group from one location or community to another.

State A society's set of political structures.

Status A socially defined position in a group.

Status group A category of people within a social class, defined by how much honor or prestige they receive from the society in general.

Status symbols Material objects or behaviors that indicate social status or prestige.

Stereotype An inflexible image of the members of a particular group that is held without regard to whether it is true.

Stigma An attribute or quality of an individual that is deeply discrediting.

Structural mobility Movement of an individual or group from one social stratum to another that is caused by the elimination of an entire class as a result of changes in the means of existence.

Subculture A group of people who hold many of the values and norms of the larger culture but also hold certain beliefs, values, or norms that set them apart from that culture.

Subjective class In Marxian theory, the way members of a given social class perceive their situation as a class.

Superego According to Freud, the part of the human personality that internalizes the moral codes of adults.

Technologies The products and the norms for using them that are found in a given culture.

Technology Tools, procedures, and forms of social organization that increase human productive capacity.

Territorial community A population that functions within a particular geographic area.

Theism A belief system that conceives of a god or gods as separate from humans and from other living things on the earth.

Theoretical perspective A set of interrelated theories that offer explanations for important aspects of social behavior.

Theory A set of interrelated concepts that seek to explain the causes of an observable phenomenon.

Total institution A setting in which people undergoing resocialization are isolated from the larger society under the control of a specialized staff.

Totalitarian regime A regime that accepts no limits to its power and seeks to exert its rule at all levels of society.

Traditional authority Authority that is hereditary and is legitimated by traditional values, particularly people's ideas of the sacred.

Transsexuals People who feel very strongly that the sexual organs they were born with do not conform to their deep-seated sense of what their sex should be.

Triad a group consisting of three people.

Unobtrusive measures Observational techniques that measure behavior but intrude as little as possible into actual social settings.

Upward mobility Movement by an individual or group to a higher social stratum.

Urbanization A process in which an increasing proportion of a total population becomes concentrated in urban settlements.

Values The ideas that support or justify norms.

Variable A characteristic of an individual, group, or society that can vary from one case to another.

Voluntary association A formal organization whose members pursue shared interests and arrive at decisions through some sort of democratic process.

Welfare capitalism An economic philosophy in which markets determine what goods will be produced and how, but in which the government regulates economic competition.

REFERENCES

Abbott, A. D. (1993). The sociology of work and occupations. *Annual Review of Sociology, 19,* 187–209.

Abbott, A. D. (1988). *The system of professions: An essay on the division of expert labor.* Chicago: University of Chicago Press.

Addams, J. (1895). *Hull House maps and papers.* New York: Crowell.

Adler, M. (1979). *Drawing down the moon: Witches, druids, goddess-worshippers, and other pagans in America today.* Boston: Beacon.

Aganbegyan, A. (1989). *Perestroika 1989.* New York: Scribner.

Ahlburg, D. A., & De Vita, C. J. (1992, August). New realities of the American family. *Population Bulletin,* vol. 47, no. 2. Washington, DC: Population Reference Bureau.

Ahmad, O., Lopez, A., & Inoue, M. (2000, November 13). Decrease worldwide reaches target, but many countries still lagging. *AIDS Weekly.*

Aldous, J. (1982). From dual-earner to dual-career families and back again. In J. Aldous (Ed.), *Two paychecks: Life in dual-career families.* Newbury Park, CA: Sage.

Alland, A. (1973). *Human diversity.* Garden City, NY: Doubleday.

Allen, F. L. (1931). *Only yesterday: An informal history of the 1920s.* New York: Harper.

Allen, J. L., Jr. (2000, June 2). Key principles of liberation theology. *National Catholic Reporter,* p. 16.

Altman, L. K. (2001, January 30). The AIDS questions that linger. *New York Times, Science Times,* pp. 1, 6.

Ammerman, N. T. (1990). *Baptist battles: Social change and religious conflict in the Southern Baptist Convention.* New Brunswick, NJ: Rutgers University Press.

Andersen, M. L. (1993). *Thinking about women: Sociological perspectives on sex and gender* (3rd ed.). New York: Macmillan.

Angier, N. (1995, September 26). At a conference on crime's links to heredity, a calm after the storm. *New York Times,* p. C8.

Ariès, P. (1962). *Centuries of childhood.* New York: Vintage Books.

Armstrong, S. (2000, March). Honor's victims. *Chatelaine,* pp. 54–61.

Arnold, G. B. (2000). *The politics of faculty unionization.* Westport, CT: Bergin and Garvey.

Aron, R. (1955). *The century of total war.* Boston: Beacon.

Aronowitz, S. (1984). *Working class hero.* New York: Schocken.

Aronowitz, S., & Giroux, H. A. (1985). *Education under siege.* South Hadley, MA: Bergin and Garvey.

Asch, S. E. (1966). Effects of group pressure upon the modification and distortion of judgments. In H. Proshansky & B. Seidenberg (Eds.), *Basic studies in social psychology.* Fort Worth, TX: Holt, Rinehart and Winston.

Aspaklaria, S. (1985, September 12). A divorced father, a child, and a summer visit together. *New York Times,* p. C1.

Atchley, R. C. (1991). *Social forces and aging: An introduction to social gerontology.* Belmont, CA: Wadsworth.

Austin, G. A. (1978). *Perspectives on the history of psychoactive substance abuse.* Rockville, MD: National Institute on Drug Abuse.

Ayala, V. (1996). *Falling through the cracks: AIDS and the urban poor.* Putnam Valley, NY: Social Change Press.

Bakke, E. W. (1933). *The unemployed man.* London: Nisbet.

Balch, E. G. (1910). *Our Slavic fellow citizens.* New York: Charities Publication Committee.

Baldassare, M. (1986). *Trouble in paradise: The suburban transformation in America.* New York: Columbia University Press.

Bales, R. F., & Slater, P. E. (1955). Role differentiation in small decision-making groups. In T. Parsons & R. F. Bales (Eds.), *Family, socialization, and interaction process.* New York: Free Press.

Bales, K. (1999). *Disposable people: New slavery in the global economy.* Berkeley: University of California Press.

Ballantine, J. H. (1993). *The sociology of education: A systematic analysis* (3rd ed.). Englewood Cliffs, NJ: Prentice Hall.

Barker, K., & Christensen, K. (Eds.). (1998). *Contingent work: American employment relations in transition.* Ithaca, NY: ILR Press.

Barnet, R. J. (1980). *The lean years.* New York: Simon & Schuster.

Baron, J. N., & Bielby, W. T. (1980). Bringing the firms back in: Stratification, segmentation, and the organization of work. *American Sociological Review, 45,* 737–765.

Barringer, F. (1989, June 9). Divorce data stir doubt on trial marriage. *New York Times,* pp. A1, A28.

Bashi, V. (1991). Mentoring of at-risk students. *Focus* (University of Wisconsin, Institute for Research on Poverty), 13, 26–32.

Bassuk, E. L. (1984, July). The homelessness problem. *Scientific American,* 40–45.

Bates, F. L., & Murray, V. K. (1975). The school as a behavior system. *Journal of Research and Development in Education, 2,* 23–33.

Battaglia, L. (1999). *Photographs of Sicily.* New York: Aperture.

Baum, S. (1987). Financial aid to low-income college students: Its history and prospects (Institute for Research on Poverty Discussion Paper No. 846–87). Madison: University of Wisconsin.

Beale, C. L. (2000, January–April). A century of population growth and change. *Food Review,* p. 16.

Becker, G. S. (1973). A theory of marriage. In T. W. Schultz (Ed.), *Economics of the family: Marriage, children, and human capital.* Chicago: University of Chicago Press.

Becker, H. S. (1961). *Boys in white: Student culture in medical school.* Chicago: University of Chicago Press.

Becker, H. S. (1963). *The outsiders: Studies in the sociology of deviance.* New York: Free Press.

Beer, F. A. (1981). *Peace against war: The ecology of international violence.* San Francisco: Freeman.

Bell, D. (1962). Crime as an American way of life. In D. Bell, *The end of ideology.* New York: Free Press.

Bell, D. (1973). *The coming of post industrial society: A venture in social forecasting.* New York: Basic Books.

Bellah, R. N., Madison, R., Sullivan, W. M., Swidler, A., & Tipton, S. M. (1985). *Habits of the heart: Individualism and commitment in American life.* Berkeley: University of California Press.

Benavot, A., Cha, K-Y, & Kamen, D. G. A. (1991). Knowledge for the masses: World models and national curricula, 1920–1986. *American Sociological Review, 56,* 85–91.

Bendix, R. (1969). *Nation-building and citizenship.* Garden City, NY: Doubleday.

Bendix, R. (1974). *Work and authority in industry.* Berkeley: University of California Press.

Bendle, M. F. (1999). The death of the sociology of deviance? *Journal of Sociology, 35,* 42.

Bennett, W. (1996, June 3). Leave marriage alone. *Newsweek,* p. 27.

Benokraitis, N. V., & Feagin, J. R. (1986). *Modern sexism: Blatant, subtle, and covert discrimination.* Englewood Cliffs, NJ: Prentice Hall.

Bensman, D., & Lynch, R. (1987). *Rusted dreams: Hard times in a steel community.* New York: McGraw-Hill.

Bentham, J. (1789). *An introduction to the principles of morals and legislation.* London: T. Payne.

Berger, B. (1968). *Working class suburb: A study of auto workers in suburbia.* Berkeley: University of California Press.

Bergmann, B. R., & Hartmann, H. (1995, May 1). A program to help working parents. *The Nation,* pp. 592ff.

Berke, R. L., & Elder, J. (2000, November 6). Poll shows either candidate within reach of victory. *New York Times,* p. A1.

Bernard, J. (1982). *The future of marriage* (2nd ed.). New Haven, CT: Yale University Press.

Berrios, A. (2000, September 25). No wealth of opportunity. *New York Times,* p. A27.

Berrueta-Clement, J. R., Schweinhart, L. J., Barnett, W. S., Epstein, A. S., & Weikart, D. P. (1984). *Changed lives: The effects of the Perry preschool program on youths through age 19.* Ypsilanti, MI: High/Scope Press.

Berry, B. (1978). Latent structure of urban systems: Research methods and findings. In L. S. Bourne & J. W. Simmons (Eds.), *Systems of cities: Readings on structure, growth, and policy.* New York: Oxford University Press.

Bhalla, A. S., & Lapeyre, F. (1999). *Poverty in a global world.* New York: St. Martin's Press.

Bianchi, S. M., & Spain, D. (1986). *American women in transition.* New York: Russell Sage.

Bianchi, S. M., & Spain, D. (1996). Women, work, and family in America. *Population Bulletin,* vol. 51, no. 3. Washington, DC: Population Reference Bureau.

Bierstedt, R. (1963). *The social order.* New York: McGraw-Hill.

Bigham, G. D., & Bigham,V. S. (2000, November). Laptops revisited. *Curriculum Administrator,* p. 35.

Bigner, J. J., & Bozett, F. W. (1989). Parenting by gay fathers. *Marriage and Family Review, 14,* 155–175.

Bishop, G. (1999, Fall). Americans' belief in God. *Public Opinion Quarterly,* p. 421.

Bissinger, B. (1990). *Friday night lights.* New York: Harper Perennial.

Bjorntorp, P. (1997). Obesity. *The Lancet, 350,* 423–426.

Blau, F. D., & Kahn, L. M. (1999). Understanding international differences in the gender pay gap. Unpublished working paper, National Bureau for Economic Research.

Blau, P. M., & Scott, R. W. (1962). *Formal organizations: A comparative approach.* San Francisco: Chandler.

Blauner, R. (1972). *Racial oppression in America.* New York: Harper.

Blauner, R. (Ed.) (1989). *Black lives, white lives: Three decades of race relations in the United States.* Berkeley: University of California Press.

Blinder, D., & Crossette, B. (1993, February 7). As ethnic wars multiply, U.S. strives for a policy. *New York Times,* p. A1.

Block, F. (1990). *Postindustrial possibilities: A critique of economic discourse.* Berkeley: University of California Press.

Block, R. (1994). The tragedy of Rwanda. *New York Review of Books, 41,* 3–8.

Bloom, B. S. (1976). *Human characteristics and school learning.* New York: McGraw-Hill.

Blumenfeld, W. J. (Ed.) (1992). *Homophobia.* Boston: Beacon.

Blumer, H. (1969a). Elementary collective groupings. In A. M. Lee (Ed.), *Principles of sociology* (3rd ed.). New York: Barnes & Noble.

Blumer, H. (1969b). *Symbolic interactionism.* Englewood Cliffs, NJ: Prentice Hall.

Blumer, H. (1978). Elementary collective behavior. In L. E. Genevie (Ed.), *Collective behavior and social movements.* Itasca, IL: Peacock.

Blumstein, P., & Schwartz, P. (1983). *American couples: Money, work, sex.* New York: Morrow.

Bobo, L., & Hutchings, V. L. (1996). Perceptions of racial group competition. *American Sociological Review, 61,* 951–972.

Bogue, D. J. (1969). *Principles of demography.* New York: Wiley.

Bogue, D. J. (1985). *The population of the United States: Historical trends and future projections.* New York: Free Press.

Bongaarts, J., & Bulatao, R. A. (1999). Completing the demographic transition. *Population and Development Review, 25,* 515.

Bonner, R. (1994, August 25). Hutu and Tutsi mill the rice and set an example. *New York Times,* p. A1.

Boocock, S. S. (1980). *Sociology of education: An introduction* (2nd ed.). Boston: Houghton Mifflin.

Bosk, C. (1979). *Forgive and remember: Managing medical failure.* Chicago: University of Chicago Press.

Bottomore, T. (Ed.) (1973). *Karl Marx.* Englewood Cliffs, NJ: Prentice Hall.

Bourdieu, P. (1988). *Homo academicus.* London: Polity.

Bowles, S., & Gintis, H. (1977). *Schooling in capitalist America.* New York: Basic Books.

Braidwood, R. S. (1967). *Prehistoric man* (7th ed.). Glenview, IL: Morrow.

Brake, M. (1980). *The sociology of youth cultures and youth subcultures.* London: Routledge and Kegan Paul.

Braudel, F. (1976/1949). *The Mediterranean and the Mediterranean world in the age of Philip II.* New York: Harper.

Braudel, F. (1984). The perspective of the world: Vol. 3. *Civilizations and capitalism: 15th 18th century.* New York: Harper.

Braverman, H. (1974). *Labor and monopoly capital.* New York: Monthly Review Press.

Bray, L. (1999, November). Consider the alternatives. *Association Management,* p. 33.

Brimelow, P. (1995). *Alien nation: Common sense about America's immigration disaster.* New York: Random House.

Brockerhoff, M. P. (2000). An urbanizing world population. *Population Bulletin,* vol. 55, no. 3. Washington, DC: Population Reference Bureau.

Bromt. S. (1998). *Schools and societies.* Thousand Oaks, CA: Pine Forge Press.

Brown, C. (1966). *Manchild in the Promised Land.* New York: Macmillan.

Brown, D. (1970). *Bury my heart at Wounded Knee.* New York: Washington Square Press.

Brown, R. (1965). *Social psychology.* New York: Free Press.

Browne, M. W. (1987, September 1). Relics of Carthage show brutality amid the good life. *New York Times,* pp. C1, C10.

Browne Miller, A. (1990). *The day care dilemma: Critical concerns for American families.* New York: Plenum Press.

Buckley, W. F., Jr. (2000, August 28). On the right—Colin Powell says it almost right. *National Review.*

Bumpass, L. L., & Sweet, J. A. (1989). Children's experience in single-parent families: Implications of cohabitation and marital transitions. *Family Planning Perspectives, 21,* 256–260.

Burawoy, M. (1980). *Manufacturing consent.* Chicago: University of Chicago Press.

Bureau of Justice Statistics, U. S. Department of Justice (2000, June 16). Serious violent crime levels continued to decline in 1998 (online).

Buruma, I. (1984). *Behind the mask: On sexual demons, sacred mothers, transvestites, gangsters, drifters, and other Japanese cultural heroes.* New York: Pantheon.

Bury, J. B. (1932). *The idea of progress: An inquiry into its origin and growth.* New York: Dover.

Butalia, U. (2000). *The other side of silence.* Durham, NC: Duke University Press.

Butler, R. (1989). Dispelling ageism: The cross-cutting intervention. *Annals of the American Academy of Political and Social Science, 503,* 138–148.

Butterfield, F. (1992a, July 19). Are American jails becoming shelters from the storm*? New York Times,* p. E4.

Butterfield, F. (1992b, January 13). Studies find a family link to criminality. *New York Times,* pp. A1, A16.

Butterfield, F. (1997, September 28). Prison population growing although crime rate drops. *New York Times,* p. A1.

Cahill, S. E. (1992). The sociology of childhood at and in an uncertain age. *Contemporary Sociology, 21,* 669–672.

Cain, L. D. (1964). Life course and social structure. In R. F. Faris (Ed.), *Handbook of modern sociology.* Chicago: Rand McNally.

Cameron, W. B. (1966). *Modern social movements: A sociological outline.* New York: Random House.

Caplan, A. L. (1978). *The sociobiology debate: Readings on ethical and scientific issues.* New York: Harper.

Castells, M. (1999). *Critical education in the new information age.* Lanham, MD: Rowman and Littlefield.

Cazden, C. B. (1988). *Classroom discourse.* New York: Heinemann.

Chadwick, B. A., & Heaton, T. B. (1999). *The statistical handbook on the American family,* (2nd ed.). Phoenix: Oryx Press.

Chambliss, W. J. (1973, December). *The Saints and the Roughnecks. Society,* pp. 23–31.

Cherlin, A. J. (1981). *Marriage, divorce, remarriage.* Cambridge, MA: Harvard University Press.

Chirot, D. (1986). *Social change in the modern era.* San Diego, CA: Harcourt Brace Jovanovich.

Chirot, D. (1994). *How societies change.* Thousand Oaks, CA: Pine Forge Press.

Chodorow, N. (1978). *The reproduction of mothering: Psychoanalysis and the sociology of gender.* Berkeley: University of California Press.

Chomsky, N. (1965). *Aspects of the theory of syntax.* Cambridge, MA: MIT Press.

Chomsky, N. (1985). *Knowledge of language: Its nature, origins, and use.* New York: Praeger.

Clausen, J. A. (1968). *Socialization and society.* Boston: Little, Brown.

Clay, G. (1980). *Close-up: How to read the American city.* Chicago: University of Chicago Press.

Clay, G. (1994). *Real places: An unconventional guide to America's landscape.* Chicago: University of Chicago Press.

Clinton, B. (1999, May 3). Remarks announcing measure to address school violence. *Weekly Compilation of Presidential Documents,* pp. 767–769.

Cloward, R. A., & Ohlin, L. (1960). *Delinquency and opportunity: A theory of delinquent gangs.* New York: Free Press.

Cloward, R. A., & Piven, F. F. (1993, May 24). The fraud of workfare: Punishing the poor, again. *The Nation,* pp. 693–697.

Clydesdale, T. T. (1997). Family behaviors among early U.S. baby boomers: Exploring the effects of religion and income change, 1965–1982. *Social Forces, 76,* 605–635.

Cockerham, W. C. (1998). *Medical sociology* (7th ed.). Englewood Cliffs, NJ: Prentice Hall.

Cohen, N. (1961). *The pursuit of the millennium.* New York: Harper.

Cohen, S. A., & Zysman, J. (1987). *Manufacturing matters: The myth of the post industrial economy.* New York: Basic Books.

Cohen, S. F. (2000). *Failed crusade.* New York: Norton.

Coleman, J. S. (1964). Research chronicle: The adolescent society. In P. E. Hammond (Ed.), *Sociologists at work.* Garden City, NY: Doubleday.

Coleman, J. S. (1976). Liberty and equality in school desegregation. *School Policy, 6,* 9–13.

Collins, E. G. C., & Scott, P. (Eds.) (1978). Everyone who makes it has a mentor. *Harvard Business Review, 56,* 89–101.

Comer, J. P. (1984). Home–school relationships as they affect the academic success of children. *Education and Urban Society, 16,* 323–327.

Comer, J. P. (1985). *Investing in our children.* New York: Committee for Economic Development.

Comer, J. P. (1987). *Children in need: Investment strategies for the educationally disadvantaged.* New York: Committee for Economic Development.

Comte, A. (1971/1854). The positive philosophy. In M. Truzzi (Ed.), *Sociology: The classic statements.* New York: Random House.

Condon, K. A. (1991). Cohabitation, marriage, marital dissolution, and remarriage, U.S., 1988. *Advance data from vital and health statistics* (No. 194). Hyattsville, MD: National Center for Health Statistics.

Connell, R. W. (1995). *Masculinities.* Berkeley: University of California Press.

Connell, R. W. (1998). Gender regimes and the gender order. In *The Policy reader in gender studies.* Cambridge, England: Polity Press.

Connelly, M. (2000, November 8). Who voted: A portrait of American politics, 1976–2000. *New York Times,* p. B1.

Conway, G. (1997). *The doubly green revolution.* Ithaca, NY: Cornell University Press.

Cooley, C. H. (1909). *Social organization: A study of the large mind.* New York: Scribner.

Coon, C. (1962). *The origin of races.* New York: Knopf.

Coser, L. A. (1966). *The functions of social conflict.* New York: Free Press.

Council of Economic Advisors (2000). The changing American family. *Population and Development Review, 26,* 617–628.

Cranz, G. (1982). *The politics of park design: A history of urban parks in America.* Cambridge, MA: MIT Press.

Cremin, L. A. (1980). *American education: The national experience 1783–1876.* New York: Harper.

Crowder, M. (1966). *A short history of Nigeria* (Rev. ed.). New York: Praeger.

Csikszentmihalyi, M., Larson, R., & Prescott, S. (1977). The ecology of adolescent activity and experience. *Journal of Youth and Adolescence, 6,* 281–294.

Csilag, C. (1999). Sex education is key to combatting AIDS in Brazil. *The Lancet,* 353, 171.

Cuban, L. (1994, March). Computers meet classroom: Who wins? *Education Digest,* pp. 50–53.

Cuban, L. (2000, January 7). Computers for all by 2002. *New York Times Educational Supplement,* p. D6.

Cuber, J. F., & Haroff, P. B. (1980). Five types of marriage. In A. Skolnick & J. Skolnick (Eds.), *Family in transition.* Boston: Little, Brown.

Cumming, E., & Henry, W. (1971). *Growing old: The process of disengagement.* New York: Basic Books.

Curtin, P. D. (1969). *The Atlantic slave trade.* Madison: University of Wisconsin Press.

Curtiss, S. (1977). *Genie: A psycholinguistic study of a modern-day "wild child."* New York: Academic Press.

Dahl, R. (1961). *Who governs?* New Haven, CT: Yale University Press.

Dahrendorf, R. (1959). *Class and class conflict in industrial society.* Stanford, CA: Stanford University Press.

Danziger, K. (1971). *Socialization.* Harmondsworth, England: Penguin.

Danziger, S., & Gottschalk, P. (Eds.). (1993). *Uneven tides: Rising inequality in America.* New York: Russell Sage.

Davis, Kimberly (2000). Making blended families work. *Ebony.*

Davis, Kingsley (1939). Illegitimacy and the social structures. *American Journal of Sociology, 45,* 215–233.

Davis, K. (1947). Final note on a case of extreme isolation. *American Journal of Sociology, 52,* 432–437.

Davis, K. (1955). The origin and growth of urbanization in the world. *American Journal of Sociology, 60,* 429–437.

Davis, K. (1968). The urbanization of the human population. In S. F. Fava (Ed.), *Urbanism in world perspective: A reader.* New York: Crowell.

Davis, K. (1991). Population and resources: Fact and interpretation. In K. Davis & M. S. Bernstam (Eds.), *Resources, environment, population: Present knowledge, future options.* New York: Oxford University Press.

Davis, K., & Moore, W. E. (1945). Some principles of stratification. *American Sociological Review, 10,* 242–249.

Davis, K. C. (1995, September 3). Ethnic cleansing didn't start in Bosnia. *New York Times,* sec. 4, pp. 1, 6.

Davis, N. J. (1975). *Deviance: Perspectives and issues in the field.* Dubuque, IA: Brown.

Delgado-Gaiton, C. (1987). Traditions and transitions in the learning process of Mexican American children. In G. Spindler & L. Spindler (Eds.), *Interpretative ethnography of education.* Hillsdale, NJ: Erlbaum.

Demerath, N.J., III (1998). Excepting exceptionalism: American religion in comparative relief. *Annals of the American Academy of Political and Social Science, 558,* 28–39.

D'Emilio, J., & Freedman, E. B. (1988). *Intimate matters: A history of sexuality in America.* New York: Harper.

Denno, D. W. (1990). *Biology and violence: From birth to adulthood.* New York: Cambridge University Press.

de Sola Pool, I. (1983). *Technologies of freedom.* Cambridge, MA: Belknap Press.

de Sola Pool, I. (1990). *Technologies without boundaries.* Cambridge, MA: Harvard University Press.

de Tocqueville, A. (1955/1856). *The old regime and the French revolution* (S. Gilbert, Trans.). Garden City, NY: Doubleday.

de Tocqueville, A. (1980/1835). *On democracy, revolution, and society* (J. Stone & S. Mennell, Trans.). Chicago: University of Chicago Press.

Dewey, R. (1948). Charles Horton Cooley: Pioneer in psychosociology. In H. F. Barnes (Ed.), *An introduction to the history of sociology.* Chicago: University of Chicago Press.

De Witt, K. (1994, August 15). Wave of suburban growth is being fed by minorities. *New York Times,* pp. A1, B6.

Dillon, S. (2000, July 6). *Familiar foe for Mexico's new leader: Corruption. New York Times.*

Djilas, M. (1982). *The new class: An analysis of the communist system.* Orlando: Harcourt Brace Jovanovich.

Dobjas, G. J., & Freeman, R. B. (Eds.) (1992). *Immigration and the work force: Economic consequences for the United States and source areas.* Chicago: University of Chicago Press.

Dobzhansky, T. (1962). *Mankind evolving.* New Haven, CT: Yale University Press.

Dodgson, R. A. (1987). *The European past: Social evolution and spatial order.* London: Macmillan.

Dollard, J. (1937). *Caste and class in a southern town.* New Haven, CT: Yale University Press.

Dollard, J., Miller, N., & Doob, L. (1939). *Frustration and aggression.* New Haven, CT: Yale University Press.

Domhoff, G. W. (1978). *The powers that be.* New York: Random House.

Domhoff, G. W. (1983). *Who rules America now?* New York: Simon & Schuster.

Dornbusch, S. (1955). The military as an assimilating institution. *Social Forces, 33,* 316–321.

Dority, B. (1999). The Columbine tragedy: Countering the hysteria. *The Humanist, 59,* 7.

Douglas, J. D., & Rasmussen, P. K. (1977). *The nude beach.* Newbury Park, CA: Sage.

Dowd, M. (1985, November 17). Youth, art, hype: A different Bohemia. *New York Times Magazine,* pp. 26ff.

Dreger, A. D. (1998). *Hermaphrodites and the medical invention of sex.* Cambridge, MA: Harvard University Press.

Dublin, T. (1996). *Becoming American, becoming ethnic: College students explore their roots.* Philadelphia: Temple University Press.

Du Bois, W. E. B. (1967/1899). *The Philadelphia Negro: A social study.* New York: Schocken.

Dugger, C. W. (1996, September 11). A refugee's body is intact but her family is torn. *New York Times,* p. A1.

Duneier, M. (1992). *Slim's table: Race, respectability, and masculinity.* Chicago: University of Chicago Press.

Dunlop, C., & Kling, R. (Eds.). (1991). *Computerization and controversy: Value conflicts and social choices.* San Diego: Academic Press.

Durkheim, E. (1964/1893). *The division of labor in society* (2nd ed.). New York: Free Press.

Eibl-Eibesfeldt, I. (1989). *Human ethology.* Hawthorne, NY: Aldine.

Eisley, L. (1961). *Darwin's century.* Garden City, NY: Doubleday Anchor.

Eisley, L. (1970). *The invisible pyramid.* New York: Scribner.

Elder, G. H. (1981). History and the life course. In D. Berteaux (Ed.), *Biography and society: The life history approach to the social sciences.* Newbury Park, CA: Sage.

Elder, G. H., Jr., Modell, J., & Parke, R. D. (Eds.). (1992). *Children in time and place: Intersecting historical and developmental insights.* New York: Cambridge University Press.

Eldridge, C. C. (1978). *Victorian imperialism.* Atlantic Highlands, NJ: Humanities Press.

Elkin, F., & Handel, G. (1989). *The child and society: The process of socialization* (5th ed.). New York: Random House.

Elkind, D. (1970). *Children and adolescents: Interpretative essays on Jean Piaget.* New York: Oxford University Press.

Ellul, J. (1964). *The technological society.* New York: Vintage Books.

Ellwood, D. (1988). *Poor support: Poverty in the American family.* New York: Basic Books.

Ellwood, D. T. (2000, Winter). Anti-poverty policy for families in the next century: From welfare to work—and worries. *Journal of Economic Perspectives,* p. 187.

England, P. (1986). *Households, employment, and gender: A social, economic, and demographic view.* New York: Aldine.

Epstein, C. F. (1985). Ideal roles and real roles. *Research in Social Stratification and Mobility, 4,* 29–51.

Epstein, C. F. (1988). Deceptive distinctions: Sex, gender, and the social order. New Haven, CT: Yale University Press.

Epstein, C. F. (1995, Fall). Affirmative action. *Dissent,* pp. 463–465.

Erikson, E. (Ed.). (1997). *Sociological visions.* Oxford, England: Rowman and Littlefields.

Erikson, K. T. (1962). Notes on the sociology of deviance. *Social Problems, 9,* 307–314.

Erikson, K. T. (1966). *Wayward puritans: A study in the sociology of deviance.* New York: Wiley.

Etzioni, A. (1997). *The new golden rule.* New York: Basic Books.

Ewald, W. (1996). *I dreamed I had a girl in my pocket.* New York: Norton.

Ezorsky, G. (1991). *Affirmative action.* Ithaca, NY: Cornell University Press.

Falah, G. (1996). Living together apart: Residential segregation in mixed Arab-Jewish cities in Israel. *Urban Studies, 33,* 823–857.

Fallows, D. (1985). *A mother's work.* Boston: Houghton Mifflin.

Fantasia, R. (1988). *Cultures of solidarity: Consciousness, action, and contemporary American workers.* Berkeley: University of California Press.

Farley, R., & Allen, W. R. (1987). *The color line and the quality of life in America.* New York: Russell Sage.

Farnsworth, C. H. (1994, August 15). With her songs, Eskimo bares her people's pain. *New York Times,* p. A4.

Farrington, K., & Chertok, E. (1993). Social conflict theories of the family. In P. G. Boss, W. J. Doherty, R. LaRossa, W. R. Schumm, & S. K. Steinmetz (Eds.), *Sourcebook of family theories and methods: A contextual approach.* New York: Plenum Press.

Fathalla, M. F. (1992). Reproductive health in the world: Two decades of progress and the challenge ahead. Discussion note, Expert Group Meeting on Population and Women, Gaborone, Botswana, June 22–26.

Feagin, J. R. (1991, November 27). Blacks still face the malevolent reality of white racism. *Chronicle of Higher Education,* p. A44.

Feagin, J. R. (2000). *Racist America: Roots, current realities, and future reparations.* New York: Routledge.

Fernandes, F. (1968). The weight of the past. In J. H. Franklin (Ed.), *Color and race.* Boston: Beacon.

Fernandez, J. (1992). *Tales out of school: Joseph Fernandez's crusade to rescue American education.* Boston: Little, Brown.

Ferree, M. M., & Hall, E. J. (2000). Gender stratification and paradigm change. *American Sociological Review, 65,* 475–481.

Finke, R., & Stark, R. (1992). *The churching of America 1776–1992.* New Brunswick, NJ: Rutgers University Press.

Fischer, C. S. (1976, 1987). *The urban experience.* Orlando: Harcourt Brace Jovanovich.

Fischer, C. S., Hout, M., Sanchez-Jankowski, M., Lucas, S. R., Swidler, A., & Voss, K. (1996). *Cracking the bell curve myth.* Princeton, NJ: Princeton University Press.

Fishman, R. (1987). *Bourgeois utopias: The rise and fall of suburbia.* New York: Basic Books.

Flink, J. J. (1988). *The automobile age.* Cambridge, MA: MIT Press.

Fortin, N., & Ecker, M. (1978). Social problems at Cape Cod National Seashore. Unpublished report to U.S. Department of the Interior, National Park Service.

Frank, A. G. (1966). The development of underdevelopment. *Monthly Review, 18,* 3–17.

Frank, R. H. (1988). *Passions within reason: The strategic role of the emotions.* New York: Norton.

Freeland, C. (2000). *Sale of the century.* New York: Crown Business.

Freeman, R., & Rogers, R. (1999). *What workers want.* Ithaca, NY: ILR Press.

Frey, W. H. (1998). The diversity myth. *American Demographics, 20,* 38–43.

Friedlander, D. (1969). Demographic responses and population change. *Demography, 6,* 359–381.

Fullan, M. (1993). *Change forces: Probing the depths of educational reform.* New York: Falmer Press.

Furstenberg, F. F., & Cherlin, A. J. (1991). *Divided families: What happens to children when parents part.* Cambridge, MA: Harvard University Press.

Fyfe, C. (1976). The dynamics of African dispersal. In M. L. Kilson & R. I. Rothberg (Eds.), *The African diaspora.* Cambridge, MA: Harvard University Press.

Gabriel, T. (1995a, February 12). A generation's heritage: After the boom, a boomlet. *New York Times,* pp. 1, 34.

Gabriel, T. (1995b, June 12). A new generation seems ready to give bisexuality a place in the spectrum, *New York Times,* p. A12.

Gaines, D. (1992). *Teenage wasteland: suburbia's dead end kids.* New York: Harper Perennial.

Galtung, J. (1985). War. In A. Kuper & J. Kuper (Eds.), *The social science encyclopedia.* London: Routledge.

Gamson, W. A. (1990). *The strategy of social protest,* (2nd ed.). Belmont, CA: Wadsworth.

Gamson, W. A., & Wolfsfield, G. (1993). Movements and media as interacting systems. *Annals of the American Academy of Political and Social Science, 528,* 114–125.

Gans, H. (1967, 1976). *The Levittowners: Way of life and politics in a new suburban community.* New York: Pantheon.

Gans, H. (1984). *The urban villagers.* New York: Free Press.

Gans, H. (1995). *The war against the poor.* New York: Basic Books.

Gardner, D. (2000, December 23). A season of ill-will and violence: India's Christian minority is fearing another. *Financial Times* (London).

Gardner, H. (1983). *Frames of mind: The theory of multiple intelligences.* New York: Basic Books.

Garfinkel, H. (1967). *Studies in ethnomethodology.* Englewood Cliffs, NJ: Prentice Hall.

Garfinkel, I., & McLanahan, S. S. (1986). *Single mothers and their children: A new American dilemma.* Washington, DC: Urban Institute.

Garreau, J. (1991). *Edge city: Life on the new frontier.* Garden City, NY: Doubleday Anchor.

Geertz, C. (1973). The growth of culture and the evolution of mind. In C. Geertz, *The interpretation of culture.* New York: Basic Books.

Gelbard, A., Haub, C., & Kent, M. M. (1999). World population beyond six billion. *Population Bulletin,* vol. 54, no. 1. Washington, DC: Population Reference Bureau.

Genevie, L. E. (Ed.) (1978). *Collective behavior and social movements.* Itasca, IL: Peacock.

Geoghegan, T. (1991). *Which side are you on?* New York: Farrar Straus Giroux.

Gerbner, G. (1990). *Violence profile.* Philadelphia: Annenberg School of Communications.

Gergen, D. R. (1997, October 20). Profiles in creativity: Smart innovations are sweeping government agencies across the country. *U.S. News & World Report,* p. 92.

Gerth, H., & Mills, C. W. (1958). *From Max Weber: Essays in sociology.* New York: Oxford University Press.

Gilligan, C. (1987). Adolescent development reconsidered. In C. E. Irwin, Jr. (Ed.),

New directions for child development: No. 37, Adolescent social behavior and health. San Francisco: Jossey-Bass.

Gilligan, C., Ward, J. V., Taylor, J. M., & Bardige, B. (Eds.). *Mapping the moral domain.* Cambridge, MA: Harvard University Press.

Gimenez, M. E. (1998, November). Looking back, looking forward: LAP in the 21st century. *Latin American Perspectives,* pp. 59–61.

Giugni, M. G. (1998). Was it worth the effort? The outcomes and consequences of social movements. *Annual Review of Sociology, 24,* 371–393.

Glazer, N., & Moynihan, D. P. (1970). *Beyond the melting pot: The Negroes, Puerto Ricans, Jews, Italians, and Irish of New York City* (2nd rev. ed.). Cambridge, MA: MIT Press.

Glenny, M. (2000). *The Balkans: Nationalism, war, and the great powers, 1804–1999.* New York: Viking.

Glick, P. C., & Parke, R., Jr. (1965). New approaches in studying the life cycle of the family. *Demography, 2,* 187–202.

Glueck, S., & Glueck, E. T. (1950). *Unraveling juvenile delinquency.* New York: Commonwealth Fund.

Goffman, E. (1958). Deference and demeanor. *American Anthropologist, 58,* 488–489.

Goffman, E. (1959). *The presentation of self in everyday life.* Garden City, NY: Doubleday.

Goffman, E. (1961). *Asylums.* Garden City, NY: Doubleday.

Goffman, E. (1963). *Stigma: Notes on the management of spoiled identity.* Englewood Cliffs, NJ: Prentice Hall.

Goffman, E. (1965). *Interaction ritual: Essays on face-to-face behavior.* Garden City, NY: Doubleday.

Goffman, E. (1972). Territories of the self. In E. Goffman (Ed.), *Relations in public.* New York: Harper.

Gold, S. (2000, May 4). Here we go again—ILOVEYOU virus sweeps the world. *Technology Information* (online).

Goldberg, J. (1997, March 2). Their Africa problem, and ours. *New York Times Magazine,* pp. 34–39.

Golden, T. (2000, July 6). Fungus considered as a tool to kill coca in Colombia. *New York Times,* p. 1.

Goldfarb, W. (1945). Psychological privation in infancy and subsequent adjustment. *American Journal of Orthopsychiatry, 15,* 247–253.

Goldhamer, H., & Shils, E. (1939). Types of power and status. *American Journal of Sociology, 45,* 171–182.

Goodall, J. V. L. (1968). A preliminary report on expressive movements and communications in Gombe Stream chimpanzees. In P. Jay (Ed.), *Primates: Studies in adaptation and variability.* Fort Worth, TX: Holt, Rinehart and Winston.

Goode, Eric. (1992). *Collective behavior.* Fort Worth, TX: Harcourt Brace.

Goode, Eric. (1994). *Deviant behavior* (4th ed.). Englewood Cliffs, NJ: Prentice Hall.

Goode, Erica (2000, August 27). Hey, what if contestants gave each other shocks? *New York Times,* p. 3.

Goode, W. J. (1963). *World revolution and family patterns.* Glencoe, IL: Free Press.

Goode, W. J. (1964). *The family.* Englewood Cliffs, NJ: Prentice Hall.

Goode, W. J. (1993) *World changes in divorce.* New Haven, CT: Yale University Press.

Gordon, M. (1964). *Assimilation in American life.* New York: Oxford University Press.

Goslin, D. A. (1965). *The school in contemporary society.* Glenview, IL: Scott, Foresman.

Gottmann, J. (1978). Megalopolitan systems around the world. In L. S. Bourne & J. W. Simmons (Eds.), *Systems of cities: Readings on structure, growth, and policy.* New York: Oxford University Press.

Gould, S. J. (1981). *The mismeasure of man.* New York: Norton.

Gould, S. J. (1995, February). Ghosts of bell curves past. *Natural History,* pp. 12–19.

Gramlich, E. M., Kasten, R., & Sammartino, F. (1993). Growing inequality in the 1980s: The role of federal taxes and cash transfers. In S. Danziger & P. Gottschalk (Eds.), *Uneven tides: Rising inequality in America.* New York: Russell Sage.

Greeley, A., Michael, R. T., & Smith, T. (1990). Americans and their sexual partners. *Society, 27,* 36–42.

Greenberg, D. S. (1999). Delete the revolution. *The Lancet, 353,* 764.

Greenman, C. (1999, November 24). V-chip veto? *Technology Information,* p. 15.

Gross, J. (1992, March 29). Collapse of inner-city families creates America's new orphans. *New York Times,* pp. 1, 20.

Grossman, C. L. (2000, December 27). T.D. Jakes spiritual salesman. *USA TODAY,* p. 8D.

Grunwald, L. (1995, February). Do I look fat to you? *Life,* pp. 58–71.

Gusfield, J. (1987). *Symbolic crusades: Status politics and the American temperance movement,* 2nd ed. Urbana: University of Illinois Press.

Gusfield, J. (1981). *The culture of public problems: Drinking, driving, and the symbolic order.* Chicago: University of Chicago Press.

Gutman, H. (1976). *The black family in slavery.* New York: Pantheon.

Gutwirth, J. (1999, Fall). From the Word to the televisual image: The televangelists and Pope John Paul II. *Diogenes,* p. 122.

Haber, L. D., & Dowd, J. E. (1994, January). A human development agenda for disability: Statistical considerations (Working Paper for the Statistical Division of the United Nations Secretariat). New York: United Nations.

Hacker, A. (1997). *Money: Who has how much and why.* New York: Simon & Schuster.

Hagestad, G. O., & Neugarten, B. L. (1985). Age and the life course. In R. H. Binstock & E. Shanas (Eds.), *Handbook of aging and the social sciences* (2nd ed.). New York: Van Nostrand Reinhold.

Halevy, E. (1955). *The growth of philosophic radicalism* (M. Morris, Trans.). Boston: Beacon.

Hall, J. (1987). *Gone from the Promised Land.* New Brunswick, NJ: Transaction Books.

Halsey, M. (1946). *Color blind.* New York: Simon & Schuster.

Hamilton, C. V. (1969). The politics of race relations. In C. U. Daley (Ed.), *The minority report.* New York: Pantheon.

Hanna, S. L. (1994). *Person to person: Positive relationships don't just happen.* Englewood Cliffs, NJ: Prentice Hall.

Hansen, W. L., & Stampen, J. O. (1987, April). Economics and financing of higher education: The tension between quality and equity (Rev.). Paper presented at the annual meeting of the Association for the Study of Higher Education, San Diego.

Harbison, F. H. (1973). *Human resources as the wealth of nations.* New York: Oxford University Press.

Hare, A. P., Blumberg, H. H., Davis, M. F., & Kent, V. (1994). *Small group research: A handbook.* Norwood, NJ: Ablex.

Harmon, A. (2000, July 3). As computers idle in classrooms, training for teachers is the next challenge. *New York Times,* pp. A17, B1.

Harper, D. A. (1982). *Good company.* Chicago: University of Chicago Press.

Harrington, M. (1973). *Socialism.* New York: Bantam.

Harrington, M. (1987). *The next lift: The history of a future.* New York: Holt.

Harrison, B. (1990). *The great U-turn: Corporate restructuring and the polarizing of America.* New York: Basic Books.

Harrison, B. B., Tilly, C., & Bluestone, B. (1986, March–April). Wage inequality takes a great U-turn. *Challenge,* pp. 26–32.

Hauser, P. M. (1957). The changing population pattern of the modern city. In P. K. Hatt & A. J. Reiss (Eds.), *Cities and society.* New York: Free Press.

Hawkins, G. (1976). *The prison: Policy and practice.* Chicago: University of Chicago Press.

Hechter, M. (1987). *Principles of group solidarity.* Berkeley: University of California Press.

Hechter, M. (2000). *Containing nationalism.* New York: Oxford University Press.

Heilbroner, R. (1990, February 15). Seize the day. *New York Review of Books,* pp. 30–31.

Helmreich, W. B. (1982). *The things they say behind your back.* Garden City, NY: Doubleday.

Helvarg, D. (1994). *The war against the greens.* San Francisco: Sierra Club Books.

Henry, D. O. (1989). *From foraging to agriculture.* Philadelphia: University of Pennsylvania Press.

Henslin, J., & Briggs, M. (1971). Dramaturgical desexualization: The sociology of the vaginal examination. In J. Henslin (Ed.), *Studies in the sociology of sex.* New York: Appleton-Century-Crofts.

Herrnstein, R., & Murray, C. (1994). *The bell curve.* New York: Free Press.

Hills, L., & Trapp, R. (2000, October 20). African-Americans and Latinos, in a San Diego study, represent 28 percent of the driving population, but are 40 percent of those stopped and 60 percent of those searched. *San Diego Union-Tribune,* p. B–11.

Hochschild, A. (1989). *The second shift: Working parents and the revolution at home.* New York: Viking.

Holt, J. (1965). *How children fail.* New York: Dell.

Holt, J. (1967). *How children learn.* New York: Pitman.

Holt, T. C. (1980). Afro-Americans. In *Harvard encyclopedia of American ethnic groups.* Cambridge, MA: Belknap Press.

Holy, L. (1985). Groups. In A. Kuper & J. Kuper (Eds.), *The social science encyclopedia.* London: Routledge.

Homans, G. (1950). *The human group.* Orlando: Harcourt Brace Jovanovich.

Homans, G. (1961). *Social behavior: Its elementary forms.* Orlando: Harcourt Brace Jovanovich.

Hopper, R. D. (1950). The revolutionary process: A frame of reference for the study of revolutionary movements. *Social Forces, 25,* 270–279.

House, J. S., Lepkowski, J. M., Kinny, A. M., Mero, R. P., Kessler, A., & Herzog, R. (1994). The social stratification of aging and health. *Journal of Health and Social Behavior, 35,* 213–234.

Howe, I. (1983). *1984 revisited: Totalitarianism in our century.* New York: Harper.

Huang, C. (1994). Immigration and the underclass. Unpublished doctoral dissertation, City University of New York Graduate School.

Huber, J., & Spitz, G. (1988). Trends in family sociology. In N. Smelser (Ed.), *Handbook of sociology.* Newbury Park, CA: Sage.

Hughes, E. (1945). The dilemmas and contradictions of status. *American Journal of Sociology, 50,* 353–359.

Hughes, E. (1958). *Men and their work.* New York: Free Press.

Hughes, E. (1959). The study of occupations. In R. K. Merton, L. Broom, & L. S. Cottrell, Jr. (Eds.), *Sociology today: Problems and prospects.* New York: Basic Books.

Hummin, D. (1990). *Commonplaces: Community ideology and identity in American cities.* Albany: State University of New York Press.

Hunt, M. (1985). *Profiles of social research: The scientific study of human interactions.* New York: Russell Sage.

Hunter, F. (1953). *Community power structure: A study of decision makers.* Chapel Hill: University of North Carolina Press.

IBRD (International Bank for Reconstruction and Development) (2000). *World Bank indicators.* Washington, DC: IBRD.

Inglehart, R., & Baker, W. E. (2000). Modernization, cultural change, and the persistence of traditional values. *American Sociological Review, 65,* 19–51.

Ingraham, L. (2001, January 7). Bush's first task: Handle McCain factor. *Los Angeles Times,* part M, p. 5.

Jackman, M. R., & Jackman, R. W. (1983). *Class awareness in the United States.* Berkeley: University of California Press.

Jackson, B. (1972). *In the life: Versions of the criminal experience.* New York: NAL.

Jackson, K. (1985). *The crab grass frontier: The suburbanization of the United States.* New York: Oxford University Press.

Jacot, M. (1999, October). The death penalty: Abolition gains ground. *UNESCO Courier,* p. 37.

Jahoda, M. (1982). *Employment and unemployment: A social-psychological analysis.* London: Cambridge University Press.

Jahoda, M., Lazarsfeld, P., & Zeisel, H. (1971). *Marienthal: A study of an unemployed community.* Hawthorne, NY: Aldine.

Janowitz, M. (1968). Political sociology. In D. Sills (Ed.), *The international encyclopedia of the social sciences.* New York: Free Press.

Janowitz, M. (1978). *The last half century: Societal change and politics in America.* Chicago: University of Chicago Press.

Janowitz, M. (1991). Aspects of social control. In J. Burk (Ed.), *Morris Janowitz on social organization and social control.* Chicago: University of Chicago Press.

Jargowsky, P. A. (1996). Take the money and run: Economic segregation in U.S. metropolitan areas. *American Sociological Review, 61,* 984–998.

Jaynes, G. D., & Williams, R. M., Jr. (1989). *A common destiny: Blacks and American society.* Washington, DC: National Academy Press.

Jencks, C. (1993). *Rethinking social policy: Race, poverty and the underclass.* Cambridge, MA: Harvard University Press.

Jencks, C. (1994). *The homeless.* Cambridge, MA: Harvard University Press.

Jencks, C., Smith, M., Acland, H., Bane, M. J., Cohen, D., Gintis, H., Heyns, B., & Michelson, S. (1972). *Inequality: A reassessment of the effect of family and schooling in America.* New York: Basic Books.

Jennings, F. (1975). *The invasion of America: Indians, colonialism, and the cant of conquest.* New York: Norton.

Johnson, J. H., Jr., & Farrell, W. C., Jr. (1995, July 7). Race still matters. *Chronicle of Higher Education,* p. A48.

Johnson, M. A. (Ed.). (1989). *The many faces of Hull House.* Urbana: University of Illinois Press.

Johnson, W. Y. (1999). *Youth suicide: The school's role in prevention and response.* Bloomington, IN: Phi Kappa Delta Educational Foundation.

Johnston, C. (2000). Computers for all by 2002. TES online supplement.

Johnston, D. K. (1988). Adolescents' solutions to dilemmas in fables: Two moral orientations—two problem solving strategies. In C. Gilligan, J. V. Ward, J. M. Taylor, & B. Bardige (Eds.), *Mapping the moral domain.* Cambridge, MA: Harvard University Press.

Jones, R. C. (1998). Remittances and inequality: A question of migration stage and geographic scale. *Economic Geography, 74,* 8–25.

Kalleberg, A. L., Reskin, B. F., & Hudson, K. (2000). Bad jobs in America: Standard and nonstandard employment relations and job quality in the United States. *American Sociological Review, 65,* 256–278.

Kammerman, S., & Kahn, A. (1981). *Child care, family benefits, and working parents: A study in comparative policy.* New York: Columbia University Press.

Kane, T. J. (1996). Lessons from the largest school voucher program. In Fuller, B., Elmore, R. F., & Orfield, G. (Eds.), *Who chooses? Who loses? Culture, institutions, and the unequal effects of school choice.* New York: Teachers College Press.

Kapur, A. (2000, December 10). Subcontinental divide. *New York Times,* sec. 7, p. 38.

Karabel, J., & Halsey, A. H. (Eds.) (1988). *Power and ideology in education.* New York: Oxford University Press.

Kasarda, J. D. (1988). *Metropolis era.* Newbury Park, CA: Sage.

Kasarda, J. D. (1989). Urban industrial transition and the underclass. *Annals of the American Academy of Political and Social Science, 501,* 26–47.

Katz, F. E. (1964). The school as a complex social organization. *Harvard Educational Review, 34,* 428–455.

Katz, M. (1983). *Poverty and policy in American history.* New York: Academic Press.

Katznelson, I. (1981). *City trenches: Urban politics and the patterning of class in the United States.* Chicago: University of Chicago Press.

Keely, C. B. (2000, June 3). Nation's religious makeup shifting with immigrant tide. *America,* p. 4.

Keith, J. (1982). *Old people, new lives: Community creation in a retirement residence* (2nd ed.). Chicago: University of Chicago Press.

Kempe, H., & Helfer, R. E. (Eds.) (1980). *The battered child* (3rd ed.). Chicago: University of Chicago Press.

Kemper, T. D. (1990). *Social structure and testosterone: Explorations of the sociobiosocial chain.* New Brunswick, NJ: Rutgers University Press.

Keniston, K. (1977). *All our children.* Orlando: Harcourt Brace Jovanovich.

Keyfitz, N. (1986). The population that does not reproduce itself. In K. Davis, M. S. Bernstam, & R. Ricardo-Campbell (Eds.), *Below replacement fertility in industrial societies: Causes, consequences, policies. Population and Development Review,* 12 (Suppl.).

Kilson, M. (1995, Fall). Affirmative action. *Dissent,* pp. 469–470.

Kim, I. (1981). *New urban immigrants: The Korean community in New York.* Princeton, NJ: Princeton University Press.

Kimmel, M. S. (1990). *Revolution: A sociological interpretation.* Philadephia: Temple University Press.

Kinsey, A. C., Pomeroy, W. B., & Martin, C. E. (1948). *Sexual behavior in the human male.* Philadelphia: Saunders.

Kinsey, A. C., Pomeroy, W. B., & Martin, C. E. (1953). *Sexual behavior in the human female.* Philadelphia: Saunders.

Kitano, H. H. I. (1980). *Race relations.* Englewood Cliffs, NJ: Prentice Hall.

Kitsuse, J. I. (1962). Societal reaction to deviant behavior; problems of theory and method. *Social Problems, 9,* 247–257.

Kleinberg, O. (1935). *Race differences.* New York: Harper.

Kohlberg, L., & Gilligan, C. (1971). The adolescent as a philosopher: The discovery of the self in a post-conventional world. *Daedalus, 100,* 1051–1086.

Kolata, G. (1992, November 11). New views on life span alter forecasts on elderly. *New York Times,* p. A1.

Kornblum, W. (1974). *Blue collar community.* Chicago: University of Chicago Press.

Kornblum, W., & Julian, J. (2001). *Social problems.* Englewood Cliffs, NJ: Prentice Hall.

Kunen, J. S. (1995, July 10). Teaching prisoners a lesson. *New Yorker,* pp. 34–39.

Kutner, B., Wilkins, C., & Yarrow, P. R. (1952). Verbal attitudes and overt behavior involving racial prejudice. *Journal of Abnormal and Social Psychology, 47,* 649–652.

Kwamena-Poh, M., Tosh, J., Waller, R., & Tidy, M. (1982). *African history in maps.* Burnt Hill, England: Longman.

Lai, H. M. (1980). The Chinese. In S. Thornstrom (Ed.), *Harvard encyclopedia of ethnic groups.* Cambridge, MA: Harvard University Press.

Lambert, H. H. (1987). Biology and equality: A perspective on sex differences. In S. Harding & J. F. O'Barr, *Sex and scientific inquiry.* Chicago: University of Chicago Press.

LaPiere, R. (1934). Attitudes vs. actions. *Social Forces, 13,* 230–237.

Laslett, P. (1972). Introduction. In P. Laslett & R. Wall (Eds.), *Household and family in past time.* Cambridge, Eng.: Cambridge University Press.

Laslett, P. (1983). *The world we have lost* (3rd ed.). London: Methuen.

Laslett, P., & Wachter, K. W. (1978). *Statistical studies of historical social structure.* New York: Academic Press.

Lasswell, H. (1936). *Politics: Who gets what, when, and how.* New York: McGraw-Hill.

Lasswell, H. D. (1941). The garrison state. *American Journal of Sociology, 46,* 455–468.

Latané, B., & Darley, J. (1970). *The unresponsive bystander: Why doesn't he help?* New York: Meredith.

Laumann, E. O., Gagnon, J. H., Michaels, R. T., & Michaels, S. (1994). *The social organization of sexuality: Sexual practices in the United States.* Chicago: University of Chicago Press.

Lazreg, M. (1994). *The eloquence of silence: Algerian women in question.* New York: Routledge.

LeBon, G. (1947/1896). *The crowd.* London: Ernest Bonn.

Lee, V. E., & Smith, J. B. (1993). Effects of school restructuring on the achievement and engagement of middle-grade students. *Sociology of Education, 66,* 148–163.

Levitan, S., & Belous, R. S. (1981). *What's happening to the American family?* Baltimore: Johns Hopkins University Press.

Lewin, T. (1989, November 14). Aging parents: Women's burden grows. *New York Times,* pp. A1, B12.

Lewis, M., & Fiering, C. (1982). Some American families at dinner. In L. M. Laosa & I. E. Sigel (Eds.), *Families as learning environments for children.* New York: Plenum Press.

Lewontin, R. C. (1992). *Biology as ideology.* New York: Harper Perennial.

Lieberson, S. (1980). *A piece of the pie: Blacks and white immigrants since 1880.* Berkeley: University of California Press.

Liebow, E. (1967). *Tally's corner: A study of Negro streetcorner men.* Boston: Little, Brown.

Liebow, E. (1995). *Tell them who I am.* Baltimore: Penguin.

Lim, L. L. (Ed.) (1998). *The sex sector.* Geneva: International Labor Organization.

Lipset, S. M. (1979). *The first new nation.* New York: Norton

Lipset, S. M. (1981). *Political man.* Baltimore: Johns Hopkins University Press.

Lipset, S. M. (1994). The social prerequisites of democracy revisited. *American Sociological Review, 59,* 1–22.

Livernash, R., & Rodenberg, E. (1998). Population change, resources, and the environment. *Population Bulletin,* vol. 33, no. 1. Washington, DC: Population Reference Bureau.

Loewy, E. H. (1993). *Freedom and community: The ethics of interdependence.* Albany: State University of New York Press.

Lofland, J. F. (1981). Collective behavior: The elementary forms. In N. Rosenberg & R. H. Turner (Eds.), *Social psychology: Sociological perspectives.* New York: Basic Books.

Lombroso, C. (1911). *Crime: Its cause and remedies.* Boston: Little, Brown.

London, B. (1992). School-enrollment rates and trends, gender, and fertility: A cross-national analysis. *Sociology of Education, 65,* 305–318.

Lorber, J. (1994). *Paradoxes of gender.* New Haven, CT: Yale University Press.

Lowry, I. S. (1990). World urbanization in perspective. In K. Davis & M. S. Bernstam (Eds.), Resources, environment, and population. *Population and Development Review,* 15 (Suppl.).

Luhmann, N. (1982). *The differentiation of society.* New York: Columbia University Press.

Lundahl, M., & Silie, R. (1998). Economic reform in Haiti: Past failures and future success? *Comparative Economic Studies, 40,* 43–71.

Lynch, M. W. (2000, June). Walk on by. *Reason,* p. 16.

MacLeod, J. (1987, 1995). *Ain't no makin' it: Leveled aspirations in a low-income neighborhood.* Boulder, CO: Westview Press.

Majundar, R. C. (Ed.) (1951). *The history and culture of the Indian people.* London: Allen and Unwin.

Malinowski, B. (1927). *Sex and repression in savage society.* London: Harcourt Brace.

Malson, L. (1972). *Wolf children and the problem of human nature* (E. Fawcett, P. Aryton, & J. White, Trans.). New York: Monthly Review Press.

Malthus, T. (1927–1928/1798). *An essay on population.* New York: Dutton.

Mannheim, K. (1941). *Man and society in an age of reconstruction.* Orlando: Harcourt Brace Jovanovich.

Mansnerus, L. (1992, November 1). Should tracking be derailed? Education Life, *New York Times,* pp. 14–16.

Mariani, M. (2000, Fall). Telecommuters. *Occupational Outlook Quarterly,* p. 10.

Marks, J. (1994, December). Black, white, other. *Natural History,* pp. 32–35.

Marshall, T. H. (1964). *Class, citizenship, and social development.* Garden City, NY: Doubleday.

Martinson, R. (1972, April 29). Planning for public safety. *New Republic,* pp. 21–23.

Marx, K. (1962/1867). *Capital: A critique of political economy.* Moscow: Foreign Languages Publishing House.

Marx, K., & Engels, F. (1955). Wage labor and capital. In *Selected works in two volumes.* Moscow: Foreign Languages Publishing House.

Marx, K., & Engels, F. (1969/1848). *The communist manifesto.* New York: Penguin.

Mascie-Taylor, C. G. N. (1990). *Biosocial aspects of social class.* New York: Oxford University Press.

Mason, K. O. (1997). Gender and demographic change: What do we know? In G. W. Jones et al. (Eds.), *The continuing demographic transition.* Oxford, Eng.: Clarendon Press.

Massey, D. S., & Denton, N. A. (1993). *American apartheid: Segregation and the making of the underclass.* Cambridge, MA: Harvard University Press.

Massey, D. S., & Fischer, M. J. (1998, December 14). Where we live, in black and white. *The Nation,* p. 25.

Massing, M. (2000, June 15). The narco-state? *New York Review of Books,* pp. 24–29.

Matras, J. (1973). *Populations and societies.* Englewood Cliffs, NJ: Prentice Hall.

Maxwell, M. (1991). *The sociobiological imagination.* Albany: State University of New York Press.

Mayfield, L. (1984). Teenage pregnancy. Unpublished doctoral dissertation, City University of New York.

Mayo, E. (1945). *The social problems of an industrial civilization.* Boston: Harvard University, Graduate School of Business Administration.

McAdam, D., McCarthy, J. D., & Zald, N. (1988). Social movements. In N. J. Smelser (Ed.), *The handbook of sociology.* Newbury Park, CA: Sage.

McAndrew, M. (1985). Women's magazines in the Soviet Union. In B. Holland (Ed.), *Soviet sisterhood.* Bloomington: Indiana University Press.

McFalls, J. A., Jr. (1998). Population: A lively introduction. *Population Bulletin,* vol. 53, no. 3. Washington, DC: Population Reference Bureau.

McGaw, J. A. (1987). Women and the history of American technology. In S. Harding & J. F. O'Barr (Eds.), *Sex and scientific inquiry.* Chicago: University of Chicago Press.

McGoldrick, M., & Carter, E. A. (1989). *The family life cycle in normal family processes,* 2nd ed. London: Guilford Press.

McGuire, M. B. (1987). *Religion: The social context* (2nd ed.). Belmont, CA: Wadsworth.

McKendrick, N., Brewer, J., & Plump, J. H. (1982). *The birth of a consumer society.* Bloomington: Indiana University Press.

McKinlay, J. B., & McKinlay, S. M. (1977). The questionable contribution of medical measures to the decline of mortality in the United States in the twentieth century. *Health and Society, 53,* 405.

McLanahan, S. S. (1984). Family structure and stress: A longitudinal comparison of two-parent and female-headed families. *Journal of Marriage and the Family, 24* 347–357.

McNeill, W. (1963). *The rise of the West: A history of the human community.* Chicago: University of Chicago Press.

McNeill, W. H. (1982). *The pursuit of power: Technology, armed force, and society since A.D. 1000.* Chicago: University of Chicago Press.

Mead, M. (1950). *Sex and temperament in three primitive societies.* New York: NAL Mentor.

Mehan, H. (1992). Understanding inequality in schools: The contribution of interpretative studies. *Sociology of Education, 65,* 1–20.

Meier, D. (1995). *The power of their ideas: Lessons from a small school in Harlem.* Boston: Beacon.

Merton, R. K. (1938). Social structure and anomie. *American Sociological Review, 3,* 672–682.

Merton, R. K. (1948). Discrimination and the American creed. In R. M. MacIver (Ed.), *Discrimination and national welfare.* New York: Institute for Religious and Social Studies; dist. by Harper.

Mĕstrovíc, S. G. (1991, September 25). Point of view: Why East Europe's upheavals caught social scientists off guard. *Chronicle of Higher Education,* p. A56.

Meyer, J. (1990). Guess who's coming to dinner this time? A study of gay intimate relationships and the support for those relationships. *Marriage and Family Review, 14,* 59–82.

Meyer, J. W., Ramirez, F. O., & Soysal, Y. N. (1992). World expansion of mass education, 1870–1980. *Sociology of Education, 65,* 128–149.

Milgram, S. (1970). The experience of living in cities. *Science, 167,* 1461–1468.

Milgram, S. (1974). *Obedience to authority: An experimental view.* New York: Harper.

Miller, D. C., & Form, W. H. (1964). *Industrial sociology* (2nd ed.). New York: Harper.

Miller, N. (1989). *In search of gay America: Women and men in a time of change.* New York: Atlantic Monthly Press.

Miller, W. B. (1958). Lower class culture as a generating milieu of gang delinquency. *Journal of Social Issues, 14,* 5–19.

Mills, C. W. (1956). *The power elite.* New York: Oxford University Press.

Mills, C. W. (1959). *The sociological imagination.* New York: Oxford University Press.

Modell, J., & Hareven, T. K. (1973). Urbanization and the malleable household: An examination of boarding and lodging in American families. *Journal of Marriage and the Family, 35,* 466–479.

Mollenkopf, J. (1985). *The contested city.* Princeton, NJ: Princeton University Press.

Money, J., & Tucker, P. (1975). *Sexual signatures.* Boston: Little, Brown.

Monnier, M. (1993). *Mapping it out: Expository cartography for the humanities and social sciences.* Chicago: University of Chicago Press.

Montagna, P. D. (1977). *Occupations and society.* New York: Wiley.

Moore, B. (1968). *The social origins of dictatorship and democracy: Lord and peasant in the making of the modern world.* Boston: Beacon.

Moore, H. (1998). The cultural constitution of gender. In *The polity reader in gender studies.* Cambridge, England: Polity Press.

Moscos, C. A. (1988). *Call to civic service.* New York: Free Press.

Mosley, W. H., & Cowley, P. (1991, December). *The challenge of world health.* Washington, DC: Population Reference Bureau.

Mulkey, L. M. (1993). *Sociology of education. Theoretical and empirical investigations.* Fort Worth, TX: Harcourt Brace.

Muller, W. (2000, July 24). From data to wisdom. *Forbes,* p. 384.

Murdock, G. (1983). *Outline of world cultures* (6th ed.). New Haven: Human Relations Area Files.

Murray, J. F. (2000). *Intensive care: A doctor's journal.* Berkeley: University of California Press.

Murti, M., Guio, A., & Dreze, J. (1995). Mortality, fertility, and gender bias in India. *Population and Development Review, 21,* 745–782.

Myers, D., & Wolch, J. R. (1995). In R. Farley (Ed.), *The state of the union.* New York: Russell Sage.

Nagel, T. (1994, May 12). Freud's permanent revolution. *New York Review of Books,* pp. 34–39.

Nasrin, T. (2000, June). No progress without a secular society. *UNESCO Courier,* p. 17.

National Center for Educational Statistics. (1995). *The educational progress of black students.* Washington, DC: U.S. Department of Education.

National Commission on America's Urban Families. (1993). *Families first.* Washington, DC: U.S. Government Printing Office.

Neihardt, J. G. (1959/1932). *Black Elk speaks: Being the life story of a holy man of the Oglala Sioux.* New York: Washington Square Press.

Nelson, G. K. (1984). Cults and new religions: Towards a sociology of religious creativity. *Sociology and Social Research, 68,* 301–325.

Nesterenko, A. (2000, September). Godfather of the Kremlin: The modernization challenge facing President Putin. *Finance & Development,* p. 20.

Neugarten, B. L. (1969). Continuities and discontinuities of psychological issues into adult life. *Human Development, 12,* 121–130.

Newman, K. S. (1988). *Falling from grace.* New York: Free Press.

Niebuhr, G. (1995, October 3). With every wave of newcomers, a church more diverse. *New York Times,* p. B6.

Niebuhr, H. R. (1929). *The social sources of denominationalism.* New York: Meridian.

Ning, P. (1995). *Red in tooth and claw: Twenty-six years in communist Chinese prisons.* New York: Grove Press.

Nisbet, R. A. (1969). *Social change and history.* New York: Oxford University Press.

Nisbet, R. A. (1970). *The social bond.* New York: Knopf.

NORC (National Opinion Research Center) [Various years]. General social survey, cumulative codebook. Chicago: University of Chicago Press.

Oakes, J. (1985). *Keeping track: How schools structure inequality.* New Haven, CT: Yale University Press.

Oberschall, A. (1973). *Social conflict and social movements.* Englewood Cliffs, NJ: Prentice Hall.

O'Hare, W. P. (1996). A new look at poverty in America. *Population Bulletin,* vol. 51, no. 2. Washington, DC: Population Reference Bureau.

Okun, B. F. (1996). *Understanding diverse families: What practitioners need to know.* New York: Guilford Press.

Oldenburg, R. (1989). *The great good place.* New York: Paragon.

Oliver, M. L., & Shapiro, T. M. (1990). Wealth of a nation: A reassessment of asset inequality in America shows at least one third of households are asset-poor. *American Journal of Economics and Sociology, 49,* 129–150.

Olojede, D. (1995, June 5). Chaos lurks in Nigeria. *Newsday,* pp. A6, A22.

O'Malley, E. (1996). Post-social criminologies. *Current Issues in Criminal Justice, 8,* 26.

Oppenheimer, V. K. (1994). Women's rising employment and the future of the family in industrial societies. *Population and Development Review, 20,* 293–342.

Orfield, G. (1999, Fall). The resegregation of our nation's schools: A troubling trend. *Civil Rights Journal,* p. 8.

Orr, D. W. (1998, November-December). Technological fundamentalism. *The Ecologist,* pp. 329–332.

Osborne, L. (1999, October 24). A linguistic big bang. *New York Times Magazine,* pp. 84–89.

Osmond, M. W. (1985). Comparative marriage and the family. In B. C. Miller & D. H. Olson (Eds.), *Family studies review yearbook* (Vol. 3). Newbury Park, CA: Sage.

Palmore, E. (1981). *Social patterns in normal aging.* Durham, NC: Duke University Press.

Parclius, A. P., & Parclius, R. J. (1987). *The sociology of education* (2nd ed.). Englewood Cliffs, NJ: Prentice Hall.

Park, R. E. (1967/1925a). The city: Suggestions for the investigation of human behavior in the urban environment. In R. E. Park & E. W. Burgess (Eds.), *The city.* Chicago: University of Chicago Press.

Park, R. E. (1967/1926). The urban community as a spatial pattern and a moral order. In R. H. Turner (Ed.), *Robert E. Park on social control and collective behavior.* Chicago: University of Chicago Press.

Park, R. E., & Burgess, E. W. (1921). *Introduction to the science of sociology* (Rev. ed.). Chicago: University of Chicago Press.

Parkin, F. (1971). *Class inequality and political order.* New York: Praeger.

Parsons, T. (1937). *The structure of social action.* New York: McGraw-Hill.

Parsons, T. (1940). An analytic approach to the theory of social stratification. *American Journal of Sociology, 45,* 841–862.

Parsons, T. (1951). *The social system.* New York: Free Press.

Parsons, T. (1960). *Structure and process in modern societies.* New York: Free Press.

Parsons, T. (1966). *Societies: Evolutionary and comparative.* Englewood Cliffs, NJ: Prentice Hall.

Parsons, T. (1968). The problem of polarization along the axis of color. In J. H. Franklin (Ed.), *Color and race.* Boston: Beacon.

Parsons, T., & Bales, R. F. (1955). *Family, socialization, and interaction process.* New York: Free Press.

Partin, M., & Palloni, A. (1995, Spring). Accounting for the recent increases in low birth weight among African Americans. *Focus,* pp. 33–38.

Patterson, O. (1982). *Slavery and social death.* Cambridge, MA: Harvard University Press.

Pavlov, I. (1927). *Conditioned reflexes: An investigation of the physiological activity of the cerebral cortex* (G. V. Anrep, Trans. & Ed.). London: Oxford University Press.

Pedersen, E., & Faucher, T. A., with Eaton, W. W. (1978). A new perspective on the effects of first-grade teachers on children's subsequent adult status. *Harvard Educational Review, 48,* 1–31.

Perez-Lopez, J., & Diaz, S-B. (1998). The determinants of Hispanic remittances: An exploration using U.S. census data. *Hispanic Journal of Behavioral Sciences, 20,* 320–348.

Perin, C. (1977). *Everything in its place: Social order and land use in America.* Princeton, NJ: Princeton University Press.

Peterson, D., & Goodall, J. (1973). *Visions of Caliban: On chimpanzees and people.* Boston: Houghton Mifflin.

Pfeiffer, E., Verwoerdt, A., & Davis, G. (1972). Sexual behavior in middle life. *American Journal of Psychiatry, 128,* 1262–1267.

Philips, S. (1982). *The invisible culture: Communications in classroom and community on the Warmsprings Indian Reservation.* White Plains, NY: Longman.

Piaget, J., & Inhelder, B. (1969). *The psychology of the child.* New York: Basic Books.

Pinkewr, S. (1994). *The language instinct.* New York: Morrow.

Piore, M. J., & Sabel, C. F. (1984). *The second industrial divide: Possibilities for prosperity.* New York: Basic Books.

Piven, F. F., & Cloward, R. A. (1996). Northern Bourbons: A preliminary report on the National Voter Registration Act. PS: Political Science 7 *Politics, 29,* 39–42.

Polansky, N. A., Chalmers, M. A., Buttenweiser, E., & Williams, D. P. (1981). *Damaged parents, an anatomy of child neglect.* Chicago: University of Chicago Press.

Polanyi, K. (1944). *The great transformation.* Boston: Beacon.

Pope, L. (1942). *Millhands and preachers.* New Haven, CT: Yale University Press.

Popenoe, D. (1994). Family decline and scholarly optimism. *Family Affairs, 6,* 9–10.

Popenoe, D. (1996). *Life without father: Compelling evidence that fatherhood and marriage are indispensable for the good of children and society.* New York: Free Press.

Population Reference Bureau. (2000). 2000 world population data sheet. Washington, DC: U.S. Government Printing Office.

Population Today (1997, August). Infectious diseases continue to threaten world health.

Portes, A. (Ed.) (1996). *The new second generation.* New York: Russell Sage.

Portes, A., & Rumbaut, R. G. (1990). *Immigrant America: A portrait.* Berkeley: University of California Press.

Pressman, J. L., & Wildavsky, A. (1984). *Implementation* (3rd ed.). Berkeley: University of California Press.

Preston, S. H. (1984). Children and the elderly: Divergent paths for America's dependents. *Demography, 21,* 435–457.

Putnam, R. (1995). Bowling alone: America's declining social capital. *Journal of Democracy, 6,* 65–78.

Quadagno, J. S. (1979). Paradigms in evolutionary theory—Sociobiological models of natural selection. *American Sociological Review, 44,* 100–109.

Quinney, R. (1978). The ideology of law: Notes for a radical alternative to legal oppression. In C. E. Reasons & R. M. Rich (Eds.), *Sociology of law: A conflict perspective.* Toronto: Butterworths.

Quinney, R. (1980). *Class, state, and crime.* White Plains, NY: Longman.

Radelet, M. L., & Bedau, H. A. (1992). *In spite of innocence: Erroneous convictions in capital cases.* Boston: Northeastern University Press.

Rathje, W. L. (1993). Less fat? Aw, baloney. Garbage Project studies indicate misreporting of fatty meat consumption. *Garbage, 5,* 22–23.

Ratzen, S. C., Filerman, G. L., & LeSar, J. W. (2000, March). Attaining global health: Challenge and opportunity. *Population Bulletin,* vol. 55, no. 1. Washington, DC: Population Reference Bureau.

Ravenholt, R. G. (1990). Tobacco's global death march. *Population and Development Review, 16,* 213–240.

Reich, R. B. (1992, Winter). Training a skilled work force. *Dissent,* pp. 42–46.

Richards, E. (1989). *The knife and gun club: Scenes from an emergency room.* New York: Atlantic Monthly Press.

Richie, M. F. (2000, June). America's diversity and growth. *Population Bulletin,* Vol. 55, no. 2. Washington, DC: Population Reference Bureau.

Richmond-Abbott, M. (1992). *Masculine and feminine: Gender roles over the life cycle* (2nd ed.). New York: McGraw-Hill.

Riggs, F. W. (1997). Modernity and bureaucracy. *Public Administration Review, 57,* 347–353.

Riis, J. A. (1890). *How the other half lives: Studies among the tenements of New York.* New York: Scribner.

Riley, M. W., Foner, A., & Waring, J. (1988). The sociology of age. In N. E. Smelser (Ed.), *The handbook of sociology.* Newbury Park, CA: Sage.

Riley, M. W., Kahn, R. L., & Foner, A. (1994). *Age and structural lag.* New York: Wiley.

Rist, R. C. (1973). *The urban school: A factory for failure.* Cambridge, MA: MIT Press.

Roberts, S. (1995, April 27). Women's work: What's new, what isn't. *New York Times,* p. B6.

Robinson, P. (1994). *Freud and his critics.* Berkeley: University of California Press.

Robinson, W. C. (1963). Urbanization and fertility: The non-western experience. *Millbank Memorial Fund Quarterly, 4,* 291–308.

Rock, P. (1985). Symbolic interactionism. In A. Kuper & J. Kuper (Eds.), *The social science encyclopedia.* London: Routledge.

Rockett, I. R. H. (1999). Population and health: An introduction to epidemiology. *Population Bulletin,* vol. 54, no. 4. Washington, DC: Population Reference Bureau.

Rose, M. (1995). *Possible lives: The promise of public education.* Boston: Houghton Mifflin.

Rosenhan, D. L. (1973). On being sane in insane places. *Science, 179,* 250–258.

Rosow, I. (1965). Forms and functions of adult socialization. *Social Forces, 44,* 38–55.

Ross, H. L. (1963). *Perspectives on the social order.* New York: McGraw-Hill.

Ross, S. (2000, September 18). Lower income users get online as gap in digital divide closes. *Brandweek,* p. 56.

Rossi, A. (1977). The biosocial basis of parenting. *Daedalus, 106,* 1–31.

Rossi, A. (1980). Aging and parenthood in the middle years. In P. B. Baltes & O. G. Brim, Jr. (Eds.), *Life-span development and behavior* (Vol. 3). Orlando: Academic Press.

Rossi, A. S., & Rossi, P. H. (1990). *Of human bonding: Parent–child relationships across the life course.* Hawthorne, NY: Aldine.

Roszak, T. (1969). *The making of a counterculture.* Garden City, NY: Doubleday.

Rubington, E., & Weinberg, M. S. (1996). *Deviance: The interactionist perspective* (6th ed.). Needham Heights, MA: Allyn & Bacon.

Russo, J. (1999). Strategic campaigns and international collective bargaining: The case of the IBT, FIET, and Royal Ahold NV. *Labor Studies Journal, 24,* 23.

Rutter, M. (1974). *The qualities of mothering: Maternal deprivation reassessed.* New York: Jason Aronson.

Ryan, A. (1992). Twenty-first century limited. *New York Review of Books, 39,* 20–24.

Rymer, R. (1992a, April 13). A silent childhood. *New Yorker,* pp. 41–53.

Rymer, R. (1992b, April 20). A silent childhood, part II. *New Yorker,* pp. 43–47.

Sachs, S. (2000, June 10) Egyptian's arrest seen as penalty for criticism. *New York Times.*

Sadik, N. (1995). Decisions for development: Women, empowerment and reproductive health. In *The state of the world population 1995.* New York: United Nations Population Fund.

Sagarin, E. (1975). *Deviants and deviance: A study of disvalued people and behavior.* New York: Praeger.

Sahlins, M. D. (1960, September). The origin of society. *Scientific American,* pp. 76–87.

Salisbury, R. F. (1962). *From stone to steel.* Parkville, Australia: Melbourne University Press.

Samuelson, P. (1980). *Economics* (11th ed.). New York: McGraw-Hill.

Sanchez-Jankowski, M. S. (1991). *Islands in the street: Gangs and American urban society.* Berkeley: University of California Press.

Sandefur, G. D., & Tienda, M. (Eds.) (1988). *Divided opportunities: Minorities, poverty, and social policy.* New York: Plenum Press.

Sassen, S. (1991). *Global city.* Princeton, NJ: Princeton University Press.

Savage, M. (1997). Social mobility, individual ability, and the inheritance of class inequality. *Sociology, 31,* 645–672.

Scanzoni, L., & Scanzoni, J. (1976). *Men, women, and change.* New York: McGraw-Hill.

Scarf, M. (1995). *Intimate worlds: Life inside the family.* New York: Random House.

Schaeffer, R. (1999). Civilization and its discontents: Essays on the new mobility of people and money (Saskia Sassen, New Press, 1998) (Review). *Social Forces, 77,* 1197–1198.

Schaller, G. B. (1964). *The year of the gorilla.* Chicago: University of Chicago Press.

Schuman, H., & Scott, J. (1989). Generations and collective memories. *American Sociological Review, 54,* 359–381.

Schumpeter, J. A. (1950). *Capitalism, socialism, and democracy* (3rd ed.). New York: Harper.

Schur, E. M. (1973). *Radical nonintervention: Rethinking the delinquency problem.* Englewood Cliffs, NJ: Prentice Hall.

Schur, E. M. (1984). *Labeling women deviant: Gender, stigma, and social control.* New York: Random House.

Schwartz, F. (1989, January–February). Management women and the new facts of life. *Harvard Business Review,* pp. 65ff.

Sconing, J. (1999). Cracking the bell curve myth (review). *Journal of the American Statistical Association, 94,* 335–337.

Scull, A. T. (1988). Deviance and social control. In N. Smelser (Ed.), *The handbook of sociology.* Newbury Park, CA: Sage.

Seel, N., & Casey, N. (2001). *Teens and technology.* New York: Russell Sage.

Seiden, P. A. (2000, Summer). Bridging the digital divide. *Reference & User Services Quarterly,* p. 329.

Selznick, P. (1952). *The organizational weapon: A study of Bolshevik strategy and tactics.* New York: McGraw-Hill.

Sen, A. (1981). Poverty and famines: An essay on entitlement and deprivation. New York: Oxford University Press.

Sen, A. (1993, May). The economics of life and death. *Scientific American,* pp. 40–47.

Sen, A. (2000). *Development as freedom.* New York: Knopf.

Sennett, R., & Cobb, J. (1972). *The hidden injuries of class.* New York: Random House.

Serfaty, S. (2000, Autumn). Europe 2007: From nation-states to member states. *Washington Quarterly,* p. 15.

Shanks, B. (1991). The endangered ranger. *National Parks, 65,* 32–36.

Shavelson, L. (1998). *A chosen death.* Berkeley: University of California Press.

Shaw, C. R. (1929). *Delinquency areas: A study of the geographic distribution of school truants, juvenile delinquents, and adult offenders in Chicago.* Chicago: University of Chicago Press.

Shaw, J. A. (1979). The child in the military community. In J. D. Call, J. D. Noshpitz, R. L. Cohen, & I. N. Berlin (Eds.), *Basic handbook of child psychiatry.* New York: Basic Books.

Sheehan, J. P. (2000, March). Caring for the deaf. *RN,* p. 69.

Shibutani, T., & Kwan, K. M. (1965). Ethnic stratification. New York: Macmillan.

Shils, E. (1962). The theory of mass society. *Diogenes, 39,* 45–66.

Shils, E. (1963). On the contemporary study of the new states. In C. Geertz (Ed.), *New societies and old states: The quest for modernity in Asia and Africa.* New York: Free Press.

Shils, E. (1970). Tradition, ecology, and institution in the history of sociology. *Daedalus, 99,* 760–825.

Shorris, E. (1992). *Latinos: A biography of the people.* New York: Norton.

Shweder, R. H. (2000, Summer). Rethinking the object of anthropology. *Items: Social Science Research Council, 1,* 7–9.

Siegal, L., & Sullivan, D. (1997). Temporary services employment durations: Evidence from state UI data. Working Paper WP-97–23, Federal Reserve Bank of Chicago.

Silberman, C. (1980). *Criminal violence, criminal justice.* New York: Random House.

Silk, L. (1989, May 12). Rich and poor: The gap widens. *New York Times,* p. D2.

Simmel, G. (1904). The sociology of conflict. *American Journal of Sociology, 9,* 490ff.

Simmons, J. L. (1985). The nature of deviant subcultures. In E. Rubington & M. S. Weinberg (Eds.), *Deviance: The interactionist perspective.* New York: Macmillan.

Simpson, C. E., & Yinger, J. M. (1953). *Racial and cultural minorities: An analysis of prejudice and discrimination.* New York: Harper.

Simpson, J. H. (1983). Moral issues and status politics. In R. C. Liebman & R. Wuthnow (Eds.), *The new Christian right: Mobilization and legitimation.* Hawthorne, NY: Aldine.

Sizer, T. (1995, January 8). What's wrong with standard tests. *New York Times,* Education Life, p. 58.

Skinner, B. F. (1976). *Walden two.* New York: Macmillan.

Skocpol, T. (1979). *States and social revolutions.* New York: Cambridge University Press.

Skolnick, A. S. (1991). *Embattled paradise: The American family in an age of uncertainty.* New York: Basic Books.

Smelser, N. J. (1962). *Theory of collective behavior.* New York: Free Press.

Smelser, N. J. (1966). The modernization of social relations. In M. Weiner (Ed.), *Modernization.* New York: Basic Books.

Smelser, N. J. (1984). *Sociology.* Englewood Cliffs, NJ: Prentice Hall.

Smith, A. (1910/1776). *The wealth of nations.* London: University Paperbacks.

Smith, C. D. (1994). *The absentee American: Repatriates' perspectives on America.* Putnam Valley, NY: Aletheia.

Smith, D. (1989). Promises (Letter from Madurai). Oberlin, OH: Oberlin Shansi Memorial Association.

Smith, H. (1952). *The religions of man.* New York: Mentor.

Smith, H. (1976). *The Russians.* New York: Ballantine.

Smith, R. A. (1968). Los Angeles, prototype of supercity. In S. F. Fava (Ed.), *Urbanism in world perspective: A reader.* New York: Crowell.

Snipp, C. M. (1991). *American Indians: The first of this land.* New York: Russell Sage.

Sorenson, A. B., Weinert, F. E., & Sherrod, L. R. (Eds.). (1986). *Human development and the life course: Multidisciplinary perspectives.* Hillsdale, NJ: Erlbaum.

Sorokin, P. (1937). *Social and cultural dynamics: Vol. 3. Fluctuation of social relationships, war, and revolution.* New York: American Book.

Spain, D. (2000). *How women saved the city.* Minneapolis: University of Minnesota Press.

Spencer, H. (1874). *The study of sociology.* New York: Appleton.

Stacey, J. (1990). *Brave new families: Stories of domestic upheaval in late twentieth century America.* New York: Basic Books.

Stack, C. (1974). *All our kin.* New York: Harper.

Staggenborn, S. (1998). *Gender, family and social movements.* Thousand Oaks, CA: Pine Forge Press.

Stark, R., & Bainbridge, W. S. (1979). Of churches, sects, and cults: Preliminary concepts for a theory of religious movements. *Journal for the Scientific Study of Religion, 18,* 117–133.

Stark, R., & Bainbridge, W. S. (1985). *The future of religion: Secularization, revival, and cult formation.* Berkeley: University of California Press.

Stark, R., & Bainbridge, W. S. (1997). *Religion, deviance, and social control.* New York and London: Routledge.

Starr, P. (1982). *The social transformation of American medicine.* New York: Basic Books.

Statistical abstract of the United States [Various years]. Washington, DC: U.S. Bureau of the Census.

Steinberg, J. (2000, December 30). Roderick Raynor Paige. *New York Times,* p. A10.

Steltzer, U. (1982). *Inuit: The north in transition.* Chicago: University of Chicago Press.

Stevens, W. K. (1988, December 20). Life in the Stone Age: New findings point to complex societies. *New York Times,* pp. C1, C15.

Stevenson, H. W. (1992, December). Learning from Asian schools. *Scientific American,* pp. 70–76.

Stevenson, H. W. (1998, March). A study of three cultures: Germany, Japan, and the United States–An overview of the TIMSS Case Study Project. *Phi Delta Kappan,* pp. 524–529.

Stockard, J., & Johnson, M. M. (1992). *Sex and gender in society.* Englewood Cliffs, NJ: Prentice Hall.

Stouffer, S. A., Suchman, E. A., DeVinney, L. C., Star, S. A., & Williams, R. A., Jr. (1949). *Studies in social psychology in World War II: Vol. 1. The American soldier: Adjustment during army life.* Princeton, NJ: Princeton University Press.

Sudnow, D. (1967). *Passing on: The social organization of dying.* Englewood Cliffs, NJ: Prentice Hall.

Sullivan, A. (1996, June 3). Let gays marry. *Newsweek,* p. 26.

Sullivan, W. M. (1999, March). What is left of professionalism after managed care? *Hastings Center Report,* p. 7.

Sumner, W. G. (1940/1907). *Folkways.* Boston: Ginn.

Sumner, W. G. (1963/1911). *Social Darwinism: Selected essays.* Englewood Cliffs, NJ: Prentice Hall.

Sutherland, E. H. (1940). White collar criminality. *American Sociological Review, 5,* 1–12.

Suttles, G. (1972). *The social construction of communities.* Chicago: University of Chicago Press.

Swartz, D. (1997). *Culture and power: The sociology of Pierre Bourdieu.* Chicago: University of Chicago Press.

Szelenyi, I. (1983). *Urban inequalities under state socialism.* New York: Oxford University Press.

Tabb, W. (1986). *Churches in struggle: Liberation theologies and social change.* New York: Monthly Review Press.

Taeuber, K. E., & Taeuber, A. F. (1965). *Negroes in cities.* Hawthorne, NY: Aldine.

Tarrow, S. G. (1994). *Power in movement: Social movements, collective action, and politics.* Cambridge, England: Cambridge University Press.

Taylor, J. M., Gilligan, C., & Sullivan, A. M. (1995). *Between voice and silence: Women and girls, race and relationship.* Cambridge, MA: Harvard University Press.

Thomas, L. (1979). *The medusa and the snail.* New York: Bantam.

Thomas, W. I. (1971/1921). *Old world traits transplanted.* Montclair, NJ: Patterson Publishers.

Thorne, B. (1993). *Gender play: Boys and girls in school.* New Brunswick, NJ: Rutgers University Press.

Thornton, R. (1987). *American Indian holocaust and survival: A population history since 1492.* Norman: University of Oklahoma Press.

Thrasher, F. M. (1963/1926). *The gang: A study of 1,313 gangs in Chicago* (Abridged ed. by J. F. Short, Jr.). Chicago: University of Chicago Press.

Tienda, M., & Singer, A. (1995). Wage mobility of undocumented workers in the United States. *International Migration Review, 29,* 112–138.

Tienda, M., & Wilson, F. D. (1992). Migration and the earnings of Hispanic men. *American Sociological Review, 57,* 661–678.

Tobach, E., & Rosoff, B. (Eds.) (1994). *Challenging racism and sexism: Alternatives to genetic explanations.* New York: Feminist Press at the City University of New York.

Tobin, J. J., Wu, D. Y. H., & Davidson, D. H. (1989). *Preschool in three cultures: Japan, China, and the United States.* New Haven, CT: Yale University Press.

Tönnies, F. (1957/1887). *Community and society* (C. P. Loomis, Trans. & Ed.). East Lansing: Michigan State University Press.

Touraine, A. (1971). *The post industrial society: Tomorrow's social history: Classes, conflicts and culture in the programmed society.* New York: Random House.

Troeltsch, E. (1931). *The social teachings of the Christian churches* (O. Wyon, Trans.). New York: Macmillan.

Trow, M. (1966). The second transformation of American secondary education. In R. Bendix & S. M. Lipset (Eds.), *Class, status, and power* (2nd ed.). New York: Free Press.

Truzzi, M. (1971). *Sociology: The classic statements.* New York: McGraw-Hill.

Tumin, M. M. (1967). *Social stratification: The forms and functions of inequality.* Englewood Cliffs, NJ: Prentice Hall.

Turner, R. H. (1974). The theme of contemporary social movements. In R. E. L. Faris (Ed.), *Handbook of modern sociology.* Chicago: Rand McNally.

Uchitelle, L. (1993, May 14). Pay of college graduates is outpaced by inflation. *New York Times,* pp. A1, B12.

UCR (Federal Bureau of Investigation) [Annual]. *Crime in the United States* (Uniform Crime Reports). Washington, DC: Government Printing Office.

Uhlenberg, P. (1980). Death and the family. *Journal of Family History, 5,* 313–320.

UN Chronicle (2000, Spring). Digital divide to digital dividend. P. 46.

United Nations (1987). *The prospects of world urbanization.* New York: United Nations.

United Nations (1995a, March). *Adoption of the declaration and programme of action of the World Summit for Social Development.* New York: United Nations.

United Nations (1995b). *Gender, population & development: The role of the United Nations Population Fund.* New York: United Nations, UNFPA.

United Nations (1995c, March). *Summary of the programme of action of the International Conference on Population and Development.* New York: United Nations, Department of Public Information.

UNDP (United Nations Development Programme) (Various years). *Human development report.* Geneva: United Nations.

U.S. Department of Education. (1994). *High school students ten years after "A Nation at Risk": Findings from the condition of education 1994.* Washington, DC: U.S. Department of Education, Office of Educational Research and Development.

van den Berghe, P. L. (1975). *Man in society: A biosocial view.* New York: Elsevier.

Van Gennep, A. (1960/1908). *The rites of passage.* Chicago: University of Chicago Press.

Verbrugge, L. M. (1985). Gender and health: An update on hypotheses and evidence. *Journal of Health and Social Behavior, 26,* 156–182.

Vilas, C. M. (1993). The hour of civil society. *NACLA Report on the Americas, 27,* 38–43.

Vining, D. R., Jr. (1985, April). The growth of core regions in the Third World. *Scientific American,* pp. 42–49.

Vondra, J. (1996, April). Resolving conflicts over values. *Educational Leadership,* pp. 76–79.

Waldrop, J. (1994). More than a typist. *American Demographics, 16,* 4.

Wallace, B. (1999, October 18). Quebec separatism dominated International Conference on Federalism. *Maclean's,* p. 38.

Wallbank, T. W. (1996). *Civilization past and present.* New York: HarperCollins.

Wallerstein, I. (1974). *The modern world system: Capitalist agriculture and the origins of the European world-economy in the sixteenth century.* Orlando: Academic Press.

Wallerstein, J., & Blakeslee, S. (1989). *Second chances: Men, women & children a decade after divorce.* New York: Ticknor & Fields.

Walton, J. (1970). A systematic survey of community power research. In M. Aiken & P. Mott (Eds.), *The structure of community power.* New York: Random House.

Walton, J. (1992). *Western times and water wars: State, culture, and rebellion in California.* Berkeley: University of California Press.

Waltzer, M. (1980). Pluralism. In *Harvard encyclopedia of American ethnic groups.* Cambridge, MA: Belknap Press.

Warner, W. L., Meeker, M., & Calls, K. (1949). *Social class in America: A manual of procedure for the measurement of social status.* Chicago: Science Research Associates.

Washburn, P. C. (1982). *Political sociology.* Englewood Cliffs, NJ: Prentice Hall.

Watson, J. B. (1930). *Behaviorism.* New York: Norton.

Webb, E., Campbell, D. T., Schwarz, R. D., & Sechrest, L. (1966). *Unobtrusive measures: Nonreactive research in the social sciences.* Chicago: Rand McNally.

Weber, M. (1922). *Gesammelte aufsatze zur Religionssoziologie.* Tubingen, Germany: Mohr.

Weber, M. (1947). *The theory of social and economic organization* (A. M. Henderson & T. Parsons, Trans.). New York: Free Press.

Weber, M. (1958/1922). Economy and society. In H. Gerth & C. W. Mills (Trans. & Eds.), *From Max Weber: Essays in sociology.* New York: Oxford University Press.

Weber, M. (1962/1921). *The city.* New York: Collier.

Weber, M. (1963/1922). *The sociology of religion* (E. Fischoff, Trans.). Boston: Beacon.

Weber, M. (1968). The concept of citizenship. In S. N. Eisenstadt (Ed.), *Max Weber on charisma and institution building.* Chicago: University of Chicago Press.

Weber, M. (1974/1904). *The Protestant ethic and the spirit of capitalism* (T. Parsons, Trans.). New York: Scribner.

Weinberg, H. (1994). Marital reconciliation in the United States: Which couples are successful? *Journal of Marriage and the Family, 56,* 80–88.

Weinstein, H. (2000, April 12). Court bars border stops based on ethnicity. *Los Angeles Times,* p. 1.

Weisberger, B. A. (1997, September). What made the government grow? *American Heritage,* pp. 34–45.

Wellins, S. (1990). *Children's use of television in England and the United States.* Doctoral dissertation, City University of New York.

Wells, M. J. (1984). The resurgence of sharecropping: Historical anomaly or political strategy? *American Journal of Sociology, 90,* 1–30.

Wells, M. J. (1996). *Strawberry fields: Politics, class and work in California agriculture.* Ithaca; Cornell University Press.

West, C. (1992, February 8). Learning to talk of race. *New York Times Magazine,* pp. 24–25.

Westin, A. (1967). *Privacy and freedom.* New York: Atheneum.

White, M. J. (1987). *American neighborhoods and residential differentiation.* New York: Russell Sage.

WHO (World Health Organization) (1994). *Women and AIDS: Agenda for action.* Geneva: World Health Organization.

Whorf, B. L. (1961). The relation of habitual thought and behavior to language. In J. B. Carroll (Ed.), *Language, thought, and reality: Selected writings of Benjamin Lee Whorf.* Cambridge, MA: MIT Press.

Whyte, W. F. (1943). *Street corner society.* Chicago: University of Chicago Press.

Whyte, W. F. (1949). The social structure of the restaurant. *American Journal of Sociology, 54,* 302–310.

Whyte, W. F. (1984). *Learning from the field.* Newbury Park, CA: Sage.

Wickens, B. (1996, February 19). Fretting over the undercover nest egg. *Maclean's,* p. 13.

Wilhelm, A. G., & Thierer, A. D. (2000, September 4). Symposium. *Insight on the News,* p. 40.

Williams, T. (1989). *The cocaine kids.* Reading, MA: Addison-Wesley.

Williams, T. (1992). *Crack house.* Reading, MA: Addison-Wesley.

Willis, P. (1983). Cultural production and theories of reproduction. In L. Barton & S. Walker (Eds.), *Race, class, and education.* London: Croom-Helm.

Willmott, P., & Young, M. (1971). *Family and class in a London suburb.* London: New American Library.

Wilson, B. L. (1993). *Mandating academic excellence: High school responses to state curriculum reform.* New York: Teachers College Press.

Wilson, E. O. (1975). *Sociobiology.* Cambridge, MA: Belknap Press.

Wilson, E. O. (1979). *On human nature.* New York: Bantam.

Wilson, J. Q. (1977). *Thinking about crime.* New York: Vintage Books.

Wilson, W. J. (1978). *The declining significance of race: Blacks and changing American institutions.* Chicago: University of Chicago Press.

Wilson, W. J. (1984). The urban underclass. In L. W. Dunbar (Ed.), *The minority report.* New York: Pantheon.

Wilson, W. J. (1987). *The truly disadvantaged: The inner city, the underclass, and public policy.* Chicago: University of Chicago Press.

Wilson, W. J. (1996). *When work disappears: The world of the urban poor.* New York: Knopf.

Winik, M. F. (1996, November 13). Sharing a legacy of rescue. *The Christian Century,* pp. 1112–1116.

Wirth, L. (1945). The problem of minority groups. In R. Linton (Ed.), *The science of man in the world crisis.* New York: Columbia University Press.

Wirth, L. (1968/1938). Urbanism as a way of life. In S. F. Fava (Ed.), *Urbanism in world perspective: A reader.* New York: Crowell.

Wittfogel, K. (1957). *Oriental despotism: A comparative study of total power.* New Haven, CT: Yale University Press.

Wixen, B. N. (1979). Children of the rich. In J. D. Call, J. D. Noshpitz, R. L. Cohen, & I. N. Berlin (Eds.), *Basic handbook of child psychiatry.* New York: Basic Books.

Wolf, E. R. (1984a, November 4). The perspective of the world. *New York Times Book Review,* pp. 13–14.

Wolf, E. R. (1984b, November 4). Unifying the vision. *New York Times Book Review,* p. 11.

Wolff, E. N. (1995). *Top heavy: A study of the increasing inequality of wealth in America.* New York: Twentieth Century Fund.

Wolfgang, M. E., & Riedel, M. (1973). Race, judicial discretion, and the death penalty. *Annals of the American Academy of Political and Social Science, 407,* 119–133.

Woodall, J. (1997, August). Infectious diseases continue to threaten world health. *Population Today,* pp. 36–40.

Woodman. S. (1998). *Last rights: The struggle over the right to die.* New York: Plenum.

Wren, C. S. (1998, January 9). Drugs or alcohol linked to 80% of inmates. *New York Times,* p. A14.

Wren, D. J. (1997). Adolescent females' "voice" changes can signal difficulties for teachers and administrators. *Adolescence, 32,* 463–470.

Wright, E. O. (1989). *The debate on classes.* New York: Verso.

Wright, R., & Jacobs, J. A. (1994). Male flight from computer work: A new look at occupational resegregation and ghettoization. *American Sociological Review, 59,* 511–536.

Wrigley, E. A. (1969). *Population and history.* New York: McGraw-Hill.

Wu, H., & Wakeman, C. (1995). *Bitter winds: A memoir of my years in China's gulag.* New York: Wiley.

Wuthnow, R. (1988). Sociology of religion. In N. E. Smelser (Ed.), *The handbook of sociology.* Newbury Park, CA: Sage.

Zerubavel, E. (1986). *Hidden rhythms: Schedules and calendars in social life.* Chicago: University of Chicago Press.

Zill, N. (1995, Spring). Back to the future: Improving child indicators by remembering their origins. *Focus* (University of Wisconsin, Institute for Research on Poverty), pp. 17–24.

Zimmerman, E., & Newman, J. D. (1995). *Current topics in primate vocal communication.* New York: Plenum Press.

CREDITS AND ACKNOWLEDGMENTS

Credits

02 all	© PhotoDisc
04	AP / Wide World Photos
08	Courtesy of Manjula Giri
11	AP / Wide World Photos
16	Courtesy of Joyce Ladner
26	AP / Wide World Photos
33	© PhotoDisc, © Digital Vision
34	© Susan Meiselas / Magnum Photos
37	© Seth Mydans / NYT Pictures
45	AP / Wide World Photos
46	© Macduff Everton / CORBIS
51	© AFP / CORBIS
52	©Susan Meiselas / Magnum Photos
53	AP / Wide World Photo
58 all	© PhotoDisc
60	© Reuters / TimePix
63 l	© Araldo de Luca / CORBIS
63 r	AP / Wide World Photos
67 l	© Archive Photos
67 r	© Bob Daemmrich / The Image Works
70	© Chris Heflin / Venice Arts Mecca
80 l	©1988 by Muzafer Sherif.
80 r	©1988 by Muzafer Sherif.
86 all	© PhotoDisc
88	© Eugene Richards
94 l	© David Hiser / Photographers Aspen / Picture Quest
94 r	© Lowell Georgia / CORBIS
95 l	© Phil Schermeister / CORBIS
95 r	© James A. Sugar / CORBIS
96 tl	© Nicholas Sapieha / Art Resource, NY
96 tr	© John Neubauer
96 bl	© Culver Pictures / Picture Quest
96 br	© Ed Kashi / CORBIS
112	© David Kampfner / Liaison International
115 t	© Eugene Richards
115 b	© Eugene Richards
122 all	© PhotoDisc
124	near the church at Santa Chiara. Palermo, 1982 © Letizia Battaglia, Letizia Battaglia: Passion, Justice, Freedom—Photographs of Sicily, New York: Aperture, 1999.
126	AP / Wide World Photos
131 l	Stock Montage, Inc.
131 r	AP / Wide World Photos
133	© David & Peter Tumley / CORBIS
142	AP / Wide World Photos
145 t	Letizia Battaglia. © Franco Zecchin, Letizia Battaglia: Passion, Justice, Freedom—Photographs of Sicily, New York: Aperture, 1999.
145 b	the wife and daughters of Benedetto Grado at the site of his murder, Palermo. November 11, 1983. © Franco Zecchin, Letizia Battaglia: Passion, Justice, Freedom—Photographs of Sicily, New York: Aperture, 1999.
150	© PhotoDisc, © Corel, © Pictor International / Pictor International, Ltd. / PictureQuest
152	© Lynn Johnson / Aurora / PictureQuest
159	© Beryl Goldberg
168 r	© Bill Varie / CORBIS
168 l	AP / Wide World Photos
176	© Nathan Benn / Woodfin Camp & Associates
180	© Annie Griffiths Belt / CORBIS
186 all	© PhotoDisc
188	AP / Wide World Photos
193 l	© Bettmann / CORBIS
193 r	© David & Peter Tumley / CORBIS
200	© AFP / CORBIS
209	© Martha Stewart
210 l	© Reuters NewMedia, Inc. / CORBIS
210 r	© Liaison / Newsmakers / OnlineUSA
216 all	© PhotoDisc
218	AP / Wide World Photos
223	AP / Wide World Photos
224	AP / Wide World Photos
228 l	© P. Lisenkin / Sovfoto / Eastfoto
228 r	© 1996 The Hearst Corporation
231	© Catherine Ursillo / Photo Researchers
246 l	AP / Wide World Photos
246 r	© David & Peter Tumley / CORBIS
252	© PhotoDisc, © Digital Visions, © Joseph Sohm; Chromo Sohm, Inc. / CORBIS, © Paul Saltzman / Contact Press Images / PictureQuest
254	© Frank Simonetti / Index Stock Imagery / PictureQuest
257 l	© FPG International
257 r	© Ron Chappel / FPG
261 r	© Owen Franken / Stock, Boston
261 l	© Bob Daemmrich / Stock, Boston
263	© Reuters / STR / Archive Photos
271 l	© Margo Smith
271 r	© Rick Reinhard / Impact Visuals / PictureQuest
276 all	© PhotoDisc
278	© Reuters / Sunil Malhotra / Archive Photos
284 l	© Bruno Barbey / Magnum Photos
284 r	© Richard Elliott / Stone
285	© A. Keler / Sygma
286	© Bruno Barbey / Magnum Photos
287	© A. Ramey / Stock, Boston
289	© Oliver Martel / Liaison Agency
291	AP / Wide World Photos
295	AP / Wide World Photos
297 l	© Bruno Barbey / Magnum Photos
297 r	© Earl & Nazima Kowall / CORBIS
302	© PhotoDisc, © Bob Daemmrich / Stock,Boston / PictureQuest, © Naoki Okamoto / Black Star Publishing / PictureQuest
304	AP / Wide World Photos
310	© Liba Taylor / CORBIS
314 t	© UPI / Bettmann / CORBIS
314 b	© Piet Van Lier

315 Copyright © 1954 by The New York Times Co. Reprinted by permission.
327 tl © Robert Clark
327 bl © Robert Clark
327 r © Robert Clark
330 all © PhotoDisc
332 AP / Wide World Photos
346 l © Robert Brenner / PhotoEdit
346 r © Mark Richards / PhotoEdit
348 Jenny Matthews / Network Photographers Ltd.
349 t International Labour Organization: Photograph by Kevin Bales
349 b © Mark Edwards / Still Pictures
362 l © John Parker
362 r © AFP / CORBIS
368 © PhotoDisc, © CORBIS
370 "Tidi and her neighbors" by Wendy Ewald
376 Reuters / Bettmann / CORBIS
389 t Wallace Kirkland Papers (JAMC neg. 613), Jane Addams Memorial collection, Special Collections, The University Library, University of Illinois at Chicago
389 b Wallace Kirkland Papers (JAMC neg. 839), Jane Addams Memorial collection, Special Collections, The University Library, University of Illinois at Chicago
394 © Liba Taylor / CORBIS
396 r "Hasmukh, Chandrakant, Harshad & Dasrath learning to use the camera" by Wendy Ewald
396 l "Grandfather is smoking on the ledge with the dog" by Chandrakant
400 © PhotoDisc, © Tomas Muscionico / Contact Press Images / PictureQuest, © David and Peter Tumley / CORBIS
402 © Betty Press / Woodfin Camp & Associates
404 l The Granger Collection, New York
404 r © Jon Anderson / Sygma Photo News
413 © Brian Willer / NYT Pictures
414 t © Mark Ludak
414 b © Jacques Jangoux / Stone
418 AP / Wide World Photos
425 l Independent Picture Service
425 r photograph by Eliot Elisofon, 1969, Slide No. TIVC2, Eliot Elisofon Photographic Archives, National Museum of African Art

INDEX